Discovering Design with Earth Science

Published by
Berean Builders
Muncie, IN
www.bereanbuilders.com

Manufactured in the United States of America
Second Printing 2023
Printed by Jostens

ISBN: 978-0-9962784-3-0

Cover layout by Kim Williams

Cover Art from www.shuttterstock.com, © Albert Russ, Deni Sugandi, Naeblys

Discovering Design with Earth Science

Introduction

You have lived on the earth all your life, but you probably don't know very much about it. As a child, you probably enjoyed digging in the dirt. But what is dirt? How is it different from rocks? How are rocks different from fossils and gems, which are usually found in rocks? You have sometimes enjoyed the weather and sometimes complained about it. But what makes the different kinds of weather you have experienced? You generally get up after the sun rises, and you have probably gazed at the stars after the sun has set. But what makes the sun rise and set? What are the stars? You will find the answers to these questions through a study of earth science, which is what I will cover in this book.

The earth is a marvel of design and complexity, because God made it. Psalm 24:1 tells us, "The earth is the Lord's, and all it contains, The world, and those who dwell in it." Psalm 111:2 also tells us, "Great are the works of the Lord; They are studied by all who delight in them." I am sure that there have been times you have been delighted by the earth. I know that every time I sit on a beach or scuba dive in the ocean, I am delighted. The same thing happens when I gaze at a beautiful mountain or see constellations of stars in the night sky. Because I have been so delighted by the earth, I want to study it. I hope that this course makes you take even more delight in this planet that you call home, and I hope that it encourages you to continue studying the earth, even once you are finished with this book.

How to Use This Book

This book is made up of 16 chapters. Each chapter contains reading and experiments that you need to complete as well as questions you must answer. You are supposed to perform the experiments when they come to you in the reading, because right after the experiment, I will discuss what the experiment means.

The questions that you answer while you are reading are called "Comprehension Check" questions, and they represent important milestones in your work. Each time you reach a "Comprehension Check" box, you are at the end of the day's lesson. You need to answer the questions and then check your answers against the answers that appear right before the Chapter Review at the end of the chapter. Once you check your answers and understand anything you answered incorrectly, you are done with science for the day!

The chapters have seven "Comprehension Check" boxes, which means you will use seven school days to work through each chapter. Some of those days will consist entirely of reading, and some of them will consist of less reading and an experiment to do. At the end of those seven days, you need to spend a day or two answering the questions in the Chapter Review that appears at the end of each chapter. Your parent/teacher has the answers to those questions, but you should not use them until you have completed the entire review. Feel free to use the book to help you with the review.

Once you have finished answering and checking all your answers for the Chapter Review, you are ready to take the test that covers the chapter you have been working on. You cannot use your book for the test, but you can use a calculator if the test has any math-related questions in it.

As you read, you will see some statements and equations that are centered and surrounded by pink boxes. You must memorize any information that you see in those pink boxes. In addition, there are definitions that are centered in the text. They also need to be committed to memory. Finally, there are some words in boldface type scattered through the text. Those are terms with which you need to be familiar.

Most students should try to cover this course in one year of school. If you think about it, all the Chapters have seven "Comprehension Check" boxes, which means it will take seven days to get through each chapter. After that, suppose you spend two days working on the Review and studying for the test. Then on the next day, you take the test. That means it would take ten school days (two weeks of school) to cover the chapter, which tells you that it will take 32 weeks to finish the entire book. Most school years are 36 weeks long, so you have some built-in "flex time" in case some chapters take longer than others.

Experiments and Activities

There are several experiments that are scattered throughout the course. They will help you get serious, "hands on" experience with the concepts that you are learning.

********* **Please Do The Experiments With Common Sense And Adult Supervision**. *********

You will not find these experiments to be any more dangerous than cooking or cleaning, but that doesn't mean students can't get hurt. Your parent/teacher should supervise you to make sure you aren't doing anything dangerous, and you must ***always*** wear safety goggles while doing experiments!

Experiment Supplies

The experiments you do in this course use a kit that has been made specifically for this course. In addition, they use household items or items that are easy to get at a grocery store, drug store or hardware store. Before you start each chapter, you should look at Appendix B near the end of the book. It contains a list of all the materials you will need to do the experiments in that chapter. Check and make sure you have all of them, and if you don't, ask your parents to get them for you so that you can do the experiments when you get to them in the reading.

The kit that has been designed for this course contains several mineral, rock, sediment, and fossil specimens as well as certain materials you will need to analyze them. You need to use the specific kit made for this course, because I need to know that you have the specific specimens I want you to analyze. Some will have numbers on them, and I will refer to them by numbers. Others will be in labeled bags, and I will refer to those labels. **It is impossible to do many of the experiments in the course without the kit**. You can order it at:

bereanbuilders.com/lk/ddes

One supply you will need for every experiment is your laboratory notebook. You should keep a record of all your experiments. If you want guidance on how to do that, there are directions on the course website, which is discussed in the next part of this introduction. You can also find samples of how I would record the first few labs on the course website. The notebook can be anything you want it to be – a spiral-bound notebook, a cloth-bound notebook, a diary, etc. It just needs to have lots of pages for you to write in. Honestly, the best kind of pages are blank pages, but you will be doing a lot of writing on those pages. If you have a hard time writing without lines, get a notebook that has lined paper.

Course Website

There is a website that goes along with this course, and it contains some helpful resources. The address is:

bereanbuilders.com/olc/ddes

This is where you will find direction on how to record your labs and samples of what that looks like for the first few experiments in the course. You will also find videos that relate to the material in the book as well as links to other resources. If you find them helpful, the course website also contains worksheets that you can print out to help you study.

Question/Answer Service

One of the most important things you need to remember while using this course is that there is a way to get help if you are stuck or confused. Just go to:

www.askdrwile.com

and register. Registration is free, and once you have registered, you can log in to use the website. Your information will not be shared with others. The website allows you to ask questions about the material, and it also allows you to search for the answers to questions that have been asked previously. Please feel free to ask as many questions as you like!

The earth has been fashioned lovingly by God to be a haven for life. As you learn the ways in which He has done this, I pray that you are filled with awe!

Discovering Design with Earth Science

Table of Contents

Chapter 1: Basic Concepts Required to Study Earth Science

If you have not read the introduction to the book (pp. i-vii), do so now. You can count it as your first day of science, and you can start reading what's below tomorrow.

Introduction

This picture, taken by astronaut William Anders, shows the earth as seen from a spacecraft orbiting the moon. It is rotated by 90 degrees so that it looks like it was taken from the moon's surface.

Have you ever seen the picture shown on the right? It's usually called "Earthrise," and it was taken by astronaut William Anders on Christmas Eve in 1968, when he and his team were orbiting the moon in a spacecraft. The bottom of the picture shows the surface of the moon, and the gorgeous blue-and-white ball above the moon's surface is the earth. While you can find lots of color pictures of the earth as seen from space, this is one of the first that was taken, and it had a profound effect on many people, because it showed the earth from a completely new perspective. On Christmas Eve fifty years after taking the picture, Anders declared, "We set out to explore the moon and instead discovered the earth." (https://www.space.com/42848-earthrise-photo-apollo-8-legacy-bill-anders.html, retrieved on 5/27/2020). In this course, I hope to show you the earth from a completely new perspective so that you can discover truths about its design and the One who designed it.

Now there are a lot of things to learn about this picture. Why is only half of the earth visible? Why don't you see any stars, even though the "sky" above the moon's surface is black? In a later chapter, I will answer those questions. For right now, just notice the shape of the earth. Only a portion of the earth is visible, but based on that portion, the earth is clearly a sphere. Actually, it's not a perfect sphere. Technically, it's an **oblate** (oh blayt') **spheroid** (sfear' oyd), which means it looks like a ball that has been squashed by something pushing down on its top and up on its bottom.

The Earth's Shape

We all know that the earth is a sphere, but you might not be aware that people have known this for more than 2,500 years. While you may have heard that people in ancient times believed the earth was flat, that's really not true. I am sure if you go back far enough in history, you can find some people who believed the earth was flat, but the earliest historical records we have indicate that most people understood it is a sphere.

For example, Aristotle (air' ih stot' uhl) was a natural philosopher (an older term that essentially means "scientist") who lived from 323 to 385 BC. He concluded that the earth was a sphere based on several different observations that he could make. He was considered one of the greatest minds of his time, so pretty much everyone agreed with him. Around 200 BC, another natural philosopher, Eratosthenes (air' uh tas' thuh neez'), measured the shadow cast by a stick in one location on a specific day at a specific time. He then walked about 500 miles southeast and measured the same stick's shadow on the same day the next year at the same time. Using geometry based on the fact that the earth is a sphere, he actually determined the distance around the earth, and it turns out that his value was very close to the satellite-measured value we have today!

Of course, Aristotle and Eratosthenes were well-educated people, so it's not surprising that they could figure out that the earth is a sphere. However, to the average, poorly-educated person back then, it looked pretty flat. Thus, you might think that most people believed the earth was flat, right? Probably not. We can't say for sure, of course, but we know that sailors and people who lived near the ocean, most of whom were poorly educated, understood that the earth is a sphere. Why? Because of something they saw regularly. The following experiment will help you understand what they saw.

Experiment 1.1: Hull Down

Supplies:
- Five sheets of plain paper
- Tape
- Many books of different thicknesses
- A flat table or desk that is at least as long as four of the sheets of paper laid end-to-end
- Play-Doh or modeling clay
- Aluminum foil
- A toothpick
- Someone to help you

Instructions:
1. Make a rough cube of clay that is a little over 1 cm (about half an inch) on each side. It doesn't have to be a really good cube. Just make sure the bottom is flat.
2. Tear off a small piece of aluminum foil and wrap it around the Play-Doh.
3. Put the bottom on the flat surface and stick the toothpick into the clay so it stands on its own.
4. Tear off a small piece from one of the sheets of paper and tape it near the top of the toothpick so that you have a pretty sad-looking model of a boat (see the top picture below).
5. Tape the other four sheets of paper end-to-end.
6. Lay the long strip of paper you just made on the flat table or desk.
7. Have your helper hold the model of the boat near the center of the long strip of paper.
8. Position your head so that one eye is level with the flat surface, and you are looking down the long strip of paper. Close the other eye. You should see something like the picture on the top right.
9. Hold the paper strip to the surface so it doesn't move.
10. Have your helper move the boat model close to you and then slowly pull it away from you until it reaches the end of the paper strip. How did the model's appearance change?
11. Arrange stacks of books on the floor as shown in the picture on the bottom right. They should slowly decrease in height so that the long strip of paper forms a slope. The first stack of books should be about twice as tall as the boat model.
12. Lay the long strip of paper on the stack of books.

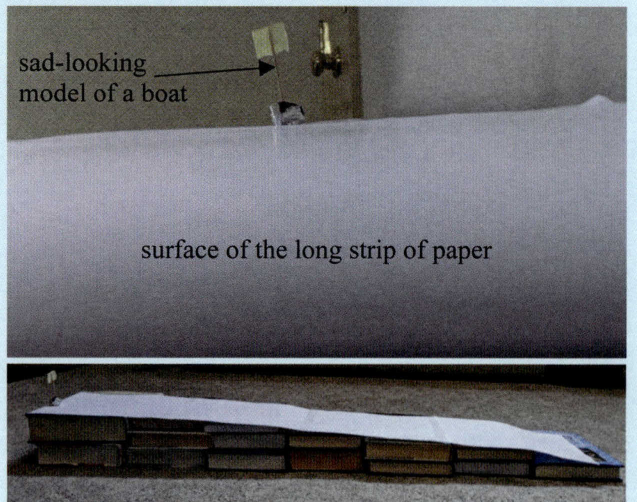

13. Have your helper hold the model of the boat.
14. Position your head so that one eye is level with the highest stack of books and you are looking down the long strip, just like you did in step 8. Once again, close the other eye.
15. Hold the edge of the paper strip against the stack of books so that the strip doesn't move.
16. Have your helper move the boat model close to you and then slowly move it down the slope of the paper strip. How does the model's appearance change? The change should be noticeably different from what you saw when the paper strip was on the flat surface. If you don't see the difference, repeat both parts of the experiment.
17. Clean up your mess.

What difference did you see in how the boat model's appearance changed? You should have seen that the model appeared to get smaller as it moved away from you on the flat surface, but that's about it. However, on the sloped surface, things should have been rather different. The boat appeared to get smaller, but also, you eventually stopped seeing the foil part of the boat. It "disappeared" as the boat moved away. However, you could still see the sail on top of the toothpick, even after you couldn't see the foil part of the boat. That's because the boat was following the slope made by the books, and the slope hid the bottom of the boat before it hid the sail of the boat.

This is something sailors and people who lived near the ocean have seen since people first traveled on the ocean. As a ship traveled away from the shore, the bottom of the ship (the hull) would disappear before the top (the sails). As a ship travelled towards the shore, the sails would appear first, and the hull would appear later. This is because the ships are traveling along a slope. The ocean looks flat, but if it were, a boat would appear smaller and smaller as it traveled away from shore, but you would continue to see the entire boat, just like you did in the first part of the experiment. But the ocean is actually sloped, because it follows the curve of the earth's sphere. As a result, boats traveling toward or away from shore appear like the model did in the second part of the experiment.

Because of this, sailors have a term called "hull down." When a ship is "hull down," it is far enough away that the curve of the earth's sphere hides its hull but not its sails (or smokestacks and antennas in modern boats). This is something even ancient sailors experienced, so even though they didn't have the benefit of being taught by people like Aristotle, they knew that the earth is a sphere.

But wait a minute. Wasn't Christopher Columbus's plan to sail around the world opposed because people thought he would sail off the edge of the earth? No! While that's a popular myth, it is completely false. In Christopher Columbus's day, European people knew the earth was a sphere because the church taught that the earth is a sphere, and the people believed the church's teachings. If you study actual history, you will find that Christopher Columbus's plan was opposed because Eratosthenes's measurement of earth was known. People knew how far Columbus would have to travel to go around the world, and they didn't think such a long journey was possible.

Comprehension Check

1.1 If the ocean is curved because it follows the earth's sphere, why does it appear to be flat?

1.2 If you look at a flat map of the world, you will see that Greenland looks almost as big as Africa. On a globe, however, Africa looks a lot bigger than Greenland. In reality, is Africa larger than or roughly the same size as Greenland?

Chemicals

Once again, you probably knew that the earth is spherical, but I want to make sure we are all on the "same page," so I want to cover a few more basic concepts that will be necessary to understand the rest of the course. Some of them will be review for you, but some might not be. Be sure to study this chapter carefully even if you recognize the material, because throughout the rest of the course, I will be assuming that you understand everything presented in this chapter.

Everything that makes up the earth is composed of **chemicals**. In fact, almost all of creation (except light) is made up of chemicals, so you will need to understand chemicals if you want to understand earth science. Chemicals are made up of one or more **atoms**, which you can think of as tiny building blocks. Some chemicals, like the carbon found in diamonds, are made up of individual atoms. Most chemicals, however, are made up of atoms that are joined together.

Consider a chemical with which you are very familiar: water. It is formed when one oxygen atom attaches to two hydrogen atoms. The attachments are called **chemical bonds**, and they keep the atoms together. When atoms are joined together by chemical bonds, we usually call them **molecules** (mol' ih kyoolz). The drawing below, for example, represents a molecule of water.

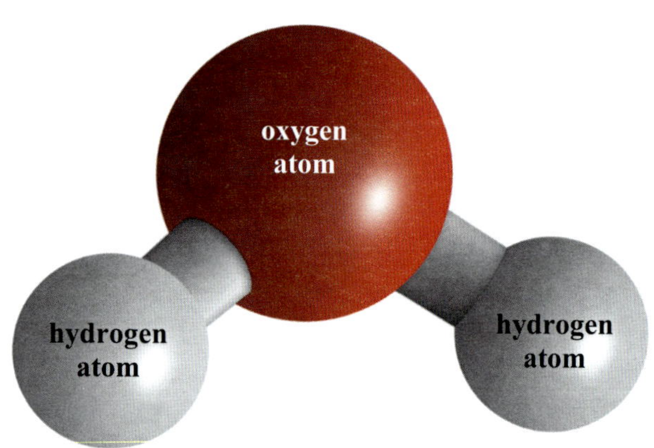

A molecule of water is made of two hydrogen atoms attached to an oxygen atom with chemical bonds.

Under certain circumstances, the chemical bonds in a water molecule can be broken down, making hydrogen and oxygen. This is called a **chemical reaction**. Interestingly enough, hydrogen and oxygen are nothing like water. At room temperature, for example, hydrogen is a gas that burns violently, and oxygen is a gas that makes fire possible. At room temperature, however, water is a liquid that can be used to put out many kinds of fire! When atoms join together to make a molecule, then, the molecule they make behaves differently from the atoms themselves. I want to emphasize this by having you do an experiment.

Experiment 1.2: Very Similar but Very Different!

Supplies:
- Hydrogen peroxide from the laboratory kit made for this course
- Active dry yeast (Yeast for a bread machine will also work.)
- Water
- Two smaller (like "snack sized") plastic bags that can be zipped shut (like Ziploc bags)
- Two larger (like "quart sized") plastic bags that can be zipped shut (like Ziploc bags)
- A measuring cup
- A measuring tablespoon
- Somewhere outside that you can get messy

Instructions:
1. Add a cup of water to one of the small bags and zip it shut.
2. Put the small bag of water into one of the larger bags.
3. Add a tablespoon of yeast to the larger bag that you just put the small bag into.

4. Zip the big bag shut. You should now have a larger, zipped-shut bag that holds a smaller, zipped-shut bag of water and some yeast.
5. Repeat steps 1-4 with the other bags, but use hydrogen peroxide in the small bag instead of water.
6. Take both bags outside to the place that you can get messy.
7. Set the bag that holds the small bag of hydrogen peroxide down carefully, making sure that you don't break or open the small bag inside.
8. Hold the bag that contains the small bag of water in both hands and use your fingers to unzip the small bag of water inside while keeping the larger bag zipped.
9. Shake the larger bag so that the water and yeast mix well.
10. Drop the larger bag on the ground and move a couple of meters (a few feet) away. Watch the bag to see if anything happens.
11. Repeat steps 8-10 with the larger bag that has hydrogen peroxide and yeast in it.
12. What's the difference?
13. Empty the bags that aren't already empty and throw them away. Also, clean up your mess inside.

What did you see in the experiment? Nothing exciting should have happened with the bag that had water in it. The water and yeast mixed, but nothing else should have happened. However, when you used the hydrogen peroxide, you should have seen a lot of bubbles. When you dropped the bag and moved away, you should have seen the bag inflate. Eventually, it should have inflated so much that it popped open, and some liquid probably sprayed out.

What's the difference between these situations? It's the hydrogen peroxide, which is a fairly unstable molecule. Over time, it will break down, and one of the things this breakdown makes is oxygen. Yeast speeds up the process quite a bit. So, when the yeast and hydrogen peroxide mixed, oxygen started being produced. Under the conditions of the experiment, oxygen is a gas, so it started inflating the bag. It inflated the bag so much that the bag became highly pressurized, and eventually, it had to pop open.

Now here's the truly remarkable thing: A molecule of water and a molecule of hydrogen peroxide are very, very similar. How similar? The drawing on the previous page tells you that a water molecule is made of one oxygen atom bonded to two hydrogen atoms. In chemistry, we can abbreviate this with a **chemical formula**. First, we abbreviate the atoms. Instead of saying "hydrogen," we just use an "H." Instead of saying oxygen, we just use an "O." Then, we put a subscript after each atom's abbreviation to indicate how many of that atom are in the molecule. If there is only one atom, we don't write any number. So, the chemical formula of water is H_2O. The subscript of "2" after the H tells us there are two hydrogen atoms, and the fact that there is no subscript after the O tells us there is only one oxygen atom.

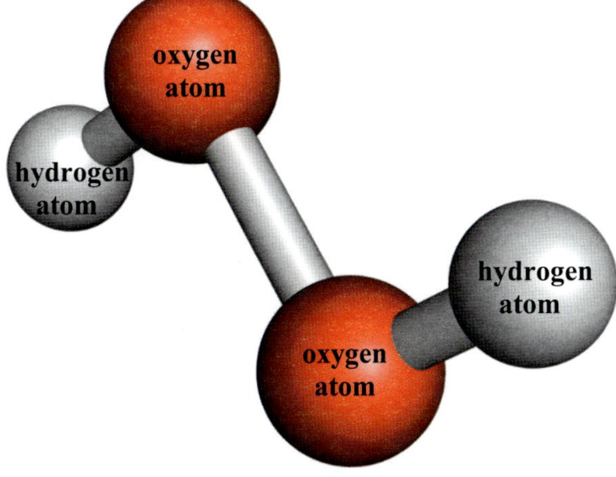

This drawing shows you a molecule of hydrogen peroxide.

Compare that to a molecule of hydrogen peroxide, which is shown in the drawing on the right. As you can see, it is made of two hydrogen atoms and two oxygen atoms, so it has a chemical formula of H_2O_2. Water is made up of hydrogen atoms and oxygen atoms, but so is hydrogen peroxide. Both water and hydrogen peroxide molecules have two hydrogen atoms. The only difference is that while a water molecule has only one oxygen atom, a hydrogen peroxide molecule has two oxygen atoms. Nevertheless, these molecules that have so many similarities act very differently when exposed to yeast, all because of the difference in the number of oxygen atoms in each molecule! When molecules form, then, *each atom is important.* If you change just one atom in a molecule, the properties of the molecule can change completely. Because of this, the 98 naturally-occurring kinds of atoms in creation can combine together to make an unimaginable number of chemicals.

Each of these 98 types of atoms forms its own **element**, which has its own abbreviation. So far, I have talked about two: H stands for hydrogen, and O stands for oxygen. Those make a lot of sense. However, some elements have two letters in their abbreviation. Calcium, for example, is abbreviated as Ca. If an element has two letters in its abbreviation, the first one is always capitalized, and the second one is always lower case. That way, when you look at a list of element abbreviations, you know that you have come to a new element when you come to a new capital letter. For example, the molecule CaO is composed of one calcium atom and one oxygen atom. You know this because the O is capitalized, telling you it is a new atom. Since there is no subscript after the Ca, you know there is one Ca atom. Similarly, since there is no subscript after the O, you know that there is one O atom.

While H, O, and Ca all make sense. Some element abbreviations are a bit less obvious. Sodium atoms, for example, are abbreviated as Na. Why? Because sodium's Latin name is *natrium.* Similarly, potassium atoms are abbreviated with a K, since potassium's Latin name is *kalium.* In general, we can say that the abbreviation of an atom is one or two letters from its English or Latin name. The first letter is always capitalized, and if there is a second letter, it is lower case.

Suppose, then, I told you that a certain rock contained the chemical $KAlSi_3O_8$. Even though you probably don't know all the symbols, you could tell me that this molecule contains one K atom, one Al atom, three Si atoms, and eight O atoms. You can figure this out because each capital letter starts a new atom, and you know the subscript after the atom tells you how many atoms there are in the molecule. If there is no subscript, there is only one of that atom.

Comprehension Check

1.3 Automobile exhaust contains CO. We exhale CO_2. A student reasons that since the chemical formulas are similar, the chemicals must be similar as well. Is he correct? Why or why not?

1.4 Rust has the chemical formula Fe_2O_3. How many atoms (total) are in a molecule of rust?

1.5 You might have some Epsom salt in your medicine cabinet. Each molecule has one atom of magnesium (Mg), one atom of sulfur (S), and four atoms of oxygen (O). What is its chemical formula?

It's Just a Phase

I have been using terms like "liquid" and "gas," and I am sure that you understand what they mean. However, I need to make sure you understand that those terms refer to different forms of matter which we call **phases**. In the conditions found on earth, chemicals are typically in one of three phases: **solid**, **liquid**, or **gas**. If you lower the temperature enough, you can make water a solid. This process is called **freezing**, and we call the solid water ice. If you heat up ice enough, it will turn back into a liquid. That process is called **melting**. If you heat the liquid enough, it will **boil**, changing into a gas. If you cool the gas down, it will turn back into a liquid, and we call that **condensing**.

Given the right conditions, it is possible for any chemical to attain any of the three phases listed above. For example, the photograph on the right shows a beaker that holds liquid oxygen. As a gas, oxygen has no color, but if you cool it down to a very cold temperature, it condenses and becomes a faintly blue liquid. The "smoke" you see around the beaker isn't really smoke. The liquid oxygen is so cold that it is causing gases in the air to condense into liquids as well, forming tiny droplets that look like smoke. While I don't have a picture of it, if you cooled liquid oxygen down even more, you could make solid oxygen, which is also has a blue color.

This beaker contains oxygen that is so cold it is in its liquid phase.

There are a couple of things you need to realize about the phases in which matter can exist. First, the phase doesn't change the chemical nature of the molecules. The molecules that make up liquid oxygen are the same as the molecules that make up oxygen gas. Anything that oxygen gas can do chemically, liquid oxygen can do as well. For example, oxygen gas is necessary for fire, because a fire is caused by a chemical reaction between oxygen gas and whatever fuel that is being burned. Well, despite the fact that it is very, very cold, liquid oxygen can also react with a fuel to make fire.

What's the difference between solids, liquids, and gases? In most chemicals, the molecules are closest in a chemical's solid phase, farther apart in its liquid phase, and even farther apart in its gaseous phase. In addition, the mobility of the molecules is different. In solids, molecules are only allowed to vibrate. They cannot move around. In liquids, the molecules can move around, but their movement is a bit limited. In gases, the molecules move around like crazy. We will be discussing chemical formulas and phases quite a bit in this course, and if you understand everything you have read, you will be able to grasp that upcoming material. If not, you should go to the course website, which is given in the introduction to the book. There, you will find links that explain these things in more detail.

Units

When we study the earth, we need to measure things. We might need to know how tall something is, how much space it takes up, or how much material is in it. Measurements include numbers, but the numbers are meaningless by themselves. If I told you I am 67 tall, would that mean anything to you? No. I have to tell you what I used to measure my height so that you could figure out whether I am a tall, short, or medium-height guy. You might be able to guess what I used to measure myself based on your experience with the height of adults. Nevertheless, if I don't tell you what I used, you won't know for sure.

The units of the measurement tell you what I used to measure myself. I am 67 *inches* tall. Now you know that I am a short guy, since the average adult man is 68.9 inches tall. Of course, I could use other units to measure my height. I could use feet and inches, for example. I am 5 feet, 7 inches tall. I could also use a unit called **meters**. If you have ever seen a meterstick, it has a length of one meter. I am 1.7 meters tall. I could also use centimeters, which is another unit you have probably seen before. I am 170 centimeters tall. All those measurements: 67 inches, 5 feet 7 inches, 1.7 meters, and 170 centimeters represent the same height. The numbers are different because the units are different, but all the heights they represent are the same.

When reporting measurements, then, you must report your units. A measurement without units means nothing. I remember the first time I was taught this. I was in high school, and my teacher had just gotten done explaining how important units are in science. He said that all students must report their units with each measurement, or their answer will be counted as incorrect. I wanted to earn high grades in my classes, so I took his warning to heart.

The first test required us to use some equations, but he didn't want us to memorize them, so he said that while taking the test, we could use a 3x5 card and write any equations we thought we might use on it (something I *will not* allow you to do in this course!). I raised my hand to confirm this. I asked, "A 3x5 card, right?" He replied, "yes." For the first test, I showed up with a poster board that was 3-feet tall and 5-feet wide. It had all sorts of equations (some that I didn't even understand) copied on it. Of course, he wanted us to use a 3-inch by 5-inch card, but he had to let me use my poster, because he had not listed his units. When I saw him years after I had graduated, he said that since then, he told his students the story of me and my poster every year to emphasize how important units are!

The Metric System

Since units are so important, all scientists try to use the same set of units so that we can easily compare our results to one another. In this class, we will use the **metric system** as our system of units. You might be familiar with the metric system already, but if not, that's fine. You just need to know the basic units of the metric system and how to modify them.

A guitar is about a meter long.

Let's start with the metric unit I already mentioned: the meter. It is abbreviated with "m" and measures distance. The distance between the bottom of my feet to the top of my head is my height: 1.7 meters. As a way of visualizing a meter, consider a guitar. The distance from one end of a guitar to the other is about a meter. If that doesn't help, most standard doorways are just under a meter wide. Most countertops are one meter high.

Besides learning how tall, wide, or high something is, we might want to know how much **matter** is in it. If you aren't familiar with that term, "matter" can refer to any substance. This book is made of matter, you are made of matter, everything you see around you (except light) is made of matter. The amount of matter in a substance is called its **mass**.

<u>Mass</u> – A measure of how much matter is in an object

This book has a certain amount of matter in it (you might think it is a lot!), but a car has more matter in it. A house has even more matter in it. Thus, a house's mass is larger than a car's mass, which is larger than this book's mass. In general, the more mass something has, the heavier it is. Mass and weight are not the same thing, but they are related. So, the more something weighs, the more mass it has, which means the more matter it has in it.

Mass is often measured using a balance, like the one pictured on the right. The silvery objects in the pan on the far right each have a known mass. The mass of the yellow material in the left pan is being measured. If the mass of the yellow material were larger than the total mass of the silvery objects, the balance would tilt towards the yellow material. If the total mass of the silvery objects was larger, it would tilt towards the silvery objects. Since the balance is not tilted either way, we know that the mass of the yellow material is the same as the mass of the silvery objects. Even though they are very different, this tells you the total amount of matter in the yellow material is equal to the total amount of matter in the two silvery objects.

The fact that this balance is not tilted towards either side tells you that the mass of the things in each pan are the same.

The metric unit for mass is the **gram**, which is abbreviated "g." An American dollar bill has a mass of about one gram. So does a standard metal paper clip. That means even though one is made of paper and ink, while the other is made of metal, the total amount of matter in a dollar bill is the same as the total amount of matter in a standard metal paper clip.

You might also need to know how much time it takes for something to happen. The metric unit for time is one with which you are already familiar: the **second**, which is abbreviated with an "s." There are other ways to measure the passage of time, such as hours, minutes, days, etc. However, they are not considered part of the metric system.

This gives us three things we can measure: length (in meters), mass (in grams), and time (in seconds). But what happens when I am measuring something for which those units don't make much sense? For example, suppose I want to measure the thickness of a page in this book. That's pretty small – a lot less than a meter. Isn't there a better unit? Yes, there is. In the metric system, when we want to measure something really big or really small, we modify the unit with a prefix. If we are measuring something that is really small, we use a prefix that makes the unit small. If we want to measure something that is really big, we use a prefix that makes the unit big.

In the metric system, the prefixes are all based on powers of ten. If I wanted to measure the thickness of this page of paper, for example, I would probably choose the millimeter. The prefix "**milli**" means one thousandth (0.001), so a millimeter is one thousandth of a meter. If I used that unit, I would find that this page is 0.1 millimeters thick. On the other hand, if I wanted to measure the distance between cities, I would need a much bigger unit, so I might choose the kilometer. The prefix "**kilo**" means one thousand (1,000), so a kilometer is 1,000 meters. The distance between New York City and Washington, DC, for example, is 370 kilometers.

There are lots and lots of prefixes that can be used to make a metric unit meet your measurements needs. The table on the next page lists several of them.

Metric Prefixes and Their Meanings

Prefix	Abbreviation	Meaning		Prefix	Abbreviation	Meaning
mega	M	1,000,000		**centi**	c	**0.01**
kilo	**k**	**1,000**		**milli**	m	**0.001**
hecto	H	100		micro	μ	0.000001
deca	Da	10		nano	n	0.000000001

Don't worry. You don't need to remember all of these. I only want you to remember the three in boldface type: **kilo** (1,000), **centi** (0.01), and **milli** (0.001). If you don't already know them, don't worry about memorizing them right now. The next time you do science, you will see these prefixes in use, and that should help make them more familiar to you.

Suppose, then, you want to measure the mass of a car. If you were limited to the three prefixes I want you to know (kilo, centi, milli), what measurement unit would you use? Well, since you are measuring mass, you would use some form of the gram, because the gram is the metric unit for mass. However, the mass of a dollar bill is a gram, and a car has *a lot* more mass than a dollar bill. Thus, you would use the kilogram. The mass of the car that I drive (a Jeep Liberty) is 1,900 kilograms. Notice that the abbreviation for "kilo" listed in the table is "k." That means I could abbreviate the unit and say that my car has a mass of 1,900 kg.

Suppose you wanted to measure something that happens in a really short amount of time, like a lightning strike. Since you are measuring time, you would use the second, but a lightning strike is a lot shorter than a second. Using the three prefixes I want you to know, you would choose milliseconds. A lightning strike usually lasts 0.03 milliseconds, or 0.03 ms.

Math With Units

We need to measure a lot of things besides just mass, length, and time. Interestingly enough, however, most of the other things we need to measure are the result of mathematics being done with measurements of mass, length, and time. Consider, for example, comparing the areas of two different rooms. One of the rooms is 10 meters long and 3 meters wide. The other is 7 meters long and 5 meters wide. Which is the bigger room? You might think that's a hard question to answer. One room is longer, but the other is wider. Does the increased width of the second room make up for the decreased length? You might consider adding the measurements. The first room's measurements add to 13, while the second room's measurements add to 12. Does that mean the first room is bigger? No. The first room is actually smaller.

How do I know? Because I can determine the **area** of the room, which is a measure of how much space exists within the room's boundaries. The area of a rectangle is the rectangle's length times its width:

$$\text{Area} = (\text{length})\cdot(\text{width})$$

You have probably seen that formula before. If not, it is one that you will have to remember and be comfortable using. That's why it is in a pink box. As you were told in the introduction to the book, you need to memorize anything you see in pink boxes. If you haven't seen it before, the "·" in the box is one way to abbreviate multiplication. You are probably used to abbreviating it with "x," but scientists often use "x" for other things, so we typically abbreviate multiplication using a "·".

If we use that formula to determine the first room's area, we end up with this equation:

$$\text{Area} = (\text{length})\cdot(\text{width}) = (10\ \text{meters})\cdot(3\ \text{meters})$$

I have to keep the units with the number, because a measurement without units means nothing. But how do I deal with them in the equation? I do the numbers separately from the units. 10 times 3 is 30, so the number I get is 30. But what happens to the units? I multiply them together as well. I can't get a number for an answer, but I can say the units is meters·meters. Thus, my answer is:

$$\text{Area} = (\text{length})\cdot(\text{width}) = (10\ \text{meters})\cdot(3\ \text{meters}) = 30\ \text{meters}\cdot\text{meters}$$

But what does that mean? Well, what happens when I multiply 2·2? I could just get the answer, which is 4, or I could say that 2·2 is the same as 2^2, which is read as "two squared." The superscript tells me I multiply 2 by itself. Well, if 2·2 is two squared, what is meters·meters? It's meters squared, which we can write as m^2. Remember, meters can be abbreviated with an "m," and the superscript just means multiply it by itself. So m^2 is really meters·meters. That means my answer is:

$$\text{Area} = (\text{length})\cdot(\text{width}) = (10\ \text{meters})\cdot(3\ \text{meters}) = 30\ \text{meters}\cdot\text{meters} = 30\ \text{m}^2$$

So, if I want to measure the area of something, I multiply length times width, and if I am using meters, the unit for area is m^2. Compare this area to the area of the second room:

$$\text{Area} = (\text{length})\cdot(\text{width}) = (7\ \text{meters})\cdot(5\ \text{meters}) = 35\ \text{meters}\cdot\text{meters} = 35\ \text{m}^2$$

This tells me that the second room is bigger, because it has more area. More importantly, we now know that we can express the unit for area in terms of meters. It's m^2.

It turns out that we will be using units in math from time to time, so you need to get comfortable doing it. If you aren't quite comfortable yet, that's fine. You will see examples throughout the rest of this chapter.

Comprehension Check

1.6 In my research, I often have to heat solid metal. If I heat it long enough, what phase will it become? If I heat it a lot more, what phase will it become?

1.7 Suppose you want to measure how much time you work on this course to finish it. If you want to use a metric unit with one of the three prefixes you need to know, what unit would you use?

1.8 A page from this book has a length of 28 cm and a width of 22 cm. In case you don't recognize it, "cm" stands for "centimeters." What is its area, including the units?

Volume

Area tells you how much space exists within a specific boundary, but something that scientists measure much more frequently is **volume**, which indicates how much total space an object takes up. If I want to put an object into a container, for example, the only way the object will fit into the container is if the object's volume is smaller than the container's volume. For example, in the late 1800s, a Russian woodcarver made a set of hollow dolls that are now called "nesting dolls." A modern example is shown in the photo below. Notice that starting from the far left, each doll gets smaller. In other words, the volume of each doll decreases. Because of this, the dolls fit inside one another, so they can all be "nested" inside the largest one. Nesting dolls work because each doll's volume is smaller than the previous one, which ensures that each doll fits inside the previous one.

The volume of each doll decreases from left to right. That means each smaller doll can fit in the doll on its left. Since they are all hollow and can be opened, you can put each doll inside the doll on its left, so that all of them will be "nested" inside the largest one.

What do you need to know to determine an object's volume? It's not enough to know how long the object is and how wide it is. That will give you the area, but not the volume. The total amount of space that an object takes up depends on its length, its width, and its height. When you know those three measurements, you can determine the volume by multiplying them together according to the formula:

$$\text{Volume} = (\text{length})\cdot(\text{width})\cdot(\text{height})$$

Let's suppose you have a box that measures 1.2 m long, 1.5 m wide, and 2.5 m high. The volume of the box would be:

$$\text{Volume} = (\text{length})\cdot(\text{width})\cdot(\text{height}) = (1.2\text{ m})\cdot(1.5\text{ m})\cdot(2.5\text{ m}) = 4.5\text{ m}\cdot\text{m}\cdot\text{m}$$

Notice once again that I did the numbers separately from the units. I multiplied the numbers together to get 4.5, and then I multiplied the units together to get m·m·m. How do we abbreviate that? Well, what is 2·2·2? It's 2^3, right? That means m·m·m is m^3, which is usually referred to as "cubic meters" or "meters cubed," because when you multiply something by itself three times, mathematicians say that you are "cubing" it. So, the box has a volume of 4.5 m^3.

Suppose I have another box that is 1 m long, 2 m wide, and 2.2 m high. It's shorter in length and height than the previous box, but it's wider. Which box can hold more stuff? All you have to do is compare the volumes. The first box has a volume of 4.5 m^3, and the second box has a volume of:

$$\text{Volume} = (\text{length})\cdot(\text{width})\cdot(\text{height}) = (1\text{ m})\cdot(2\text{ m})\cdot(2.2\text{ m}) = 4.4\text{ m}^3$$

The first box can hold a bit more stuff, because its volume is a bit larger.

Units like m^2 and m^3 might seem strange to you, but they are very common in science. In fact, they are so common that they have a name – **derived units**.

Derived unit – A unit of measurement produced by the mathematical combination of simpler units

The meter, for example, is a simple unit telling you distance. When you multiply length and width, you get the derived unit for area, m^2. When you multiply length, width, and height, you get the derived unit for volume, m^3.

While cubic meter is the standard unit for volume, it isn't used nearly as much as another metric volume unit with which you are familiar – the **liter**, which is abbreviated as "L." Many consumable liquids like bottled water, juice, and soda are sold in ½-liter, 1-liter or 2-liter bottles. Where does this unit come from? Believe it or not, it comes from the formula for volume that is on the previous page. However, instead of using meters (m) as the unit in the equation, it uses centimeters (cm). What unit would you get if you measured length, width, and height in cm and then multiplied them together to get volume? You would get cm·cm·cm, or cm^3. This is called a **cubic centimeter**, and it is often abbreviated as "cc." The volume of medicines is often measured in cm^3, so you might hear a doctor say something like, "administer 10 cc's of Penicillin." That means the patient is supposed to be injected with a dose of Penicillin whose volume is 10 cm^3.

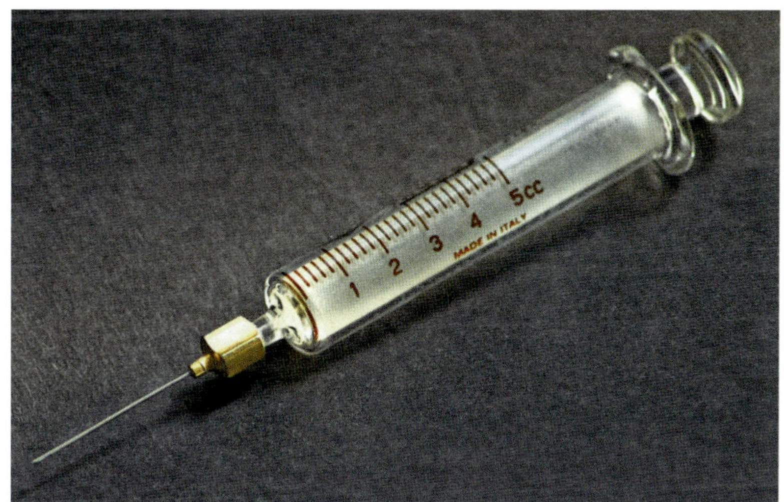

This needle could be used to inject medicine into a patient. Notice that the unit listed is "cc." That means the marks measure the volume of the medicine in cubic centimeters.

But what does this have to do with liters? Well, the liter is defined as 1,000 cubic centimeters. So even though the metric unit of liter doesn't look like a derived unit, it is. It is based on the cubic centimeter, which is what results when you measure length, width, and height in cm and then multiply them together to get volume. Of course, since liter is a metric unit, it can have prefixes as well, so you can measure very large volumes in kiloliters (kL) and very small volumes in centiliters (cL) or milliliters (mL), with mL being the most common.

Converting Between Metric Units

Now you know the metric units for length (m), mass (g), time (s), and volume (m^3 or L). But since they are metric units, you can modify them with prefixes. While this is convenient, it can also lead to issues when comparing one measurement to another. For example, suppose you want to put curtains on your windows. You need to know the length and width of your windows, so you measure them to be 1.2 m high and 1.2 m wide. You then go shopping for curtains, but you find that all the lengths and widths listed are in cm, not m. Do you have to go back home and remeasure the windows using cm as the unit? No. Since centimeters and meters are both units of length, they can easily be converted into one another.

How do you do the conversion? Different people do it in different ways, but in this course, I want to teach you a very specific way. While it might seem a little annoying, it is important to learn the method, because it is used quite frequently in science. In addition, once you understand the method, you can apply it to any set of units, whether or not you are familiar with them. This method is called the **factor-label method**, and it is based on a very simple mathematical fact. Suppose you are faced with the math problem below:

$$\frac{17}{2} \cdot \frac{5}{17} =$$

How would you solve it? You *could* multiply 17 by 5 to get the numerator, multiply 2 by 17 to get the denominator, and then simplify. However, it's much easier to recognize that when multiplying fractions, like terms in the numerator and denominator cancel. In other words, you can just cancel the 17's:

$$\frac{\cancel{17}}{2} \cdot \frac{5}{\cancel{17}} = \frac{5}{2}$$

Now remember, in equations, you treat the units separately from the numbers, but the units still go through the same mathematical steps. We can use this to our advantage.

Remember, your windows are 1.2 m high and 1.2 m wide. However, the measurements on the curtains in the store are in cm. You can convert between meters and centimeters by starting with the definition of the prefix "centi." It means "0.01." That tells you:

$$1 \text{ cm} = 0.01 \text{ m}$$

Notice what I did. I started with the unit that has the prefix: 1 cm. After the equal sign, I replaced the "c" in "cm" with the number it represents, 0.01. This is called a **conversion relationship**. It tells you how these two units of distance compare to one another.

Now that I have the conversion relationship, I can convert between units by setting things up like I am multiplying fractions. Remember that I can make any number a fraction by just putting it over 1. So a measurement of 1.2 m can be turned into a fraction this way:

$$\frac{1.2 \text{ m}}{1}$$

This is still the same measurement: 1.2 m. I have just turned it into a fraction. Now I need to multiply this fraction by another fraction that will convert the meters into centimeters. Not surprisingly, I use the conversion relationship to build that second fraction. I will do this by putting one side of the equation over the other side. The way I do that is very important, however. I need to make the fraction so that "m" cancels and "cm" ends up being the final unit. I can do that if I put the "1 cm" on the top and the "0.01 m" on the bottom:

$$\frac{1.2 \text{ m}}{1} \cdot \frac{1 \text{ cm}}{0.01 \text{ m}}$$

Remember, we do the numbers and the units separately, but they each go through all the mathematical steps. Before doing the numbers, then, look at the units. There is an "m" in the numerator and an "m" in the denominator. Those are the same, so they cancel each other out:

$$\frac{1.2\ \cancel{m}}{1} \cdot \frac{1\ cm}{0.01\ \cancel{m}}$$

That leaves us with "cm" as our only unit. That's good, because we want to know the height and width in cm. Now we just need to deal with the numbers. Once you have cancelled everything you can, you multiply the numerators together, multiply the denominators together, and then divide the numerator by the denominator:

$$\frac{1.2\ \cancel{m}}{1} \cdot \frac{1\ cm}{0.01\ \cancel{m}} = \frac{1.2\ cm}{0.01} = 120\ cm$$

We now know that 1.2 m is the same as 120 cm. You might be able to do that in your head. However, in this course, I want you to learn how to do it this way because later on, you will deal with units like parts per million (ppm) and carats (ct), and you will have a hard time converting those kinds of units in your head. However, this method always works, regardless of whether or not you know the units. Study the following example problem so you can get more experience with this method.

Example 1.1

The highest mass a regulation golf ball can have is 46 grams. How many kilograms is that?

First, we use the prefix to determine the conversion relationship. The prefix "kilo" means "1,000," so we put a "1" next to the unit with the prefix, and then on the other side of the equal sign, we replace the prefix with its meaning: 1,000:

$$1\ kg = 1{,}000\ g$$

Now we set up the conversion like we are multiplying fractions. The original measurement we have goes over 1:

$$\frac{46\ g}{1}$$

Now we multiply by a fraction made from the conversion relationship. However, we need to get rid of the "g," so the "1,000 g" needs to go on the bottom of the fraction. That way, the g's will cancel:

$$\frac{46\ \cancel{g}}{1} \cdot \frac{1\ kg}{1{,}000\ \cancel{g}} = \underline{0.046\ kg}$$

To get the number part of the answer, we multiply 46 by 1 and divide the result by 1 times 1,000. In other words, it's 46 divided by 1,000, which is 0.046.

I strongly recommend that you use a calculator when you do problems like these. There are so many other things you need to think about, I don't want you to have to think about the multiplication and division. You should even use a calculator on the tests.

In case you are wondering why we can take a measurement and multiply by the conversion relationship in fraction form, remember that the two sides of the conversion relationship are the same. That's what the equal sign means. In the example, the conversion relationship was 1 kg = 1,000 g. That means 1 kg and 1,000 g are equal. When you take two equal things and divide one by the other, what do you get? You always get 1. So, when you multiply by the conversion relationship in fraction form, you are really just multiplying by 1, which doesn't change anything. The original measurement remains the same value, just in the new unit. Let's do one more just so you have some more practice.

Example 1.2

The volume of a regulation golf ball is at least 40.7 milliliters. How many liters is that?

Remember, we start with the conversion relationship. The prefix "milli" means "0.001," so we put a "1" next to the unit with the prefix, and then on the other side of the equation, we replace the prefix with 0.001:

$$1 \text{ mL} = 0.001 \text{ L}$$

Now we set up the conversion like we are multiplying fractions. The original measurement we have goes over 1:

$$\frac{40.7 \text{ mL}}{1}$$

Now we multiply by a fraction made from the conversion relationship. However, we need to get rid of the "mL," so the "1 mL" needs to go on the bottom of the fraction. That way, the mL's will cancel:

$$\frac{40.7 \cancel{\text{mL}}}{1} \cdot \frac{0.001 \text{ L}}{1 \cancel{\text{mL}}} = \underline{0.0407 \text{ L}}$$

To get the number part of the answer, we multiply 40.7 by 0.001 and divide the result by 1 times 1. In other words, it's 40.7 times 0.001 divided by 1, which is 0.0407.

Make sure you understand what you learned today by answering the questions that follow.

Comprehension Check

1.9 What is the volume of a tiny box that is 16 mm long, 20 mm wide, and 15 mm tall? Don't forget to include the unit!

1.10 The maximum mass of a regulation bowling ball is 7.3 kg. How many grams is that?

1.11 Light can travel once around the earth in 13.4 centiseconds. How many seconds is that?

Converting Between Unit Systems

Scientists tend to use metric units, but depending on the country in which you live, they may not be familiar to you. If you live in the United States, for example, you don't measure distances in meters, millimeters, centimeters, and kilometers. Instead, you use inches, feet, and miles. Those are distance units, but they are part of the **imperial unit system**.

Imperial units – Measurement units defined by Great Britain in 1825

Obviously, imperial units are different from metric units. For example, an inch is longer than a centimeter, but it is shorter than a meter. Nevertheless, inches (abbreviated with "in") and centimeters measure the same thing: distance. Thus, there must be a relationship between them. It turns out that an inch is 2.54 times longer than a centimeter. In other words, it takes 2.54 cm to make 1 in. We could express this mathematically as follows:

$$1 \text{ in} = 2.54 \text{ cm}$$

Based on what you have learned so far, what is that? It's a conversion relationship! That equation allows us to convert between centimeters and inches (or inches and centimeters), as shown in the example below:

Example 1.3

The length of a piece of string is 16.51 cm. How many inches is that?

Remember, we start with the conversion relationship, which was given above:

$$1 \text{ in} = 2.54 \text{ cm}$$

Now we set up the conversion like we are multiplying fractions. The original measurement we have goes over 1:

$$\frac{16.51 \text{ cm}}{1}$$

Now we multiply by a fraction made from the conversion relationship. However, we need to get rid of the "cm," so the "2.54 cm" needs to go on the bottom of the fraction. That way, the cm's will cancel:

$$\frac{16.51 \cancel{\text{cm}}}{1} \cdot \frac{1 \text{ in}}{2.54 \cancel{\text{cm}}} = \underline{6.5 \text{ in}}$$

To get the number part of the answer, we multiply 16.51 by 1 and divide the result by 1 times 2.54. In other words, it's 16.51 divided by 2.54.

To make the example above a bit easier, I chose the string measurement so that it worked out to a fairly simple number of inches. However, when you are converting between systems, that doesn't always happen. For example, the average adult male in the United States is 175 cm tall. How many inches is that? Well, if we followed the same procedure given above, we would find that it is:

$$\frac{175 \cancel{\text{cm}}}{1} \cdot \frac{1 \text{ in}}{2.54 \cancel{\text{cm}}} = 68.897637795 \ldots$$

Those three dots mean that the answer continues with a lot more digits. Obviously, you don't want to write all those digits, so you need to stop somewhere. When you take high school chemistry, you will find that there are rules for when to stop reporting the digits in an answer like that one. Fortunately, you don't have to worry about that here. For this class, just keep three digits in your answer.

Because you will be dropping digits, there is one more thing you need to worry about: rounding your answer. When you drop digits, you look at the first digit you are going to drop. If it is 0-4 just drop it and all the remaining digits and don't do anything. That's called **rounding down**, because the actual answer is just a bit bigger than the answer you are reporting. However, if it is 5-9, you need to add one to the last digit you are not dropping. That's called **rounding up**, because the answer you are reporting is just a bit larger than the actual answer. Rounding up and rounding down ensures that the answer you give is closer to the actual answer given by the calculator. In this case, the calculator answer was 68.897637795..., and we need to keep three digits, so we need to drop the first "9" in the number. Because we are dropping a "9," we round up, and the answer is 68.9 cm.

Of course, each unit in the metric system has a unit in the imperial system to which it can be related. Inches, for example, can be related to cm, as I just showed you. In fact, any length unit in the metric system can be related to any length unit in any other system, including the system used in the United States, which is similar to the imperial system. This is true for mass units, volume units, etc. Below you will find a few of those relationships.

Relationships Between Some Metric and U.S. Units

Physical Quantity	Metric Unit	U.S. Unit	Relationship
Distance	centimeter (cm)	inch (in)	1 in = 2.54 cm
Mass	gram (g)	slug (sl)	1 sl = 14,594 g
Volume	liter (L)	gallon (gal)	1 gal = 3.785 L

You will not need to know any of these relationships. You won't even need to know the names of the U.S. units. If you need them to solve a problem or answer a question, they will be given to you. Let's do one more conversion between metric and U.S. units before we move on.

Example 1.4

A regulation basketball has a volume of 1.88 gallons. How many liters is that?

Remember, we start with the conversion relationship, which is given in the table:

$$1 \text{ gal} = 3.785 \text{ L}$$

Now we set up the conversion like we are multiplying fractions. The original measurement we have goes over 1:

$$\frac{1.88 \text{ gal}}{1}$$

Now we multiply by a fraction made from the conversion relationship. However, we need to get rid of the "gal," so the "1 gal" needs to go on the bottom of the fraction. That way, the gals will cancel:

$$\frac{1.88 \cancel{\text{gal}}}{1} \cdot \frac{3.785 \text{ L}}{1 \cancel{\text{gal}}} = \underline{7.12 \text{ L}}$$

The calculator's answer is 7.1158. However, I told you to keep three digits, so you need to drop the "58." Since the first digit you are dropping is a 5, you must round the answer up.

The Importance of the Factor-Label Method

Hopefully, the previous example gives you an idea of why it is important to learn the factor-label method. When you are familiar with the unit, you can often do conversions in your head. Many people have no problem realizing that 150 cm is the same as 1.5 m, because they know that a cm is 100 times smaller than a meter, so if you divide 150 by 100 (which many people can do in their heads), you convert from cm to m. Of course, if you aren't thinking really clearly, you might end up multiplying by 100, which would give you the wrong answer. More importantly, the factor-label method works regardless of whether or not you are familiar with the unit.

In the previous example problem, you know that 1 gal = 3.785 L, so you know that in order to convert from gallons to liters, you need to do something with 3.785. But what do you do? Do you divide or multiply? Well, if you are familiar with the units, you can usually reason it out. However, the more unfamiliar the unit, the harder it is to figure out. Consider, for example, the way God instructed Noah to build the ark: "This is how you shall make it: the length of the ark three hundred cubits, its breadth fifty cubits, and its height thirty cubits" (Genesis 6:15, NASB). Since God is listing length, breadth (width), and height, you know that the cubit is a unit for measuring distance. It was defined as the distance from a man's elbow to the end of his fingertips. While that varied from person-to-person, a commonly-used value is 0.517 m. So that tells us:

$$1\text{ cubit} = 0.517\text{ m}$$

You probably aren't very familiar with the cubit, but you can still convert the ark's measurements into meters, because you know the factor label method. For example, the length of the ark is 300 cubits. You put that over 1 to make it a fraction, and then you multiply by a fraction made from the conversion relationship. That fraction is made so that cubits cancel. When you do that, you get:

$$\frac{300\ \cancel{\text{cubits}}}{1} \cdot \frac{0.517\text{ m}}{1\ \cancel{\text{cubit}}} = 155\text{ m}$$

The answer is really 155.10 meters, but remember, you only keep three digits. So the ark was 155 m long. An American football field is 91.4 m long, so the ark was more than 1.5 times as long as an American football field. If you do this with the other measurements, you find that the ark was also 25.9 m wide and 15.5 m tall.

That gives you a better idea of how big the ark was, doesn't it? In fact, there is an organization in the United States, Answers in Genesis, that has built a life-sized model of Noah's ark, which is pictured below:

So that you have an idea of how big it is, look at the people in the foreground or the cars on the right! Of course, when they had the ark built, Answers in Genesis didn't use the unit cubits to guide its construction. They converted cubits to units that the builders knew how to use.

The point I am trying to make is that you don't have to be familiar with what a cubit is in order to use the factor-label method to convert it into a different unit, as long as you know the relationship between them. This is important, because in this course, you will be using some units with which you are not familiar, like parts per million (ppm) and carats (ct). When you study chemistry and physics in high school, you will use even less-familiar units, like moles. If you learn the factor label method now, it will make dealing with those units *much* easier.

Identifying Measurements Based on Units

Units are important for many reasons. Not only are they necessary for the measurements themselves, but they also tell you what is being measured. For example, consider the basic metric unit for mass – the gram. If you read a measurement whose unit is grams, (or mg, cg, kg, etc.), then you know it is a mass measurement, because that's what grams measure. Thus, I don't have to say, "An object whose mass is 15 g." I could just say, "A 15-g object." Because of the unit, you would know that I am telling you the mass of the object. In the same way, if I said "A 2.5-kL object," you would know that I am giving you the volume of the object. The metric unit liter measures volume, so any unit based on liters (mL, cL, kL, etc.) measures volume.

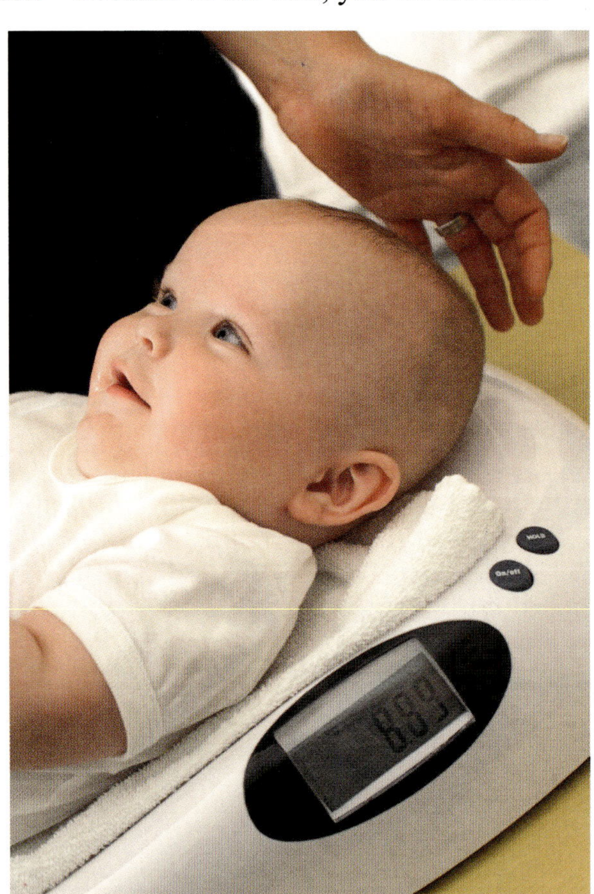
This baby is being weighed, but the scale is indicating kg, which is mass, not weight.

While scientists are very careful about this, people often aren't. In everyday language, for example, people often confuse units. For example, a doctor tends to track a patient's weight, since a person's weight is related to his or her health. This is especially important for babies, so there are special scales used to weigh them. On the right, you see a picture of a baby being weighed. Notice what the scale says the weight is: 8.89 kg. What's wrong with that? The unit kg does not measure weight; it measures mass.

As I told you already, mass and weight are related, but they are not the same thing. Mass is a measure of how much matter is in something. You will learn more about this when you take high school physics, but weight is a measure of how strong the planet's gravity pulls on an object. Since they are different things, they are measured with different units. In the metric system, the unit for weight is the Newton. Thus, since the scale is measuring the baby's weight, it should give an answer in Newtons, not kg. The U.S. unit for mass (as shown in the table on page 18) is the slug. What's the U.S. unit for weight? It's the pound. So, if I report something in pounds or Newtons, I am reporting weight. If I report it in grams (or kg, mg, etc.) or slugs, I am reporting mass.

Why does this confusion exist? Because as long as you stay near the surface of the earth, a person's mass can be calculated from his or her weight, and vice-versa. For example, I can tell you that this baby's actual weight is 87.2 Newtons (which is also 19.6 pounds). Since nearly everyone stays near the surface of the earth, the distinction between mass and weight is not all that important.

Nevertheless, for the metric system, you need to know what is being measured simply by the unit. So when you see one of the units I want you to recognize (gram, meter, seconds, liter, along with any of the three prefixes you are supposed to know), you have to be able to tell me what is being measured.

Comprehension Check

1.12 As you determined earlier, the maximum mass of a regulation bowling ball is 7,300 g. What is that mass in slugs? (1 sl = 14,594 g)

1.13 A furlong is another imperial unit for distance. If the distance between Washington, DC and New York, NY is 1,840 furlongs, what is it in meters? (1 furlong = 201 m)

1.14 You are reading a laboratory notebook, and you see several pieces of data: 14.5 mL, 16.2 kg, and 1.2 cm. For each piece of data, indicate (based on the unit) what was being measured.

It's Not the Heat

When studying the earth, another very important thing you need to measure is temperature, and while we tend to monitor temperature regularly, most people don't understand what it measures. Many students (and unfortunately, even some textbooks) say that temperature is a measure of heat. After all, when the temperature goes up, we say that it is "hot." Thus, temperature measures heat, right? Wrong! Heat has a very specific definition:

Heat – Energy that is exchanged because of a difference in temperature or a change in phase

As I discussed already, when you "heat up" ice, it melts. The reason we say "heat up" is because you are adding energy to change the phase. That means energy is being given to the ice (being exchanged) to change its phase. That's heat. If you cool something down, you are also causing energy to be exchanged, so once again, you are using heat. It is just being exchanged in the opposite direction.

So, heat is energy that is being exchanged. One reason it might be exchanged is to change phase. Another reason is because of a difference in temperature. If you have something hot (like a fire) and something cooler (like your body), energy will be exchanged. It will leave the fire and be absorbed by your body, as shown in the diagram on the right. The flow of energy illustrated by the yellow arrow is heat. Temperature, then, does not measure heat. However, it does determine the direction in which energy travels. Heat is the flow of energy from something at a higher temperature to something else at a lower temperature.

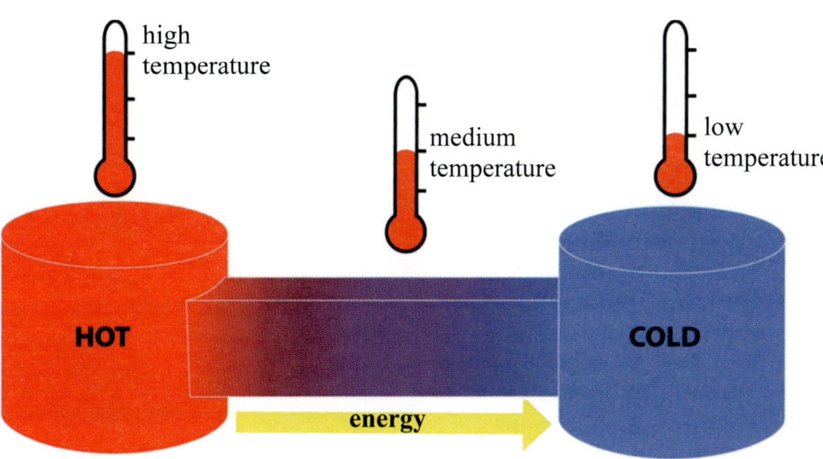

When energy is exchanged because of a temperature difference, it always flows from the hotter object to the cooler object.

So far, I have only told you what temperature *does not* measure. It doesn't measure heat. What does it measure? To help you understand the answer to that question, I want you to perform an experiment.

Experiment 1.3: Movement You Do Not See

Supplies:

- M&M candies (You need two each of four different colors. See the photo below.)
- Water
- Ice
- A pan for boiling
- A bowl that holds roughly as much water as the pan (They just need to be close.)
- Two plates that are either metal or ceramic and can sit comfortably on top of the pan and bowl (The closer to white they are, the better.)
- A large serving spoon
- A stove or hotplate

Instructions:

1. Fill the pan about half full of hot water from the tap. Put it on the stove and get the water boiling.
2. Fill the bowl with ice and water until it is almost completely full.
3. Put the two plates on a flat surface and arrange four M&M candies, each with a different color, in a square at the center, as shown in the pictures below.
4. Once the water is boiling, turn off the stove and move the pan to a part of the stove that is no longer hot so the water stops boiling.
5. Use the serving spoon to take ice-cold water out of the bowl.
6. Gently add the water from the spoon to one of the plates, far from the M&Ms. You want the water to flow to the candies, but you don't want them to move. You need to have a thin layer of water in the plate. It should not cover the candies, but it should completely surround them. If you need more than one spoonful of cold water, that is fine.
7. Put the plate on top of the bowl so that it rests there.
8. Use the serving spoon to take hot water from the pan. **Be careful, it is hot!**
9. Add that water to the other plate, just like you added the cold water to the first plate. Once you have a thin layer of water that surrounds the candies but doesn't cover them, put the plate on top of the pan. **Once again, be careful! The plate will be warm, and the pan is hot!**

10. Your experiment should now look something like the picture on the right.
11. Watch how things change over the next few minutes. What differences do you note?
12. You can leave this experiment set up for a while if you want to come back later and see what happens after a long time. However, whenever you are finished, clean up your mess.

What did you see in your experiment? In both plates, you should have seen colors spreading out away from the candies. That should probably make sense to you. The candy coating dissolved in the water, so the water started taking on the colors of the candy coatings. As time went on, what differences did you see in each plate? You should have seen that the colors spread much farther from the candies on the hot plate than they did on the cold plate. Also, while the colors stayed partly separated from each other, after a while, they did begin to mix. However, they mixed to a much greater extent on the hot plate than on the cold plate.

What caused the difference? The temperature of the water. Even though the water wasn't moving on the plate, the *water molecules* were. As you learned earlier, molecules in liquids can move around. In your experiment, the candy coating on the M&Ms dissolved in water because the water molecules were moving around and slamming into the candies. The water molecules and the molecules that make up the candy coating were attracted to one another, so after the collision, some of the molecules from the candy coating "wanted" to stay close to the water molecules that slammed into them, so they followed the water molecules. As a result, the molecules from the candy coating started spreading out into the water.

Well, at higher temperatures, water molecules move faster. In your experiment, then, the water molecules in the hot water slammed into the candies harder and pulled the molecules from the coating off more quickly. As a result, the colors spread out more quickly. That's what temperature is really a measure of. It is a measure of the energy associated with the motion of a substance's molecules.

Temperature – A measure of the energy associated with the random motion of a substance's molecules

But wait a minute. How does that make you hot or cold? If you had put your hand in the ice water, it would have gotten uncomfortably cold. If you had put your hand in the hot water, you would have gotten burned. How does the motion of the molecules do that? Well, think about it. The water molecules collide with whatever is in the water, so if you put your hand in the hot water, they would collide with your skin. Since the water molecules in the hot water move very quickly, the collision would be violent enough to damage cells in your skin, causing a burn.

Now remember, the molecules in your skin cells are moving as well. So, if you put your hand in cold water, the molecules in your skin cells would be moving faster than the molecules in the water. As a result, when a collision occurred, energy would be transferred from your skin to the water. As your skin loses energy, the molecules in your skin cells slow down, which reduces the temperature of your skin. That makes your skin cold.

In the end, then, temperature is *not* a measure of heat. It is a measure of how quickly a substance's molecules are randomly moving around. Now remember, even the molecules of a solid move. They don't move from place to place, but they do vibrate back and forth. Thus, regardless of whether it is a solid, liquid, or gas, the higher the temperature, the more quickly the molecules that make up the substance are moving.

You might still be wondering about one thing you observed in the experiment. Why did it take so long for the colors to start mixing? Well, the molecules that make up the dyes that give the candies their colors are much larger than water molecules. In order to spread, the dye molecules must move through other molecules, and it is easier for them to move through the smaller water molecules than the other dye molecules. At first, then, the color didn't mix much. However, as time went on, the dye molecules did move a bit through the other dye molecules, so there was some color mixing. However, once again, the colors mixed more in the hot water, because those dye molecules were moving faster, which made it easier for them to move through the other dye molecules.

Comprehension Check

1.15 You are watching the molecules of a substance and notice that over time, their random motion gets slower and slower. Is the substance increasing in temperature, decreasing in temperature, or remaining at the same temperature?

Measuring Temperature

Now that you know what temperature measures, we can discuss how we measure it and what units we will use. Let's start with how we measure temperature. As you already know, we use **thermometers**, to measure temperature. But how? Well, there are different kinds of thermometers, but the most common kind is illustrated in the drawing below. It consists of a tube of glass that contains a colored liquid. When you put the thermometer in something, the molecules of the substance start colliding with the molecules in the glass. If the molecules in the substance have more energy than the molecules in the glass, energy goes from the substance to the glass. If the molecules of the substance have less energy than the molecules in the glass, energy goes from the glass to the substance.

As the energy moves, the liquid inside the thermometer warms up or cools down, depending on whether the thermometer is gaining energy or losing energy. If the liquid warms up, it expands. This causes it to rise in the tube. If it cools down, the liquid contracts, which causes the liquid to lower in the tube. Based on the final height of the liquid, you can read the temperature. The higher the liquid has risen, the higher the temperature.

But how does the height of liquid in a tube tell us the temperature? We define a temperature scale and then mark a thermometer so it reads that scale. For example, the temperature scale that we will use in this course is the Celsius (sell' see us) temperature scale, named after Anders Celsius, a natural philosopher who lived in the first half of the 18th century. In this scale, water freezes at 0 degrees and boils at 100 degrees. To make the thermometer read that scale, you put it in water that is in the process of freezing, and you make a mark that represents 0. Then, you put it in boiling water and make a mark that represents 100. You then divide up the rest of the thermometer so that there are an even number of marks between 0 and 100. You now have a thermometer that reads the temperature in degrees Celsius.

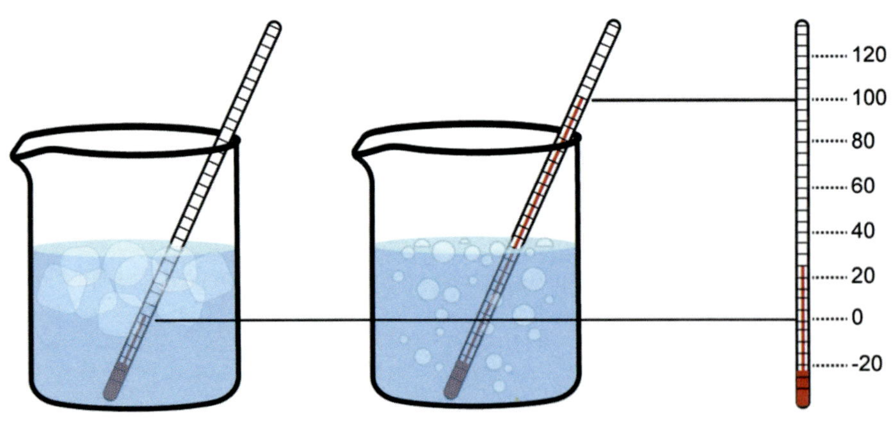

The Celsius temperature scale is defined so that water freezes at 0 degrees and boils at 100 degrees.

There are other kinds of thermometers. Some use the flow of electricity through a metal to determine the temperature. Others use the way solids expand and contract to measure temperature. In addition, there are different temperature scales. The one with which you might be more familiar is the Fahrenheit scale, named after Daniel Gabriel Fahrenheit (fair' uhn height'), a natural philosopher who was 15 years older than Anders Celsius. On that scale, water freezes at 32 degrees and boils at 212 degrees. To distinguish between the two, we used "°C" to represent temperatures in the Celsius scale and "°F" to represent temperatures in the Fahrenheit scale.

Because both scales measure temperature, you can convert between them. However, I don't want you to worry about doing that, because the conversion doesn't use the factor-label method, and I want to concentrate on that in this course. However, I do want you to get a general idea of how the scales relate to one another, so study the following example problems.

Example 1.5

Which is colder: 4 °C or 31 °F?

You might be tempted to just choose the smaller number, assuming it represents the lower temperature. However, think about the phase of water. At 4 °C, water is a liquid, because it freezes at 0 °C. However, at 31 °F, water is frozen, because it freezes at 32 °F. The temperature at which water is frozen will be lower than the temperature at which it is still a liquid, so 31 °F is the colder temperature.

Which is warmer: 200 °F or 101 °C?

You might be tempted to just choose the larger number, but once again, think about the phase of water. At 101 °C, water is a gas, because it boils at 100 °C. However, at 200 °F, water is a liquid, because it boils at 212 °F. The temperature at which water is a gas will be higher than the temperature at which it is still a liquid, so 101 °C is the warmer temperature.

Obviously, you can't compare all Celsius and Fahrenheit temperatures that way. For example, water is a solid at -1 °C and 5 °F. Thus, unless you convert one temperature to the other scale, it would be hard for you to tell that 5 °F is the colder temperature. However, if I ask you to compare Celsius and Fahrenheit temperatures, there will always be a difference in the phase of water. Thus, all you need to know for this course is how both scales are defined. If you know that, you will be able to answer any questions I ask using the same reasoning found in the two example problems above.

Measuring Density

You have probably learned about density already, but it is an important thing to measure in earth science, so I need to make sure you understand it. Density tells us how tightly-packed the matter in an object is. There is a simple equation that allows you to calculate the density of any object:

$$\text{density} = \frac{\text{mass}}{\text{volume}}$$

If you know the mass of an object, you know how much matter it has in it. If you then divide the mass by the volume, you have a number that tells you how tightly-packed that matter is.

Think about the units for density. Mass can be measured in grams, for example, while volume can be measured in liters. Now remember, when units are in an equation, you do the math with them as well. Some of them might cancel, like they do in conversions, but if they don't cancel, the results of the math are the units. If you divide grams by liters, you get $\frac{\text{grams}}{\text{liter}}$ ($\frac{\text{g}}{\text{L}}$), which is called "grams per liter." That's one unit for density. In earth science, however, it is much more common to measure volume in milliliters (mL), so the more common density unit is what results when grams are divided by milliliters: $\frac{\text{grams}}{\text{milliliter}}$ ($\frac{\text{g}}{\text{mL}}$).

This looks like gold, but its density indicates it's not.

The density of an object is important because it is different for different substances. Consider, for example, the picture on the left. What does that look like to you? It looks like gold, doesn't it? Well, at room temperature, gold has a density of 19.3 $\frac{\text{g}}{\text{mL}}$. If that really is gold, it would have the same density at room temperature. If I take the mass of that sample and divide by its volume, however, I get 4.9 $\frac{\text{g}}{\text{mL}}$, which tells me it's not gold. In fact, that's a picture of iron pyrite, which is often called "fool's gold" because it looks like gold but isn't.

Density can be useful in another way as well. It can tell you whether or not something will float in water. At room temperature, water has a density of 1.0 $\frac{\text{g}}{\text{mL}}$. Anything that has a lower density than that will float in water, and anything that has a higher density will sink. For example, the iron pyrite pictured above will sink in water, because its density (4.9 $\frac{\text{g}}{\text{mL}}$) is higher than 1.0 $\frac{\text{g}}{\text{mL}}$. However, a cork has a density of 0.2 $\frac{\text{g}}{\text{mL}}$, which is lower than water's density. That means it will float. In the same way, ice has a density of 0.9 $\frac{\text{g}}{\text{mL}}$, which explains why ice cubes float in your drink. Make sure you can use this kind of reasoning by studying the following example.

Example 1.6

A 45-gram object takes up 65 mL of space. Will it float or sink in water?

To determine whether something floats or sinks in water, you must calculate its density. Now remember, you can recognize what is being measured just by the units. So even though I didn't tell you that the mass is 45 grams, you see that it is, because a measurement with a unit of grams must be a mass measurement. In the same way, L (or any prefix with L, like mL) measures volume, so the 65 mL measurement must be the volume. Now that we know mass and volume, we simply stick them into the equation:

$$\text{density} = \frac{\text{mass}}{\text{volume}} = \frac{45\ \text{g}}{65\ \text{mL}} = 0.69\ \frac{\text{g}}{\text{mL}}$$

Now remember, we do the numbers and the units separately. When you divide 45 by 65, you get 0.69230769…. However, you only need to keep three digits (even when the one left of the decimal is 0), so that means you should drop the "2" and everything after it. Since you are dropping a two, you round down, so the number is 0.69. For the units, you do the same math, so g divided by mL is $\frac{\text{g}}{\text{mL}}$. Since the density is less than 1.0 $\frac{\text{g}}{\text{mL}}$, <u>it will float in water</u>.

I need to make two more points about this. First, I told you that an object will sink in water if its density is higher than that of water, and it will float if its density is lower. But what happens if its density actually equals the density of water? That doesn't happen very often, because each substance has its own density. However, if you are clever, you can make an object with a density equal to that of water. If you do, the object will stay wherever you place it in water. If you put it on the surface, it will completely submerge, but it will stay on the surface. If you put it 10 cm below the surface, it won't rise or sink; it will continue to stay 10 cm below the surface.

The balloons that are floating are less dense than air, while the balloons that aren't floating are denser than air.

The other point is more important. This works for any substance in which floating is possible, not just water. Consider the picture on the right. Why are some of the balloons floating, while the others rest on the floor? You know the answer – the balloons that are floating are filled with helium, while the balloons that are not floating are filled with air. But what makes helium-filled balloons float? The mass of helium is low, so when you take the total mass of the helium plus the mass of the balloon itself and divide by the volume of the balloon, the density is lower than the density of the air around it, so the helium balloon floats. When you fill a balloon with air, the total mass of the air plus the mass of the balloon itself divided by the volume is greater than the density of air, so it sinks.

Measuring Concentration

Suppose you are hosting a party. You make some snacks so that your guests can have something to eat, and you decide to make lemonade for them to drink. You get out the lemonade mix and read the instructions. It says to add one packet of lemonade mix to two liters of water and stir so that it all dissolves. The problem is that you have only one packet left, and you know that two liters will not be enough for all your guests. You decide to add one packet of lemonade mix to four liters of water. That way, there is plenty for everyone. What's wrong with that reasoning? You will, indeed, have four liters of "lemonade" to drink, but it won't taste right, will it? The "lemonade" will taste too weak.

But why will it taste weak? Because in order for it to taste right, the chemicals in the lemonade mix must have a specific **concentration.**

Concentration – The amount of a substance in a defined volume

In the scenario I just described, you took chemicals that were supposed to go into two liters of volume and instead put them into four liters of volume. That spread the chemicals over too large a volume, reducing their concentration, which reduced the strength of their taste.

It turns out that concentration is incredibly important when it comes to understanding how chemicals behave. For example, you know that you are supposed to have a diet that is rich in vitamins, right? Your body needs vitamins, and if you don't get enough of them, their concentration becomes too low, and you can get sick. However, you might not realize that for some vitamins, too high a concentration can also make you sick! Think about the lemonade. If the concentration of the mix is too low, it tastes weak. What if you added an entire packet to only one liter of water instead of the recommended two liters? It would taste too strong. For the lemonade to taste right, it needs to be at the right concentration. In the same way, for vitamins to work right, they have to be at the right concentration.

There are lots of different ways to measure concentration, but one of the most commonly-used units is **percent (%)**. When you hear that word, you probably think about buying something on sale and getting a discount. However, the word "percent" actually means "per hundred." Thus, it can be a concentration unit as well. For example, suppose I dissolve salt in water. Suppose further that I count the molecules of salt and the molecules of water after I am done. If I count 100 total molecules and find that 15 of them are salt, I could say that the concentration of salt is 15 molecules of salt per 100 molecules in the mixture. That means it is 15% salt. The "%" symbol actually means "per hundred." If there were 40 molecules of salt in 100 molecules of the mixture, the mixture would be 40% salt, which is more concentrated.

The colors tell you the concentration of drink mix in these cups.

Thinking about percent as a concentration unit, then, consider the picture on the left. Each drink was made from the same mix. Which cup holds the highest percentage of mix in it? Which holds the lowest? You can tell from the color. The darker the color, the more concentrated the mix is in the drink, so the percent of drink mix in the middle cup is the highest, and the percent of drink mix in the cup on the right is the lowest. As we study things like rocks, soil, and the air we breathe, we will discuss the concentration of chemicals in terms of percent as well as other units that I will introduce later on in the course.

Comprehension Check

1.16 Which is warmer: 215 °F or 99 °C?

1.17 At room temperature, silver has a density of $10.5\,\frac{\text{g}}{\text{mL}}$. If you find a 20-mL, 210-g object that looks like it is made of silver. Is it?

1.18 Will a 125-mL, 90-g object float in water, which has a density of $1.0\,\frac{\text{g}}{\text{mL}}$?

1.19 You are blindfolded and asked to taste two drinks. Each one is made by mixing lemon juice and water. Neither is as sour as pure lemon juice, but the first one is a lot more sour than the second. Which was made with the higher percent of lemon juice?

You are now done covering the material in the first chapter of the course. In order to prepare for the test, answer the Chapter Review questions on pages 33 and 34. Then have your parent/teacher check your answers. Correct and make sure you understand anything you got wrong, and then take the test. Recall that for the test, you need to remember anything that is in pink boxes in the chapter, including equations. You must also remember the definitions presented in the chapter.

Answers to the Comprehension Check Questions

1.1 It appears to be flat because the curve is very gentle. The bigger the sphere, the more gently its surface curves. The earth is a big sphere, so its curve is very gentle.

1.2 Africa is larger than Greenland. Since the earth is a globe, a flat representation distorts it. Things near the equator are less distorted than things far from the equator. Since Greenland is far from the equator, it is distorted a lot. If you can't picture this in your mind, that's fine. Just realize that a globe is a better representation of the earth, since the earth is actually a sphere. Thus, the globe will have a better representation of the continents than something that is flat.

1.3 The student is not correct. A change of just one atom makes a profound difference in the chemistry of the molecule. Since we breathe out CO_2, it is only bad for us when there is a lot of it. Small amounts of CO, however, can be deadly because of the way it interacts with our blood cells.

1.4 There are five atoms. "Fe" is one atom, because it has one capital letter. There is a "2" subscript after it, so there are two Fe atoms. O is the only other atom, because it is the only other capital letter. There is a "3" subscript next to it, so there are three O atoms. That makes a total of five atoms.

1.5 $MgSO_4$. Since we don't write 1's, there is nothing after Mg and nothing after S. Since there are four atoms of oxygen, there is a "4" subscript after the O.

1.6 It will become a liquid and then a gas. Yes, you can make metals into gases. You have to be careful, and it must be done without any oxygen present, but it can be done. In fact, that's why I heat metals in my research. I have to make them gases. Like all chemicals, metals can reach all three phases, if the conditions are correct.

1.7 You would use kiloseconds. A second is a short amount of time compared to the length of time you will be working on this course. The only prefix you need to know to measure large things is kilo. If you are like most students, you will spend about 700 kiloseconds on this course. That doesn't sound like much, does it?

1.8 The area is $616\ \text{cm}^2$. Area is length times width, so:

$$\text{Area} = (\text{length})\cdot(\text{width}) = (28\ \text{cm})\cdot(22\ \text{cm}) = 616\ \text{cm}^2$$

Remember, you do the numbers and units separately. $28\cdot 22 = 616$, and $\text{cm}\cdot\text{cm}$ is cm^2. If you are worried about why the unit isn't m^2 like it was in the room example I gave, remember that a unit with a prefix measures the same thing as the unit without a prefix. It is just smaller or larger than the basic unit. Since cm is a way of measuring smaller lengths than m, cm^2 is a way of measuring smaller areas than m^2. If you had measured the page in meters, you would have gotten an answer of $0.0616\ \text{m}^2$.

1.9 The volume is $4{,}800\ \text{mm}^3$. All we have to do here is plug the measurements into the formula for volume and then do the numbers and units separately:

$$\text{Volume} = (\text{length})\cdot(\text{width})\cdot(\text{height}) = (16\ \text{mm})\cdot(20\ \text{mm})\cdot(15\ \text{mm}) = 4{,}800\ \text{mm}^3$$

That may sound like a lot, but a mm^3 is a very small volume unit, so $4{,}800\ \text{mm}^3$ isn't a lot of volume.

1.10 It is 7,300 g. First, we use the prefix to determine the conversion relationship. The prefix "kilo" means "1,000," so we put a 1 next to the unit with the prefix, and then on the other side of the equal sign, we replace the prefix with 1,000:

$$1 \text{ kg} = 1{,}000 \text{ g}$$

Now we set up the conversion like we are multiplying fractions. The original measurement we have goes over 1:

$$\frac{7.3 \text{ kg}}{1}$$

Now we multiply by a fraction made from the conversion relationship. However, we need to get rid of the "kg," so the "1 kg" needs to go on the bottom of the fraction. That way, the kg's will cancel:

$$\frac{7.3 \cancel{\text{kg}}}{1} \cdot \frac{1{,}000 \text{ g}}{1 \cancel{\text{kg}}} = 7{,}300 \text{ g}$$

To get the number part of the answer, we multiply 7.3 by 1,000 and divide the result by 1 times 1. In other words, it's 7.3 times 1,000, divided by 1.

1.11 It is 0.134 s. First, we use the prefix to determine the conversion relationship. The prefix "centi" means "0.01," so we put a "1" next to the unit with the prefix, and then on the other side of the equal sign, we replace the prefix with 0.01:

$$1 \text{ cs} = 0.01 \text{ s}$$

Now we set up the conversion like we are multiplying fractions. The original measurement we have goes over 1:

$$\frac{13.4 \text{ cs}}{1}$$

Now we multiply by a fraction made from the conversion relationship. However, we need to get rid of the "cs," so the "1 cs" needs to go on the bottom of the fraction. That way, the cs's will cancel:

$$\frac{13.4 \cancel{\text{cs}}}{1} \cdot \frac{0.01 \text{ s}}{1 \cancel{\text{cs}}} = 0.134 \text{ s}$$

To get the number part of the answer, we multiply 13.4 by 0.01 and divide the result by 1 times 1. In other words, it's 13.4 times 0.01, divided by 1.

1.12 It is 0.50 sl. First, we write down the conversion relationship:

$$1 \text{ sl} = 14{,}594 \text{ g}$$

Now we set up the conversion like we are multiplying fractions. The original measurement we have goes over 1:

$$\frac{7{,}300 \text{ g}}{1}$$

Now we multiply by a fraction made from the conversion relationship. However, we need to get rid of the "g," so the "14,594 g" needs to go on the bottom of the fraction. That way, the g's will cancel:

$$\frac{7{,}300\ \cancel{g}}{1} \cdot \frac{1\ \text{sl}}{14{,}594\ \cancel{g}} = 0.50\ \text{sl}$$

The calculator says the answer is 0.50020556…, but keeping only three digits means dropping the second "0" and everything that comes after it. Since it's 0-4, we round down, making the answer 0.50.

1.13 <u>It is 370,000 m</u>. First, we write down the conversion relationship:

$$1\ \text{furlong} = 201\ \text{m}$$

Now we set up the conversion like we are multiplying fractions. The original measurement we have goes over 1:

$$\frac{1{,}840\ \text{furlongs}}{1}$$

Now we multiply by a fraction made from the conversion relationship. However, we need to get rid of the "furlong," so the "1 furlong" needs to go on the bottom of the fraction. That way, the furlongs will cancel:

$$\frac{1{,}840\ \cancel{\text{furlongs}}}{1} \cdot \frac{201\ \text{m}}{1\ \cancel{\text{furlong}}} = 370{,}000\ \text{m}$$

The calculator says the answer is 369,840, but keeping only three digits means dropping the "8" and everything that comes after it. Since that's between 5 and 9, we round up. That means turning the "9" into a "10," which makes the answer 370,000. If it bothers you that there are actually six digits in the answer, not three, remember that you can't drop any of those zeroes, because it would change the size of the number. When you drop digits, you can keep zeroes if you need to in order to keep the size of the number correct.

1.14 <u>14.5 mL is volume, 16.2 kg is mass, and 1.2 cm is distance</u>. The units tell you what is being measured. Liters (with or without prefixes) measure volume, grams (with or without prefixes) measure mass, and meters (with or without prefixes) measure distance.

1.15 <u>It is decreasing in temperature</u>. Temperature measures how quickly a substance's molecules are moving. The slower molecules move, the lower the temperature.

1.16 <u>215 °F is warmer</u>. It's not because the number is higher. It's because water boils at 212 °F and 100 °C. At 215 °F, water is a gas, but ag 99 °C, it is a liquid.

1.17 <u>Yes, it is made of silver</u>. You need to check its density. To do that, you use the formula:

$$\text{density} = \frac{\text{mass}}{\text{volume}}$$

Even though you aren't told which is mass and which is volume, you know from the units. Grams measure mass, and mL measures volume:

$$\text{density} = \frac{\text{mass}}{\text{volume}} = \frac{210\text{ g}}{20\text{ mL}} = 10.5\ \frac{\text{g}}{\text{mL}}$$

Since that equals the density of silver, it is probably made of silver.

1.18 Yes, it will float. Remember, you compare the density of the object to that of water. If its density is lower than water's density, it floats. If it is higher, it sinks. You know that 90 g must be the mass and 125 mL must be the volume because of the units:

$$\text{density} = \frac{\text{mass}}{\text{volume}} = \frac{90\text{ g}}{125\text{ mL}} = 0.72\ \frac{\text{g}}{\text{mL}}$$

Since that is lower than the density of water ($1.0\ \frac{\text{g}}{\text{mL}}$), it floats.

1.19 The first one is made with a higher percent of lemon juice. Remember, percent is a way to measure concentration. Since pure lemon juice is sour, the one that is more sour has more lemon juice in it, so it has the higher concentration and therefore the higher percent.

Chapter Review

1. Define the following terms:

a. Mass
b. Derived unit
c. Imperial units
d. Heat
e. Temperature
f. Concentration

2. Why did ancient sailors and people who lived near the ocean understand that the earth is spherical in shape?

3. What made people think that Christopher Columbus couldn't sail around the world?

4. N_2O gas is often called "laughing gas," because it can be used to help people ignore pain that occurs during medical procedures. NO_2 gas is a pollutant found in the air. Does NO_2 have the same effect on people as laughing gas?

5. Sometimes, coal can be contaminated with a chemical whose formula is $CuFeS_2$. How many of each atom is in a molecule of this chemical?

6. The main chemical in limestone is made of one calcium (Ca) atom, one carbon (C) atom, and three oxygen (O) atoms. What is its chemical formula?

7. A chemical is in its liquid phase. Are its molecules closer together or farther apart compared to when it is in its gas phase? Do the atoms move around more or less when it is a liquid as compared to when it is a gas?

8. If you have a gas and want to turn it into a liquid, do you need to heat it up or cool it down?

9. You see the following measurements: 1 kg, 34 ms, 17%, 3 L, 5 g/mL, and 14 cm. Identify each as a measurement of mass, distance, time, volume, concentration, or density.

10. What is the area of a room that measures 3 meters wide and 2 meters long?

11. What is the volume of a cube that is 12 cm long, 10 cm wide, and 5 cm high?

12. On the surface of the earth, an object that weighs 1 pound has a mass of 454 g. How many kilograms is that?

13. An adult human finger is about 20 mm wide. How many meters wide is it?

14. A regulation fencing sword is 90 cm long. If 1 in = 2.54 cm, how many inches long is it?

15. You are watching the molecules in an object move. Suddenly, they start moving faster than before. Was object cooled down or heated up?

16. Water is at a temperature of 95 °C. Someone tells you that's the same as 230 °F. Should you believe that person? Why or why not?

17. You see an object (190 g, 30 mL) that looks like it is made of copper. If copper has a density of 8.96 g/mL, is the object made of copper?

18. The density of air at 25 °C is 0.01 g/mL. You let go of a 500-mL balloon whose total mass is 10 g. Will it float away or fall to the floor?

19. If you could count the molecules in air, you would find that out of 100 molecules, 21 of them are oxygen, 78 are nitrogen, and 1 is another chemical. What percent of air is nitrogen?

20. Another common temperature scale in science is the Kelvin scale. On this scale, water freezes at 273 K and boils at 373 K. Which is warmer: 300 K or 110 °C?

Chapter 2: It's a Little Crusty

Introduction

Now that we have the basics out of the way, it's time to start concentrating on the earth and how it is designed. I started the course discussing the fact that the earth is an oblate spheroid. As I said then, it's like a ball that someone is squashing by pushing on the top and bottom. It is easy to understand how we can confirm that today with satellites monitoring the earth from space, but believe it or not, scientists figured out that the earth is an oblate spheroid back in the early 1700s!

How did they do that? First let's make sure you know some basic terminology. In order to define position on earth's sphere, Greek natural philosopher Hipparchus (hih' par kus – 190-120 BC) suggested that we should use a series of imaginary lines that are shown in the drawings on the right. One set should wrap around the earth going east to west, and we now call them **lines of latitude** (lad' uh tood). The middle line of latitude is now called the **equator**. The other set should run from the top of the sphere (the north pole) to the bottom of the sphere (the south pole). They are called the **lines of longitude** (lahn' jih tood). A single line of latitude and a single line of longitude cross at only one spot on the earth, so if you list your position with a line of latitude and a line of longitude, everyone knows where you are on the earth's sphere.

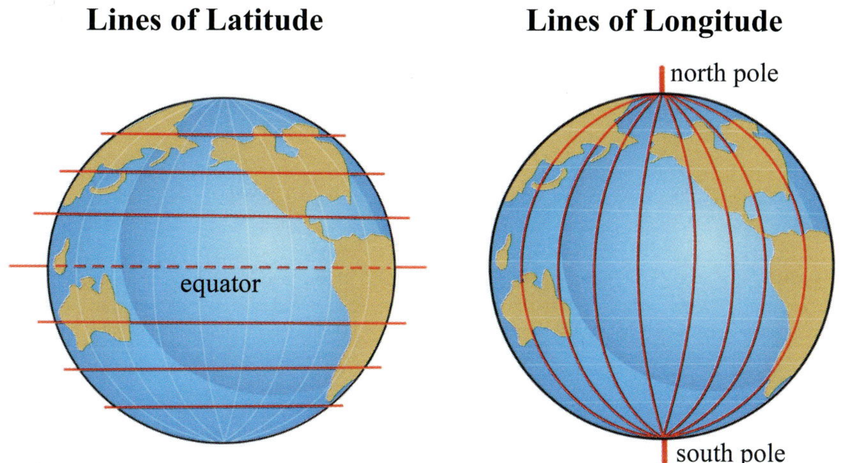

Lines of latitude wrap around the earth east to west (left), while lines of longitude run from the north pole to the south pole (right).

Comparing the positions of the stars at night to the position of the sun at specific times during the day, explorers could roughly determine the lines of longitude and latitude where they were, Thus, as long as the stars were visible at night, they could determine where they were on the earth. This, of course, allowed them to navigate.

Lines of latitude are defined by degrees. The equator, for example, is 0°. These aren't the degrees used to determine temperature. They measure angles, but you don't need to worry about that. You just need to know that the lines of latitude increase as you move away from the equator. The north pole is at a latitude of 90° north (N), while the south pole is at a latitude of 90° south (S). The larger the degrees on the lines of latitude, the farther you are from the equator. Miami, Florida, for example, is at a latitude of 26° N, while the center of Alaska is at a latitude of 65° N. Lines of longitude are also defined by degrees. The line of longitude that runs through Greenwich, England is called the **prime meridian**, and it has a longitude of 0°. Lines of longitude run east (E) and west (W) from the prime meridian, ranging up to 180° E or 180° W. Israel, for example, is at a longitude of 35° E, while Perth, Australia is at a longitude of 116° E. To specify a point on the earth, you include both its latitude and its longitude. Sydney, Australia, for example, is at 34° S and 151° E, while New York City is at 41° N and 74° W.

What does this have to do with the shape of the earth? In the 1700s, there was a big debate among natural philosophers about the specific shape of earth's sphere. Some thought the earth was egg-shaped, with the poles stretched out and the equator thinner than it would be in a perfect sphere. Others thought it was flattened at the poles, and the equator was fatter than it would be in a perfect sphere. In late 1730s, the French Academy of Sciences and the King of France (Louis XV) decided to settle the debate once and for all. They sent out two expeditions. One went to the equator in Ecuador, while the other went to Lapland, Finland, which is closer than France to the north pole.

Both expeditions used navigation techniques to measure the distance they had to travel in order to move a total of one degree in longitude. A similar measurement was made in France, so in the end, they had three measurements of how far you had to travel to move one degree in longitude. Those three measurements confirmed that the earth is flatter at the poles and fatter at the equator, which is why we say it is an oblate spheroid. Of course, modern-day measurements made by satellites in space confirm this, but it is amazing that natural philosophers figured it out almost 300 years ago!

Spheres Within Spheres

This picture was taken at night when the International Space Station was over the Middle East.

While most people think of the earth as one big oblate spheroid, in fact, it is composed of several spheres that are all put together. Look, for example, at the picture on the left. It was taken from the International Space Station (ISS), which is in space, traveling around the earth. You can see a part of the ISS in the upper right corner of the picture. The long strip of blue in the center is the Red Sea, and the curvy line of light on the left comes from cities along the Nile River. The lands in the picture include parts of Egypt, Israel, and Saudi Arabia. In other words, this is part of the earth's sphere, as seen from the ISS at night.

But what is the green glow hovering above the earth? That's the earth's **atmosphere** (at' muh sfear).

Atmosphere – The collection of gases that surround a planet

Because of chemical reactions that are constantly occurring in the upper parts of the atmosphere, gases that are found there end up getting extra energy. They need to get rid of that energy, and one way they do it is by emitting light. Thus, the upper parts of the atmosphere are always glowing. You can't see it during the day because the sun is so bright it overwhelms the glow, and it is difficult to see from the surface of the earth, even at night. However, it is easy to see at night from space.

The oblate spheroid of the earth, then, is surrounded by a spheroid of gases called the atmosphere. To distinguish the two, we often refer to the **geosphere** (jee' uh sfear).

Geosphere – The solid components of the earth

In the picture on the previous page, then, the greenish glow comes from the atmosphere, while the land is part of the geosphere. But what about the water in the Red Sea, the Nile, and the Mediterranean Sea, which can all be seen in the picture? The water is not solid, so it's not part of the geosphere. It's also not a gas, so it's not part of the atmosphere. It's part of the **hydrosphere** (hi' druh sfear).

Hydrosphere – All the water on the earth, regardless of its phase

While the atmosphere and geosphere are determined by the phase of their components, the hydrosphere is determined by the chemical. Regardless of phase, any water on the earth is part of the hydrosphere. In this course, you are going to learn about all three of these spheres, but I want to start with the geosphere. I want to spend several chapters on it so that you become familiar with all its components.

The Geosphere

Not surprisingly, the geosphere can be split into several more spheres, as shown in the drawing below. The only part of the geosphere that we have directly investigated is the outer layer of rock, which is called the earth's **crust**. Depending on where you are, it is 5-70 km thick. That's obviously a lot of rock (5 km is roughly 3.1 miles), but it's really just a thin "coating" around the earth. If the earth were an apple, for example, the crust's thickness would be roughly the same as the skin on the apple. Despite the fact that it is so "thin," no one has ever drilled all the way through it. However, scientists have been trying. In fact, the first attempt to drill through the crust of the earth started in 1961. Since then, we have been to the moon, but despite the fact that many groups have been making many attempts over the past several decades, we still have not been able to drill through just the crust of the earth!

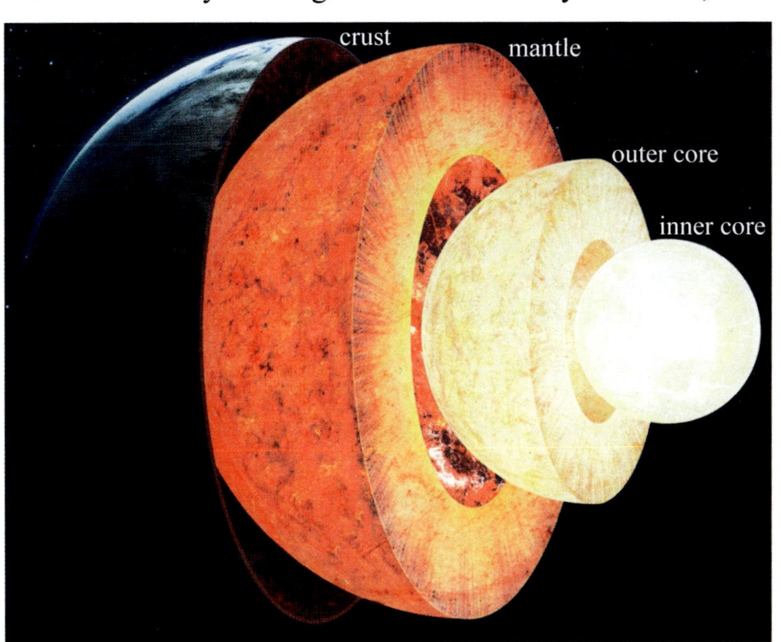

The geosphere can be divided into several smaller spheres.

If we were able to drill through the crust, the next thing we would encounter is the **mantle**, which is the thickest part of the earth. It makes up about 85% of the planet's volume, and its physical properties change dramatically as depth increases. Initially, the mantle is composed of solid rock, like the crust. However, the deeper you go in the mantle, the hotter it becomes. As rocks get really hot, of course, they melt. At the same time, however, the crust and the upper parts of the mantle push down on everything below them, and since they are pushing in from all sides, the lower parts of the mantle get squeezed by the upper parts of the mantle and the crust. In other words, the lower parts of the mantle are under a lot of pressure.

Think about what happens when you squeeze a marshmallow. It gets smaller, right? Well, the only way for an object to get smaller is for its molecules to move closer together. So, in general, when something is under pressure, its molecules get closer together. The higher the pressure, the closer the molecules get. Well, in the mantle, the rocks are hot enough to melt, but they are also being squeezed by the crust and the material above them. The heat makes the molecules in the mantle move faster and get farther apart, making them more like a liquid. However, the pressure pushes the molecules together, making them more like a solid. As a result, the mantle is an interesting "compromise" between solid and liquid. It is often called **plastic rock**, because it tends to behave like a soft plastic. In many ways, it behaves like Silly Putty.

The layer underneath the mantle is called the **outer core**. It is so hot that the pressure can't force the molecules very close together. As a result, the outer core is liquid. Below the outer core, there is an **inner core**. It is even hotter than the outer core, but the pressure produced by the crust, mantle, and outer core squeezing it pushes the molecules close together, making it a solid. If you look again at the drawing on the previous page, the colors provide an indication of the temperature. The mantle is hot, so it is red. However, the deeper parts of the mantle are hotter, so they are yellow. The outer core is hotter, so it is a brighter yellow, and the inner core is even hotter, so it is almost white.

While there are important things to learn about each of these parts of the earth, I will spend most of my time discussing the crust, because that's the part of the earth we have directly investigated, so we understand it the most. After I have covered the crust of the earth in detail, I will then discuss what we know about the mantle, outer core, and inner core.

But Wait A Minute

What you just read might have brought up a question in your mind. We have never been able to drill through the crust, so how do we know what is below it? How do we know the mantle, outer core, and inner core are real? How do we know their phases and temperatures? Although we have never observed anything below the crust *directly*, we can learn a lot by studying them *indirectly*.

What does it mean to study something indirectly? Well, suppose your parents give you a nicely-wrapped present for a special occasion, and they tell you that you can't open it right away. You have to wait. However, you really want to know what's inside. What could you do? You could look at the size of the present. The size is determined by the packaging, not the gift itself. Nevertheless, you know that the gift must fit inside the packaging, so you know that it must be at least a bit smaller. You could also lift and shake the package. That will give you an idea of how heavy it is, and you could listen to find out if it makes any noise when you shake it. You can't see what's in the package, but a rough idea of its size, its weight, and the noise it makes when shaken could help you determine what it is. In other words, you might be able to figure out what your gift is by studying it indirectly.

Scientists must study a lot of things indirectly. In the previous chapter, for example, I told you what molecules are. However, we cannot see molecules. We know what they look like because we study them indirectly. We heat them up, for example, and see what happens. We shine light of different energies on them and see what happens. We can use those observations (and others) to learn what molecules look like.

We can do the same thing with the parts of the earth we cannot study directly. When volcanoes erupt, for example, the lava that spews out comes from the upper mantle. If we study the lava, we can learn a bit about the composition of the mantle. Of course, we aren't really observing the mantle, because lava isn't under pressure like the material in the mantle. Nevertheless, we can learn about the chemicals that are found in the mantle.

More importantly, scientists called **seismologists** (size mol' uh jists) study vibrations caused by earthquakes, explosions, and other high-energy events. These vibrations produce waves, called **seismic waves**, which travel all the way through the earth. If lots of different seismologists in lots of different places study the waves that reach them as a result of a single earthquake, they can learn a lot about the structure of the earth. Consider, for example, the illustration on the right. An earthquake occurs near the north pole, and seismologists around the world study specific kinds of seismic waves that reach them. The seismologists at points A, B, and C end up seeing waves that are very similar. However, the seismologist at point D sees no waves at all. The seismologists at points E and F see fewer waves than the seismologists at points A, B and C.

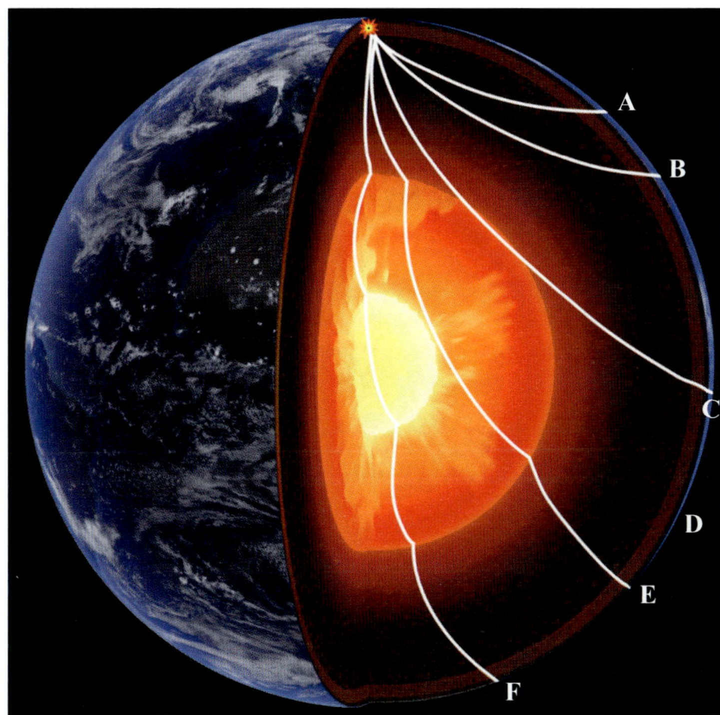

Differences in seismic waves around the world help us to understand what's below the crust.

Why do the seismologists see differences in the waves that reach them? Because the waves traveled through different materials to reach them. For seismologists A, B, and C, for example, the waves traveled through the mantle to reach them, so they can conclude that the material between them and the earthquake is very similar. However, seismologist D got no waves at all. Thus, there must be some other material between that seismologist and the earthquake, and it must be arranged so that it blocks the waves from traveling from the earthquake to her location. Seismologists E and F get waves, but not as many as the ones at A, B, and C, so the waves must have traveled through material that blocked some of them. Obviously, that material must be different from the material through which the waves traveled to reach A, B, and C.

As seismologists have made many observations like this, they have been able to map out the interior of the earth. The structure that I presented to you in the previous section is the only structure that is consistent with all those observations. Thus, while we have never been below the crust of the earth, we have a good idea of what is underneath.

Comprehension Check

2.1 You are given two positions. The first is 24° N and 17° E. The second is 49° N and 56° W. Which is closer to the north pole?

2.2 Which sphere (geosphere, atmosphere, hydrosphere) do scientists say gaseous water is a part of?

2.3 Of the crust, mantle, inner core, and outer core, which is the hottest? Which is the coolest?

2.4 You learn that the sun is roughly 75% hydrogen and 25% helium. A friend tells you that it is impossible for us to know that, since we have never been to the sun. What should you tell your friend?

Crusty and Dirty

When it comes to the structure of the earth, you are familiar with just the top part of the earth's crust. In fact, most of your experience comes from the very top of the earth's crust, which is usually made up of **soil**. While you might think of soil as "dirt," it is a lot more than that! In fact, soil is typically composed of several layers, which are called **soil horizons**. While soil horizons change from place to place, the illustration on the left is a general guide to the soil horizons that can exist in any given area.

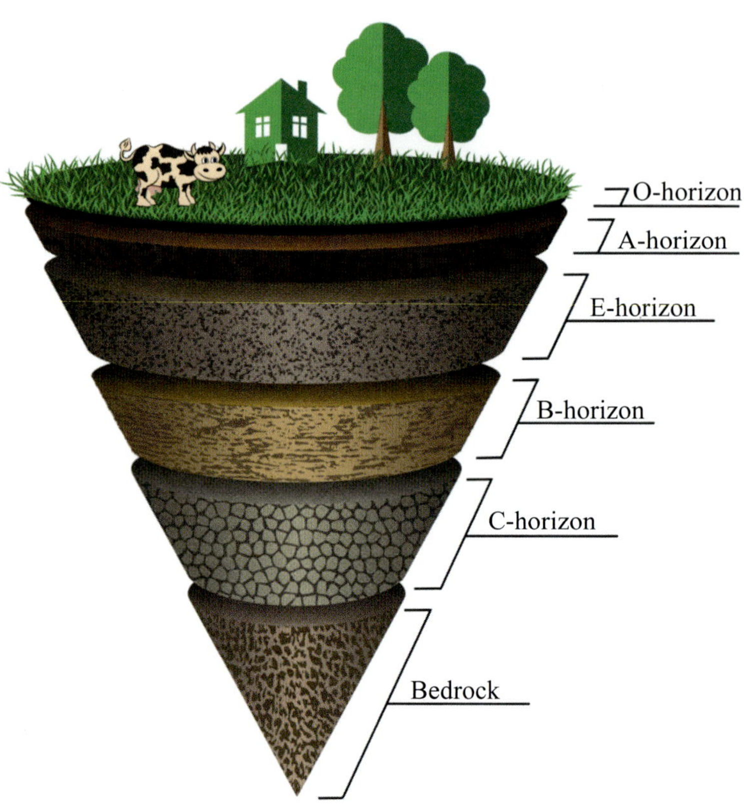

All the layers except the bottom one are soil horizons. Not all soils have all the horizons shown here.

The soil horizon found at the very top is called the **O-horizon**. It contains matter that is or was found in living things. We typically refer to it as **organic** matter. In fact, that's where the "O" comes from. The grass and tree roots, for example, are organic matter. However, some of the living things that were in and on the soil are now dead and in the process of decaying away. The decayed or decaying remains of once-living creatures are also organic matter, so the O-horizon contains those things as well. As shown in the drawing, the O-horizon is usually very thin, but it is crucial, because the decay that is going on there produces chemicals that are important to the health of the plants that are growing in the soil.

The next layer down is the **A-horizon**, and it contains a mixture of the organic material from the O-horizon as well as many minerals. While you might have heard the term "mineral" before, it is important to know its scientific definition:

Mineral – A naturally-occurring, inorganic substance usually found in rock that has its own chemical and physical properties, such as crystal structure, color, and hardness

You will learn a lot more about minerals soon, because they are important when it comes to understanding rocks. For right now, however, just think of mineral matter as another component of the soil that is very different from the organic matter found in the O-horizon.

The definition for mineral might be a bit confusing, because you might not understand how something can be naturally occurring but inorganic. After all, in today's society, "organic" typically means "natural." However, that's not the way scientists use those terms. Scientists classify all chemicals as either organic or inorganic, depending on the atoms that are found in the chemical's molecules. For this course, however, you can think of an organic substance as something that is in a living organism or produced by the decay of a once-living organism. Any other chemical is inorganic. That isn't a perfect distinction, but it is good enough for this course.

How do the organic materials from the O-horizon become a part of the A-horizon? A lot of it is the result of the activity of the organisms living in the soil. As roots grow, they push down from the O-horizon to the A-horizon. This allows the materials in the two horizons to mix. Also, there are animals, like earthworms, that burrow in both horizons. As they burrow from the O-horizon to the A-horizon, they mix the organic matter and inorganic matter.

The A-horizon is often referred to as **topsoil**, even though it isn't really the topmost soil horizon. However, it is the topmost horizon that contains both organic and inorganic matter. More importantly, it is where plants get most of the chemicals they need to be healthy. You probably know that plants make their own food through a process called photosynthesis, but they need more than just food in order to survive. They also need certain chemicals, some of which are organic, and some of which are inorganic. Both are found in the A-horizon, so it is the most important part of the soil when it comes to supplying plants with the chemicals they need. That's really why it is called the topsoil.

Below the A-horizon, you find the **E-horizon**. It is important for draining the soil, but it doesn't exist everywhere. Plants need water to survive, but they can drown in too much water. As a result, if a location gets a lot of rainfall or is exposed to a lot of water from flooding, excess water must be drained away. That's what the E-horizon does. It allows water to travel down out of the topsoil to keep the roots of the plant from drowning. If an area doesn't get a lot of rain, it might not have an E-horizon. The E-horizon is usually lighter in color than the A-horizon, and it doesn't hold on to organic matter well. As a result, it is mostly composed of minerals and plant roots.

Below the E-horizon (or below the A-horizon if there is no E-horizon), you find the **B-horizon**. It is sometimes called the "zone of accumulation" or the **subsoil**. Plant roots can be found in this horizon, but they aren't typically as plentiful as what is found in the A-horizon. Anything that has traveled down through the A-horizon and E-horizon ends up here, so it is once again a mixture of minerals and organic matter. It is usually lighter than the horizons above it.

The **C-horizon** is the last layer of soil. It consists of broken rocks, usually similar to the rocks found in the **bedrock**, which is below the soil. The bedrock isn't part of the soil. However, the broken rocks in the C-horizon are part of the "parent material" of the soil, because as you will learn later, much of the material in soil comes from rocks that break down into tiny bits.

Now remember, this discussion is for typical soil. However, soil varies from region to region. The soil horizons you find in a forest will be different from the soil horizons found in a desert. Each area might have the horizons mentioned, but the specific makeup of those horizons can be different. Also, some horizons might not be present in some regions. To become familiar with the topsoil in your area, perform the following experiment.

Experiment 2.1: It's More Than Just Dirt

Supplies:

- A place in a yard or park where you can dig. (If you don't have access to such a place, find a potted plant and use some of the soil in the pot, skipping steps 1-3. Soil from outside is better.)
- Two clear, plastic bottles, one of which has a lid (They should both be the size of a standard water bottle – 500 mL. Some sports drink bottles have wide mouths, which make the experiment easier.)
- A hand spade or a strong spoon that you can use to dig
- The funnel from the laboratory kit made for this course
- A paper towel

- Water

Instructions:
1. Use the spade or spoon to dig up a clump of grass. You don't need to dig up a lot of grass. Just pull up a small area so that you expose the soil below.
2. Shake the grass so that anything clinging to the roots falls down into the area of exposed soil.
3. Set the clump of grass off to one side.
4. Put the funnel in the bottle's opening, unless the bottle has a wide mouth.
5. Use the spade or spoon to dig up soil from the area you exposed, including the stuff you just shook off the grass. Try to go down at least a couple of centimeters (about an inch).
6. Holding the bottle over the area of soil you exposed, pour soil from the spade or spoon into the funnel or the bottle's wide mouth. If the soil doesn't fall through the funnel into the bottle, shake the funnel a bit to loosen the soil so it falls into the bottle. If the funnel makes the job too difficult, just add the soil directly to the bottle.
7. Repeat steps 5 and 6 several times, until the soil fills up about ¼ of the bottle.
8. Put the lid on the bottle and wipe the bottle with the paper towel to get rid of any dirt that is clinging to the outside of the bottle.
9. Bring the bottle to a sink.
10. Open the lid and add water until the bottle is about two-thirds full.
11. Put the lid back on the bottle and shake it vigorously so the soil and water mix well.
12. Set the bottle down and fill the other bottle with water to about the same level. Don't worry about making it exactly the same. Just get it close. If this bottle has a lid, go ahead and put it on, but it isn't necessary.
13. Set that bottle down and pick up the bottle that has the water and soil.
14. Examine that bottle for a minute or so. What seems to be happening inside?
15. Put the bottles somewhere they will not be disturbed.
16. Clean everything else up. You will look at the bottles when you start science tomorrow.

Even though my main discussion of the experiment's results will start tomorrow, I want to point out what you should have already seen happening. Most likely, you saw some material floating on top of the water and other material sinking. What does that tell you? The water is sorting the things found in the soil according to their density. The materials with low density float, and the materials with a density greater than water sink. What other sorting might take place? You will see tomorrow.

Comprehension Check

2.5 You are shown samples of two different soil horizons. They both contain organic matter and mineral matter, but the first is lighter than the second. Which horizon did each sample come from?

Components of Topsoil

The bottles you made the last time you did science have been sitting for a while. Let's see what has happened to their contents.

Experiment 2.1 Continued

Supplies:
- The bottles that have been sitting overnight. (If you must move them, do so gently.)

- A well-lit room (If it is a sunny day, outside would be even better)
- The magnifying glass from the laboratory kit made for this course

Instructions:
1. Examine the bottle that has a mixture of soil and water. What do you see? You should see some things floating on top of the water, a layer of water, and then a collection of soil underneath the layer of water.
2. Examine the soil underneath the water. Do you see any differences between the soil that is right under the water and the soil that is near bottom of the bottle?
3. Use the magnifying glass to look at the soil, starting right under the water and slowly moving to the bottom of the bottle. What differences do you see?
4. Put down the magnifying glass and pick up the other bottle so you are holding both bottles.
5. Hold them side-by-side and compare the water layer in the bottle that contains soil to the water in the other bottle. What differences do you see?
6. Keep both bottles for now, because you might want to look at them as I discuss the results of the experiment.

The results of your experiment depend on the kind of soil you got when you dug into the ground. However, there definitely should have been some things floating on top of the water. As you already know, those things are less dense than water. That's why they float. There might have been some bits of grass or roots in the floating material, but a lot of the material was either black or brown. Most of that material is made up of things that were once alive but are now dead and have decayed away. This is called **humus**, and because it is made from decayed things that were once alive, it is part of the O-horizon. However, because of the mixing that occurs due to roots and animals, it ends up becoming part of the A-horizon as well.

The next thing you saw in your bottle was a layer of water. Now, most likely, some things were floating in that layer. They weren't at the top or bottom of the water layer. Instead, they were suspended in water. As you learned in the previous chapter, that can happen when something has the same density as water. It tends to stay in the water wherever it found itself, neither rising nor sinking. Most of the suspended material was probably made from bits of grass and roots. If they get soaked with the right amount of water, they can end up having the same density as water.

Ignoring the materials that were suspended in it, how did the water layer in the bottle with soil compare to the water in the other bottle? The water layer in the bottle with soil wasn't clear like the water in the other bottle, was it? You might be tempted to blame that on the material suspended in the water layer. It was making the water "dirty." However, hopefully you could see that even if you removed all the suspended material, the water layer still wouldn't have been as clear as the water in the other bottle.

Why? Because the soil you put in the water contained all sorts of different minerals. Some of them were able to dissolve in water, and as a result, we call them **water-soluble** minerals. Other minerals are not water-soluble, and they were mostly contained in the soil that was under the water layer. When you mixed the water and the soil, any water-soluble minerals dissolved in the water, while the rest did not. Some of those water-soluble minerals ended up giving some color to the water, the same way a powdered drink mix gives color to water when it is dissolved. Thus, a lot of the differences between the water layer in the bottle with soil and the water in the other bottle were caused by those water-soluble minerals.

What did you see in the layer of soil that was underneath the water layer? Depending on the soil you had, you might have seen some obvious layers, with one part of the soil lighter or darker than another part of the soil. If nothing else, when you looked at soil with the magnifying glass, you should have seen a difference between the soil that was right under the water layer and the soil at the bottom of the bottle: The soil at the bottom shouldn't have been as smooth-looking as the soil right under the water. In other words, the soil right under the water was packed very tightly together, while the soil at the bottom of the bottle had small spaces or gaps in it. Why are there differences in the soil? Perform the following experiment to see.

Experiment 2.2: Water Sorting

Supplies:
- A raw potato (It doesn't have to be big. In fact, it could be just a piece of a potato.)
- A sharp or serrated knife
- A cutting board
- Paper towels
- A tall glass
- Water

Instructions:
1. Fill the glass nearly full of water.
2. Cut the potato so that you have a cube that is about 2 cm (¾ of an inch) tall, 2 cm wide, and 2 cm high. It doesn't have to be exactly a cube or exactly that big. However, it cannot have any part of the peel on it. It just be composed of just the starchy (white) part of the potato.
3. Repeat step one, but this time make a tiny cube. It needs to be *a lot* smaller than the cube you just made, and don't worry if it's not really a cube. Just a tiny piece of potato will do.
4. Use the paper towel to thoroughly dry both cubes of potato and your hands. You need to get rid of all the water, since water will make it hard to do step 7, especially with the tiny cube.
5. Hold one cube between the thumb and index finger of one hand, and the other cube between the thumb and index finger of the other hand.
6. Hold both cubes over the opening of the glass so that when you release them, they both fall into the glass from the same height.
7. At the same time, drop both cubes in the water and see which one hits the bottom of the glass first. If you couldn't get one or both of them to drop when you released them, you might not have dried them or your hands thoroughly enough. Try again, concentrating on making the potato cubes and your hands very dry.
8. Clean up your mess.

If things went well, you should have seen that the large potato cube reached the bottom of the glass before the small one did, but why? Remember that density determines whether or not something will sink in water. Both cubes were made out of the same thing (the starchy part of the potato), so they had the same density, which was greater than that of water. Thus, they both sank. However, the big one clearly sank faster.

Well, in order to sink, both potato cubes had to keep moving water molecules out of the way. The bigger cube had more weight, so moving water molecules out of the way was easier for it. The smaller cube had the same density, but because it was smaller, it had less weight. As a result, it had a harder time pushing its way through the water molecules. Thus, it sank more slowly.

The same thing happened to the soil in your experiment. There were a lot of things in the soil that were denser than water, so they all had to sink after you shook everything up. However, the smaller bits of soil couldn't push their way through the water as well as the bigger bits, so the larger the bits of soil, the more likely they were to reach the bottom of the bottle first. As a result, the water "sorted" the soil, with particles of larger size closer to the bottom of the soil layer, and particles of smaller size closer to the top.

The larger the particles, the larger the spaces in between them.

That, of course, changed the texture of what you saw in the soil, because larger particles can't pack together as tightly as smaller particles. Look, for example, at the picture on the right. The top diagonal section has gravel in small chunks, the middle section has gravel in larger chunks, and the bottom has sand, which is made of the smallest chunks. Notice that the diagonal region in the middle has lots of spaces in between the rocks, while the diagonal region at the top has smaller spaces in between the rocks. The sand has even smaller spaces in between the bits of sand. This is because the smaller the individual particles are, the closer they can get to one another.

In the end, the water did three things to separate the different components of the soil. First, it separated them by whether or not they were water-soluble. The water-soluble components dissolved in the water, but the rest did not. Then, it sorted them by density. Finally, it even partially sorted the components that were denser than water, sorting them by the size of the particles that make them up. It turns out that the size of the particles in soil can be important. You will learn about that the next time you do science.

You can now get rid of the two bottles. You can empty the bottle containing only water into the sink, but empty the bottle containing water and soil outside in a grassy area. It might be hard to get the wet soil out of the bottle. If you can't get it out, just throw the bottle away with the soil in it. Clean up any mess you make in the process.

Comprehension Check

2.6 Suppose you took handfuls of each material in the picture above and put them in a tall container of water. Then, you shook the container vigorously. After everything settled out, which of the materials (larger gravel, smaller gravel, or sand) would you expect to be at the bottom of the jar? Which would be closest to the top?

Particle Sizes in Soil

As the previous experiment demonstrated, soil is made of particles of different sizes. As you will see in a moment, those sizes are important, so earth scientists use size as one way to classify soil particles. The standard classification scheme is given in the illustration below. While you probably think of **sand** as being composed of tiny grains (particles), in terms of soil, sand is the component that has the largest particles, with diameters ranging from about 0.05 mm to 2 mm. Now even though I said "diameter," I don't mean to indicate that they are necessarily circular or spherical. The "diameter" of a particle in soil is generally thought of as the width of the particle, whether or not it is circular or spherical in shape. The next smaller particles are called **silt**, and they have "diameters" of 0.002 mm to just under 0.05 mm. Any particles with a "diameter" of less than 0.002 mm are called **clay** particles.

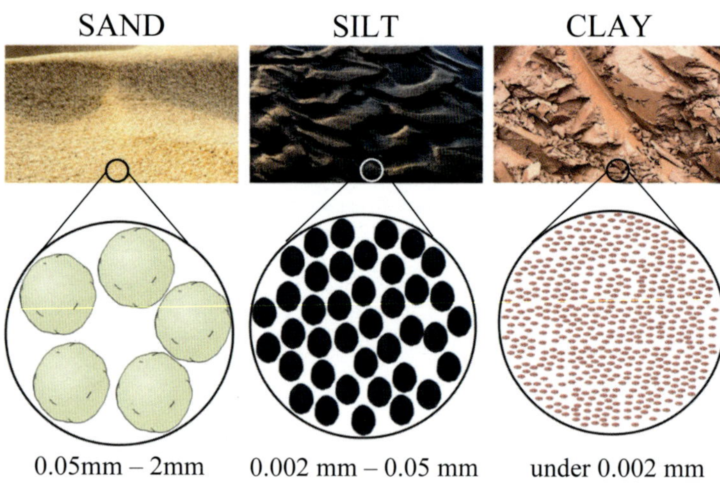

The components of soil can be classified based on the size of their particles.

You probably have experience with each of these soil components. If you have ever been to a sandy beach or played in a sandbox, you know that sand feels rough to the touch. While you might not recognize the word "silt," you have probably experienced it before. The "muck" at the bottom of a lake is mostly made of silt. It feels smoother than sand. You have probably worked with clay before. It might have been human-made (like Play-Doh) or it might have been real clay. However, it feels even smoother than silt. Part of the reason these components feel different from one another is because of the sizes of the particles that make them up, and those sizes are important to the soil.

Why? Well, look at the image above. The circles illustrate what a magnified view of each soil component might look like. Look at the white spaces between the magnified particles of sand, and compare them to the white spaces between the magnified particles of silt. The white spaces between the particles of sand are bigger than the white spaces between the particles of silt. That's because the sand particles are so big they cannot be packed really tightly together, so there is a lot of space in between them. We call the spaces between particles **pores**. Because the silt particles are smaller, they can be packed closer together, so the pores in silt are even smaller. Clay has the smallest pores of all. So what? Perform the following experiment to find out.

Experiment 2.3: Percolation

Supplies:

- Silt from the laboratory kit made for this course (It's in a labeled bag.)
- Sand (also from the kit in a labeled bag)
- Clay (also from the kit in a labeled bag)
- Three plastic water bottles (You can also use funnels or a combination of bottles and funnels.)
- Sheer hosiery that can be cut up (A couple of coffee filters will also work.)
- A ½-cup measuring cup
- A ¼-cup measuring cup

- Tape
- Scissors
- Water

Instructions:

1. Use the scissors to cut each bottle one-third of the way from the top. If you are using funnels, skip this step.
2. Cut the hosiery or coffee filter into three pieces that are big enough to wrap over the small opening of each bottle or funnel (see the picture on the right).
3. Completely cover the small opening of each bottle or funnel with a piece of the hosiery or coffee filter and use tape to hold it in place so that it will not fall off, even when weight is applied to it.
4. Invert the top of each bottle into the bottom of each bottle so that it is held in place. If you are using funnels, put the funnel into a glass that will hold it upright.
5. In the next few steps, you will be using sand, silt, and clay from the kit. They should be in plastic bags, each with a label. As you put them into the ½-cup measuring cup, rub a bit of each between your thumb and index finger so you can feel the differences among them.
6. Add ½ cup of sand to the inverted top of one bottle. It should now look like the picture on the right. Shake the inverted top so the sand is level.
7. Add ½ cup of silt to the inverted top of another bottle and shake to level it.
8. Add ½ cup of clay to the inverted top of a third bottle and shake to level it.
9. Add ¼ cup of water to the top of each bottle so that the water must fall through the sand, silt, or clay to drip through the hosiery or coffee filter.
10. Compare how quickly you see water dripping out of the hosiery or coffee filter and into the bottom of each bottle.
11. Clean up your mess. It is best to just throw the tops of the four bottles into the trash. If you are using funnels, clean out the funnels with an outdoor hose so you don't get much sand, silt, or clay in the drain of the sink.

What happened in the experiment? You should have seen that water started dripping from sand before it dripped from the other two bottles. That shouldn't surprise you if you think about the pores in sand. The water must fall through the pores, so the bigger the pores in a material, the easier it is for water to fall through it. You might not have seen water drip from the hosiery covering of the bottle that had clay in it, because the pores in clay are so small that it is hard for water to fall through it. You should have seen water drip from the bottle that contained silt, but that should have happened after the sand, because the pores in silt are larger than the pores in clay, but smaller than the pores in sand.

Now think about how this applies to soil. Every time it rains, water lands on the soil, and it will start to travel through the soil. The term we use for this process is **percolation** (pur kuh lay' shun).

Percolation – The movement of water through the pores in soil or rock

Yes, water can move through certain rocks as well, but don't worry about that right now. The key is that when water falls on soil, it must percolate through the soil, and the speed at which the percolation happens depends on the sizes of the pores in the soil.

Why is this important? Plant roots are in the soil. While they need water, they also need air. For a soil to support plant life, then, it must hold on to water, but there must be space for air as well. Clay holds on to water really well, but there is very little space for air, so most plants don't grow well in clay. Sand has lots of room for air, but water percolates through it quickly, so it doesn't hold on to water very well. As a result, most plants don't grow well in sand. For soil to promote healthy plants, it needs to be a mixture of sand, silt, and clay, which we call a **loam** (lohm).

Loam – An easily-crumbled mixture of sand, silt, and clay

The drawing below shows you different kinds of soils and how they are classified. It might be confusing at first, because it is based on a triangle, but once you get the hang of it, you can read it easily. Look, for example, at the brown area near the bottom labeled "LOAM." Near the top of the "A," there are two diagonal lines and one horizontal line that all intersect. The horizontal lines are for clay. Follow the horizontal line to the left, and you will see that it comes from 20, which means 20% clay. The diagonal lines running up and to the right are for silt, since they go to the silt side of the triangle. If you follow that diagonal line back to the "percent silt" side of the triangle, you get to 40, which means it is 40% silt. The diagonal lines running down and to the right are for sand, since they come from the sand side of the triangle. If you follow that diagonal line back to "sand," you get 40%. That intersection, then, represents a soil that is 40% sand, 40% silt, and 20% clay. Most plants grow best in a loam that is close to that mixture, because it holds water well but has room for air. However, there are certain plants that prefer a loam that is more silty, more sandy, or more rich in clay.

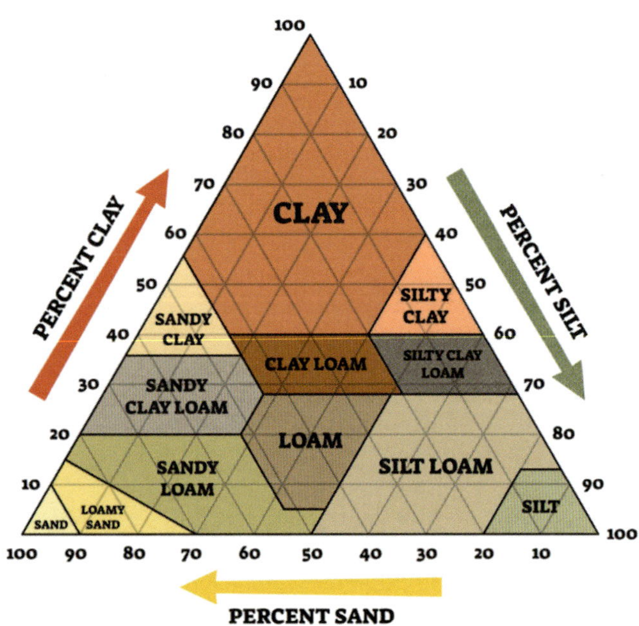

Soils can be classified based on the amount of sand, silt, and clay in them.

Comprehension Check

2.7 A loam is 30% clay, 10% silt, and 60% sand. Looking at the illustration above, how would that soil be classified?

2.8 Loam "A" is 35% clay, 55% silt, and 10% sand. Compare the size of its pores to the pores in Loam "B," which is 10% clay, 25% silt, and 65% sand.

But Where Does Soil Come From?

I've told you about the different kinds of particles that make up soil and what mixture produces healthy soil, but I haven't told you much about how soil is formed. I did give you a hint when I discussed the C-horizon of soil, which is made up of broken rocks. Do you remember what those broken rocks are called? They are called the parent material of the soil. Why? Because a large amount of the material in soil comes from broken-down rocks. So, rocks are sort of like the parents of soil, because they end up making it. How does this work? Let's start with an experiment.

Experiment 2.4: Breaking Down Rocks

Supplies:

- The rock in a small bag labeled "limestone" from the laboratory kit made for this course
- A small glass, like a juice glass
- A hammer
- White vinegar
- Aluminum foil
- A hard surface you can hit repeatedly with a hammer (A large, flat rock outside works well.)
- Eye protection, like work goggles or safety glasses

Instructions:

1. Take the rock from the plastic bag and wrap it in aluminum foil so it is completely surrounded by the foil.
2. Put on your eye protection.
3. Put the foil-wrapped rock on the hard surface and hit it twice with the hammer.
4. The foil is probably damaged, but unwrap it and examine the contents. The rock should be broken into several pieces. If not, repeat steps 1-3.
5. Choose the largest piece of rock from inside the foil and put it in the small glass.
6. Add vinegar to the glass until it completely covers the piece of rock.
7. Go back to the foil with the pieces of rock in and choose a small piece of rock.
8. Wrap that small piece of rock in another piece of foil and once again, put it on the hard surface.
9. Hit it several times with the hammer.
10. Once again, unwrap the foil and examine its contents. What do the contents look like?
11. Clean up your mess related to the foil and the pieces of rock.
12. Look at the rock that has been soaking in the vinegar. What do you see happening on the surface of the rock?
13. Leave the rock soaking in the vinegar. You will look at it again in a moment.

What did you see in your experiment? When you hit the rock with the hammer a couple of times, it broke into pieces of different sizes. When you chose the smaller piece and hit it several times with the hammer, you should have found that you had crushed the piece of rock into tiny, tiny grains that looked a bit like, well, dirt! When rocks break down into tiny grains like that, those grains become a part of the soil. If the grains are big enough, they become part of the sandy component of the soil. If they are smaller, they become part of the silty component. If they are even smaller, they are part of the clay component.

But wait a minute. There aren't hammers in nature going around and crushing rocks into tiny grains! That's true, but there are other processes that take a bit longer but produce the same results.

Roots are strong enough to grow through rocks, breaking them in the process.

Consider, for example, the picture on the left. Because of the way the side of the hill has been exposed, you can see the roots of the trees that are growing in the soil. Notice the rocks below the surface. They make up the C-horizon of the soil. They are all cracked, aren't they? Why? Well, one of the reasons can be seen near the center of the picture. A tree root has grown straight through the rock!

It might be hard to believe that roots can grow through rocks, but they can. In order to do that, however, they have to crack and break the rock. So even though you don't find hammers pulverizing rocks in nature, you do see strong things like roots exerting enough force over time to crack and break down rocks. Such processes are called **physical weathering**, because they use physical processes like strong forces to break down the rocks.

Physical weathering – The process by which physical forces break down rocks

Okay, well, the "physical" part makes sense, but why is it called "weathering?" Because weather is often responsible for the physical forces. You will learn more about that soon, but before you do, I want to finish discussing the results of the experiment.

You also soaked a piece of rock in vinegar. What did you see after it had been soaking for a while? You should have seen bubbles forming on the piece of rock. Those bubbles were caused by a chemical reaction that happened between a chemical in the vinegar, called acetic (uh' see tik) acid, and a specific mineral in the rock. You will learn more about that mineral in the next chapter.

The chemical reaction caused the acetic acid and the mineral in the rock to turn into completely different chemicals. Can you guess the phase of one of those chemicals? One of the chemicals had to be a gas, because bubbles formed, and bubbles mean that a gas is being made. The bubbles you saw were from carbon dioxide gas. The other chemical that was made dissolved in the liquid, so you didn't see it. This leads us to another way that rocks can be broken down – **chemical weathering**.

Chemical Weathering – The process by which chemical reactions break down rocks

How can a chemical reaction break down a rock? Well, look at the rock soaking in vinegar again. Swirl the glass gently so that the rock moves around at the bottom of the glass. Do you see anything else in the glass besides the vinegar and the rock? You should see tiny grains that look a bit like, well, dirt. That's because there are other chemicals in the rock that don't react with the acid in vinegar. When the mineral reacts with the acid, it leaves the rock. This breaks down the rock, and any tiny grains that were not made of that mineral fall off. If you don't see any "dirt" at the bottom of the glass, just let it sit until tomorrow, and you will see some. Once you have seen the "dirt," throw away the rock or put it outside where there is already some gravel. Also, pour the contents of the glass down the drain.

I want to talk about chemical and physical weathering more in the next section, but for now, I want you to start another experiment that needs to sit overnight.

Experiment 2.5: Physical Weathering

Supplies:

- Plaster from the laboratory kit made for this course (It's in a labeled bag.)
- A Styrofoam or paper cup
- A small balloon
- Water
- A 1-cup measuring cup
- A ¼-cup measuring cup
- A spoon for stirring.
- A freezer-safe bowl

Instructions:

1. Add enough water to the balloon so that balloon is full but not stretched out at all. It needs to be small enough to sit at the bottom of the Styrofoam or paper cup. Tie it off so the water stays inside.
2. Add ¼-cup of water to the cup.
3. Start adding plaster to the water, mixing as you go.
4. Repeat step 3 until you get a nice, thick mixture.
5. Put the tied-off balloon in the cup with the mixture. If it floats, push it down with your spoon until the plaster gets hard enough to hold it down so that it mostly stays underneath the plaster. When the plaster hardens, you want the balloon to be mostly surrounded by plaster.
6. Once the balloon looks like it will be mostly surrounded when the plaster hardens, set the cup aside so that the plaster can continue to harden. Rinse the spoon so that it is free of plaster.
7. You don't have any Comprehension Check questions for today, so you are done with science for right now. However, you need to check on the cup in a couple of hours.
8. After a couple of hours, the plaster should be hard. If not, wait another hour. Once it is hard, peel away the cup so that you now have a hard cylinder of plaster with a water balloon inside.
9. Put the cylinder of plaster in the freezer-safe bowl and put the bowl in the freezer.
10. Clean up any mess you made. You are now completely done with science for the day.

More About Physical Weathering

Yesterday, you put a cylinder of plaster in the freezer. Take it out now and look at it. What do you see? At minimum, you should see cracks in the plaster. There might even be pieces broken off the cylinder and sitting in the bowl. Once you have examined the damage done to the plaster cylinder, you can throw away the cylinder and any pieces of plaster in the bowl.

What caused the damage to the plaster cylinder? Remember, the plaster had a balloon inside, and that balloon was filled with water. When you put the cylinder in the freezer, the water inside the balloon eventually froze and became ice. That's what caused the damage. Why? As I mentioned in the previous chapter, for most chemicals, the molecules are closer together in their solid phase than in their liquid phase. However, water is an exception. Water molecules are farther apart in the solid phase than they are in the liquid phase.

Why does that make a difference? Think about it. If the molecules of a substance are close together, the substance doesn't take up a lot of volume. If they are farther apart, the substance takes up more volume. Well, when water freezes, its molecules must move farther apart from one another. That means the water must take up more volume. In other words, water expands when it freezes. This is different from most chemicals, because in most chemicals, the molecules move closer together when the chemical freezes. Thus, while most substances shrink when they freeze, water expands when it freezes. You need to remember that.

While most substances shrink when they freeze, water expands when it freezes.

How does that relate to the experiment? Think about it. The water balloon was packed inside the plaster cylinder. When the water inside the balloon froze, it had to expand. Thus, it started pushing out against the plaster. The plaster was hard, however, so it couldn't flex. As a result, it broke. The same thing can happen to rocks. Most rocks have tiny spaces (pores) in them, and if they get soaked in water, it will fill those pores. Well, if the weather gets cold enough for the water to freeze, it must expand, pushing out against the rock. Unlike your experiment, one freeze isn't enough to break most rocks, because most rocks are harder than the plaster you used. However, repeated freezing and melting will stress the rock, eventually breaking it like the balloon broke the plaster.

This rock was broken by water getting into cracks and expanding when it froze.

That's what happened to the rock pictured on the left. When the weather around it was warm, it soaked up rain, filling its pores with water. When the weather turned cold, that water froze, pushing out against the rock, making the pores slightly bigger. When the weather turned warm again, more water filled those bigger pores, so that when the weather turned cold again, even more expansion and pushing occurred. Eventually, the rock cracked and broke. This is often called "frost weathering," but you just need to recognize that it is a form of physical weathering. No chemical reactions were involved. The rock broke simply because of the physical stress produced by the freezing and melting of water inside the rock.

I am not done discussing water's role in weathering, but the fact that water expands when it freezes brings up an important point that I need to emphasize. In Chapter 1, I told you that while water has a density of $1.0 \frac{g}{mL}$, ice has a density of $0.9 \frac{g}{mL}$. You should now understand why ice has a lower density than water. Remember that density is the mass of an object divided by its volume. Well, when water freezes, its mass doesn't change. However, its volume gets bigger. So, when water freezes, the density is calculated by taking the same mass and dividing it by a larger volume. What happens when you divide by a larger number? The result is smaller. Ice's density is lower than water's density specifically because water expands when it freezes.

Why is this an important point? Think about a lake. What happens to it in the winter? The water in the lake starts to freeze. If water were like most other chemicals, it would shrink when it froze.

Its density would be calculated by taking the mass (which doesn't change) and dividing by a smaller number. What happens when you divide by a smaller number? The result gets bigger. So, if water were like most other substances, ice would be denser than water, and it would sink. As the water in the lake froze, then, the ice would sink to the bottom of the lake.

That's not what happens, however, because water expands when it freezes. That means the ice floats to the top of the lake. This is important, because the ice insulates the water below it. When it gets thick enough, the water below it can't get cold enough to freeze, so as long as the lake is deep enough, there will always be water in its liquid phase. That's good, because the organisms living in the lake need liquid water. If water shrank when it froze, the water in a lake would never get insulated, and the entire lake would freeze, killing almost all the organisms that live in it! Some might say that we are very "lucky" that water isn't like most chemicals in nature, but I think the fact that water expands when it freezes is just one of the many indications that the world we are studying was designed by God.

Other Kinds of Physical Weathering

So far, you have learned how the roots of plants and the freezing and thawing of water can produce physical weathering, but there are three more physical weathering processes that I want to discuss. Even if rocks don't have water-filled pores, temperature changes can cause damage to them. Have you ever taken something that was glass, gotten it really hot, and then put it under cold water? What happened? Depending on the glass from which it was made, it might have shattered.

Why would that happen? Well, most things expand when they warm up and contract when they cool down. When you heat up a glass object, then, the object gets bigger. When you cool it down, it gets smaller. If it cools too quickly, it shrinks too quickly, and the stress caused by the quick shrinking can break it. Rocks aren't as fragile as glass, and temperatures don't change really quickly in nature, but nevertheless, as the weather gets warm, rocks get bigger. As the weather cools down, rocks get smaller. The repeated changing in size can stress a rock, and over time, that can cause damage. This is usually referred to as **thermal stress**, since the word "thermal" refers to heat.

Another physical weathering process is caused by salt dissolved in water. When saltwater gets into the pores of a rock but doesn't freeze, the water can evaporate. However, the salt can't, so it is left behind. As this happens over and over again, the salt builds up in the pores, pushing against the rock. This happens more slowly than water freezing, and it produces a constant pressure. That makes the result very different from the pressure caused by water freezing and melting. If conditions are right, it can form patterns of pits like you see in the picture on the right. Notice the ocean in the background. This kind of weathering often occurs near an ocean, since saltwater from the ocean can spray onto the rocks. This kind of weathering is called **haloclasty** (hay' loh klas' tee).

Haloclasty can produce some very interesting weathering patterns.

The last physical weathering process I want to discuss is caused by sand and wind. Have you ever needed to make a surface smoother than it is? If so, you might have used sandpaper. If you rub sandpaper against wood, for example, the wood will usually get smoother. Once again, most rocks are harder than wood, but when they are constantly hit by bits of sand carried on the wind, the surface of the rock can be smoothed down. As the surface gets smoothed down more and more, the rock gets smaller and smaller. The speed at which this happens, of course, depends on how hard the rock is.

Wind-blown sand can also produce interesting weathering patterns.

Consider the picture on the left. It was taken in the Utah desert. There isn't a lot of water in the desert, but there is a lot of sand. When the wind blows that sand around, it hits any structures it encounters, including the rocks in the area. This will smooth down the rocks in the area, but the softer parts of the rocks will be smoothed down more than the harder parts. Looking at the structures in the picture, you can see that many of the rocks are very smooth, because they were softer and were affected strongly by the sand.

However, look at the highest piece of rock in the picture. It looks like it was just placed on top of that "tower" of smooth rock, doesn't it? But that's not what really happened. Initially, all these rocks were part of the same formation. However, as wind-blown sand started weathering the rocks, the softer parts were worn away faster than the harder parts. The parts that were much softer ended up being broken down completely. The slightly-harder parts got smoothed down, leaving what looks like towers of rock. The two rocks that look like they were placed on the towers are the hardest parts of the original rock formation, so they haven't been worn down much. That's why they look out of place.

Chemical Weathering

While chemical weathering also breaks down rocks, it is very different from physical weathering. In physical weathering, the rock gets broken down, but the chemicals in it do not change. They just are "packaged" in smaller bits. In chemical weathering, the chemicals themselves actually change. In the experiment you did, for example, a mineral in the limestone (which has the chemical formula $CaCO_3$) reacted with a chemical in the vinegar (acetic acid – $C_2H_4O_2$) to make two completely different chemicals, one of which was the gas carbon dioxide, CO_2.

To show you how the chemicals change in this process, here is the **chemical equation** for what happened in that part of the experiment:

$$CaCO_3 + 2C_2H_4O_2 \rightarrow CO_2 + H_2O + CaC_4H_6O_4$$

If you haven't seen a chemical equation before, this might look intimidating, but it isn't really that bad. The chemical formulas on the left side of the arrow indicate the chemicals that are reacting with each other, while the chemical formulas on the right side of the arrow tell you what is made in the reaction. The first chemical on the left, for example, is the mineral in the rock, which has a chemical formula of $CaCO_3$. The next chemical on the left is acetic acid, which has a chemical formula of $C_2H_4O_2$.

But wait a minute. What does the "2" to the left of $C_2H_4O_2$ mean? Well, remember, that $C_2H_4O_2$ represents a molecule, which is made of two C's, four H's, and two O's bonded together. The "2" to the left indicates that there are two of those molecules. How do I know that? Remember, if the number is a *subscript*, it's part of the chemical formula and tells you how many of the atom are found in the molecule. If there is no subscript, it means there is only one of those atoms in the molecule. If a number in a chemical reaction is *not a subscript*, it is written to the left of the chemical formula and tells you how many molecules take part in the reaction. If there is no number to the left of a chemical formula, only one molecule takes part in the reaction.

If you look at the chemical equation, then, there is no number to the left of $CaCO_3$. That means the reaction requires only one molecule of the mineral. There is a "2" next to $C_2H_4O_2$, so the reaction uses two acetic acid molecules. The chemical formulas on the right side of the arrow tell you what the reaction makes. There are no numbers to the left of those chemical formulas, so only one of each molecule is made. Thus, the reaction produces one carbon dioxide molecule (CO_2), one water molecule (H_2O), and one calcium acetate molecule ($CaC_4H_6O_4$). That means we could read the equation on the previous page like this:

"One $CaCO_3$ molecule reacts with two $C_2H_4O_2$ molecules to make
one CO_2 molecule, one H_2O molecule, and one $CaC_4H_6O_4$ molecule."

You won't ever be expected to produce chemical equations like this, but if I give you one, you should be able to tell me the chemical formulas of the chemicals that react and how many molecules of each are involved. You should also be able to tell me the chemical formulas of the chemicals that are made, along with how many of each molecule are made. You will get more experience with this the next time you do science.

The chemical equation shows you why this is chemical weathering and not physical weathering. Remember, in physical weathering, the chemicals don't change. They just get "packaged" into smaller grains. In chemical weathering, the chemicals themselves change. When $CaCO_3$ reacts with $C_2H_4O_2$, those chemicals are destroyed. In their place, three completely different chemicals (CO_2, H_2O, and $C_4H_6O_4$) are formed.

Because of chemical weathering, even something written in stone is not permanent!

While chemical weathering is different from physical weathering, it still ends up damaging the rocks, producing components that will become a part of the soil. Look at the photograph on the right, for example. All three of those old grave markers were carved out of stone, and then they were engraved with writing to identify who is buried there. However, you can't read anything on one grave marker, and the writing on the other two is very difficult to read. That's because over the years, acid has chemically weathered the stone.

You might be wondering where this acid comes from. After all, you don't normally think about acid as something you find in nature. However, acids are everywhere! In fact, every drop of rain that falls from the sky has acid in it! That's because there is carbon dioxide in the air, and when carbon dioxide reacts with water, it makes carbonic acid (H_2CO_3):

$$CO_2 + H_2O \rightarrow H_2CO_3$$

That means every drop of rain can produce chemical weathering, because every drop of rain has some acid in it. Lots of other acids are found in nature, too. Certain microscopic organisms produce acetic acid, for example. When that acid ends up on rocks which contain the right mineral, the chemical reaction that happened in your experiment occurs, weathering the rock.

Comprehension Check

2.9 You see a large rock with lots of cracks in it and what looks like dirt around its base. Suppose you collect a sample of the "dirt" and find out what chemicals are in it. Then, you analyze the rock and find out what chemicals are in it. If the "dirt" is the result of physical weathering, what can you say about how the chemicals compare. If the "dirt" is the result of chemical weathering, what can you say?

2.10 You see lots of cracks in a rock, but the area doesn't get much rain, there isn't much salt in the area, and there are no roots in the rock. Ignoring the possibility that something hard hit the rock with a lot of force, what type of physical weathering could explain the cracks?

2.11 Later on in the course, you will learn about rocks that form from lava. They often contain the chemical $HAlSi_3O_8$. It can react with water as shown in this chemical equation:

$$2HAlSi_3O_8 + 5H_2O \rightarrow Al_2O_3 + 6H_2SiO_3$$

What are the chemical formulas of the molecules that are reacting? How many of each are needed? What are the chemical formulas of the molecules that are produced? How many of each is made?

More About Chemical Weathering

Chemical weathering breaks down rocks into components that become part of the soil, but it can also produce some amazing views. Look at the photograph below, which shows part of the Rainbow Mountains in China. Believe it or not, no one painted those colors onto the rocks or added them to the picture afterwards. The coloring is a result of both the chemicals originally in the rocks and chemical weathering that took place after the rocks formed! For example, these rocks are rich in iron. Do you know what happens to iron that is exposed to rain and oxygen? It rusts. When you see red in rocks, it is often because of rust.

The colors in these mountains are natural.

But wait a minute. Rocks aren't made of iron. How can they rust? Well, they aren't made of iron, but they contain minerals that have iron atoms in them. For example, a naturally-occurring mineral called ferrosilite (fehr oh' sil eyet') has the chemical formula $FeSiO_3$, and "Fe" is the symbol for iron. When ferrosilite is exposed to oxygen (O_2), the following chemical reaction can occur:

$$4FeSiO_3 + O_2 \rightarrow 2Fe_2O_3 + 4SiO_2$$

Now remember, the numbers to the left of the chemical formulas just tell you how many molecules are involved, and if there isn't a number to the left, it means only one molecule is involved. In this chemical equation, then, four $FeSiO_3$ molecules react with one O_2 molecule to make two Fe_2O_3 molecules and four SiO_2 molecules. Guess what Fe_2O_3 is? It's rust! So some of the reds that you see in the Rainbow Mountains are the result of rust that forms because of chemical weathering.

What about the other colors? They come from more complicated chemicals in the rocks as well as more complicated reactions that occur during chemical weathering. While it's unusual to find such a wide variety of colors in the earth's crust, you can usually find some beautiful sights caused by chemical weathering, such as what you see in the photo on the right, which was taken in Reflection Canyon in Utah. The reds that you see in this photo are largely due to iron-containing minerals reacting with oxygen to make rust.

The reds you see in this photo are mostly from rust formed by chemical weathering.

Now once again, remember the long-term result of weathering, whether it is chemical or physical – it breaks down rocks, forming smaller fragments and grains that will eventually become part of the soil. Those grains, however, are only part of the mixture that ends up becoming soil. Weathering produces grains that contain minerals, but soil also contains organic chemicals as well. Those come from living things or the decaying remains of living things. As the grains produced by weathering are mixed with those organic chemicals, soil is formed.

Erosion

If you look at the photograph above, you don't really see any soil. Lots of weathering has occurred, but no soil has formed. Why? Because of another process that is closely-tied to weathering: **erosion** (uh' roz hun).

Erosion – The process by which rocks and soil components are transported to a different location

When weathering produces grains that fall out of rocks (like the grains you saw in Experiment 2.4), those grains might not stay where they are. After all, some are small enough that they can be blown around by the wind. Others can be washed away by the rain. Others can be carried away by rivers and streams. As a result, the components of soil might move around quite a bit before they become part of a soil horizon.

Erosion is a very important process, and we will discuss it a great deal in this course, because it is one of the major processes that shape the earth's crust. Think for a moment about the last two pictures I showed you. They contain rock formations that are not underground. They are exposed so that you can see them. But remember from our discussion of soil horizons that the parent material of the soil and the bedrock are found underneath the soil. Why are these rocks exposed? Because of erosion.

Consider, for example, the photo below. It shows one of the most amazing sights in the United States, the Grand Canyon. Off in the distance, you see green plants, which are growing in soil. The tree in the top left of the photo is also growing in soil. The rocks that you see here are underneath the soil, but their sides have been exposed. What exposed those sides? A combination of weathering and erosion. Erosion carried away the top soil horizons, exposing the parent material and bedrock, which was subjected to weathering. As the weathering broke down the rocks, the materials produced were carried away. As this process continued, it "carved" through the bedrock, making the incredible canyon you see today.

Canyons are formed by weathering and erosion.

But what is responsible for all that weathering and erosion? After all, *a lot* of rock had to be broken down and carried away. If you look at the bottom of the canyon, you will see a river – the Colorado River. It has a lot of water running through it. Water can cause physical weathering and can carry other chemicals that perform chemical weathering. Thus, it's tempting to think that perhaps the river "carved" the canyon. In fact, that's generally the story you will hear if you take a tour of the Grand Canyon or go to one of the visitor centers that have been set up there, because most **geologists** (scientists who study the geosphere) believe that's what happened.

If that explanation is correct, there is one thing you can conclude right away: It took *a long, long time* for the Grand Canyon to form. After all, we see lots of rivers like the Colorado River with lots of water running through them, but we don't see them cutting through the bedrock below them. Also, when we observe weathering, it generally takes a long time to make a noticeable impact. On page 55, for example, I showed you some grave markers that had been weathered to the point that you couldn't read much of what was originally carved into them. However, that's because those grave markers have been sitting outside for more than 100 years. If it takes more than 100 years to obscure the carvings on a gravestone, it must have taken *a lot longer* for the weathering required to make the Grand Canyon. Indeed, most geologists will tell you it took millions of years for the Colorado River to carve the Grand Canyon.

But there is another possibility. We have actually watched some canyons form, and they are not carved out slowly by rivers. Instead, they are carved out very quickly by floods. For example, there is a lake in Texas called Canyon Lake. It is an important source of water for the area, so it is managed by the state. One thing the state did was cut a small ditch to catch water that might flow out of the lake if the water level got too high. This is called a "spillway," and it causes any water that might "spill" out of the lake to stay away from areas where it might do damage.

Well, in 2002, there was a lot of rain in that part of the state, and it produced a lot of flooding. The lake "spilled over" into the ditch, but there was way too much water, so it just continued flowing in that direction. It destroyed trees, took out a bridge, removed the soil, and then started eroding the bedrock. In a matter of only three days, the flowing water carved a canyon that was 2.2-km (1.4 miles) long, hundreds of meters (hundreds of yards) wide, and 7 meters (23 feet) deep. A portion of that canyon, which is now called Canyon Lake Gorge, is shown in the picture on the right.

This canyon was carved by a flood in three days. The river formed after the canyon was formed.

Now just like the Grand Canyon, you might be tempted to think that the river you see in the picture carved the canyon over a long period of time. However, that's not what happened. A flood caused the weathering and erosion that carved the canyon, and it happened in just three days. After the canyon formed, water started naturally collecting in the canyon, making the river. So, the river did not make the canyon. Instead, the canyon made the river.

Not surprisingly, the canyon is a tourist attraction, and the state has made it convenient for people to go down into the canyon by adding handrails and other safety features. The picture below shows you some people going down into the canyon. The reason I am showing you this picture is so that you can appreciate the size of the boulders. They are huge, aren't they? You certainly couldn't lift one and move it, could you? However, they were lifted and moved...by water from the flood! The water knocked those boulders out of the bedrock and actually pushed them to where they are today. That tells you just how powerful a force erosion can be when there is a lot of water involved!

In the end, then, we know that at least some canyons are carved quickly as a result of floods. Was the Grand Canyon carved that way as well? The short answer is that we don't know. However, the long answer is a lot more interesting, and it will come up again and again in this course. Since we weren't around to watch how most features of the earth's crust were made, we have to speculate about what happened to form them. I want to end this chapter with a discussion of two very different views that affect such speculation.

The large boulders in this picture were carried by the flood waters that carved Canyon Lake Gorge.

Uniformitarianism and Catastrophism

When we study the earth's crust, we are analyzing things that already exist. In a few cases, like Canyon Lake Gorge, we are fortunate enough to have historical records about what happened. However, we usually aren't that fortunate. There are no historical records that tell us how the Grand Canyon formed. Even in the case of Canyon Lake Gorge, we have records about how the bedrock was carved out, but we don't have records about how the bedrock was formed to begin with. Nevertheless, as scientists, we would like to figure that out. Is there some way to do that? Of course!

We have never been able to get below the crust of the earth, but that hasn't stopped us from figuring out that there is a mantle, an outer core, and an inner core. In fact, we can even tell you about the chemicals you find in those parts of the earth. As I explained previously, we can do that by studying the interior of the earth indirectly. In the same way, we cannot directly study how the Grand Canyon and the bedrock through which it was cut was formed, but we can study it indirectly.

How do we do that? We use the scientific method. Hopefully, you have learned this before, but very briefly, we can study the details of the canyons or rocks that we are interested in, and we use our observations to come up with an idea of how they were formed. That idea is called a **hypothesis** (hy pah' thuh sis). We can then use that hypothesis to make a prediction about some aspect of the rocks or canyon that hasn't been observed yet. If we look at the rocks or canyon again and confirm that our prediction was correct, we have evidence that our hypothesis is true. If the hypothesis has several predictions confirmed, it can be considered a **theory**, which is considered a reasonable scientific explanation for how the bedrock or canyon formed.

One important thing you have to realize, however, is that every scientist looks at things a bit differently. As a result, the hypothesis that one scientist makes could be very different from the hypothesis another scientist makes, even though they are studying the same thing. This can be especially true when studying the earth. Scientists approach earth science with very different viewpoints, which can result in very different hypotheses.

For example, many scientists think that the best way to explain how most features of the earth's crust came to be is through the same things we see happening today. We see rivers flowing through areas, causing slow weathering and erosion. Thus, when these scientists see a canyon with a river on the bottom, it makes sense to them that the same processes we see happening in other rivers is what formed the canyon. They just took a long, long time to do it. This view is often called **uniformitarianism** (yoo' nuh form' uh tair' ee uhn iz uhm).

> Uniformitarianism – The view that most of the crust's features are the result of slow, gradual processes that have been at work for millions or even billions of years

If you start with the assumption of uniformitarianism, then, your hypotheses will generally involve things forming over long timespans.

While uniformitarianism is the way most geologists today prefer to study the earth, it didn't use to be. Throughout most of the history of science, scientists preferred to think that most features of the earth's crust were formed quickly, as a result of catastrophic processes. Not surprisingly, this is called **catastrophism** (kuh ta' stroh fiz uhm).

Catastrophism – The view that most of the crust's features are the result of large-scale catastrophes such as floods, volcanic eruptions, etc.

What caused these scientists to think catastrophism was a reasonable assumption? Because many (but not all) of them believed that there was a worldwide Flood as described in the Bible. Such a catastrophe would clearly have a lasting impact on the earth's crust, so it made sense to them to start with that perspective.

While most scientists throughout the course of history didn't have the opportunity to study catastrophes in detail, over the past few decades, many have. Lake Canyon Gorge is one example. A small flood caused by excess rainfall in Texas was able to carve a canyon and move huge boulders around. Imagine what a catastrophic flood that covered the entire world could do!

Catastrophism uses catastrophes, like the global Flood, to explain most features of the earth's crust.

Not surprisingly, studying the earth's crust from the viewpoint of catastrophism produces hypotheses that are very different from the ones that are produced by the uniformitarianism viewpoint. Which one makes the better hypotheses? I have my own view, and it will probably become apparent as the course progresses, but I think it is important for you to look at the evidence and come to your own view. Thus, you will see uniformitarianism and catastrophism compared throughout this course. In each instance, you will see the different hypotheses that come from these views and the evidence that relates to them. Hopefully, that will start you on an intellectual journey to discover which is the better way of studying the earth.

Comprehension Check

2.12 While we know it would be impossible, think about an earth with no oxygen in its atmosphere. Would the rainbow mountains still be so colorful? Why or why not?

2.13 While I concentrated on water as the main agent of erosion, there is another very important part of the weather that can also cause a lot of erosion. What is it? (HINT: The products of weathering are often very tiny grains.)

2.14 In the definitions of catastrophism and uniformitarianism, I use the word "most." It's important for both, but based on what you learned already, you should see that it is important for the definition of uniformitarianism. Why?

Answers to the Comprehension Check Questions

2.1 The second is closer to the north pole. Longitude (identified with either E or W) measures your position east or west, so it doesn't tell you anything about how close you are to either pole. Only the latitude (identified with either N or S) matters for that, and the bigger the number, the farther you are from the equator, so the closer you are to a pole.

2.2 It is part of the hydrosphere. The atmosphere is the collection of gases around the planet, but all water is in the hydrosphere.

2.3 The inner core is the hottest, and the crust is the coolest. For the earth, the deeper you go, the hotter it gets.

2.4 You should tell your friend that scientists can measure things indirectly, so they don't have to visit the sun to know its composition. Gases emit specific kinds of light, and by studying the light that comes from the sun and comparing it to the kinds of light emitted by chemicals on earth, we can determine the amount and types of chemicals in the sun.

2.5 The first is from the B-horizon, and the second from the A-horizon. Since there is mineral matter in both, they cannot be from the O-horizon. Since there is organic matter, they can't be from the E-horizon or the C-horizon. Thus, they must be from the A- and B-horizons. B-horizons are usually lighter, so the lighter one is from the B-horizon.

2.6 The larger gravel would be at the bottom, and the sand would be closest to the top. The sorting wouldn't be perfect, because you would need a *very* tall container to get perfect separation. Nevertheless, the largest particles would be more likely to be at the bottom, and the smallest particles would be more likely to be at the top, because the smallest particles would sink the slowest, while the largest particles would sink the fastest.

2.7 It is a sandy clay loam. Remember, the horizontal lines are for clay. So, you look at the horizontal line that comes from the 30% clay mark. The lines running diagonally to the silt side (up and to the right) are the silt lines. So you are looking at the diagonal line running up to 10% silt. The lines running down and to the right are for sand. So you look at the line running to 60% sand. Those three lines intersect right next to the first "A" in "SANDY CLAY LOAM."

2.8 Loam A has smaller pores. Clay has the smallest pores, and sand has the largest. Silt is in between. Notice that Loam A has more clay and more silt than Loam B. Also, Loam A has less sand. So Loam A has more of the small-pore material and less of the large-pore material. Thus, its pores will be smaller.

2.9 If physical weathering made the "dirt," the chemicals will be the same. If chemical weathering made it, the chemicals will be different.

2.10 Thermal stress caused the cracks. You could just say lots of temperature changes. Since there isn't a lot of rainfall, there wouldn't be much of an opportunity for freezing and melting to cause physical weathering, and haloclasty can't cause it, because there is no salt. No plants means roots couldn't cause it, either.

2.11 $HAlSi_3O_8$ and H_2O are the chemicals that react. Two molecules of $HAlSi_3O_8$ and five molecules of H_2O are required. Al_2O_3 and H_2SiO_3 are made. One molecule of Al_2O_3 is made, while six molecules of H_2SiO_3 are made. Now remember, only the numbers that are subscripted are part of the chemical formulas, so the 2, 5, and 6 in the equation are not part of any chemical formula. They tell you how many of each molecule is needed or produced. The molecules on the left are the ones doing the reacting, and the ones on the right are being made.

2.12 No, they would not be so colorful. Look at the chemical equation for the formation of rust. It requires oxygen, since O_2 is on the left-hand side of the equation. Without the rust, there would be no red. In fact, most of the colors you see are caused by chemical weathering processes that depend on oxygen.

2.13 Wind can also cause erosion. Remember, erosion is a process by which things are carried to a different location. If the products of weathering are small enough, they can be blown away by the wind.

2.14 Even those who follow uniformitarianism agree that some things, like Lake Canyon Gorge, were formed as a result of catastrophes. Remember, I said that most geologists are uniformitarians, but they don't think that Lake Canyon Gorge formed slowly, because we saw it form. As you will see later, scientists who follow catastrophism agree that some things we see are the result of slow, gradual processes. Thus, these two views don't exclude each other. One just emphasizes slow, gradual processes over catastrophes, while the other does the opposite.

Chapter Review

1. Define the following terms:

a. Atmosphere
b. Geosphere
c. Hydrosphere
d. Mineral
e. Percolation
f. Loam
g. Physical weathering
h. Chemical Weathering
i. Erosion
j. Uniformitarianism
k. Catastrophism

2. The latitude and longitude of city A is 12° S, 77° W. For city B, the latitude and longitude are 35 °S, 58 °W. Which is closer to the equator? Which is closer to the prime meridian?

3. Is the earth's crust part of the atmosphere, geosphere, or hydrosphere?

4. What are the spheres into which the geosphere can be divided? List them in order of their depth.

5. Of the spheres listed above, which is the coolest? Which is the hottest? Which is liquid? Which is solid? Which can be described as being made of "plastic" rock?

6. List the soil horizons in order of their depth. Which horizon might not exist in an area?

7. For each horizon listed above, describe its contents.

8. Why is broken bedrock sometimes called the "parent material" of soil?

9. List the three kinds of particles found in a loam, from largest to smallest.

10. Why is it important for a loam to have a mixture of those three types of particles?

11. A loam doesn't allow water to percolate through it. If it has the right amount of silt, what kind of particle does it have too much of? What kind of particle would you mix in to make it better for plants?

12. You learned about five things that produce physical weathering. What are they?

13. A geologist describes a chemical weathering process this way: Two molecules of lactic acid ($C_3H_6O_3$) made by bacteria react with one molecule of magnesium carbonate ($MgCO_3$) to make one molecule of water (H_2O), one molecule of carbon dioxide (CO_2), and one molecule of magnesium lactate ($MgC_6H_{10}O_6$). Write the chemical equation that describes this process.

14. Oxygen and water can chemically weather a rock containing pyrite (FeS_2) this way:

$$4FeS_2 + 15O_2 + 8H_2O \rightarrow 2Fe_2O_3 + 8H_2SO_4$$

List the molecules that react and the molecules that are made, including how many of each is involved.

15. Which would usually happen to a rock first: weathering or erosion?

16. Explain how a scientist guided by uniformitarianism would explain how the Grand Canyon was formed. How would a scientist guided by catastrophism explain it?

17. What catastrophe inspired many of the scientists in history to follow catastrophism?

Chapter 3: Minerals

Introduction

In the previous chapter, I defined the term "mineral" and used it several times to refer to chemicals found in rocks. In this chapter, I am going to tell you a lot more about minerals. Let me start by reminding you that earth scientists use the term rather differently from the way it is used in everyday language. For example, you have probably been told that to be healthy, you need to eat foods that are rich in vitamins and minerals. In that context, a mineral is a specific element that is important to your body. Nutritionists list calcium, magnesium, phosphorus, and ten other elements as minerals that are necessary for life.

However, that's not how the term is used in earth science. In earth science, a mineral is an inorganic substance found in rock that has its own chemical and physical properties. While elements can be minerals (gold, copper, and silver, for example), most minerals are compounds made up of many different elements. The picture on the right, for example, shows emeralds as they are often found in nature. They represent one form of a mineral known as **beryl** (bear' uhl), which has the chemical formula $Be_3Al_2Si_6O_{18}$. Believe it or not, there are more than 5,400 different named minerals, and new ones are still being discovered!

These emeralds are a form of the mineral beryl.

Minerals and Rocks

As you learned in the previous chapter, minerals and rocks are related. When you put the piece of rock in vinegar and saw the bubbles, I told you that those bubbles were the result of a chemical reaction between a chemical in vinegar (acetic acid) and a mineral in the rock. I want you to get a bit of experience with the relationship between minerals and rocks by performing the following experiment.

Experiment 3.1: Minerals and Rocks

Supplies:

- A mix of minerals and rocks from the laboratory kit made for this course, found in the "Geology Basics Kit" box
- A magnifying glass from the laboratory kit made for this course, found in a small, white box in the "Geology Testing Kit" box
- A well-lit room

Instructions:

1. Open the Geology Basics Kit. You should see a collection of specimens, each in its own space, along with a few specimens in plastic bags. The specimens not in bags are labelled with numbers,

which is important. Don't let the numbers come off, since I will refer to them by number in all the experiments that you do with them.

2. Your goal in this experiment is to make as many observations as you can regarding the differences between specimens 1-6 and specimens 7-15. Start by just looking at the 15 specimens. Can you make any statements about most of specimens 1-6 which cannot be made about most of 7-15?
3. Make sure there is plenty of light in the room so that you can see things really well.
4. Pull the first specimen out of its space and hold it between your thumb and index finger. Slowly tilt the specimen one way and then another. Write down a description of what you see. For example, is there a specific color? Is it transparent or at least partially transparent? Does it shine in the light? Does it make sparkles as you tilt it in the light?
5. Look at the specimen through the magnifying glass. Does this allow you to add more to the description? For example, are there small things in the specimen that you didn't notice until you looked through the magnifying glass? If so, describe them.
6. Repeat steps 4 and 5 for each of the next 14 specimens (2-15).
7. Reviewing your results and thinking about what you saw, are there specific observations that you mostly found with specimens 1-6 or specimens 7-15?
8. Put everything away. You need to keep this kit orderly, because you will be using the specimens a lot in the next few chapters.

Specimens 1-6 were minerals, while 7-15 were rocks. Did you notice any consistent differences between the two groups? Most of the minerals were "shinier" than most of the rocks. Some parts of the rocks might have sparkled a bit, but in general, the rocks shouldn't have been nearly as shiny as the minerals. More importantly, the minerals should have been a lot more consistent in their appearance. There were imperfections, to be sure, but in general, most of the minerals should have had roughly the same appearance throughout, and they should have reflected light more consistently throughout. That's because aside from contaminants, a mineral has the same chemical and physical properties throughout the sample.

A rock rarely has the same chemical and physical composition throughout. That should have been apparent when you looked at most of the specimens with the magnifying glass. In general, the rocks should have looked more like mixtures of many different things, while the minerals looked more like they were made of a single substance. The pictures on the left, for example, show two red specimens, but they are clearly different. That's because the one on the far left is a rock, while the one next to it is a mineral. Notice how the mineral is shinier than the rock and is more uniform throughout the specimen. Not all minerals are completely uniform. There can be other substances found in and around the mineral. Nevertheless, a mineral is a lot closer to being uniform than a rock.

A red rock (left) compared to a red mineral (right).

In general, then, rocks are like mixtures, while minerals are composed of a single substance, along with maybe a few contaminants. Because minerals are mostly a single substance, they have specific properties that can be used to differentiate them from other minerals. Let's go over a few of those properties so you can start to learn how to distinguish one mineral from another.

Color: An Optical Property of Minerals

One of the first things you notice about any object is its color. The same can be said for minerals. I am sure that each time you looked at one of the mineral samples, you took note of its color. Well, color is an **optical property**, because it comes from the way that light reflects off the mineral. You have probably learned this already, but the light that comes from the sun (or a white light bulb) is made up of all the colors of the rainbow. When white light hits an object, different things can happen, which affects the object's appearance.

If the object absorbs all the light that hits it, we don't see any light coming off the object. Thus, it appears black to us. If it reflects all the light that hits it, we see all the colors of the rainbow coming from it, so it appears white. If it absorbs some colors and reflects others, the color we see is a mixture of the colors that are reflected. The red mineral pictured on the previous page, for example. absorbs all colors of light except red. It reflects red light. As a result, we see only red light coming from the mineral, so we see it as red. Thus, the color of a mineral is based on how that mineral interacts with different colors of light. As you will see the next time you do science, a mineral's color can be dependent on how we observe it.

The color of a mineral is important, because it is sometimes the main thing that distinguishes one mineral from another. Consider the four minerals pictured on the right. They are all versions of beryl, which you saw on page 65. They all have the same chemical formula, $Be_3Al_2Si_6O_{18}$. Why do they look so different? Well, pure $Be_3Al_2Si_6O_{18}$ is colorless, so the mineral on the top left is mostly just $Be_3Al_2Si_6O_{18}$. However, when $Be_3Al_2Si_6O_{18}$ is contaminated by small amounts of another chemical, it will have a color. Each of the other samples of beryl, then, have different contaminants in them. Since they are all basically the same chemical, we call them all beryl, but because of the differences caused by the different impurities, we say they are each their own **variety** of beryl.

This image shows four different varieties of beryl: mostly pure (top left), rose (top right), blue (bottom left), and yellow (bottom right).

Comprehension Check

3.1 You are looking at a sample of something from the earth's crust. It is gray with irregular red stripes in it. Is it more likely to be a rock or a mineral?

3.2 You will learn about shape and luster in the next section. Just based on the names, which is an optical property?

Other Optical Properties of Minerals

While the color of a mineral is easy to observe, as I mentioned previously, it can change depending on how you observe it. After all, the color of a mineral (or any other object) depends on how it interacts with light. If you change the mineral so that it interacts with light differently, the color will change. One very simple way to change a mineral is to turn it into a powder. That can end up changing its color, so earth scientists have another term for the color of a mineral when it is turned into a powder. It's called the **streak** of a mineral.

Streak – The color of a mineral when it is a powder

Why is it called "streak"? You will see when you do the next experiment.

In the previous experiment, one of the things you looked at was how light reflects off the surface of the mineral and rock specimens you had. Well, for minerals, that's another important optical property, which is called **luster**.

Luster – The way a mineral shines when light hits it

As you might already have guessed from the experiment, minerals have many different lusters. There are two broad categories, however: **metallic luster** and **non-metallic luster**. Not surprisingly, a metallic luster refers to a mineral that shines like a metal under the light. Now remember, some metals can get dull, and some minerals with a metallic luster can also get dull. Thus, if a mineral would shine like a metal once it is polished, it has a metallic luster.

If a mineral has a non-metallic luster, it can be classified in several different ways. If it shines like glass, we say it has a "vitreous" (vih' tree us) luster. If it shines like it is covered with oil, it has a "greasy" luster. Other non-metallic lusters include waxy, pearly, porcelain, and silky. If it doesn't shine at all, it has a **dull** luster. In a moment, you will try to categorize the luster of the minerals in your lab kit. Don't worry if you don't do well. Lusters can be difficult to identify. Also, the only lusters you need to remember are metallic, non-metallic, and dull.

In the previous experiment, you probably noticed another important optical property: **transparency**, which describes how light goes through the mineral. If it is mostly transparent, light passes through the mineral easily, and we say that the mineral is **transparent**. As a result, the mineral looks pretty clear. Most minerals are **translucent**, which means light can pass through them, but not well. Often, only certain colors of light make it through the mineral. Some minerals are **opaque** (oh payk'), which means light doesn't pass through them at all.

Get some experience exploring the optical properties of your minerals by performing the following experiment.

Experiment 3.2: Optical Properties of Minerals

Supplies:

- Specimens 1-6 from the laboratory kit made for this course
- The white and black tiles from the laboratory kit made for this course
- A penny (There is one in the laboratory kit made for this course.)

Instructions:

1. Pick up one of the minerals labeled specimen 1. Classify it as transparent, translucent, or opaque.
2. Identify the color of the specimen. (red, yellow, blue, clear, etc.)
3. Hold the black streak plate in your other hand, and push the specimen into the plate hard, near the top. As you continue to push, drag the specimen across the plate. It should leave a streak on the plate. The color tells you the specimen's streak. It may be different from the color you wrote down in step 2.
4. Try using different sides of the mineral to make the streak to see if the color changes.
5. Tilt the specimen back and forth and turn it around so that you can see all sides, noting how it shines in the light. Compare its luster to the penny's luster. The penny is coated in a metal (copper), so the penny has a metallic luster. Identify your specimen's luster as metallic or non-metallic.
6. If you classified the mineral as having non-metallic luster, try to identify its luster according to the words I used on the previous page.
7. Repeat steps 1-6 for the other sample of specimen 1. Label it as "specimen 1b." For step 3, make the streak just below the streaks from the first specimen.
8. Repeat steps 1-6 for the other mineral specimens (2-6). Once again, do the streaks below the ones you already made so you can fit all the streaks onto one tile. Also, if you don't see the streak with the black tile, use the white tile instead.
9. Don't clean up your mess yet. Keep everything out as you read the following discussion, and feel free to look at the specimens, make more streaks, etc., while you are reading.

For the color and streak of the minerals, you can compare your answers to the ones on the sheet that is in your kit. You might not have exactly the same answers, because there might be some specific contaminant in one of your specimens that isn't common. For example, your kit says that specimen 6 (hematite) should have a brick red streak. You might have gotten that color when you used one side of the mineral, but the streak might have been brown or black when you used a different side.

Regardless of how your answers compared to the answers on the paper in the kit, you now know why it's called the "streak" of a mineral. Because of the rough nature of the two tiles, when you pulled the mineral across the tile, bits of the mineral broke off, forming a streak of powder. Thus, you can see the color of a mineral in its powdered form by making a streak like that. Not surprisingly, it is called a **streak test**.

Did you classify any of your minerals as having a metallic luster? Since you don't have a lot of experience with this, you might not have. However, the hematite (6) has a metallic luster. You might have called it dull, but remember, if it looks like it could be polished so that it shines like a metal, then it has a metallic luster. The other minerals had non-metallic lusters.

Were you able to further classify the non-metallic lusters? If not, don't worry about it. However, specimens 4 (calcite) and 5 (fluorite) were probably vitreous, because they looked a lot like glass. They were probably tinted, but they were probably glassy. The transparent form of specimen 1 (quartz) was also vitreous, but the opaque form might have been waxy or porcelain. Specimen 2 (muscovite [muh' skuh vight] mica) was probably pearly, and specimen 3 (potassium feldspar) was probably porcelain as well.

But what about transparency? Most likely, specimens 4, 5, and 2 were translucent, while specimens 3 and 6 were opaque. But what about specimen 1, the quartz? Why was one sample transparent, while the other was opaque? Well, think about the streak test. What did the powdered form of the transparent quartz look like? It looked white, didn't it? Imagine turning the entire transparent quartz sample into a powder and then mashing the powder together. What would it look like? It would look opaque, wouldn't it?

This image shows four different minerals that had room to grow and contain few impurities: iron pyrite (top left), quartz (top right), fluorite (bottom left), and calcite (bottom right).

Minerals actually grow as their molecules collect together. If they have plenty of room to grow, the molecules form a repeating pattern, which is called the mineral's **crystal form** or **habit**. When there is plenty of room for the crystal to grow, and when there aren't any impurities, the molecules keep forming that repeating pattern, and the result is a uniform crystal with a very specific geometry. Look, for example, at the four minerals on the left. You can see that each has a distinct shape and looks fairly uniform throughout. This indicates there was plenty of room to grow, and there were few impurities. Why didn't your transparent quartz specimen look like the one pictured in the upper right photo? Because your specimen was just one piece that was broken out of a bigger crystal like that one. The flat surfaces you see are called the faces of the crystal.

If there isn't enough room for the crystal to grow in one direction, it will stop growing in that direction but continue to grow in other directions, deforming the shape of the crystal. If there are impurities, the crystals become less uniform throughout. In the case of your opaque quartz specimen, there were water molecules that disrupted the growth of the crystal. This resulted in lots of tiny crystals that have water molecules in between. As a result, it looks more like powdered quartz, which is what was formed in the streak test.

It might seem odd that a crystal can "grow," but remember, a crystal is a collection of molecules with some impurities. If mineral molecules are brought into an area, they can collect together, and the crystal can start to grow. How could mineral molecules be brought into an area? Well, think about water percolating through the soil. Mineral molecules can dissolve into it and be carried to wherever the water is going.

Comprehension Check

3.3 You see a sample of quartz shaped like the one in the picture above, but it is opaque. What can you say about the room it had to grow and the level of impurities in the area where it grew?

3.4 Using the same specimen, one student claims a mineral's streak is brown, and another student says it is green. Is it possible that both are accurately reporting the results of a streak test? Why or why not?

You are done with science for the day, so clean up and put all the experiment materials away.

Mechanical Properties of Minerals

The optical properties of minerals are important, but they aren't enough to completely describe a mineral. Minerals also have **mechanical properties**, which can be investigated by seeing how the crystal can be damaged. Suppose, for example, you wanted to break a mineral, like you broke the limestone rock in Experiment 2.4. Many minerals will break along flat surfaces, leaving relatively smooth faces where the break occurs. For example, the mineral halite is NaCl – the salt you use to season food. When it is found in nature, it is often in the form of large box-like shapes, as shown on the right. All those samples of halite were broken from a larger halite crystal, but they broke so that they have flat faces that form a "box" of halite. When minerals break like this, we call it **cleavage** (klee' vij).

When halite is broken, it breaks along flat surfaces, which is called cleavage

Cleavage – The tendency of a mineral to break along parallel flat surfaces

In cleavage, there are two flat surfaces, and they are parallel to one another. The best way to test for cleavage is to try to pinch the material with your thumb and index finger each on a flat surface. If your thumb and index finger are parallel, the mineral has cleavage.

While it is often enough to just indicate whether or not a mineral has cleavage, there are more details that can be added. For example, the cleavage can be perfect, good, poor, or indistinct. Perfect cleavage results in almost perfectly flat surfaces, while good cleavage will have some roughness to the surface. Poor cleavage results in noticeably rough surfaces, and indistinct cleavage exists when you know the mineral should have cleavage, but you can't really find it. As a sort of reference, the halite crystals above have perfect cleavage. They may not look like they have perfectly flat surfaces to you, but in the mineral world, they are.

In addition, you can specify the directions in which the cleavage can happen. In the case of the halite crystals above, we say that the cleavage is perfect in three directions. Why? Think about how you could pinch those crystals with your thumb and index finger and end up having them parallel. You could pinch the top and bottom, you could pinch the sides, or you could pinch the ends. Since there are three sets of parallel flat surfaces, we say that the cleavage is in three directions. While many minerals have cleavage, some do not. If they don't have cleavage, we say that they **fracture**. You will get some experience determining whether or not a mineral has cleavage or fracture in a moment.

Another important mechanical property is the hardness of a mineral. As you have been handling the minerals in your kit, you have probably noticed that some are harder than others. In fact, there is a scale that is used to measure a mineral's hardness, and it was developed in 1812 by Friedrich Mohs, a German **mineralogist**. If you haven't seen that term before, it refers to a scientist who studies minerals. Of all the minerals he had studied, the softest one he could find was talc. If you have heard of "talcum powder," it is made by crushing talc and is useful for many applications, including absorbing water. The hardest mineral he could find was diamond. He decided to define talc's hardness as 1 and the diamond's hardness as 10. He then found other minerals in between, and defined the scale shown in the table below, which is called the **Mohs Hardness Scale**.

While his scale was defined by specific minerals, Mohs wanted mineralogists and other earth scientists to be able to use his scale with very simple materials, so he found an easy way to estimate how hard a mineral is on his scale. Essentially, you see whether or not the mineral can make a scratch on glass. If so, it is considered a hard mineral, with a hardness of 5.5 or higher. If it can scratch one of the plates you used for the streak test, it is even harder, with a hardness of 6.5 or higher.

If the mineral can't scratch glass, it is softer than 5.5. To determine its hardness, you try to scratch the mineral. If you can scratch it with your fingernail, its hardness is 2.5 or lower. If you can scratch it with a penny, its hardness is 3.0 or lower. If you can scratch it with an iron nail, its hardness is 4.5 or lower. For example, suppose a mineral cannot scratch glass. You find that you can scratch it with an iron nail, but you can't scratch it with a penny. Thus, it is harder than 3.0, because the penny couldn't scratch it. However, it is 4.5 or softer, because the iron nail could scratch it. Thus, it is between 3.0 and 4.5 on the Mohs Hardness Scale.

The Mohs Hardness Scale

DEFINITION

Mineral	Hardness	Mineral	Hardness
Talc	1	**Potassium feldspar**	6
Gypsum	2	**Quartz**	7
Calcite	3	**Topaz**	8
Fluorite	4	**Corundum**	9
Apatite	5	**Diamond**	10

SCRATCH TESTING

Scratch Tool	Hardness
Fingernail	2.5
Penny	3.0
Iron Nail	4.5
Glass plate	5.5
Streak plate	6.5

Get some experience identifying cleavage/fracture and the hardness of minerals by performing the following experiment.

Experiment 3.3: Mechanical Properties of Minerals

Supplies:

- Specimens 1-6 from the laboratory kit made for this course
- The black and glass tiles from the laboratory kit made for this course
- A penny (There is one in the laboratory kit made for this course.)
- An iron nail (There is one in the laboratory kit made for this course.)

Instructions:

1. Pick up specimen 1 from the Geology Basics Kit.
2. Examine it for parallel, flat sides. Once again, do this by trying to pinch two flat sides between your thumb and index finger. If your thumb and index finger are parallel, you have a cleavage direction.
3. If you did find cleavage, see how many cleavage directions you can find and determine whether or not the cleavage is perfect, good, or poor. If you didn't find any cleavage, the mineral has fracture.
4. Now it's time to do scratch tests to measure hardness. Start with the glass plate. Try to scratch the glass plate with the mineral. Essentially, act like you are doing a streak test, but move it back and forth a couple of times to see if you can leave scratches in the glass. Hold the glass up to the light

to look for scratches. You aren't looking for a streak of powder. You are looking for a scratch that has damaged the glass. If you see a mark, rub it with your thumb to see if you can rub it off. If you can rub it off, it is a streak, not a scratch.

5. If your mineral did scratch the glass, see if it scratches the black streak plate. Once again, you are looking for a mark on the plate that you can't rub off.
6. If your mineral didn't scratch the glass, try to scratch it with the nail. If you leave a mark on the mineral, it has been scratched.
7. If the nail scratched your mineral, try to scratch it with the edge of the penny. Once again, if you leave a mark on the mineral, it has been scratched.
8. If the penny left a scratch, see if you can scratch it with your fingernail.
9. Based on the results of the scratch test, determine the range in hardness of the mineral.
10. Repeat steps 1-9 with the other five minerals (specimens 2-6).
11. Keep everything out while you check your results as discussed below.

To see how you did, compare your determination of fracture/cleavage with the paper that came in your kit. When it comes to cleavage, don't worry if you didn't find all the directions. It's possible that your specimen didn't cleave exactly the way that it should have. However, you should have been able to correctly identify that all of your minerals except the quartz and hematite did have some cleavage. If not, try doing the test again, seeing if you can understand where you went wrong. Technically, the quartz has a "conchoidal" (kahng koy' dul) fracture, which just means the fracture surface can be curved. It is not smooth and parallel to another surface, however, so it is just a special kind of fracture. Also, don't worry if your judgement of perfect, good, or poor isn't the same as what is on the page. What's more important is that you got experience trying to make that judgement.

Now look at the range for hardness. Is it consistent with what is on the paper? For example, you should have been able to scratch the fluorite with the nail but not the penny, so you should have gotten a hardness between 3 and 4.5. The paper says 4, which is in that range. If one or more of your ranges are not consistent with what is on the paper, try again. Sometimes, determining whether or not scratches are left on the glass plate or the streak plate can be tricky.

While I expect you to be able to use the Mohs hardness scale, you do not need to remember which mineral has which hardness. You also don't need to memorize the scratch tools and what they mean. However, if you are given the tables on the previous page, you should be able to use them to interpret the results of a scratch test, like you did in the experiment. Comprehension Check question 3.6 also gives you an idea of how I want you to use the Mohs Hardness scale.

Comprehension Check

3.5 The pictures on the right are of two different minerals. Which has cleavage and which has fracture?

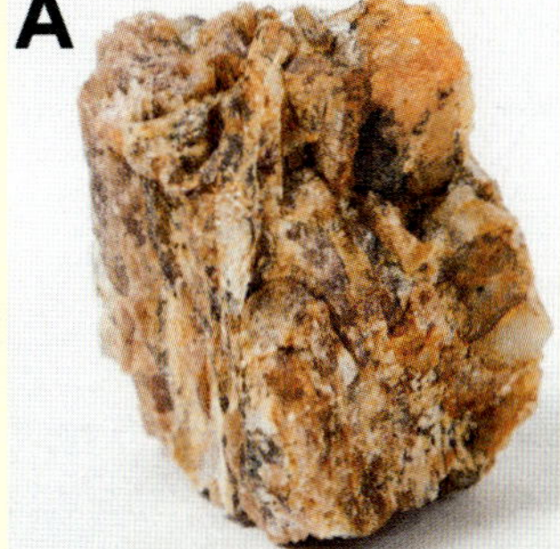

3.6 An iron nail cannot scratch a mineral, but that mineral will not scratch a glass plate. If you are told it is one of the minerals used to define the Mohs Hardness Scale, which is it?

You are done with science for the day, so clean up and put all the experiment materials away.

Other Properties of Minerals

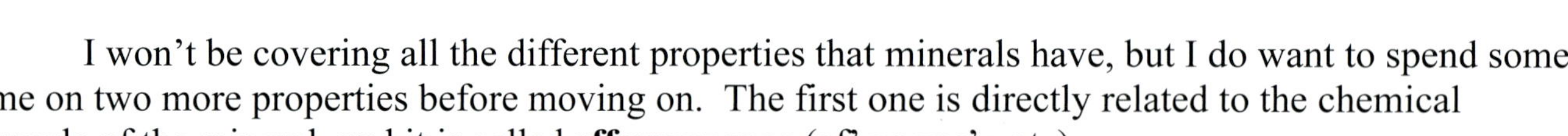

I won't be covering all the different properties that minerals have, but I do want to spend some time on two more properties before moving on. The first one is directly related to the chemical formula of the mineral, and it is called **effervescence** (ef' ur ves' untz).

Effervescence – The bubbling of a mineral when it is exposed to acid

In Experiment 2.4, you saw that limestone has effervescence, because it bubbled when it was placed in vinegar, which is acidic. That's because the limestone contains a mineral with that property. In a moment, you will find out what mineral that is.

These two samples of magnetite are attracting bits of iron.

Some minerals can also have **magnetic properties**. For example, look at the photo on the left. The two objects being held are samples of the mineral known as magnetite. Based on just its name, you could probably guess that magnetite is magnetic. You can see that in the picture, because little bits of iron are clinging to the minerals, making them look "hairy." This mineral has been known for quite some time. For example, an ancient historian known as Pliny the Elder discusses it in a book he wrote in AD 77. Of course, he didn't call it magnetite. That name wasn't coined until 1845. However, his description makes it clear that he was talking about magnetite.

The magnetic properties of minerals can be a bit less obvious than what you see in magnetite. That's because there are different levels of magnetism. For example, magnetite is considered a **ferromagnetic** (fehr' oh mag neh' tik) mineral.

Ferromagnetic mineral – A mineral that is strongly attracted to a magnet and can become a magnet if exposed to a magnet for a sufficient amount of time

However, a mineral can have magnetic properties and not be ferromagnetic. You will learn about those properties after you do the following experiment.

Experiment 3.4: Effervescence and Magnetic Properties in Minerals

Supplies:

- Specimens 1-6 from the laboratory kit made for this course
- The magnifying glass from the laboratory kit made for this course
- The nail from the laboratory kit made for this course
- The magnet from the laboratory kit made for this course
- The dropper bottle with the blue top from the laboratory kit made for this course
- White vinegar

- A metal paper clip (It must be bare metal, not covered in plastic.)
- A paper towel

Instructions:

1. Add white vinegar to the dropper bottle until it is about ¾ full and replace the lid.
2. Hold the magnet between your thumb and index finger and touch one of its flat sides to specimen 1 in your Geology Basics Kit.
3. Lift the magnet up and see if the mineral comes up with it or is at least attracted enough to the magnet to cling to it for a moment.
4. Repeat step 2 for the other 5 mineral specimens (2-6).
5. Repeat step 2 using the nail instead of a mineral specimen.
6. You should have found one mineral that was attracted to the magnet, just as the nail was attracted to the magnet. See if there is a difference in the amount that they are attracted to the magnet. Put the magnet on the mineral and see how hard it is to pull the two apart. Do the same with the nail. Which was harder to pull off the magnet, the mineral or the nail?
7. Touch the mineral to the nail. Are they attracted to one another?
8. Put the nail on a flat side of the magnet so that as much of the bottom part of the nail as possible (including the tip) is resting against the magnet.
9. Wait for one full minute.
10. Pull the nail off the magnet and place the magnet off to the side.
11. Touch the paper clip with the bottom part of the nail (the same part that was resting on the magnet) and gently try to pull the paper clip along the surface upon which you are working. Is the paper clip attracted to the nail?
12. Place the nail to one side and put a flat side of the magnet on the mineral that was attracted to it. Position it so that the magnet is on one edge of the mineral and is in contact with as much of the mineral as possible.
13. Wait for one full minute.
14. Pull the mineral off the magnet and place the magnet to one side.
15. Touch the paper clip with the same edge of the mineral that was in contact with the magnet and try to pull it across the surface like you did with the nail. Is the paper clip attracted to the mineral?
16. Turn each mineral specimen over in its container so the numbers are facing down. That way, what you do next will not mess up the numbers in any way.
17. Position each mineral so that some part of it is level.
18. Use the dropper bottle to add a single drop of vinegar to the level part of each mineral.
19. Wait a full minute.
20. Use the magnifying glass to examine the drop of vinegar on each mineral. You are looking for several bubbles inside the drop. You should see them in only one specimen.
21. Use a paper towel to wipe the vinegar off the specimens.
22. Put the lid back on the dropper bottle securely. That way, you can keep the vinegar in it when you put it away.
23. Clean up your mess.

Let me discuss the second part of the experiment first. You should have seen lots of bubbles in the drop that you put on the calcite sample (specimen 4), which tells you that calcite has effervescence. That's because calcite's chemical formula is $CaCO_3$, and as you learned in the previous chapter, $CaCO_3$ reacts with acid to form carbon dioxide, which is a gas. The bubbles that you saw contained the carbon dioxide produced. Calcite is the mineral found in limestone, which is why you saw bubbles in Experiment 2.4. You shouldn't have seen bubbles produced in any of the other specimens, because the chemicals that make up the other specimens don't react with acid.

Now what about the first part of the experiment? You should have found that the only mineral attracted to the magnet was hematite (specimen 6). Since the hematite was attracted to the magnet and the nail was attracted to the magnet, you might have thought that the hematite would be attracted to the nail. However, you should have found that it wasn't. Why? Because while the hematite responds to magnets, its response is weak. You should have noticed that when you compared the difference between how the mineral was attracted to the magnet and how the nail was attracted to the magnet. The nail should have been harder to pull away from the magnet because it was more strongly attracted to the magnet.

There was another difference between the magnetic properties of the nail and the mineral. You should have found that after the nail had been sitting on the magnet for a full minute, the paper clip was attracted to it, at least a bit. However, after the mineral had been sitting on the magnet for a full minute, the paper clip was not attracted to it. That's because the nail is made of a ferromagnetic substance (iron). Go back and look at the definition of a ferromagnetic substance. It is strongly attracted to a magnet, and it can become a magnet if exposed to a magnet. That's what happened. The nail became temporarily magnetic after being exposed to the magnet you were using.

Since the hematite's attraction to the magnet is weaker, and since it cannot become a magnet after being exposed to one, we say that the hematite is **paramagnetic** (pehr' uh mag neh' tik).

Paramagnetic mineral – A mineral that is weakly attracted to a magnet and cannot become a magnet when exposed to one

Most minerals that are attracted to a magnet are paramagnetic. However, there are some minerals, like certain forms of graphite, that are repelled by magnets. We call them **diamagnetic** (dy' uh mag neh' tik) minerals.

Diamagnetic mineral – A mineral that is repelled by a magnet, no matter how the magnet is oriented

You probably already know that the north pole of a magnet is repelled by the north pole of another magnet. That's not diamagnetism. That's just the basic Law of Magnetism: like poles repel, while opposite poles attract. A diamagnetic substance is repelled by *both* poles of a magnet.

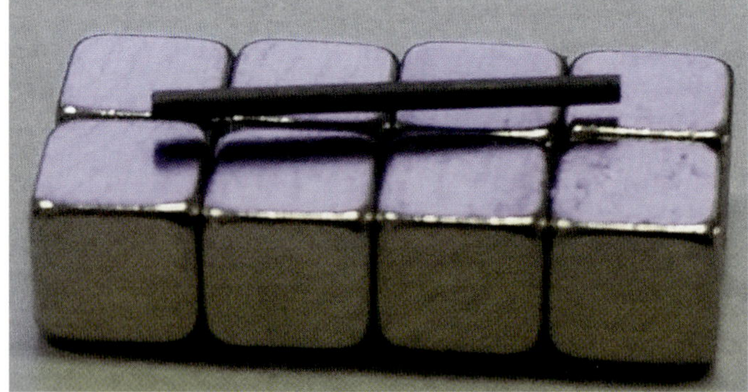
The bar in this photo is made of graphite. It is diamagnetic, so it is repelled by the magnets below and floats above them.

Why are some minerals ferromagnetic, others paramagnetic, and one diamagnetic? It depends on the way the atoms in their molecules are connected to one another. I don't want you to worry about it now, but if you take high school physics, you will learn about it. For now, just understand the definitions and be able to use them to answer questions like the Comprehension Check question below.

Comprehension Check

3.7 A mineral doesn't attract any metal to it. However, after being put next to a strong magnet, the mineral attracts bits of iron to itself. Is this mineral ferromagnetic, diamagnetic, or paramagnetic?

Chemical Properties of Minerals

You have learned about several properties that can help us determine the identity of a mineral. Those properties depend on the chemical nature of the mineral. If a mineral is made up of one type of molecule (or element), its properties will be different from a mineral made up of a different type of molecule (or element). Thus, we need to spend some time discussing the chemical makeup of minerals. Well, since the minerals we have been studying are all in the crust of the earth, we need to see what elements are available to make minerals.

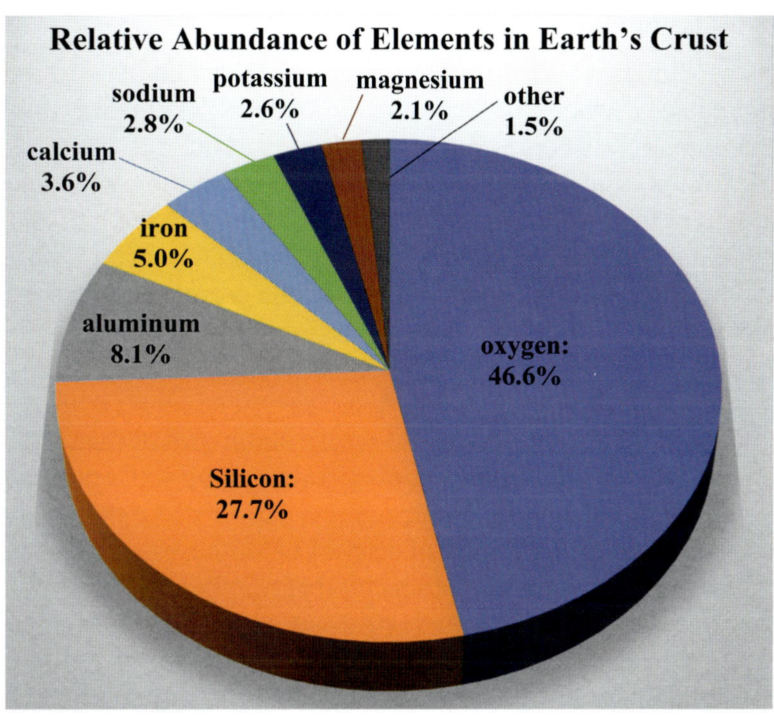

The chart on the right shows you the elements found in the earth's crust along with their relative abundance, which means how common they are compared to one another. Notice that 46.6% of all atoms in the earth's crust are oxygen atoms (O). That makes it the most abundant element. Now you must remember that the contents of earth's crust are solid, so the oxygen I am discussing is not oxygen gas, which is actually O_2. Oxygen gas is mostly in earth's atmosphere. The oxygen in the earth's crust refers to oxygen atoms that are a part of the molecules which make up the rocks and minerals of the crust. The next most abundant element is silicon (Si). Those two elements make up almost ¾ of the mass in the earth's crust! As a result, it shouldn't come as a surprise to you that the most common minerals in the crust are ones that contain both silicon and oxygen. We call them the **silicate minerals**, and they represent about 92% of all the minerals in the crust. The other 8% are called **non-silicate minerals**.

The quartz that you have been testing (specimen 1) is an example of a silicate mineral. In fact, the chemical formula of quartz is SiO_2, so silicon and oxygen are the only two elements in the molecules that make up quartz. Now remember, even though a mineral is quartz, there can be things in it other than silicon and oxygen. After all, you have both transparent quartz and opaque quartz in your lab kit. The transparent one is mostly pure SiO_2, while the opaque quartz has water molecules in it. Those water molecules are contaminants, which change the optical properties of the quartz. Thus, when you read the chemical formula for a mineral, remember that it is for the molecules that define the mineral, but there may be contaminating chemicals that can affect the mineral's properties.

Of course, there are a lot of different silicate minerals, and even though they all contain silicon and oxygen atoms, they can have very different properties. Think, for a moment, about the quartz (specimen 1) and potassium feldspar (specimen 3) in your kit. They are both hard minerals, but quartz is harder. Quartz has fracture, while potassium feldspar has cleavage. Their optical properties are quite different as well. Nevertheless, they are both silicate minerals. The chemical formula for potassium feldspar is $KAlSi_3O_8$. It is clearly a silicate mineral, since it has silicon and oxygen atoms in it. However, the number of silicon and oxygen atoms are different from what you find in quartz, and there are also potassium (K) and aluminum (Al) atoms in the molecule. The minerals are different because their chemical formulas are different.

By now, you have been handling the muscovite mica (specimen 2) quite a bit, so you are probably used to it. The first time you held it, however, you might have thought that it was manufactured, not natural. After all, it is very different from the other minerals you have been testing. Its chemical formula is very complicated: $KAl_2(Si_3AlO_{10})(OH)_2$. You can see that it is a silicate mineral, because there are Si and O atoms in it. However, you've probably never seen parentheses in a chemical formula, so let me explain a little chemistry before I go any further.

In chemical formulas, parentheses can be used to indicate how atoms are arranged in a molecule. Notice, for example, the "$(OH)_2$" at the end of the chemical formula. The "OH" means an oxygen atom is connected to a hydrogen atom, and the parentheses followed by the subscripted "2" means that this arrangement is found twice in the molecule. However, there are ten *other* oxygen atoms in the molecule, and they aren't connected to a hydrogen atom. Instead, they group with three silicon atoms (Si) and one aluminum atom (Al). But there are also two more aluminum atoms in the molecule that aren't part of the Si_3AlO_{10} group. So the parentheses point out subgroups that exist within a molecule. All those atoms are in the molecule, but they are arranged in groups that are indicated by the parentheses.

Now you don't have to worry about any of that. It means something important to a chemist, and when you take high school chemistry, you will learn more about that. What you need to know is how to count the atoms in a chemical formula like that. First, you must realize that any subscript coming right after the end of a set of parentheses applies to all the atoms inside the parentheses. Thus, you must multiply those atoms by that number. So, for the "$(OH)_2$" at the end of the chemical formula, you multiply the O and the H by 2. Well, there is only one O and one H in the parentheses, so when you multiply them both by 2, you know there are two O atoms and two H atoms in that group. There is no subscript after "(Si_3AlO_{10})," but remember, that represents a group, so the Al that is in that group is different from the other two Al's in the molecule. Thus, this molecule has one K atom, three Al atoms – two from the "Al_2" and one from the "(Si_3AlO_{10})", three Si atoms, twelve O atoms – ten from the "(Si_3AlO_{10})" and two from the "$(OH)_2$", and two H atoms. Now that was a bit complicated, so let me give you another example.

Example 3.1

The mineral shown on the left is called "lizardite," which is sometimes called "Norwegian jade." Its chemical formula is $Mg_3(Si_2O_5)(OH)_4$. List the number of each type of atom in one molecule of lizardite.

You read this chemical formula like any other chemical formula, but if you see a subscript after the end of a parentheses, you must multiply everything inside the parentheses by that number. Thus, the "$(OH)_4$" tells you that there are four O's and four H's represented there. Those are the only H's in the chemical formula, so that means each molecule has <u>four H atoms</u>. There is a subscript of "3" after the Mg, and Mg doesn't appear anywhere else, so there are <u>three Mg atoms</u>.

The "Si_2O_5" is in parentheses, but there is no subscript afterwards, so you don't have to worry about multiplying them by anything. Since there are no other Si's in the chemical formula, each molecule has two Si atoms. There are two groups with O atoms, however. The "(Si_2O_5)" group has five, and the "$(OH)_4$" group has four more, so there are nine O atoms.

To make matters more confusing, there is one other reason you might see parentheses in a mineral's chemical formula. It's when there is a choice between the atoms. For example, the lizardite that you see on the previous page has the chemical formula $Mg_3(Si_2O_5)(OH)_4$. The mineral pictured on the right is called "kaolinite," and it has the chemical formula $Al_2(Si_2O_5)(OH)_4$. Notice that the only thing different is the first atom. In lizardite, there are three magnesium (Mg) atoms, while in kaolinite, there are two aluminum (Al) atoms. Everything else is the same. Even though these minerals are different, their chemical formulas are so similar that they are often discussed together.

The molecules that make up this mineral, kaolinite, are surprisingly similar to molecules in the lizardite pictured on the previous page.

When that happens, their chemical formulas are often expressed as $(Mg_3, Al_2)(Si_2O_5)(OH)_4$. Do you see what this means? Notice the comma between the "Mg_3" and the "Al_2". That means this formula represents two chemicals. You get the formula of one chemical by choosing one of the things separated by the comma. Choosing the first thing (Mg_3) gives you lizardite, $Mg_3(Si_2O_5)(OH)_4$. The other chemical's formula comes from choosing the other thing in the parentheses, which gives you kaolinite, $Al_2(Si_2O_5)(OH)_4$. This happens quite a lot in the study of minerals, so let's do an example that combines the two notations you just learned.

Example 3.2

The chemical formula for a group of minerals can be given as $(Mg, Fe, Mn)_2SiO_4$. How many different chemical formulas does this represent? What are they?

Remember, the comma inside the parentheses indicates that you have choices. Since there are three atoms separated by commas, you have three choices. That makes three different chemical formulas. You get the first one by choosing the first atom. Now remember, any subscript outside the parentheses applies to all atoms inside the parentheses, so once you choose one of the atoms inside the parentheses, you must multiply by the subscript, if there is one. In this case, the subscript after the parentheses is 2. That means the first possibility is Mg_2SiO_4, the second one is Fe_2SiO_4, and the third one is Mn_2SiO_4.

Now that you know how to interpret the chemical formulas of minerals, I want you to recognize something important. In the example you just read, there were three chemical formulas.

The only difference between them, however, was the first atom. After that, all three of the chemical formulas ended in "SiO_4." These three minerals each have their own name. Mg_2SiO_4 is called forsterite (for' stuh right), Fe_2SiO_4 is called fayalite (fay' uh light), and Mn_2SiO_4 is called tephroite (teh' froh ight). However, because the group containing silicon and oxygen are the same, the minerals are similar. In fact, when you find one of those three minerals, you often find one or both of the others mixed in with it. Because of this, it is useful to have another name that can refer to any of these three minerals, or even a mixture of two or more of them. That name is olivine (ah' luh veen), a sample of which is pictured on the left.

This sample of olivine could be a mixture of up to six different minerals that are similar enough to belong to the olivine group.

In fact, there are other minerals that can be included in olivine. Their chemical formulas are $CaMgSiO_4$, $CaFeSiO_4$, and $CaMnSiO_4$. Notice that once again, they all end in SiO_4, and the only differences are the first two atoms. That means the olivine pictured on the left could possibly have up to six different minerals in it, but the chemical formulas are all so similar that they can be grouped together. As a result, we can say that all six of those minerals are part of the "olivine group."

But why do we group all these different chemicals together? Remember, we are talking about *silicate* minerals. How do we define a silicate mineral? It's a mineral with silicon and oxygen. Well, if the silicon and oxygen parts are the same, then their silicate natures are the same. That means when it comes to their mineral properties, they should be similar. As a result, it makes sense to put them together in a group.

Let me give you one more example. You have been testing muscovite mica, which I told you has the chemical formula $KAl_2(Si_3AlO_{10})(OH)_2$. Remember why we use parentheses in chemical formulas. They tell you how the atoms group together. So, what group of atoms are the most important in this silicate mineral? The group that has silicon and oxygen. That group is in parentheses: Si_3AlO_{10}. There is another kind of mica as well. It's called biotite (by' uh tight) mica. Its chemical formula is $K(Mg, Fe)_3(Si_3AlO_{10})(F,OH)_2$. Notice that it has the same group of silicon and oxygen atoms, along with aluminum (Al). Thus, you would expect biotite mica to be very similar to muscovite mica. In fact, it is. It has the same kind of cleavage as muscovite mica and roughly the same hardness.

Now don't get lost in all these chemical formulas and strange names. You don't have to know any of the names, and you don't have to remember any of the chemical formulas. However, you do need to be able to interpret the chemical formulas of minerals if I give them to you. You should be able to do problems like the ones I gave in Examples 3.1 and 3.2. You should also be able to look at a few chemical formulas for silicate minerals and identify which ones are similar enough to belong to a group. You will get some experience doing that when you answer the Comprehension Check problems that are coming up soon.

So far, I have been giving you lots of examples of minerals, but they have all been silicate minerals. That makes sense, since they are the most abundant minerals. However, there are non-silicate minerals as well. In fact, half of the minerals in your kit are non-silicate. The calcite ($CaCO_3$), fluorite (CaF_2), and hematite (Fe_2O_3) in your kit are all non-silicate minerals. Two of them have oxygen, but none of them have silicon. A few other non-silicate minerals are pictured on the right. Halite ($NaCl$) is a mineral that you might use a lot, but you probably don't think about it as a mineral. As I told you already, it's the salt you use to season food. Galena (PbS) is another non-silicate mineral. Most of the lead found in the earth's crust is found in galena. Baryte ($BaSO_4$) is a mineral that holds most of the earth's barium.

Halite Galena

Baryte Silicon

Also, any solid, pure element in nature is also a non-silicate mineral. Gold and silver are precious metals, but from a mineralogist's point of view, they are non-silicate minerals. Interestingly enough, while most of the silicon in the earth's crust is found in silicate minerals, you can sometimes find pure silicon. Pure silicon is a non-silicate mineral. That might sound odd at first, but remember that a silicate mineral must have both silicon *and* oxygen. Pure silicon has only silicon in it, so it is a non-silicate mineral!

Comprehension Check

3.8 Horneblend's chemical formula can be given as $Ca_2(Fe, Mg)_5Si_8O_{22}(OH)_2$.

a. Is this a silicate or non-silicate mineral?
b. How many different chemical formulas can horneblend have?
c. Give the possible chemical formulas of horneblend.
d. List the number and type of atom in each chemical formula.

3.9 There is a mineral group called augite. One member is $MgSiO_3$. Which of the following would also be a member of the augite group: $KAlSi_3O_8$, $FeSiO_3$, or $MgSiO_2$?

3.10 Which of the following are non-silicate minerals: SiO_2, Cu, $FeSiO_2$, $CaSO_4$, KCl, Al_2O_3?

How Do Minerals Form?

Now that you've learned about minerals and some of their properties, you might be wondering where they come from. Yes, you can dig them up from the earth, but how do they form to begin with? Well, have you ever made something like what is pictured below? Some people call them "sugar stirrers," others called them "sugar sticks." They are basically sticks with big crystals of sugar on them. In the picture, the sticks have brown sugar on them, but you can make them out of any kind of sugar.

The crystals on the sugar sticks (top) form when lots of brown sugar (bottom) is dissolved in boiling-hot water and then the water is allowed to cool.

If you've never made them before, they are pretty easy to make, but they do take some patience. Basically, you heat up some water until it is boiling, and then you add a huge amount of brown sugar, like the brown sugar shown in the picture below the sugar sticks. Most recipes call for adding two cups of brown sugar to every cup of boiling-hot water. The sugar dissolves in the water, making a thick syrup. If you want, you can add something to give the sugar a flavor. You then put just a little bit of sugar on the sticks and put them in the thick syrup. As the syrup cools, sugar crystals start growing on the sticks!

How does this work? When you heat water, you can dissolve more solid into it. If you tried to dissolve two cups of brown sugar in one cup of water at room temperature, it wouldn't work. You would get some of the sugar to dissolve, but you wouldn't get it all to dissolve. To get all the sugar to dissolve, you would need to heat up the water. In fact, to get two cups of brown sugar to dissolve in one cup of water, the water has to be boiling-hot!

As long as the water stays boiling-hot, the brown sugar stays dissolved, and you have a **solution** of brown sugar water. However, think about what happens as it cools down. The cooler the water gets, the less brown sugar it can dissolve. So what happens? Some of the brown sugar comes out of the solution and turns back into a solid. This process is called **precipitation** (pre sip' uh tay' shun).

<u>Precipitation</u> – (chemistry) The process by which a solid forms from a solution

I have to add "chemistry" to the definition, because when it comes to weather, "precipitation" means something different.

Now you might ask yourself why the brown sugar on the sticks looks so different from the brown sugar that was dissolved into the boiling-hot water. That's because when precipitation occurs slowly, the molecules that make up the solid can slowly add to the molecules that are already a solid. This "slow building" tends to produce large crystals, and that's the only difference between the brown sugar on the sticks and the brown sugar pictured below them. The molecules are exactly the same, but because the solid was produced slowly, it could form large crystals. The brown sugar pictured below the sticks is made up of the same kind of crystals as the brown sugar on the sticks; they are just smaller. As a result, they have a different appearance.

Those sugar sticks should have reminded you of minerals, like the fluorite shown on the right, which is also specimen 5 in your kit. Why? Because that's how this sample of fluorite was formed. Very hot water had CaF_2 dissolved in it, but that water eventually cooled. It cooled slowly, so the CaF_2 slowly precipitated out, forming the crystals you see on the right. In general, then, the molecules that make up minerals start out in some kind of solution. Then, something happens to force them to precipitate out of solution. If they precipitate out quickly, they form small crystals. If they precipitate out slowly, they form large crystals.

This sample of fluorite was formed when CaF_2 that was dissolved in hot water slowly precipitated as the solution cooled.

There are a couple of things you have to realize about minerals precipitating out of solution. First, it's not always caused by the solution cooling. Sometimes, it's caused by water evaporating. After all, a molecule will precipitate out of solution when it cannot stay dissolved. If there are a lot of molecules dissolved in water, and the water starts to evaporate, there is less room for the dissolved molecules. If enough water evaporates, there won't be room for all the dissolved molecules, and they will have to precipitate. Minerals that form this way are called **evaporites.**

The second thing you have to realize is that for minerals, the solution is not always formed by hot water. In fact, the silicate minerals (which are the most common) form by precipitating out of solutions formed by molten rock! Remember from Chapter 2 that the deeper you go in the geosphere, the hotter it gets. In the mantle, the temperature gets hot enough to melt rock. When it is under the crust, that molten rock is called **magma**.

Magma – Extremely hot liquid or semi-liquid rock located beneath the surface of the earth

What do I mean by "semi-liquid"? Remember that the mantle is under a lot of pressure, which makes the molten rock behave a bit differently than the liquids that you are used to. As I told you in Chapter 2, the rock in the mantle is often called "plastic rock" because of that.

But wait a minute. Minerals are found in the earth's crust. How do they go from being dissolved in magma to precipitating in the earth's crust? There are cracks in the earth's crust, and the plastic rock of the mantle is under so much pressure that it gets pushed up into the cracks, rising to fill them. If the magma gets pushed up so far that it gets released onto the earth's surface, we no longer call it magma. Instead, we call it **lava**.

Lava – Magma that emerges onto the earth's surface as a liquid

Of course, you know that lava comes out of volcanoes, so a volcano is a place where magma rises all the way to the surface of the earth and flows out as a liquid.

To give you an idea of how this works, look at the drawing below. It shows a volcano and the magma underneath. The magma, of course, is coming from the mantle below. It rises up through cracks in the crust, but most of them don't lead to the surface. As a result, the magma never becomes lava. However, if it can get to the surface, it flows over the surface, and we call it lava.

The yellow and orange colors represent molten rock rising up into the crust. If it stays underground and remains molten, it is magma. If it reaches the surface, it is lava.

Now remember, when liquids cool, they can turn into solids. Thus, when lava cools, it eventually becomes solid and becomes rock. You will learn more about that in the next chapter. However, magma can cool as well. It won't cool as fast as lava, since the geosphere gets warmer the deeper you go. If the magma is deep enough, it will continue to be magma. However, if it rises high enough, it will also cool to the point where it solidifies and becomes rock. In both cases, as the molten rock cools, dissolved chemicals will precipitate, forming mineral deposits. But think about the differences between the minerals formed in lava and those formed in magma. Since the lava cools quickly, the size of the mineral crystals will be small. Since the magma cools more slowly, the size of the mineral crystals will be larger. In general, the deeper the magma, the more slowly it will cool, so the larger the crystals.

There is one other important thing to consider in this process. When I told you about making the sugar sticks, I told you that most recipes call for two cups of brown sugar for each cup of water. As long as the water is boiling-hot, all the brown sugar will dissolve. Suppose you tried to dissolve two cups of salt into boiling-hot water. Could you get it all to dissolve? No. You can't dissolve nearly as much salt in boiling-hot water as you can sugar.

In the same way, some minerals are easier to dissolve in magma than others. The minerals that are harder to dissolve will precipitate at higher temperatures, while the minerals that are easier to dissolve will precipitate at lower temperatures. Because of this fact, minerals tend to form at different times during the cooling process. Olivines, which I discussed on page 80, are very hard to dissolve in magma. As a result, they tend to precipitate at high temperatures. Quartz, on the other hand, is rather easy to dissolve in magma, so it doesn't precipitate until the temperature decreases quite a bit.

So let's suppose there is some magma that has both Mg_2SiO_4 (one form of olivine) and SiO_2 (quartz) in it. As it rises through the crust, it cools. Which mineral precipitates first? The Mg_2SiO_4, because it is harder to dissolve in molten rock. By the time the magma cools to the point where SiO_2 precipitates, all the Mg_2SiO_4 will have precipitated. Thus, even if the minerals are mixed together in the magma, their crystals won't necessarily be mixed together once the magma cools, because the crystals form at different temperatures, which usually happens at different times.

The problem with the example I just gave you is that it is overly simplified. In fact, magma rarely has just two chemicals dissolved in it. Usually, there are all sorts of chemicals, including the chemicals needed to make various forms of a mineral. The magma might have some iron in it as well. If so, then in addition to Mg_2SiO_4 forming at a high temperature, there will be Fe_2SiO_4 forming as well. If that's the case, the olivine that is made will be a mixture of those two different forms of olivine.

Of course, in addition to the fact that different forms of a mineral can be produced by the same magma, there are other chemicals dissolved as well. As a mineral precipitates out, it might pull some other chemicals out of solution with it. Do you remember the four different colors of the mineral beryl that I showed you on page 67? They are all made of the same basic molecule, $Be_3Al_2Si_6O_8$. However, they have different colors because as they precipitated, they pulled some other chemicals out of solution. Green beryl, which is commonly called "emerald," results when the precipitating $Be_3Al_2Si_6O_8$ molecules end up pulling some chromium out of solution. Yellow beryl results when iron is pulled out of solution. If the beryl is clear, it is pure $Be_3Al_2Si_6O_8$, which means no other chemicals were pulled out of solution.

Polymorphism

There is one other consideration we must make when thinking about how minerals form. It's possible for the same chemical to form radically different minerals depending on the conditions of the environment. Consider, for example, the two minerals pictured below. While they look very different, they are both made entirely of the element carbon. The one on the top is called **graphite**, and it is the most common mineral form of carbon. The part of the pencil that you use to write with is usually graphite. The picture on the bottom is of a **diamond**. It doesn't look like the diamonds you are used to seeing, because it is the natural form of the mineral that comes out of the earth's crust. Diamonds used in jewelry are made by cutting and polishing pure samples of the mineral.

So what's the difference between graphite and diamond? Surprisingly, it's not the presence or absence of impurities, which is what causes some minerals made of the same chemical to look so different. In this case, it's the way the carbon atoms are arranged. In graphite, the atoms are arranged in layers, while in diamond, the atoms are arranged in triangular pyramids, which are also called tetrahedrons (tet' ruh hee' druns). This different arrangement causes the different optical properties, but it also changes the hardness. Graphite has a hardness of 1-2 on the Mohs Hardness Scale, but diamond is used to define a hardness of 10. Thus, while graphite is a very soft mineral, diamond is the hardest naturally-occurring mineral on earth. When the same chemical can produce radically different minerals, we say that the minerals are **polymorphs**. "Poly" means "many," and in Greek, the word "morph" refers to the form or shape of something. So "polymorph" means "many forms."

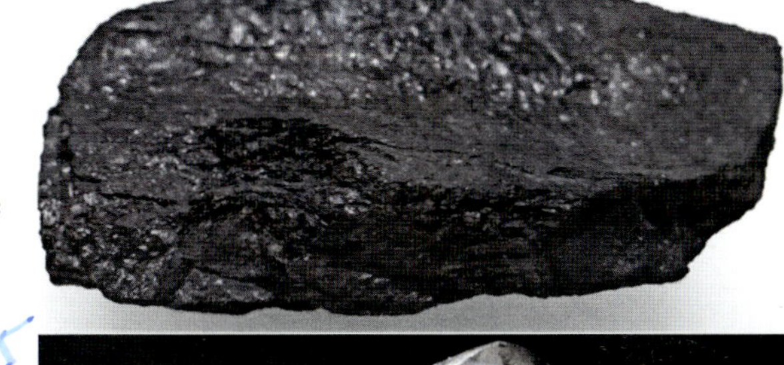

Graphite (top) and diamond (bottom) are both made of the element carbon, but the difference in the way the carbon atoms are arranged causes them to be very different minerals.

But what is responsible for these two polymorphs of carbon? It's the different conditions under which they form. Graphite forms in the crust or the upper mantle. At the temperatures and pressures found there, carbon precipitates out of magma in layers, which leads to the formation of graphite. In the mantle, temperatures are much hotter and pressures are much higher. Under those conditions, carbon precipitates out of the magma as tetrahedrons, making diamonds. Now, of course, even though diamonds are formed in the mantle, we can't get them until they make their way to the crust. For that to happen, they must be carried up to the crust by magma.

You might think that since the mineral diamond is formed at higher temperatures and pressures than graphite, it would be possible to turn graphite into diamond by simply heating it and applying a lot of pressure. However, that's not correct. Diamonds aren't formed *from* graphite. They are formed at higher temperatures and pressures *instead of* graphite. Thus, if you want to make artificial diamonds, you must essentially melt a sample of carbon, recreate the conditions found in the lower mantle, and then allow diamonds to form. Surprisingly, modern technology allows us to do that, so it is possible to purchase synthetic diamonds. They are indistinguishable from natural diamonds without some specific laboratory instruments, because they are formed in nearly the same way that we think natural diamonds are formed in the mantle.

Comprehension Check

3.11 You are given two samples of a silicate mineral that both come directly from the earth's crust. The first contains large crystals, while the second contains small crystals. If you are told one of them formed from the cooling of lava and the other from the cooling of magma, which is which?

3.12 A sample of magma is cooling. The first mineral you see forming is biotite mica. Later on, muscovite mica is formed. Which mineral is easier to dissolve in magma?

3.13 "A diamond is a chunk of coal that did well under pressure" is an inspirational quote you might read from time to time. Given that coal is a form of carbon found in the earth's crust, is this quote accurate? Why or why not?

How We Use Minerals

From the earliest of times, people have been using minerals in many practical (and some not-so-practical) ways. In Genesis 2:12-13, the region of Havilah is described as having gold and onyx, both of which are minerals. The great-grandson of Cain, Tubal-Cain, is described in Genesis 4:22 as a forger of bronze and iron. Bronze is a mixture of metals, usually copper and tin. Iron is also a metal. Since copper, tin, and iron are all elements, when they are found in the earth's crust, they are minerals. Before they are used, however, these metals had to be mined from the crust.

When we observe what's in the earth's crust, we rarely find deposits of pure iron, copper, tin, and other metals. Instead, we find the metals mixed together with other chemicals. When a mixture like that is found in the earth's crust, it is called an **ore.**

Ore – A naturally-occurring solid material from which a useful mineral can be extracted

Miners generally extract ores from the ground, and those ores are then treated so that the useful mineral can be separated from all the other chemicals found in the ore. The picture below, for example, is of some gold ore. You might be able to see that there is gold in it, but there are clearly other things besides just gold. The gold must be separated from those other things before it can be used.

Gold, iron, copper, and tin are all examples of what are called **metallic resources** that come from the earth's crust. Some of their uses are obvious. We can use iron to build structures, and when we mix iron with a small amount of carbon, we get steel, which can be an even more useful building material. Of course, different metals have different characteristics, so they are used to build different things. Steel, for example, is harder than aluminum, but aluminum is easier to shape than steel. In addition, steel is generally less expensive than aluminum, but it is also much heavier. When buildings, cars, airplanes, etc., are being built, such characteristics end up determining what metal (or group of metals) is used.

Gold ore like this is the most common way gold is found in the earth's crust.

Metals have other uses that you might not think about right away. They conduct heat very well, so they can be used to make cookware that will pass heat from the oven, stove, or fire to cook food efficiently. They also conduct electricity. Most wires are made of copper, for example. It is an inexpensive metal that can conduct electricity from one place to another. While copper is used in most wires, there are metals that do better when it comes to conducting electricity. Silver, for example, is the very best conductor of electricity. However, it is expensive because it is rare. Thus, it is only used in electrical applications when its superior conductive property is necessary.

Gold, silver, and platinum are often called **precious metals**, because they are used in the making of jewelry and sometimes used as currency. But because of their chemical properties, they are used in other applications as well. For example, today's automobiles are equipped with catalytic converters that turn carbon monoxide (a gas that can be toxic) produced by the burning of fuel into carbon dioxide (a gas that is much less toxic). It turns out that platinum makes that process much more efficient, so catalytic converters contain platinum. In fact, catalytic converters contain a mixture of several metals, one of which is palladium. It costs more than gold, but a small amount of it is usually necessary for a catalytic converter to do its job well.

While metallic resources are important, there are many other useful minerals that are not metallic. Not surprisingly, they are considered **nonmetallic resources**, and some of them are rocks. We mine a lot of rocks from the earth's crust and use them for building and other purposes. Remember, rocks are mixtures that contain minerals, and you will learn a lot more about them in the next chapter. For right now, then, I don't want to spend any more time talking about them. Instead, I want to discuss how we use some of the nonmetallic mineral resources in the earth's crust.

You are already familiar with the uses of some nonmetallic mineral resources. As I have already told you, halite (NaCl) is a source of table salt. Graphite (carbon) is used in pencils, but because it forms in layers that slide easily against one another, it is slippery. As a result, it is often used as a lubricant. Why use graphite as a lubricant instead of oil or some other slippery liquid? Well, most oils leave a residue behind. In many applications, that's not a problem, but in other applications it can be. In those situations, a solid lubricant (sometimes called a "dry lubricant") like graphite works best.

These days, one of the most important nonmetallic resources is silica (SiO_2), which is a common silicate mineral. You have one variety of silica in your kit: quartz. Silica is also the main component in the sand found on many beaches. While silica can be used to make things like concrete and glass, it is very important to digital devices like computers and smartphones. That's because digital devices depend on the electrical properties of the element silicon. However, pure silicon isn't common in the earth's crust. Instead, it is mostly found in silicate minerals, like SiO_2. To make digital devices, silica must be mined, and then it must go through a chemical process to separate the silicon atoms from the oxygen atoms to which they are bonded so that the element silicon can be produced.

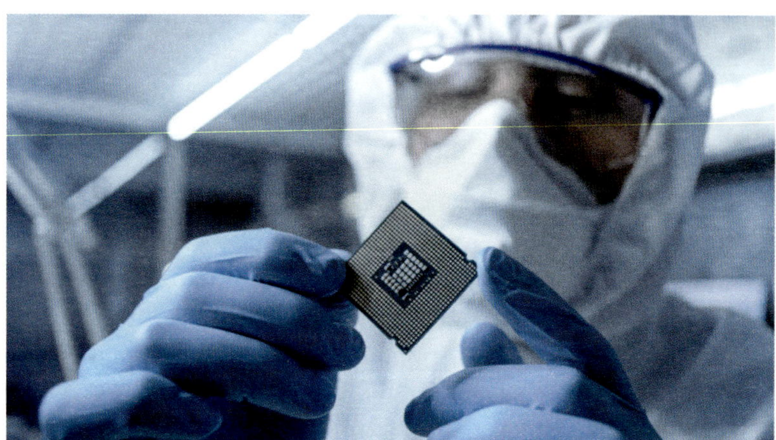

This microchip requires the element silicon, which is produced from the mineral silica. The man is dressed like that to ensure that the production process is clean.

You have been experimenting with another important nonmetallic resource: mica. The mica that you have is muscovite mica [$KAl_2(Si_3AlO_{10})(OH)_2$], but there are more than 35 different varieties of mica. What was the first thing you noticed when you started handling your mica? For me, I noticed that it felt a lot like plastic. It isn't plastic, of course, but it feels like it. Also, when you were looking at cleavage, you saw that it had perfect cleavage in one dimension. In other words, it is composed of sheets, and those sheets are very thin.

Well, even though mica is not plastic, it has some things in common with plastic. For example, it is very bad at conducting electricity. While that means it can't be used in wires and things like that, it can be used *between* electronic components that should not exchange electricity. We call that an electrical insulator, and small electronic devices need to have thin insulators. Mica is excellent for those kinds of applications. It is also very bad at conducting heat, so mica can be used as a thermal insulator as well. While the cleavage property of mica makes it excellent for those kinds of uses, mica can also be useful when it is ground into a powder. It is used along with a non-silicate mineral, gypsum (jip' sum), to make drywall, which is most likely what makes up the interior walls of your home. It is also used in paints to make them more resistant to water damage and in plastics to reinforce their strength.

As I already mentioned, there are more than 5,400 identified minerals in the earth's crust, so it isn't practical to go through all the uses that can be made of minerals. Nevertheless, I hope this discussion has helped you understand just how important minerals are to your everyday life. I also told you that new minerals are still being discovered. In 2019, for example, a student studying for her

Ph.D. discovered a completely new mineral trapped inside a diamond. Two years previously, scientists were studying a Russian site that experienced a lot of volcanic activity. Their analysis revealed another brand-new mineral.

Precious and Semi-Precious Gems

Metallic minerals are often used in jewelry, but the most expensive jewelry-related items are precious gems, which are nonmetallic minerals. However, the minerals that come from the earth don't exactly look like the gems you normally see on jewelry. For example, the picture on the right shows you a ruby that has been prepared to go on a piece of jewelry. The picture below also shows you a ruby, but as it is found in the earth's crust. What's the difference?

The ruby gemstone (top) is a cut and treated version of the mineral corundum (bottom).

Rubies come from the mineral corundum (kuh run' dum), which has a chemical formula of Al_2O_3. Their red color is a result of a small number of the aluminum atoms (Al) being replaced with chromium (Cr) atoms. When they precipitate out of cooling magma, there can be impurities pulled into the mineral as well. After it forms, its surface ends up interacting with its surroundings, and external forces can cause it to crack or deform. As a result, a natural ruby doesn't look nearly as appealing as a ruby gemstone.

To increase their beauty, rubies are polished to remove surface contaminants. They usually have imperfections inside of them, which are often treated with heat and chemicals that can remove or cover up the imperfections. Then, each ruby is cut into the shape that it needs to be for the piece of jewelry it is going on. In the cutting process, flat surfaces are made and then polished. They make the gem's **facets**, which reflect light to make it sparkle. Notice how the ruby in the top picture has bright, triangular patches. Those are caused by the light reflecting off the facets that have been cut into it.

You might be surprised to learn that another popular gemstone, the sapphire, also comes from the mineral corundum. Sapphires have the same basic characteristics of rubies, because they come from the same mineral. However, they are typically blue, yellow, purple, or green because they have small amounts of elements such as iron and titanium in them. Rubies and sapphires, along with diamonds and emeralds, are called **precious gems**, because they are generally the most expensive and sought after gemstones. If you recall what I told you in the first part of this chapter, emeralds come from the mineral beryl, which has the chemical formula $Be_3Al_2Si_6O_{18}$.

What makes a gem precious? Its rarity, which makes it expensive. How expensive? That depends on how big they are. You might have heard the term **carat** applied to jewelry. A wedding ring, for example, might have a diamond that is ½ of a carat. What does that mean? Believe it or not, it's just another unit for mass. There are a lot of factors that determine the price of a gemstone, but one of the most important is its mass. The more mass a gemstone has, the more it usually costs. A carat is defined as follows:

$$1 \text{ carat} = 0.2 \text{ grams}$$

Since a carat is simply a mass measurement, we can convert any carat measurement into grams using the above conversion relationship. See what I mean by studying the following example:

Example 3.3

An American dollar bill has a mass of 1 gram. How many carats does a diamond of the same mass have?

A diamond with the same mass as a dollar bill must also be 1 gram. However, diamonds are usually rated by carats, so to determine how many carats that is, you have to do a conversion. Remember, you put the number you know over 1 to make it a fraction, and then you multiply by the conversion relationship in the form of a fraction. You put the unit you want to cancel on the bottom of the fraction, and the unit you want to keep on the top. Thus, the conversion would be:

$$\frac{1 \text{ g}}{1} \cdot \frac{1 \text{ carat}}{0.2 \text{ g}} = 5 \text{ carats}$$

Remember, you multiply by everything on the top and divide by everything on the bottom, so you take 1 times 1 and then divide by 1x0.2. That gives you 1 ÷ 0.2, which is 5. So a diamond that has the same mass as an American dollar bill is 5 carats, which will sell for a minimum of about 45,000 American dollars!

Please note that while I will expect you to do a conversion like this, I will not expect you to memorize the conversion relationship. It will be given to you on the test.

In the world of jewelry, the value of a gemstone is often given as the price per carat. Diamonds, rubies, sapphires, and emeralds are considered precious gems because their price per carat is very high. However, there are a lot of gemstones that are beautiful but do not have as high a price per carat. They are called **semi-precious gems**. I personally think that amethyst (am' us thust) is one of the prettiest gems. It comes from quartz that has iron in it and has also been exposed to a radioactive substance. The energy from the radiation causes a change in the iron, giving it the characteristic purple color. Notice in the picture on the near right that the bottom of the mineral is clear quartz, like one of the quartz samples in your kit. However, the quartz is purple near the top. That's where a lot of radiation was hitting the mineral.

The mineral form of amethyst is on the left, while the cut and polished form is on the right.

Minerals in the Soil

Now remember, minerals are found in rocks, and rocks are the parent material of soil. Thus, as rocks are broken down by weathering and incorporated into soil, the minerals in the rocks find their way into the soil. This is important, because while plants make their own food by photosynthesis, they need certain elements that are found in minerals. The potassium feldspar that you have in your kit, for example, is the main source of the potassium that is found in soil, and plants need potassium to stay healthy.

You also need elements that come from minerals to stay healthy. However, you don't eat minerals directly. Instead, you eat plants that have absorbed what they need from minerals in the soil, or you eat organisms that ate those plants. As a result, you are also getting the elements you need from minerals in the soil. So, while minerals are useful when it comes to making things and can also be a source of beauty and wonder, some of them are absolutely necessary for life!

Comprehension Check

3.14 Of the nonmetallic resources I discussed (halite, graphite, quartz, and mica), which are silicate minerals?

3.15 If a mineral has a lot of facets, what was probably done to it after it was pulled from the ground?

3.16 A ring has a 3.5-carat ruby in it. What is the mass of the ruby in grams? (1 carat = 0.2 grams)

3.17 A scientist starts out with a sample of quartz that is nearly clear. She puts it in a box and lets it sit there for a long time. When it comes out, it has a purple color to it, but its shape and hardness have not changed. What happened to the quartz while it was in the box?

Answers to the Comprehension Check Questions

3.1 It is more likely to be a rock. Remember, rocks are mixtures, while minerals are more uniform throughout. The red stripes are most likely made of a different chemical than the rest of the gray material. Thus, it is probably a mixture, which is characteristic of a rock.

3.2 Luster is an optical property. Shape isn't affected by the way light interacts with an object, so shape is not an optical property. You might not know the proper definition of "luster," and that's fine. You should know that shape is not an optical property, so the only other choice is luster. However, if you have heard the term before, you know it has something to do with the shininess of an object, and that is clearly affected by the way it interacts with light.

3.3 It had enough room to grow, but there were lots of impurities present. The shape tells you that there was plenty of room. However, if there were no impurities, it would be transparent. The more impurities, the less transparent it will become. The fact that it is opaque and not translucent indicates that there were a lot of impurities.

3.4 Yes, they could both be accurate in their reporting. It is possible for a mineral to have different results in a streak test, depending on the side used to make the streak.

3.5 A has fracture and B has cleavage. Only B has nice, flat surfaces.

3.6 The mineral is apatite. Since the nail can't scratch the mineral, it is harder than 4.5. However, it can't scratch the glass plate, so it is softer than 5.5. The only mineral on the table that is consistent is the one with a hardness of 5, which is apatite.

3.7 It is ferromagnetic. Remember, a ferromagnetic substance is strongly attracted to a magnet *and* can become a magnet if exposed to a strong magnet for a while. The question didn't mention that it was strongly attracted to the magnet, but since it became a magnet, the only possibility would be that it is ferromagnetic.

3.8 a. It is a silicate mineral, since it has Si and O in it.

b. There are two different chemical formulas. The comma inside parentheses tells you the choices. Two atoms are separated by a comma, so that means there are two choices.

c. The chemical formulas are $Ca_2Fe_5Si_8O_{22}(OH)_2$ and $Ca_2Mg_5Si_8O_{22}(OH)_2$.

d. Both molecules have two Ca's, eight Si's, twenty-four O's, and two H's. The first one also has five Fe's, while the other has five Mg's instead. Choosing the Fe gives the first formula, but remember, you have to multiply it by 5, because of the subscript after the parentheses. The subscripts tell you how many of each atom, but you need to multiply both the O and the H at the end by 2, because of the subscript outside the parentheses. Since O appears twice, you must add to get the total. The "O_{22}" gives twenty-two O's, but the $(OH)_2$ gives two more, for a total of twenty-four.

3.9 $FeSiO_3$ would also be a member. This is a silicate mineral, so you need to match the group that has silicon and oxygen. Thus, the chemical formula must have SiO_3 in it.

3.10 Cu, $CaSO_4$, KCl, and Al_2O_3 are all non-silicate minerals, because they do not have *both* Si and O.

3.11 The first came from magma, the second from lava. Slower cooling makes larger crystals, and magma cools more slowly than lava.

3.12 Muscovite mica is easier to dissolve in magma. The harder-to-dissolve chemicals precipitate at higher temperatures. Since the magma is cooling, the one that precipitates first is doing so at a higher temperature.

3.13 It is not accurate. You can't convert carbon from the crust into diamond. You can only form diamond if you recreate the conditions of the mantle and let the carbon precipitate under those conditions.

3.14 Quartz and mica are the silicate minerals. You need to go back and look at the chemical formulas I gave when I discussed each of them. Silicate minerals must have silicon and oxygen in them, and only quartz and mica have those atoms.

3.15 It was probably cut and polished. You don't necessarily have to say polished, but when a natural mineral has a flat surface, there are usually only a few that correspond to cleavages. Gems are usually cut to give them a lot of facets.

3.16 Its mass is 0.7 g. This is just a conversion from one mass unit to another. To do a conversion, you put the number you know over 1 to make it a fraction, and then you multiply by the conversion relationship in the form of a fraction. You put the unit you want to cancel on the bottom of the fraction, and the unit you want to keep on the top. Thus, the conversion would be:

$$\frac{3.5\ \cancel{\text{carats}}}{1} \cdot \frac{0.2\text{ g}}{1\ \cancel{\text{carat}}} = 0.7\text{ g}$$

3.17 It must have been exposed to radiation. An amethyst is purple quartz, and the color comes from being exposed to radiation.

Chapter Review

1. Define the following terms:

a. Streak
b. Luster
c. Cleavage
d. Effervescence
e. Ferromagnetic mineral
f. Paramagnetic mineral
g. Diamagnetic mineral
h. Precipitation
i. Magma
j. Lava
k. Ore

2. You are looking at two solid objects that came from the earth's crust. The first has lots of different chemicals all mixed together. The second is mostly made up of one chemical. Which is a mineral?

3. Which of the following are optical properties: hardness, luster, cleavage, color, effervescence?

4. If you cannot find parallel, flat surfaces on a mineral, does it have cleavage or fracture?

5. Refer to Tables 3.1 and 3.2 on page 72 to answer the following questions:

a. You have a sample of fluorite. Can it scratch a glass plate? Can a penny scratch it? What about an iron nail?
b. A mineral can scratch a glass plate but not a streak plate. What is its range in hardness?
c. A mineral cannot scratch a glass plate, can be scratched by a nail and a penny, but can't be scratched by a fingernail. What is its range in hardness?

6. A mineral is attracted to a magnet but cannot become a magnet. What word describes this?

7. You have two samples of quartz. Sample A is white, and if you try to look through it, you cannot see anything on the other side. Sample B is somewhat clear, and you can partially see things through it. How would you describe each sample's optical property, using a single word for each?

8. The chemical formula for the mineral anthophyllite can be given as $(Mg,Fe)_7Si_8O_{22}(OH)_2$.

a. List the possible chemical formulas of the molecules in this mineral.
b. How many of each type of atom exist in each of the molecules?
c. Which of the following molecules would be in a mineral from the same group: $Mg_3Si_4O_{10}(OH)_2$, $Fe_2Al_4Si_5O_{18}$, or $Ca_2Mg_5Si_8O_{22}(OH)_2$?

9. Which of the following are non-silicate minerals: $NaAlSi_2O_6$, PbS, $CaSiO_3$, $Ca_5(PO_4)_3(OH)$?

10. Magma is rising through a volcano. Mineral A precipitates out while it is magma, but mineral B doesn't precipitate out until it is lava. Which mineral precipitates at a higher temperature?

11. Calcite and aragonite are polymorphs. If calcite's chemical formula is $CaCO_3$, what is aragonite's chemical formula?

12. A metallic resource is mined from the ground, but it must be treated so that the metal can be extracted and used. What is the name we use to refer to what was mined?

13. The sapphire in a ring is 2.2 carats. What is its mass in grams?

14. You measure the mass of a diamond to be 1.1 grams. How many carats is it?

Chapter 4: Rocks

Introduction

In the previous chapter, you learned a lot about minerals. I want to continue our study of the earth's crust by showing how minerals come together to make rocks. Earth scientists think of a rock as a collection. For example, a rock can be a collection of different minerals. Granite is a collection of quartz crystals, feldspar crystals, and mica crystals, along with minor amounts of other mineral crystals. However, rocks can also be a collection of small crystals from a single mineral. The calcite in your kit is a large, single crystal of calcite, so it is a mineral. If you have lots of little calcite crystals held together as a collection, that's limestone. In other words, rocks are a collection of small crystals from a single mineral or small crystals from more than one kind of mineral, often with other substances mixed in.

Before I start this discussion, I need to point out something important about science in general and earth science in particular. Scientists make sense of what they study by classifying things. We classify a substance as a mineral because it is made of only one chemical. The mineral corundum, for example, is Al_2O_3. However, it's not really pure, is it? It has contaminants in it, which change its optical properties. With one set of contaminants, corundum is red, and we fashion it into ruby gemstones. If it has other contaminants, it is blue, yellow, purple, or green, and we use it to make sapphire gemstones. So even though we think of minerals as made of only one chemical, they often aren't. In the same way, we think of rocks as mixtures, but some (like limestone) can be made of only one mineral (calcite). Thus, the distinction between a rock and a mineral can be a bit "fuzzy."

Both of these are rocks, but the one on the left (granite) is made of a three different minerals. The one on the right (limestone) is made of only one mineral.

Why is that? Because in order to classify things, we must make definitions. God's creation, however, is very complex, and it doesn't always fit nicely into our human-made definitions. Because of this, it is sometimes hard for students to fully grasp the distinctions that science makes. Honestly, that's fine. The more you study any area of science, the more experience you will get learning when God's creation doesn't exactly fit our definitions and how to deal with it. For right now, learn the definitions, and just be aware that there will be times when they are a bit hard to apply. Rest assured, however, that I will never expect you to deal with such situations when it comes to answering questions or taking the tests!

Minerals in Rocks

To get you started in your study of rocks, I want you to get some experience looking for the minerals that can be found in a rock. Now, of course, to be 100% sure of what minerals are in a rock, you must do tests that determine the chemicals found in the rock. Once you know the chemicals in the rock, you can determine what minerals went into making the rock. However, there are some rocks for which the mineral content can be identified visually. Since you have been studying the minerals in your kit over the past two weeks, let's see if you can identify them when they are mixed together in a particular type of rock.

Experiment 4.1: Minerals in Granite

Supplies:
- The minerals and rocks from the laboratory kit made for this course
- The magnifying glass from the laboratory kit made for this course
- A well-lit room

Instructions:
1. Make sure there is a lot of light so that you can easily observe optical properties.
2. Pick up specimen 7 from your kit. It is an example of granite rock.
3. Examine all sides of the rock, trying to note what makes it different from the minerals that you have been studying. How many different colors do you see in this rock?
4. Even though it is different from the minerals in your kit, the rock might have some similarities to at least some of them. Consider the colors, for example. Do any of the colors you see match the color of at least one of the minerals?
5. Use the magnifying glass to study the granite more closely, tilting it as you do so in order to get light to reflect off it at different angles. Do you see that small bits of the rock look like minerals? For example, you should see that many parts of the rock have luster, but the luster is different depending on where you are looking. Can you distinguish between different bits of the rock based on the luster that you see?
6. Concentrate on the different shapes you see in the rock. Are some bits of the rock flat? Are other bits shaped like tiny crystals?
7. After you have really examined granite well, pick up specimen 1 and examine it with the magnifying glass. Does anything that you saw in the granite look like what you see in specimen 1? Consider color, luster, and shape. Feel free to look back and forth between the granite and specimen 1. If so, you might have discovered one of the minerals that make up granite.
8. Repeat step 6 for the rest of the minerals (samples 2-6).
9. At least two of the minerals in your kit are present in the granite. Given that information, try to determine which ones they are, just based on what you saw with the magnifying glass.
10. Keep everything out. You may want to look at things again when you find out what those two minerals are.

It is easy to identify two of the minerals in this granite.

If you had trouble identifying the minerals in the sample of granite, don't worry! It often takes a practiced eye to identify the minerals in a rock based on their visible characteristics alone. Nevertheless, look at the picture on the left. It is a closeup image of a sample of granite. It might look different from yours, but you ought to at least notice some similarities. As pointed out in the picture, granite has quartz and potassium feldspar in it. It can also have biotite or muscovite mica in it, so if you thought you saw muscovite mica in your sample, you probably did. In addition, granite can also contain hornblende, a mineral that is not in your kit. Even if you weren't able to identify any of the minerals, you should have at least been able to see that there is a mixture of minerals in granite.

11. If you couldn't see the potassium feldspar or quartz in the granite when you did the experiment, look at the granite specimen again. With the help of the picture on the previous page, can you see them now?
12. Put everything away and clean up your mess.

Three Basic Types of Rock

The minerals that make up a rock are important, and we will talk about them quite a bit in this chapter. However, one of the most important distinctions between rocks involves how the rock was formed. There are three basic types of rock: **igneous** (ig' nee us), **sedimentary** (sed ih men' tuh ree), and **metamorphic** (met uh mor' fik).

Igneous rock – Rock formed from the freezing of magma or lava

Sedimentary rock – Rock formed from small particles accumulating and then sticking together

Metamorphic rock – Rock formed from existing rock that was subjected to heat and/or pressure

It's easy to understand the formation of igneous rock. After all, magma and lava are both molten rock. When molten rock cools it freezes, becoming a solid. That's igneous rock. But what about sedimentary rock? Think about the sand, silt, and clay you handled in Experiment 2.3. They were each a collection of particles with different sizes. Because the particles weren't attached to each other, they could be poured into things, like the inverted top of a plastic bottle. Suppose, however, the particles of sand had all been stuck together. You couldn't pour it anymore, could you? You would have a single, solid object. In that case, you would have sandstone, which is a sedimentary rock.

Metamorphic rock usually starts out as either sedimentary or igneous rock. If one of those two types of rock is exposed to a lot of heat or pressure (or both), it can fundamentally alter the properties of the rock, making a metamorphic rock. This kind of rock gets its name from "metamorphosis," which means the process of transforming from one thing to something that is very different. Metamorphic rock is rock that has been transformed from igneous or sedimentary rock into rock that is very different. Consider, for example, the two rocks pictured on the right. The nearest one is granite, which is an igneous rock. The other is gneiss (pronouced "nice"), which is a metamorphic rock formed from granite. As you can see, heat and pressure can radically alter a rock!

Granite (left), an igneous rock, and gneiss (right), a metamorphic rock formed from granite

Comprehension Check

4.1 While it's not considered metamorphosis, there is a way that igneous rock can be turned into sedimentary rock. How?

Igneous Rock

As you already learned, igneous rock is formed when magma or lava cools and solidifies. Since magma is, by definition, underground, igneous rock formed from magma is formed underground and is called **intrusive igneous rock**. If it forms from lava, it is formed above ground and is called **extrusive igneous rock**. The pictures below show you an example of each. The large, vertical formation the man is examining in the picture on the left is called a "dike," and it is made of intrusive igneous rock. Magma ran up through a crack in the surrounding rock, which is sedimentary. When it cooled, it became a column of intrusive igneous rock. Weathering and erosion eventually removed enough rock to expose it. Otherwise, it would still be underground. The picture on the right shows a formation of rhyolite (rye' uh light), an extrusive igneous rock that formed as a result of lava being released from a volcano and then freezing. Extrusive and intrusive igneous rocks are different from one another. The following experiment will allow you to compare one intrusive igneous rock to two extrusive igneous rocks.

A dike, which is made of intrusive igneous rock (left), and a formation of extrusive igneous rock (right)

Experiment 4.2: Igneous Rock

Supplies:
- The rocks from the laboratory kit made for this course
- The magnifying glass from the laboratory kit made for this course
- A well-lit room

Instructions:
1. Make sure there is a lot of light so that you can easily observe optical properties.
2. You have already examined specimen 7 from your kit, but pick it up again and examine it with the magnifying glass to remind yourself what it looks like.
3. Pick up specimen 8 and compare it visually to specimen 7. What are the differences between them?
4. Examine specimen 8 with the magnifying glass, looking specifically for the minerals in the rock. Remember, luster is a mineral property. If one part of the rock sparkles, that's probably due to the luster of a mineral in the rock. Also, different colors in the rock usually mean different minerals as well. What can you say about the size of the minerals you are seeing in this rock compared to the size of the minerals in specimen 7?
5. Repeat steps 2 and 3 with specimen 9. Compare the color, structure, and mineral sizes to both of the other specimens.
6. Keep the specimens and magnifying glass handy as you read my discussion.

Specimen 7 (granite) is an intrusive igneous rock. Specimen 8 (basalt) and specimen 9 (pumice) are extrusive igneous rocks. What differences did you notice right away? The extrusive rocks were more uniform in color than the intrusive rock, weren't they? You might have also noticed that specimen 7 sparkled in the light when you were looking at it with your naked eyes. Specimen 8 might have sparkled a bit, but specimen 9 probably didn't sparkle at all when you looked at it with your naked eyes. When you used the magnifying glass, however, you should have seen at least some sparkles, even with Specimen 9.

What explains these differences? Well, remember what you learned about the size of mineral crystals. When the solution in which they are dissolved cools slowly, the crystals that precipitate out are bigger. When it cools quickly, the crystals are smaller. You could actually see different crystals in granite with your naked eyes, which is why it is multicolored. That's because the magma cooled slowly, allowing for larger mineral crystals to form. The other two rocks appeared more uniform, because the mineral crystals were smaller. The sparkles showed this really well. The sparkles were smaller in the extrusive rocks, because the mineral crystals with luster were smaller.

You should have noticed one other major difference. The pumice (specimen 9) was very different in structure from the granite (specimen 7) and the basalt (specimen 8). While the granite and basalt were solid throughout, the pumice had lots of holes and pits in it. Why? It's a consequence of the fact that there were a lot of gases trapped in the magma, and when it reached the surface and turned into lava, those gases formed bubbles and escaped. Since this happened while the lava was freezing into a solid, those bubbles left holes and pits in the rock.

If you are wondering how gases got into the magma, remember that there are all sorts of chemicals underground, and at high temperatures, some of those chemicals (water, for example) turn into gases. As magma flows through the ground, then, it picks up lots of chemicals. Because of the temperature, some of them turn into gases. But why don't those gases bubble out of the magma? Because the magma is under pressure.

Think about opening a bottle or can of soda. What happens? You hear a "whoosh," right? That's because the bottle or can is under pressure, which is forcing carbon dioxide gas to stay dissolved in the liquid of the soda. When you open the can or bottle, that pressure is released. What happens as a result? You see bubbles in the soda, because some of the carbon dioxide gas is escaping. If enough of it escapes, you see a froth at the top of the liquid. The same thing happens when magma reaches the surface and turns into lava. The pressure suddenly goes way down, and the gases start to escape, forming a "froth." When that froth cools, it solidifies into pumice.

7. If my discussion mentioned something you didn't observe in the experiment, look for it now.
8. Once you have observed what I discussed, put everything away and clean up your mess.

Now remember, God's creation is incredibly complex, so it is hard for us to describe it with simple definitions. In the same way, igneous rocks come in many forms. For example, the granite in your kit had large mineral crystals, so you could see different colors with your naked eye. However, not all granite is like that. After all, the size of the minerals depends on the rate at which the magma cools. While magma usually cools more slowly than lava, magma in different locations will cool at different rates. Thus, the size of the mineral crystals in different granite rocks will be different.

The mineral content, especially the content of the silicate minerals, allows us to classify igneous rocks. There are four basic categories for the mineral content of igneous rocks, and it is often based on the percentage of SiO_2 in the mineral. If more than 65% is SiO_2, the rock is called **felsic** (fell' sik) and is usually lighter in color. The granite in your kit, for example, is a felsic intrusive igneous rock. "Felsic" tells you that it has a high SiO_2 content, while "intrusive" tells you it formed from magma. The pumice in your kit is a felsic extrusive igneous rock.

If 55%-65% is SiO_2, the igneous rock is called **intermediate** and is usually gray or gray with dark spots in it. The diorite (dye' uh right) pictured below is an intermediate intrusive igneous rock. If 45-55% is SiO_2, the igneous rock is called **mafic** (maf' ik) and is usually dark. The basalt in your kit is a mafic extrusive igneous rock, and the gabbro pictured below is a mafic intrusive igneous rock. If less than 45% is SiO_2, the igneous rock is called **ultramafic**, and it is often very dark. The kimberlite pictured below is an ultramafic intrusive igneous rock.

Now I do have to make a couple of points before you finish science for the day. First, you won't need to know the percentages I just gave you. In fact, they can be different in different classifications schemes. However, you will need to know that felsic rocks have the highest percentage of silicates, intermediate rocks have a lower percentage, mafic rocks have an even lower percentage, and ultramafic rocks have the lowest percentage. Also, while color is often related to the mineral content, it sometimes is not. The obsidian pictured on the left, for example, is very dark. However, it is a felsic extrusive igneous rock. It is dark because it cooled so quickly that no mineral crystals formed, so the colors of the minerals did not affect the color of the rock.

Diorite Gabbro

Kimberlite Obsidian

Comprehension Check

4.2 You have four igneous rocks that are described as follows: (I) intermediate extrusive, (II) ultramafic intrusive, (III) felsic extrusive, (IV) mafic intrusive.

a. Which of the rocks should have smaller mineral crystals in them?
b. Order them in terms of the percentage of silicate minerals, starting with the lowest percentage and ending with the highest percentage.
c. Two of them were pulled from an underground mine. Which two?

Sedimentary Rock

While you might have recognized some of the igneous rocks I discussed (granite and obsidian, for example), you are probably much more familiar with sedimentary rocks. While they make up only a minor portion of the earth's crust, they are easy to recognize. When you see multiple layers of rock that are exposed by weathering and erosion, you are usually looking at sedimentary rocks. The beautiful rock formation on the right, for example, is composed of layers of sedimentary rock. Why are they so uniformly horizontal? Think about what happened in Experiment 2.1. When you shook up the jar, the soil settled out in layers, and those layers were horizontal. Now imagine those layers hardening. That would result in something like what you see in the picture. Get some experience investigating sedimentary rocks by performing the following experiment.

Sedimentary rock is often found in horizontal layers.

Experiment 4.3: Sedimentary Rock

Supplies:

- The rocks from the laboratory kit made for this course
- The bag of silt from the laboratory kit made for this course
- The bag of sand from the laboratory kit made for this course
- The blue-topped dropper bottle with vinegar from the laboratory kit made for this course
- The magnifying glass from the laboratory kit made for this course
- A blank white sheet of paper with no lines
- A paper towel
- A well-lit room

Instructions:

1. Make sure there is a lot of light so that you can easily observe optical properties.
2. Pull specimens 10 and 12 out of your kit.
3. Hold one specimen in one hand and the other in the other hand.
4. Rub your thumbs back and forth on the specimens. Turn the specimens around in your hands as you do this, so you feel different parts of each. Note how different they feel.
5. Lay the sheet of paper on a flat surface.
6. Stick your hand in the bag of sand and pinch some sand between your thumb and index finger.
7. Pull the pinch of sand out of the bag and hold it over the paper,
8. Rub the sand between your thumb and index finger so that it sprinkles on the paper. Don't cover the paper with sand. Just allow it to fall onto one part of the paper.
9. Repeat steps 6-8 with the silt, and make sure it lands on a part of the paper where there is no sand.
10. When you were rubbing, did the feel of the sand remind you of how one of the specimens felt? What about the silt?
11. Use the magnifying glass to examine specimen 10.

12. Use the magnifying glass to examine specimen 12.
13. Examine both the silt and the sand with the magnifying glass.
14. One of the specimens is called sandstone, and the other is called siltstone. Which is which?
15. Hold specimen 11 in your hand and rub your thumb against it. How does it compare to the others?
16. Examine specimen 11 with the magnifying glass. How does it compare to the others?
17. Specimen 11 and specimen 9 (pumice) both have pits in them. How do those pits differ?
18. Use the dropper bottle to put a drop of vinegar on specimen 11.
19. Hold the rock close to your ear. You should hear something. What is it?
20. Wipe the rock with a paper towel to dry it.
21. Repeat steps 18-20 with specimens 10 and 12. Do you hear anything with either of them?
22. Keep the specimens and magnifying glass handy as I discuss the results.

As you've already learned, sand is made of larger particles than silt. You should have felt that as you rubbed sand and silt between your thumb and index finger. You also should have seen that with the magnifying glass. You probably guessed that specimen 10 is sandstone, while specimen 12 is siltstone. The sandstone felt rougher because the particles that collected and stuck together were larger, which made the resulting rock rougher. When you examined the sandstone, you should have been able to see grains and sparkles from the larger particles. Siltstone is made when smaller particles collect and stick together. As a result, it is smoother, and it should have been much harder (if not impossible) to see any grains, and the sparkles were probably small, even with a magnifying glass.

What about specimen 11? It is limestone, and you should remember that it contains a lot of calcite. That's why it "crackled" when you added a drop of vinegar to it. The calcite reacted with the vinegar to make carbon dioxide gas, which formed bubbles like you saw in Experiment 3.4 when you put a drop of vinegar on a sample of calcite. It would be difficult to see the bubbles in this case, because the vinegar soaked into the rock. However, you could hear the bubbles as they reached the surface of the vinegar and popped. You heard no crackling with the other two specimens, because they had no calcite in them.

The limestone also had some pits in it, but as you compared those pits to the ones in pumice, you should have noticed some major differences. The pits in the pumice were generally small and round. That's because they are formed by bubbles. However, the limestone's pits were probably bigger, and they weren't all round, were they? In fact, they might have had noticeable patterns in them. That's because some of them were impressions left behind by the shells or other remains of organisms that lived in the water where the calcite and other particles in the limestone formed. Often, those remains leave obvious impressions, as shown in the picture on the left.

Limestone often has the impressions of shells in it.

Comparing the feel of the limestone to the other two specimens, you probably found that it was a bit smoother than the sandstone but rougher than the siltstone. The size of the particles that came together to make the limestone affected how it felt, but so did the pits, so just from how it feels, you probably can't say anything about the size of the particles in the limestone compared to the siltstone and sandstone. However, when you looked at the limestone with the magnifying glass, you should have been able to see clumps and sparkles more easily than you did in the siltstone, indicating that the particle size was larger than what is found in siltstone.

23. If you didn't see or feel what I discussed, try to do so now.
24. Put everything away (making sure the top of the bottle is clicked shut) and clean up your mess.

There are three basic kinds of sedimentary rock: **clastic** (klas' tik), **chemical**, and **organic**. Clastic sedimentary rock is formed by particles that come from other rocks. Weathering and erosion can bring bits of rock together, and if they harden together, the result is a clastic sedimentary rock. The sandstone and siltstone you experimented with are both clastic sedimentary rocks because the particles that formed them came from weathered rocks. Now remember, there is a third classification based on particle size. Clay is made of even smaller particles than silt. When a clastic sedimentary rock is made of only clay-size particles, it is called **shale**, and an example is shown below. You can also have a mixture of particle sizes, and the resulting clastic sedimentary rock is called **mudstone**. There are also clastic sedimentary rocks that form from large bits of rock. They are called **conglomerates**, an example of which is shown below.

A chemical sedimentary rock is formed by particles that either precipitate out of water or are chemically altered in the presence of water. Many limestones are chemical sedimentary rocks, forming from calcite particles that precipitated out of water. However, another chemical sedimentary rock is dolomite, an example of which is shown on the right. The particles in dolomite are calcite minerals that were altered in water so that magnesium was inserted. While calcite is $CaCO_3$, dolomite is $CaMg(CO_3)_2$.

Shale Conglomerate

Dolomite Coal

As its name implies, an organic sedimentary rock forms from particles that come from something that was once alive. Coal, which is also pictured above, is an example of an organic sedimentary rock. Coal is made mostly of carbon particles that formed when dead plants decayed. However, some limestones are also organic sedimentary rocks. That's because many organisms that live in the ocean make shells out of $CaCO_3$, and when bits of those shells collect and harden, the result is also limestone. Once again, then, sometimes God's creation is a bit too complex to fit nicely within our definitions.

Comprehension Check

4.3 If you had shale to use in your experiment, how would it have felt compared to the sandstone and siltstone?

4.4 Some sedimentary rocks form as a result of water evaporating. Are they clastic, chemical, or organic?

Metamorphic Rock

In many ways, metamorphic rocks are the most interesting of all three types. The word "metamorphic" comes from the Greek word *metamorphoo*. It is used in 2 Corinthians 3:18 to describe what happens when someone understands who Christ is and follows Him. That person starts transforming, becoming more Christlike. In other words, the power of Christ takes a person who already exists and changes him or her into a new and better person.

Similarly, metamorphic rocks are the result of existing rocks being changed. The rocks are changed by heat and/or pressure, and that change produces a new type of rock. Now where do you think such a thing would happen? Deep in the crust, of course. As you go deeper in the geosphere, it gets hotter. Also, the weight of all the rocks above pushes down on the rocks below, producing pressure. Despite the fact that metamorphic rock is often formed deep in the crust, we can see some metamorphic rocks on the surface of the earth. For example, the picture on the left shows a slab of a metamorphic rock found in the Georgian Bay of Canada. It formed deep in the crust, but then it was pushed up by processes that you will learn about later on in this course. The rocks above it were then weathered, and erosion exposed the beautiful metamorphic rock that was below.

The two boulders in the picture are sitting on a slab of metamorphic rock that has some water-filled depressions in it.

Of course, we can't actually see metamorphic rock form, but if we compare metamorphic rock to sedimentary and igneous rock, we can learn a lot about how the metamorphic rock must have been made. Perform the following experiment to see what I mean.

Experiment 4.4: Metamorphic Rock

Supplies:

- The rocks from the laboratory kit made for this course
- The blue-topped dropper bottle with vinegar from the laboratory kit made for this course
- The red-topped dropper bottle from the laboratory kit made for this course (It contains 10% hydrochloric acid, which is caustic. Handle it with care.)
- The magnifying glass from the laboratory kit made for this course
- Cleaning gloves
- Paper towels
- Two small glasses, like juice glasses
- A well-lit room
- A sink that is **not** stainless steel and has a faucet

Instructions:
1. Make sure there is a lot of light so that you can easily observe optical properties.
2. Pull specimens 11 (limestone) and 13 (marble) out of the kit.
3. Examine each with your naked eye and then the magnifying glass. Look specifically for mineral crystals. Try to compare the size of crystals in the limestone to those in the marble. Remember, sparkles come from the minerals' luster. The bigger the crystals, the easier it is to see sparkles.
4. Using the blue-topped bottle, add a drop of vinegar to the limestone like you did before. Once again, hold it close to your ear and hear the crackling.
5. Repeat step 4 with the marble. Can you hear anything?
6. Use the magnifying glass to look for bubbles in the drop of vinegar on the marble.
7. Squirt enough vinegar in the small glass so it covers the bottom of the glass in a thin layer.
8. Put the marble specimen in the glass with the labeled side up so the unlabeled side is soaking in the thin layer of vinegar.
9. Let it sit for a full minute and then look through the bottom of the glass for bubbles or small sediments in the liquid. Do you see any?
10. Pull the marble specimen out of the glass and use the paper towel to wipe off both specimens.
11. Put on the cleaning gloves.
12. Carefully add one drop of acid from the red-topped bottle to the limestone. Do you see bubbles? They should form quickly, and there should be a lot of them. If you don't see bubbles, look for them with the magnifying glass. **Be careful**! Don't spill the acid!
13. Repeat step 12 for the marble.
14. Repeat steps 7-9 with the marble, using the acid in the red-topped bottle and the other small glass. Do you see a difference from when you used vinegar?
15. Rinse off both specimens (and your gloves) to get rid of the acid and dry them with paper towels.
16. Pull specimens 7 (granite) and 15 (gneiss) out of the kit.
17. Repeat step 3 with these two specimens.
18. Keep everything handy as I discuss the experiment.

What did you see in the experiment? The limestone and marble looked very different, didn't they? Nevertheless, marble is metamorphic rock that is made from limestone! Why do they look so different? One reason is that the calcite crystals in marble are larger than the calcite crystals in limestone. That's one reason the marble should have sparkled more.

Also, despite the fact that the marble contains calcite, it probably didn't effervesce with the vinegar. If it did, the effervescence was much gentler than it was with the limestone, which is why you heard little or no crackling when you put vinegar on the marble. Even when you let the marble soak in vinegar, you probably didn't see bubbles. When you put hydrochloric acid (which is a stronger acid) on the limestone, you should have seen lots of bubbles. When you put the hydrochloric acid on the marble, you probably still didn't see any bubbles. However, when you let it soak in hydrochloric acid for a minute, you should have seen some.

Believe it or not, this is also due to the size of the crystals. When calcite crystals in a rock are large, they react with acid slowly. The reaction is so slow, in fact, that it is very hard to see or hear bubbles when the acid is weak, like vinegar. When the acid is stronger, like hydrochloric acid, you can see the bubbles, but it takes a while. That's because the smaller calcite crystals in limestone react much more quickly than the larger crystals in marble.

The other metamorphic rock you studied was gneiss. It can be made from either sedimentary or igneous rock. The gneiss you have was most likely formed from granite, which is why I had you compare it to your granite sample. Did you notice that the different colors in the gneiss were less randomly-arranged than in the granite? The pressure at which the gneiss formed forced the mineral crystals to grow mostly parallel with each other, which tends to make bands of color. This is called **foliation** (foh' lee ay' shun), so we would say that gneiss is a foliated metamorphic rock. The picture on page 104 is also of foliated metamorphic rock.

There are three types of metamorphic rock: **regional**, **contact**, and **dynamic**. Remember, metamorphic rocks are made with lots of heat *and/or* pressure. The "and/or" is important. If both heat and pressure are involved, the result is a regional metamorphic rock. The gneiss and marble that you have in your kit, as well as the slate pictured below, are regional metamorphic rocks.

Contact metamorphic rocks are formed when there is a lot of heat, but not a huge amount of pressure. This can happen when magma fills a crack in existing rock. The magma heats up the surrounding rock. The surrounding rock doesn't melt, but some of it gets hot enough to be transformed. The quartzite pictured below, is a contact metamorphic rock that was originally sandstone.

Dynamic metamorphic rocks are formed mostly by high pressure, but not a lot of heat. They are pretty rare, because if you think about it, you typically find high pressures deep in the crust, but you also find high temperatures there. Nevertheless, there are times when rocks can experience high pressure without really high temperatures. During earthquakes, for example, masses of rock move against one another, producing a lot of pressure. If there isn't a lot of heat, that will produce a dynamic metamorphic rock, such as the mylonite (my' luh night) pictured below.

Slate Quartzite Mylonite

Comprehension Check

4.5 Which of the three rocks above is foliated?

4.6 Suppose you took sedimentary rock and used a blowtorch to get it so hot that it transforms. Which kind of metamorphic rock would you have made?

The Rock Cycle

In "Comprehension Check" question 4.1, I asked you to think about how igneous rock could be transformed into sedimentary rock. If you didn't answer the question correctly, that's okay. You are going to learn the details of that now. It's called the **rock cycle**.

Rock cycle – The process by which the three basic types of rock can be converted into one another

When you hear that something has been "set in stone," what does that mean to you? It usually means that something cannot be changed. However, the rock cycle tells us that rocks themselves can change, so setting something in stone isn't necessarily as permanent as you might first think.

The illustration on the right shows a simplified, idealized version of the rock cycle. Notice first that igneous and sedimentary rock, when subjected to heat and/or pressure, is converted into metamorphic rock. You have learned that already. Notice also that once metamorphic rock is formed, it can be heated so much that it can once again melt, forming magma. Note that it will not form lava, at least not initially, because any natural source of heat that could melt metamorphic rock will be found underground, so the metamorphic rock will melt underground. It might eventually come up to the surface as lava, but initially, melting metamorphic rock will form magma. Once the melted rock cools down and freezes, it will become igneous rock, which might also eventually become metamorphic rock again.

This is a schematic representation of the rock cycle.

Notice also that all three types of rock can be weathered to produce sediment. That sediment might just get incorporated into the soil, but it might also go through **lithification** (lith uh fuh kay shun).

Lithification – The process by which sediment becomes sedimentary rock

The process of lithification requires three distinct steps. First, the sediments must be compacted. That pushes the grains of the sediment closer together. Also, sediment usually has water in it, and the water will get in between the grains, keeping them from getting really close together. As a result, in the second step, the water must be removed. Once the grains are close enough to one another, the third step can happen: the sediments can be cemented together. This is usually done with minerals that are in the sediment or were left behind when the water was removed. Those minerals can act like "glue," holding the grains together. At that point, sedimentary rock has been formed.

You might find it hard to believe that sediments can harden into rocks, but you have already experienced it in this course. In Experiment 2.5, you made a rock out of plaster. It started out as a powder, but you added water to it. Over time, it hardened to the point where it became a single, hard object. That's essentially what happens in lithification. In fact, the plaster you used probably had a **dehydrated** (dee hi dray' ted) form of the mineral gypsum in it. That means the water had been removed from the mineral. You added water, which allowed it to go back to its natural form. As long as you didn't add too much water, the gypsum molecules could cement the powder together, forming something very similar to a sedimentary rock.

Different Scientific Perspectives of the Rock Cycle

The rock cycle has been studied in depth for quite some time. In the late 1700s, James Hutton proposed the first formal version of the rock cycle, but like all theories, it has been modified as a result of additional study. Nevertheless, Hutton proposed that the rock cycle happens continually, and that the earth's crust is constantly undergoing change. In the scientific paper in which he described the rock cycle, he wrote, "The result, therefore, of our present enquiry is, that we find no vestige of a beginning, – no prospect of an end." ("Theory of the Earth," *Transactions of the Royal Society of Edinburgh*, 1788, p. 304). In other words, the rock cycle has been happening since the time the earth was formed, and there is no reason to think that it will ever end.

Most geologists today agree with James Hutton. We see igneous rock being formed whenever lava flows from a volcano, for example. We also know that weathering is occurring, although it often takes a long time to see the results. Consider, for example, the picture below. You can clearly see that it is a wall made by people, right? The stones are much more regular in shape than what you would expect of natural stones, and they are clearly out of place in the pasture. You can also see that the wall is crumbling to some extent. When would you guess this stone wall was built? If you guessed that it was about *1900 years ago*, you were correct!

This is a section of Hadrian's Wall in the United Kingdom.

The wall in the picture is known as Hadrian's Wall, because it was built in AD 122, when Hadrian was the emperor of Rome. We don't know for sure, but it probably took about six years to complete. Some of the rocks in the foreground of the picture might have been weathered and eroded. However, they also might have been broken off by someone who needed a few rocks for building purposes. Nevertheless, it doesn't look too bad for such an old wall, does it? Most of the rocks have retained the shape into which they were cut 1,900 years ago.

James Hutton examined Hadrian's wall and compared it to the rocks that he found in the nearby countryside. He noticed that while the rocks on Hadrian's Wall were a bit different from the way they were cut due to weathering, the rocks in the surrounding countryside were very smooth. Since they probably didn't originally form with smooth surfaces, this told Hutton that they were weathered much more than the rocks of Hadrian's Wall. Consider, for example, the picture below. It shows another section of Hadrian's Wall, but there is also a rock on the ground that is not a part of the wall. Notice that while it has cracks and pits in it, the surface is smooth. That indicates it has been weathered a lot more than the rocks that make up the wall.

This is another section of Hadrian's Wall in the United Kingdom. Notice how much more weathered the natural rock on the ground is compared to the cut rocks that make up the wall.

This led Hutton to conclude that weathering happens very slowly. After all, if the rocks that make up Hadrian's wall are still mostly rectangular with sharp edges, 1900 years isn't nearly enough time to produce the weathering that he was used to seeing in the natural rock formations he studied. As a result, he concluded that weathering is a slow process, and the rocks that he found in natural formations had to be *significantly* older than Hadrian's Wall. From Hutton's (and most modern geologists') point of view, then, the rock cycle happens continuously, and it usually occurs very, very slowly. As a result, they think that the rocks are best explained by the rock cycle happening on earth over billions of years. As I mentioned in Chapter 2, this view is usually called uniformitarianism.

There are some geologists who think differently, however. They think that the rock cycle was much more active during specific times in the earth's past and less active at other times. Of course, they are the geologists who think catastrophism is a better explanation for the earth's geological features. After all, as you have already seen, lots of water flowing quickly can produce very profound geological results in a very short time. If there were, for example, a flood that covered the entire globe, the resulting weathering, erosion, and volcanic activity could cause lots of rock cycle activity in a short time. After all that was over, the rock cycle would go back to a very, very slow pace, and if we tried to judge everything based on what we observe going on now, our conclusions would be quite wrong.

In the end, then, if the rock cycle is fairly constant in terms of its activity, the best way to explain most of the rocks we see in the earth's crust is that it has been happening for billions of years. Hadrian's Wall has experienced very little weathering, but the rocks in the surrounding countryside have experienced a lot of weathering, because while Hadrian's Wall is only 1,900 years old, the rocks in the surrounding countryside are millions or billions of years old. However, if the rock cycle was very active before Hadrian's Wall was built and then ceased its high activity before the wall was built, that could explain the difference between the rocks of the wall and the rocks in the countryside.

Which explanation is best? It's hard to say from a purely scientific perspective. There are sedimentary rock structures that uniformitarianism explains well, but there are others that it doesn't explain well. For example, we know that today, weathering occurs rather slowly. We also know that over time, the sediments do accumulate, and they look similar to the kinds of sediments that make up some of the sedimentary rock we see today. In the picture below, for example, you see a stingray churning up sand near the shore of the Clarence River, where it meets the ocean on the eastern coast of Australia. We know that sand has been accumulating because of weathering and erosion, and we also know that if that river ever dried up substantially, all the ingredients needed for lithification would be there. Most likely, that sand would end up forming sandstone, and it would look similar to the sandstone that we find in the earth's crust. That supports the uniformitarian view.

The sand that this stingray is churning up was deposited here as a result of weathering and erosion, and under the right conditions, it could be turned into sandstone.

At the same time, however, there are problems with the uniformitarian view. Consider, for example, the Dolomite mountains in Italy. They get their name because they contain huge amounts of the sedimentary rock called dolomite, which is pictured on page 103. There are other large formations of dolomite rocks around the world. In fact, estimates indicate that as much as 10% of all the earth's sedimentary rocks are dolomite. However, the formation of dolomite sediment is very rare today, and that sediment doesn't have the characteristics that would end up making the dolomite rocks we see today. As a result, we know that at least in the case of forming dolomite, the rock cycle hasn't always worked the way it is working today.

We also know that while it would take a long, long time for a lot of sediment to accumulate under normal conditions, when there is a lot of flooding, sediments can accumulate very, very quickly. Remember the Canyon Lake Gorge you learned about in Chapter 2. The floodwaters eroded a lot of sediment to make that gorge, and it was carried to another location, increasing the speed of sediment accumulation there.

You might also wonder about how long it would take for the lithification process to occur. If the Clarence river pictured above were to dry up, how long would it take for the sediment to harden into rock? It's hard to say, because the speed of lithification depends on lots of factors. It is a chemical process, and chemical processes are strongly affected by the conditions under which they happen. High temperatures increase the speed of most chemical reactions, and high pressure would also speed up lithification.

More importantly, however, the chemicals and organisms present in the sediment have a strong effect on how quickly lithification can occur. At the end of the 1990s, for example, some scientists came up with a process in which chemicals are sprayed on sand, and the sand hardens into stone in just a few minutes. It is called the "calcite *in-situ* precipitation system" or CIPS for short. You should recognize the words "calcite" and "precipitation." The words "*in-situ*" come from Latin, and they simply mean "in position." So the "calcite *in-situ* precipitation system" is a system by which calcite ($CaCO_3$) precipitates in the sand (in position), making the lithification process happen very, very quickly.

Along the same lines, there is another process called the "microbially-induced calcite precipitation system" (MICP) that uses bacteria to speed up the formation of calcite in sand, which once again speeds up the lithification process. Both of these processes demonstrate that under the correct conditions, lithification can occur very quickly. Of course, lithification can occur slowly as well. In fact, if we examine the chemicals and organisms in the sediments we find accumulating today, their lithification would proceed very slowly.

So, when it comes to sedimentary rock, we know that sediments can accumulate slowly, and we know that they can accumulate quickly. We also know that lithification can occur slowly, but it can also occur quickly. Thus, we don't know for certain whether the rock cycle mostly operates constantly over long periods of time like we see happening today or quickly during specific times in the past. We know that there are specific cases, like the dolomite discussed earlier, in which the rock cycle must have acted differently in the past. However, we don't know if that is a rare exception, or it is the general rule. Hopefully, as we learn more about sedimentary rock, we can come to a more definitive conclusion.

Comprehension Check

4.7 Igneous rock becomes metamorphic rock and then eventually becomes sedimentary rock. Which of the following things had to have happened, and in which order did they happen: freezing, lithification, melting, weathering, exposure to heat and/or pressure?

4.8 Rock A has a lot of sharp corners and is really rough. Rock B is smooth and rounded. Which has probably experienced more weathering?

4.9 Can you definitively say which of the rocks in the previous question is older? Why or why not?

Sedimentary Rock Formations

I concentrated on sedimentary rock in my discussion of the rock cycle, and I want to continue discussing it a bit. Specifically, I want you to learn the basic structures that can be formed when sediments accumulate and undergo lithification. As I noted previously, one common feature of sedimentary rock is that it forms layers. The picture below, for example, shows a cliff that is composed of sedimentary rock. As you can see, the rock forms nice, horizontal layers, which are often called **strata** (the singular form is "**stratum**"). As a result, geologists will often say that sedimentary rock is **stratified**. Notice that it is easy to distinguish between many of the layers. Some are gray, some are light brown, some are dark brown, etc. Obviously, then, some of the strata are composed of rocks that are quite different from the rocks found in other strata. However, there are some strata that are composed of the same rock. The very top strata, for example, are all light gray. They are composed of the same sedimentary rock, but you can still see individual strata.

This cliff is made of horizontal strata of sedimentary rock.

If the rocks are the same, that means the sediments they formed from are also the same. So, if you think about the top strata of rock in the picture on the previous page, the same basic kind of sediment accumulated and then lithified. Nevertheless, you can still distinguish different strata. Whenever you can do that (whether the strata are made of different rocks or the same rocks), we say that you have a **bedding plane**.

Bedding plane – A distinct separation between two rock strata

When strata are parallel, as is the case for the strata in the picture on the previous page (and the picture below), we say that the rock exhibits **planar** (play' nur) **bedding**.

While they are both bedding planes, an unconformity is different from a conformable bedding plane.

A bedding plane is fairly easy to identify. If you can distinguish between two layers, they are separated by a bedding plane. However, take a look at the picture on the left, which is a closeup of a sedimentary rock formation that exhibits planar bedding. There are a lot of bedding planes in the picture, but look at the two that I have pointed out with arrows. Do you notice the difference between them? The one I labeled **conformable bedding plane** is just a line that almost looks like it was molded into the rock. However, the one I labeled **unconformity** is different. It shows a "break" between the stratum above and the stratum below.

What causes this difference? Think about what had to happen to make the rock in the picture. Sediment had to accumulate before it could undergo lithification, right? What happens if the sediment accumulates pretty continuously? The pile of sediment will be uniform. If there is a change in the way the sediment is being deposited or a change in the composition of the sediment, the sediment will look different, and once lithification occurs, the rock will look a bit different. As a result, a conformable bedding plane will be visible.

However, suppose there is a break in accumulation, and the break is long enough to allow full or partial lithification of the sediment that has already accumulated. If new sediment starts accumulating later, the bedding plane will be much more pronounced, making an unconformity.

Unconformity – A separation between two strata caused by a period of erosion or no sediment deposition

So, while a conformable bedding plane is a "smooth" separation between strata, an unconformity represents a "break" between strata. During this break, either erosion occurred or no sediment accumulated. As a point of terminology, when strata are separated by a conformable bedding plane, they are called "conformable strata." When they are separated by an unconformity, they are called "unconformable strata."

There are actually four different types of unconformities. The one pointed out in the picture on the previous page is a **disconformity**:

Disconformity – An unconformity between parallel strata of rock

But there can also be unconformities between strata that are not parallel. When one set of strata is tilted relative to another, the result is an **angular unconformity**.

Angular unconformity – An unconformity between two sets of strata that are tilted relative to each other

The picture below shows you a sedimentary rock formation that has both disconformities and an angular unconformity.

Notice how there are two distinct sets of strata in this rock formation. The rocks at the bottom form strata that are mostly parallel, but they are angled. You can see some clean "breaks" between these angled strata, and since those breaks are at the same angle as the rock strata, they are parallel to the strata. Thus, they are disconformities. However, near the middle of the formation, the strata are horizontal. Once again, there are clean "breaks" between some of the strata, and once again, since they are parallel to the layers, those are disconformities. However, where the titled strata meet the horizontal strata, you have an angular unconformity.

These sedimentary strata have disconformities and an angular unconformity.

Of course, there is also igneous and metamorphic rock in the earth's crust, and they often are a part of a formation that also has sedimentary rock in it. When sedimentary rock is found in contact with igneous or metamorphic rock, it is called a **nonconformity**, because the rock on top doesn't conform to the rock below.

Nonconformity – The contact between igneous or metamorphic rock and sedimentary rock above it

The last kind of unconformity is a bit strange. It is an unconformity that looks like a conformal bedding plane but is thought to be the result of a period in which erosion and/or no deposition occurred. This is called a **paraconformity**.

Paraconformity – A bedding plane that is thought to be an unconformity but doesn't look like one

Why would you think something that looks like a conformal bedding plane is an unconformity? That has to do with how someone interprets the rocks. As you will learn later, there are ways that geologists try to determine how old a stratum of rock is. If they find a stratum they think is very young resting right on top of a stratum that they think is very old, they assume there had to be a long time in between the accumulation of sediments. Thus, they think there should be an unconformity, even though it doesn't look like one.

Don't get lost in all these strange words and their definitions. Remember, the reason we can distinguish between different strata is because of the bedding planes that exist between them. If a bedding plane is just a line in the rocks, it is considered a conformal bedding plane and is thought to be a result of a small change in the way the sediment was deposited. If there is a "clean break" between the strata, it is an unconformity, and it is thought to represent a period of time when no sediment was deposited. During that time, the lower strata might possibly have gone through lithification and even some weathering and erosion before the sediment from the next stratum started accumulating.

If the separation between strata is an unconformity, it is one of four types. A disconformity represents a period of time between the accumulation of parallel strata of sediment. An angular unconformity represents a time between a series of tilted strata and then a series of horizontal strata being deposited on top of it. A nonconformity represents a period of time between the formation of igneous or metamorphic rock and the accumulation of sediment on top of it. Finally, a paraconformity is an unconformity "disguised" as a conformal bedding plane.

Stranger Sedimentary Rock Formations

The picture on the previous page showed you an angular unconformity, in which tilted parallel strata were underneath horizontal parallel strata. Interestingly enough, sedimentary rocks can form more elaborate patterns of strata. Consider, for example, the image below, which shows a rock formation in Crete, Greece. The rock is limestone, so it is sedimentary, and it was clearly deposited in layers. However, those layers aren't regular. It's hard to imagine how sediment could have been deposited like that, so it is assumed that the sediments accumulated horizontally, and then something happened to deform them into those curved shapes, which are called **folds**.

These limestone strata are folded.

But wait a minute. How can rocks be folded like that? Well, even though rocks are heavy and hard, there are some tremendous forces that can act on the earth's crust. Those forces can deform the rocks. However, when hard rock is deformed like that, it usually has a lot of cracks and breaks. But there are lots of folded rocks that have very little cracking and breaking. The easiest way to understand them is to assume that the sediments did not complete lithification before they were folded. If that were the case, they would still be soft and pliable, which would allow them to fold without cracking. Then, lithification could complete, forming the folded rock strata.

Back when you first learned about igneous rock, you saw a picture of a dike that was formed when magma was forced up into a crack of a rock formation. When the magma cooled, it formed an intrusive igneous rock that took on the shape of the crack. Well, sedimentary rocks can also form dikes. They don't get pushed up through cracks in the rock, however, because sediments are usually deposited on top of rocks. However, suppose you have some wet sediments accumulating on top of a rock formation, and then a crack forms in the rock below. What will happen to the sediment? It will fall down into the crack. If it then goes through lithification, there will be sedimentary rock surrounded by whatever the sediment was originally accumulating on.

The picture on the right gives you an example of what I am talking about. The column of rock pointed out in the picture is sandstone, a sedimentary rock. The surrounding rock is granite. Obviously, the granite had to form before the sandstone, because granite comes from magma. If the sandstone had formed first and then been surrounded by magma, it would not have stayed sandstone. It would have either melted and been incorporated into the granite, or it would have become quartzite, a contact metamorphic rock. Thus, the granite had to form first, a large crack must have formed, and then the sediments that make up the sandstone must have been pushed down from above. After that, lithification occurred, producing the column of rock. While this is a dike, it is also called a **sand injectite**, because the sediments were "injected" into the surrounding granite.

Sedimentary rocks can also form dikes, called sand injectites.

Now remember, we are looking at these rock formations now and trying to figure out how they formed. In some cases, it's pretty clear. As I discussed above, the sand injectite had to form after the granite. However, sometimes it's not so clear. The folded limestone I showed on the previous page is easiest to understand if the sediments had not completely undergone lithification before they were folded. However, many geologists disagree with that. They think with the right combination of conditions, it would be possible for those folds to form in hardened rock without leaving a lot of cracks. As you will see as you go through this course, because we are examining the rocks after they formed, there can be a lot of disagreement among geologists about the details regarding exactly how the formation occurred.

Comprehension Check

4.10 You are examining a rock formation that is composed of four distinct strata of sedimentary rock. How many bedding planes are present?

4.11 In the rock formation from the previous problem, the bottom layer is made of igneous rock. The other layers are all sedimentary rock. How many nonconformities are there?

4.12 In the rock formation from the previous problems, the strata all exhibit planar bedding. How many angular unconformities are present?

4.13 You are examining a dike. How can you tell if it was formed by magma being pushed up into a crack in the surrounding rock or sediments being pushed down into the crack?

More on Sedimentary Rock

In this chapter, I have devoted more discussion to sedimentary rocks than igneous or metamorphic rocks. I don't want that to make you think they are the most common rocks in the earth's crust. They are not. The earth's crust is mostly igneous rock. However, sedimentary rocks are the most common rocks found at the surface of the earth. This should make sense to you. While there are specific instances (like sand injectites) in which sediments accumulate underground, most sediments accumulate above ground. Thus, sedimentary rocks are more common on the surface of the earth.

This fossil trilobite was encased in limestone. The stone was cut to reveal the fossil and then smoothed down.

There is another reason it is important to concentrate on sedimentary rock. It is more likely to contain things other than rocks. The picture on the left, for example, shows you a limestone rock that has been cut to reveal a fossil. This fossil is from an animal that no longer exists, at least as far as we know. It's called a **trilobite** (try luh byte), and the only things we know about trilobites come from their fossils. Sedimentary rock is much more likely to contain fossils like that than igneous or metamorphic rock, so when we want to learn about the biological world during earth's past, we spend a lot of time examining the contents of sedimentary rock.

If you think about it, there are several reasons sedimentary rock is the most likely kind of rock to hold fossils. First, extrusive igneous rock forms from lava, which destroys the remains of most organisms. However, that doesn't mean you *never* find fossils in extrusive igneous rock. If lava traps an organism as it is flowing and then cools quickly, the organism can leave an impression in the rock. For example, you learned earlier that obsidian is an igneous rock. There are some samples of obsidian that contain imprints of organisms that were engulfed by the lava that formed the rock.

Second, metamorphic rock requires heat and/or pressure to form. If a sedimentary rock contains fossils and then is changed into metamorphic rock, the fossil is often destroyed in the process. Once again, however, that is not always the case. Fossils can be found in marble, which you learned is a metamorphic rock made from limestone.

Third, most organisms live above the surface of the earth. Since sediments collect there, dead organisms get encased in the sediment. This can help the fossilization process, since it reduces the chance that certain chemicals and/or living organisms can decompose the dead organism. It can also provide chemicals that facilitate the fossilization process. You will learn more about fossils and the processes that form them later on in this course. For right now, just understand why sedimentary rock is the most likely type of rock to contain them.

Fossils aren't the only interesting things found in sedimentary rock. As sediments accumulate, they can encase objects that have been made by people. Typically, we call them **artifacts**. Most artifacts are found in sediments and soil, but that's not always the case. Sometimes, they are found encased in rocks. You might think that any artifact encased in rock must be very old, since we normally think of the lithification process as taking a long time. But that's not always the case. Consider, for example, the picture below. It shows you a sedimentary rock that was cut in half to reveal something inside. The ruler gives you an idea of how big the rock and the object inside are.

If you look at the picture closely, you can tell that whatever is in the center of both halves of the rock isn't natural. In fact, based on an analysis done by several different people, it's a spark plug that was made in the 1920s! Since the rock was first discovered in 1961, it must have formed in 41 years or less. That's not as surprising as it sounds. The specific kind of rock in the picture is called a **concretion** (kahn kree' shun), which forms when sediments collect around an object and then undergo lithification. Under the right conditions, this can happen quickly. Also, the right kind of object helps the concretion form. For example, the spark plug rusted as the sediments collected, and the particles of rust sped up the lithification process.

This concretion, a type of sedimentary rock, surrounds a spark plug!

The Relative Age of Rocks

Since rocks can hold things from the earth's past, it would be nice to know how old rocks are so we know the age of the things they contain. Unfortunately, that's not easy to figure out. Nevertheless, there are some methods that earth scientists can use to infer how old a rock is. For example, sometimes when a volcano erupts, the eruption is recorded in history. For example, we know that Mt. Vesuvius erupted in AD 79, because it was documented by a Roman official, Pliny the Younger, whose works are a great source of history. Igneous rock from lava that flowed during the eruption and sedimentary rock formed from ash emitted during the eruption can be identified, so the age of those rocks are well known.

However, that doesn't happen very often. As a result, earth scientists must use other methods to estimate the age of the rocks they are studying. As time goes on, you will learn about several of those methods, but for now, let me concentrate on a general principle that most earth scientists use: the **Principle of Superposition**.

Principle of Superposition – In a series of rock strata, the oldest stratum is at the bottom, and the rocks get progressively younger the higher they are in the series.

This should probably make sense. After all, if you first consider sedimentary rock, sediments must accumulate before they can undergo lithification, and they must accumulate on something that is already there. Thus, for any stratum of sedimentary rock, whatever it is sitting on top of must have already been there. That means the rock it is sitting on top of must be older. This is also the case for extrusive igneous rock. After all, lava must flow onto something, and once it cools into rock, the underlying rocks must have already been there, so they must be older.

While it makes a lot of sense, the principle of superposition isn't always true. If cracks occur in a series of strata and magma fills those cracks and cools, the intrusive igneous rock formed is younger, even though it will be above many of the strata in the formation. More importantly, when very strong natural forces are at work (in an earthquake, for example), a series of strata can be moved so that the older layers end up on top of the younger layers. This results in **overturned strata** that go against the Principle of Superposition. But overturned strata are rare, and there are usually ways to see that they have been overturned.

A more serious challenge to the Principle of Superposition is that there are times we know strata of sediments were laid down simultaneously. For example, look at the picture below. It shows a wall of sediment that is roughly 8 meters (25 feet) high. Notice the strata you see in the wall. While the Principle of Superposition says that the lower strata are older than the upper strata, we know that's not true, because this entire wall of sediment formed in just three hours, from 9 PM to midnight on June 12, 1980. It happened when Mount St. Helens erupted in the state of Washington, but the wall is not made of igneous rock. It is made of sedimentary rock formed from the ash and water emitted by the volcano.

This wall of stratified sediment formed in a mere 3 hours, which means that all the sediments in the strata are essentially the same age.

For all practical purposes, then, the rocks in those strata are all the same age. Now, of course, we know this because we were able to see the area before and after the wall of sediment was formed. We don't normally have that luxury. When you look at a series of strata, then, it is tempting to apply the Principle of Superposition. However, you must be careful when you do that, because we know there are times when it is not true.

Really Useful Rocks

Before you finish this chapter on rocks, I need to discuss some of the ways that we use them. Remember that diamonds are minerals, but they must form deep in the earth, where there are very high temperatures and pressures. When the magma they form in rises through narrow cracks in the crust, it forms kimberlite, an intrusive igneous rock pictured on page 100. That kimberlite is mined in order to get the diamonds formed in the magma. Granite is another useful igneous rock. Because of its interesting colors, it is cut and polished to make countertops, flooring, etc.

Metamorphic rocks are also mined as a source of metals and precious gems. However, there are a lot of metamorphic rocks that are useful because of their characteristics. Marble, for example, is harder than the limestone out of which it is formed, but as rocks go, it is still pretty soft. Thus, it can be cut and shaped fairly easily. That's why lots of statues are made out of marble. Nevertheless, it is still hard enough that it can be cut into slabs and blocks and used for building.

Nowadays, you don't see a lot of blackboards, but they used to be very important in classrooms. For a long time, paper and ink were expensive, so students would use chalk (a sedimentary rock) to write on slate (a flat metamorphic rock that is often black). Each student would do his or her lessons on a small piece of slate, usually framed in wood. The slate was cheap and could be reused, because the chalk could be erased from the slate.

In 1801, West Point Military Academy mathematics teacher George Baron hung a large slab of slate at the front of his classroom and used it to teach algebra to his students. It quickly caught on, and by the middle of the 1800s, most classrooms had large slabs of slate, which were called blackboards. Teachers could write on the blackboards with chalk, and all the students could easily see what was being written. Teachers could then erase what they wrote so that they could continue with another lesson.

Blackboards were eventually replaced with cheaper surfaces made of coated metal. They were often green and became known as chalkboards. That's what I saw in most of my classes when I was growing up, and even when I started teaching at the university level, we still used green chalkboards. Of course, these days we have either digital displays or whiteboards that we write on with erasable markers. Nevertheless, it all started with slate, a metamorphic rock. Slate is still useful in other situations. For example, it is a common material used to make floors and roofs.

Sedimentary rocks are valuable in many different ways. First, they often contain useful materials. Oil and natural gas, for example, are found mostly in sedimentary rocks. Also, many of the minerals we use, like gypsum, are found in sedimentary rocks. Most of the water that we drink is flowing underground, through sedimentary rock.

Gravel is a mixture of rock pieces, most of which come from sedimentary rock. While gravel is technically a mixture of many different kinds of rocks, limestone is often mined and crushed to make a lot of gravel. Gravel walkways and driveways are the most obvious uses of gravel, but there are many others. For example, it is often used as a base. Look at the picture on the right. You might think it is a gravel walkway, but it's not. It's a paved walkway being constructed. The gravel is laid down first, and then the concrete or other paving material is poured on top of the gravel. The specific gravel in the picture is made from limestone.

The gravel in this picture is used as a base. Cement will be poured over the gravel to make a paved walkway.

One of the most important sedimentary rocks in today's economy is coal. As you already learned, it is formed when carbon particles that are produced by decaying plants accumulate and undergo lithification. It is important, of course, because it can be burned. In other words, coal is a flammable rock! The simplified chemical equation that represents the process of burning coal is:

$$C + O_2 \rightarrow CO_2$$

When you burn something, you react it with oxygen. The chemicals produced (in this case, CO_2) contain less energy than the fuel (in this case, C) and the oxygen. As a result, energy is released.

While coal is an organic sedimentary rock, it can also become a metamorphic rock. In its metamorphic form, it is called **anthracite** (an' thruh site') **coal**, and it is the most valuable. Why? For two reasons. First, it releases the most heat per gram of rock. It costs money to mine and transport rocks, so you want to get as much out of the rocks you mine as possible. Since coal is used for energy, you want to mine coal that produces the most energy per gram of rock. Also, there are very few chemicals in the rock other than carbon. Thus, when it burns, it mostly produces carbon dioxide.

Anthracite Coal | Bituminous Coal | Sub-bituminous Coal | Lignite Coal

The sedimentary forms of coal are much more common. **Bituminous** (bih tyoo' muh nus) **coal** contains the most energy per gram of all the sedimentary forms, and it is much more common than anthracite coal. Unfortunately, the amount of carbon in the coal is lower, which means there are other chemicals in it. Why is that a problem? When you burn coal, you burn those chemicals as well, and sometimes, that produces pollution. Bituminous coal contains sulfur, and when sulfur burns, it makes SO_2 or SO_3, depending on the conditions. At high concentrations, those chemicals are poisonous to people, so burning bituminous coal produces more pollution than burning anthracite coal. There are chemical processes that can remove some of the sulfur from the coal to reduce the pollution, and it is also possible to get rid of some of the SO_2 or SO_3 before it is released into the air, but there is no way to eliminate the pollution entirely. There are two lower grades of sedimentary coal: **sub-bituminous coal** and **lignite coal**. Sub-bituminous contains less carbon than bituminous, and lignite contains even less. Of course, the less carbon there is, the more sources of pollutants there are. Nevertheless, all four forms of coal are burned for energy.

Comprehension Check

4.14 While it is rare, some igneous rocks do have fossils in them. Are they extrusive or intrusive?

4.15 As mentioned in this section, we only know trilobites from their fossils. Does that mean we know for certain that there are no trilobites alive today?

4.16 Fossils of placoderm fish are found in sedimentary strata that are lower than the strata in which fossils of trees are found. Using the Principle of Superposition, which existed on earth first? Can you be certain of that conclusion?

4.17 If burning coal makes NO and NO_2, what element must also be in that coal?

Answers to the Comprehension Check Questions

4.1 Weathering of igneous rock would turn it into particles. If those particles stuck together, the result would be sedimentary rock. If you didn't get the answer to this one, don't worry. Later on, we will talk about the rock cycle, which involves this process. I just wanted to get you thinking about it now.

4.2 a. I and III would have the smaller crystals. Small crystals are formed when minerals precipitate quickly. That involves rapid cooling, which usually happens for lava. Since I and III are extrusive, they are the ones that formed from lava. The other term is not related to the size of the mineral crystals. It is related to the amount of silicate minerals in the rock.

b. II, IV, I, III. The lowest percentage of silicate minerals is found in ultramafic rock, mafic rock has a higher percentage, intermediate an even higher percentage, and felsic has the highest percentage. This applies to both extrusive and intrusive igneous rock.

c. II and IV. Intrusive igneous rocks are formed underground, because they are formed from magma, which is (by definition) underground.

4.3 It probably would have felt smoother than either of them. Since clay is made of even smaller particles, the shale would be smoother than siltstone.

4.4 They are chemical. If water is evaporating, that means the particles are precipitating out, because there is less room for them in solution.

4.5 The mylonite is obviously foliated. The bands are a dead giveaway. It turns out that the slate is foliated as well, but you can't see that from the picture.

4.6 You would have made a contact metamorphic rock. With added pressure, it would have been regional. With pressure and no heat, it would be dynamic.

4.7 Exposure to heat and/or pressure, then weathering, then lithification. Exposure to heat and/or pressure converts the igneous rock into metamorphic rock. Then you need to make sediments from it, which is what weathering does. Then the sediments need to harden, which is the process of lithification.

4.8 Rock B has probably experienced more weathering. Weathering tends to smooth rocks down.

4.9 You cannot. Rock B experienced more weathering, but that just means it was exposed to a lot more water or wind than rock A. If it is from a very different area than Rock A and was exposed to different amounts of rainfall, wind, etc., the increased weathering could have happened in a very short time.

4.10 There are three. A bedding plane separates one stratum from another. One bedding plane will separate layer 1 from layer 2, a second will separate layer 2 from layer 3, and a third will separate layer 3 from layer 4.

4.11 There is one. A nonconformity exists between either igneous or metamorphic rock and sedimentary rock. Since there is only one layer of igneous rock, there is only one separation between it and the next layer.

4.12 There are none. An angular unconformity exists between tilted strata and horizontal strata. Planar bedding means horizontal strata, so there are no tilted strata.

4.13 Determine the kind of rock that it contains. Sedimentary rock is from sediments being pushed down, while igneous rock is from magma being pushed up.

4.14 They are extrusive. Intrusive rocks are formed from magma, which is underground. It wouldn't be likely to capture living organisms. Even if it did, remember that it cools slowly, so it would destroy the organism's remains before becoming solid.

4.15 No. It only means we haven't found any living trilobites. They may be around, however. There are several examples of animals and plants that were initially known only from their fossils but were later found to still be alive. The coelacanth (a fish), the Wollemi pine (a tree), and the tuatara (a lizard) are examples.

4.16 Placoderm fish existed on earth first, but you can't be sure of that. Since their fossils are found in lower strata, The Principle of Superposition says they are older. That means they were on earth before trees. However, that principle is known to be wrong in certain cases.

4.17 It must have N (nitrogen) in it. You only needed the letter, not the name. Remember, burning coal means reacting it with oxygen. Thus, the "O" in NO and NO_2 comes from the oxygen that is reacting to make the burning process happen. However, any other element must come from the coal itself.

Chapter Review

1. Define the following terms:

a. Igneous rock
b. Sedimentary rock
c. Metamorphic rock
d. Rock cycle
e. Lithification
f. Bedding plane
g. Unconformity
h. Disconformity
i. Angular unconformity
j. Nonconformity
k. Paraconformity
l. Principle of Superposition

2. You have the following set of rocks: (I) mafic extrusive igneous rock that has holes and pits in it (II) felsic intrusive igneous rock that has no holes or pits in it, (III) intermediate extrusive igneous rock that has no holes or pits in it, and (IV) ultramafic intrusive igneous rock that has no holes or pits in it.

a. Which one has the most silicate minerals in it?
b. Which has the least silicate minerals in it?
c. Which ones do you expect to have the larger mineral crystals?
d. Which one had lots of gas in it when it was still molten rock?
e. Which ones formed above ground?

3. You have a conglomerate, a siltstone, and a sandstone. Which is probably the smoothest? Which is probably the roughest?

4. Rock A is composed mostly of chemicals that have precipitated out of water. Rock B is composed mostly of sediments that came from weathering and erosion. Rock C is composed mostly of carbon particles from the decayed remains of organisms. Identify each as clastic, chemical, or organic.

5. A layer of limestone is just under the soil, but there is a pocket of magma below it. The limestone right above the magma goes through metamorphosis. Is a regional, contact, or dynamic metamorphic rock formed? If that layer had been under many other layers of rock, how would that change the answer?

6. In the rock cycle, what must happen for a sedimentary rock to become an igneous rock?

7. If a metamorphic rock is weathered, what kind of rock can form from the remains?

8. Sediments that were originally part of an igneous rock are now in a sedimentary rock. There are two possible ways that could happen. What are they?

9. Contrast how those who are guided by uniformitarianism and those who are guided by catastrophism view the rock cycle.

10. Can we say for certain how long the lithification process for a specific sedimentary rock took? Why or why not?

11. There are two basic kinds of bedding planes. What are they and what is the difference between them?

12. If you are looking at strata with planar bedding and see an obvious unconformity, what kind of unconformity is it?

13. You are examining what looks to be a conformal bedding plane. What could it be instead of a conformal bedding plane?

14. You see folded strata with very few cracks or breaks in them. Most likely, did lithification occur before or after the folding?

15. A dike is formed when the material that makes up the rock falls down through a crack in the underlying rock. Is this dike made of sedimentary or igneous rock?

16. Which of the three types of rock is most likely to contain fossils?

17. You are looking at a series of sedimentary rock strata. The lowest strata contain fossils of trees, the middle strata have fossils from the same kind of trees as well as bird fossils, and the upper strata have only mammal fossils. According to the Principle of Superposition, which fossils are the oldest? Which are the youngest?

18. Can you be sure of the conclusion you gave for 17?

19. Which of the four types of coal produces the least amount of energy per gram when it is burned?

20. Which of the four types of coal produces the least amount of pollution when it is burned?

21. CO_2 is the only product when carbon is burned completely. Why does burning coal produce chemicals like SO_2, SO_3, NO, and NO_2?

22. Is coal always an organic sedimentary rock?

Chapter 5: The Lithosphere

Introduction

In the previous chapters, you learned about the minerals and rocks that make up the earth's crust. Now it's time to look at the crust as a whole. To do that, however, we need to back up and look at the geosphere in more detail, as shown in the illustration below. Remember that the geosphere includes the crust, mantle, outer core, and inner core. They are shown on the lower left side of the illustration in a cutaway drawing of the earth. However, in the expanded drawing to the right, there is a bit more detail. At the top, of course, is the crust. Notice, however, that the crust can be split into two different sections. If the crust is under the ocean, it is called **oceanic crust**. If it is not under the ocean, it is called **continental crust**. As you will see in a moment, it's not just their positions that distinguish one from the other.

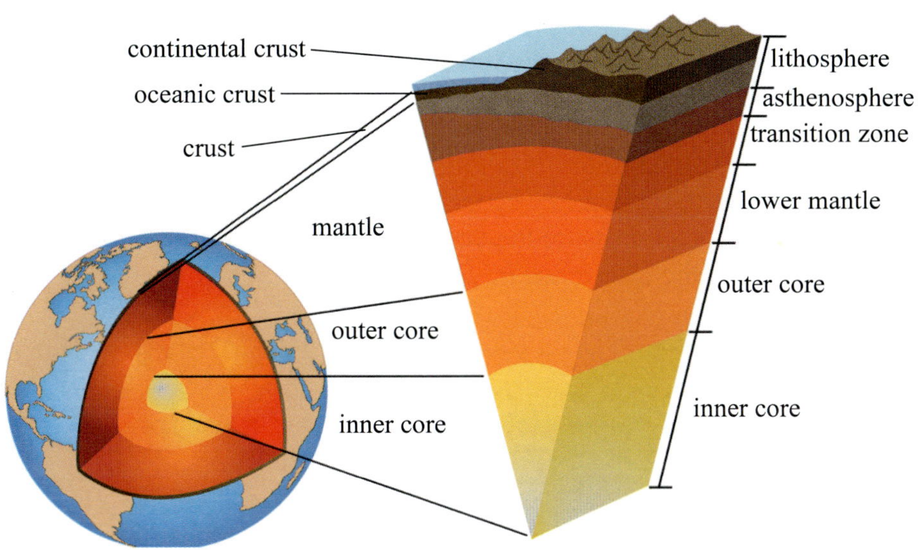

The geosphere is made up of the crust, mantle, inner core, and outer core, but the crust and mantle can be classified in more detail.

Below the crust, we find the mantle. We know it is different from the crust because the behavior of seismic waves changes abruptly when they pass into this region. This was discovered by Croatian seismologist Andrija Mohorovičić (moh huh roh' vuh chik) in 1909, so the boundary between the crust and this layer is called the "Mohorovičić discontinuity," which is mercifully referred to as the "Moho" by most geologists. The crust and the portion of the mantle directly below the Moho compose the **lithosphere** (lith' uh sfear'), which will be the focus of this chapter.

Lithosphere – The crust and the portion of the mantle directly below the Mohorovičić discontinuity

Below the lithosphere, we find the **asthenosphere** (as then' uh sfear), which is very similar to what is directly below the Moho. However, its behavior is quite different. In fact, it is sometimes referred to as "plastic rock." It is not rigid, but it is also not liquid. Have you ever used a soft, flexible putty, like Silly Putty? It is not exactly solid, but it is definitely not liquid. Instead, it can be shaped and molded. The behavior of the asthenosphere is similar.

Below the asthenosphere is a transition zone that eventually leads to the lower mantle, where the rock is not nearly as "plastic" as the upper portions of the mantle. So while it is correct to say that the geosphere is made up of four spheres (crust, mantle, outer core, and inner core), the mantle is complex. It contains at least four distinct regions: the rock just below the Moho, the asthenosphere, the transition zone, and the lower mantle.

One Big Difference Between Oceanic and Continental Crust

Before I discuss the lithosphere, I need to discuss one major difference between oceanic and continental crust – their densities. Remember that density tells you how tightly packed matter is in an object. The higher the density, the more matter is packed into a given volume. To measure an object's density, you need to know its mass and its volume. I want you to get some experience with this, so I want you to measure the densities of two rocks in your kit.

But wait a minute. How do you measure the volume of a rock? If it were shaped like a box, you could measure its length, width, and height and multiply those measurements together to get the volume. But your rocks aren't shaped like boxes. That's okay, however, because you can measure the volume by **displacement**. If an object (like a rock) sinks in water, it pushes the water molecules out of its way, taking their place. The object will displace a volume of water molecules equal to its own volume. So if you measure the volume of water, and then measure the volume of water plus the sunken object, the difference in volume will be the volume of the object.

Experiment 5.1: Density of Rocks

Supplies:

- The two rocks from the laboratory kit made for this course (They are in the plastic bag labeled "Density Rocks.")
- The graduated cylinder from the laboratory kit made for this course (The plastic cylinder with mL markings on it)
- The mass scale from the laboratory kit made for this course
- A pencil
- A small glass, like a juice glass
- A medicine dropper from the laboratory kit made for this course
- Water

There are sample calculations for this lab at the end of the chapter, right before the answers to the "Comprehension Check" questions.

Instructions:

1. Fill the glass ¾ full with water.
2. Look at the graduated cylinder. It has markings that correspond to volume in milliliters (mL). Each large line represents a multiple of 10, and each small line is a single mL.
3. Put some water in the graduated cylinder so that the water level is near the 30 mL mark. To read it properly, sit the graduated cylinder on a level surface and position your head so that your eyes are at the same level as the water level in the cylinder. That way, you are looking at it straight on. You might notice that the water level curves down slightly. If so, read the water level from the lowest point on the curve.
4. Use the medicine dropper to pull water out of the glass and put it in the graduated cylinder slowly. Do this until the level of the water (or the bottom of the curve if you see one) is right at the 30 mL mark.
5. Make sure the mass scale has batteries and that they are properly installed.
6. Set the scale on a flat surface and hit the "on/off" button to turn it on.
7. Look at the display. In the upper, right-hand corner, the unit is displayed. If it doesn't show a "g" for grams, hit the "mode" button until it does.
8. Hit the "tare" button, which resets it to zero.
9. Take the "Density Rocks" out of their bag.
10. Put one of the rocks on the scale and record the mass that the display gives you.
11. Remove that rock, hit the "tare" button to zero the scale, and repeat step 10 with the other rock.

12. Take the rock that has more mass and gently slide it into the graduated cylinder so that no water splashes out. If it doesn't fall to the bottom, use the point of the pencil to push it down until it is completely covered by water.
13. Read the new volume based on the new water level. You can do this by seeing where the bottom of the water level is. If it is 4 lines above 30, for example, the new volume is 34 mL. If it is in between two lines, try to estimate. For example, if it is halfway between the fourth and fifth lines above 30, the volume is 34.5 mL. If it is very close but not quite at the fourth line above 30, it is 33.9 mL. The difference between this reading and 30 mL is the volume of the rock. Record that.
14. Empty the graduated cylinder. If the rock is stuck, just tap the open end of the graduated cylinder against the bottom of the sink a few times. The rock should come out when you do that.
15. Repeat steps 3 and 4 so that you have the water level at the 30 mL mark again.
16. Repeat steps 12-14 for the other rock. Once again, make sure the rock is completely covered with water before you read the new volume. You may have to use the pencil to push it under the water.
17. Use the masses and volumes you recorded to calculate the density of each rock.
18. Drop each rock into the glass of water and see what happens.
19. Clean up your mess.

Did it surprise you when one of the rocks floated in water? If you had been thinking about the density, it shouldn't have. After all, the density of water is about $1\ \frac{\text{g}}{\text{mL}}$, and the density of the second rock should have been less than that. Thus, you should have expected the second rock to float. Nevertheless, you don't normally think of rocks floating in water, so it might have surprised you a bit.

Interestingly enough, both of the rocks in your experiment were igneous. The one that sank was granite, an intrusive igneous rock. The other was pumice, an extrusive igneous rock. Remember that pumice has pits and holes in it, caused by the bubbles that formed as the lava cooled. Those pits and holes make the matter in pumice pretty loosely packed, which is why its density is so low. In fact, when volcanoes under the ocean erupt, the lava can form pumice, which floats to the top of the ocean, forming a floating "island" of rock, which is often called a "pumice raft." The course website (discussed in the introduction to this book) has a couple of videos that show pumice rafts.

As the experiment shows, then, the density of a rock depends on how it was formed. It also depends on the chemicals out of which it is made. The density of any part of the earth's crust, of course, will depend on the density of the rocks it contains. However, on average, the density of oceanic crust is about $3.0\ \frac{\text{g}}{\text{mL}}$, while the density of continental crust is, on average, about $2.7\ \frac{\text{g}}{\text{mL}}$. In other words:

Oceanic crust is denser than continental crust.

As you will learn later on, that ends up being important when it comes to understanding what happens when oceanic and continental crust interact with each other.

Comprehension Check

5.1 The density of rocks varies, but on average, you can compare the densities of igneous, metamorphic and sedimentary rock. Keeping in mind that I am asking about the *average* rock, which of the three types would you think has the lowest density? Which has the highest density?

A Surprising Thing About the Continents

You might not have found it surprising that pumice floats in water. However, once it is explained to them, most students find the image below to be more surprising. It is a map of the world, as constructed by Google. Employees at NASA's Jet Propulsion Laboratory added the green dots and yellow lines to it. Each dot represents a Global Positioning System (GPS) device that is fixed in place. The GPS reads the position of each device, which is given by the green dots. The yellow lines represent *the speed and direction at which each **fixed** GPS device is traveling*! Think about what you just read. Each green dot represents a GPS device that is sitting at one place on the earth. Nevertheless, each device is moving with a speed and direction given by the yellow lines!

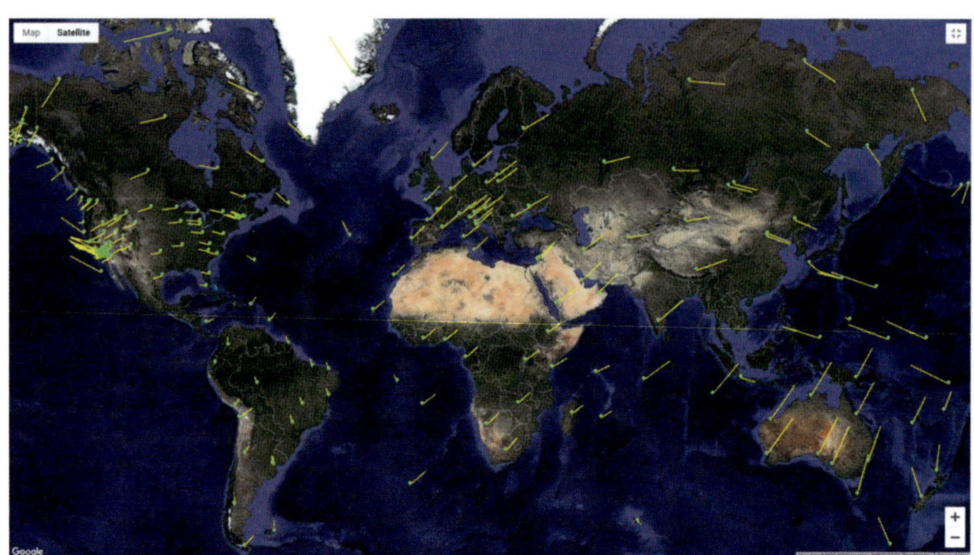

The green dots on this Google map represent fixed GPS devices, and the yellow lines indicate their motion. Note that the green dots that appear to be in the middle of the ocean are on islands. The islands are just too small to see at this scale. (Image courtesy NASA/JPL-Caltech, Google).

Why are the *fixed* GPS devices moving? Because *the land upon which they are siting is moving.* Look, for example, at North America. Notice that all the yellow lines are pointing to the left. Some are pointed up and to the left, but most are pointing down and to the left. Now look at the yellow lines on the Asian continent. They are all pointing to the right. Once again, some are pointing up and to the right, but most are pointing down and to the right. What does that tell you? The North American west coast and the Asian east coast are *moving closer to one another*!

Why don't you notice this motion? First, you are moving right along with the land. Thus, as far as you are concerned, the land isn't moving. Second, it is moving very slowly. Because of the precision of the GPS, we can actually calculate the speed of each GPS device. The GPS device that is sitting in Pittsburg, Pennsylvania, for example, is moving at a speed of $1.4 \frac{\text{cm}}{\text{year}}$. In other words, while Pittsburg, Pennsylvania is moving up and to the left, it takes a full year for it to travel a mere 1.4 centimeters. While that's *really* slow, it is measurable with the GPS!

The idea that the earth's land is moving seems outrageous, but scientists figured it out long before the GPS was made. Indeed, the first person to seriously promote it was Antonio Snider-Pellegrini, a scientist who studied geography in the mid-1800s. He noticed that the continents of South America and Africa fit together like puzzle pieces. He also noticed that distant parts of the world like North America and Europe had rocks and fossils that were very similar. He reasoned that at one time, all the continents of the world were originally part of a single continent, and that what we see today is the result of that continent being broken apart. What could have caused such a thing to happen? Pellegrini suggested it was the global Flood recorded in the Bible.

Like many revolutionary scientific ideas, Pellegrini's hypothesis was mostly ignored or ridiculed at the time. However, other scientists noticed the same things Pellegrini did and started doing more investigation. One of them was Alfred Wegener (vey'guh nuhr), a **meteorologist**, which is a scientist who studies weather. He noticed that not only did the continents seem to fit together like puzzle pieces, but there were many, many examples of rocks and fossils that lined up perfectly if you put the "puzzle" together. The rocks in Brazil, for example, matched perfectly with the rocks of southern Africa. He noticed that he could match plant fossils in the same way. Although there is no indication that he knew of Pellegrini's work, he came to the same conclusion as Pellegrini. He said that at one time, the continents were all part of a single continent that he called **Pangaea** (pan jee' uh), and over time, they drifted away from one another. In 1915, he published a paper detailing his hypothesis, which he called **continental drift**, and the evidence that supported it.

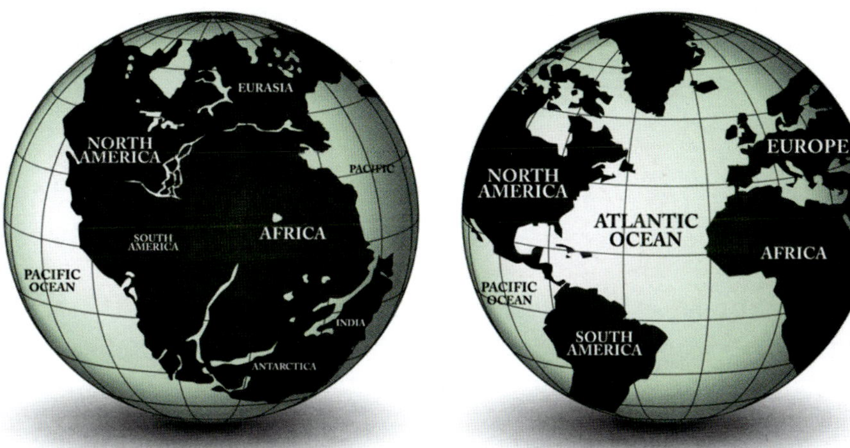

On the left is Pangaea, the supercontinent that evidence suggests was once formed by all of today's continents (shown on the right) fitting together.

While the evidence for continental drift was strong, the problem was that no one (including Wegener) had any idea what could cause the continents to move like that. Pellegrini suggested it was the catastrophic forces of the global Flood, but most scientists in the late 1800s and early 1900s didn't like the idea of using the Bible in science. Wegener himself was one of them. He thought that continental drift happened very, very slowly over millions of years. However, he couldn't come up with any explanation of how the continents would drift, so once again, the idea of the continents moving fell out of favor among scientists.

Over time, however, more evidence became available, forcing scientists to once again consider the idea that the continents might be moving. One major piece of evidence deals with the earth's magnetic field. To better understand the evidence, perform the following experiment.

Experiment 5.2: Magnetic Alignment

Supplies:

- The magnet from the laboratory kit made for this course
- A sewing needle that is strongly attracted to the magnet (most metal ones will be)
- Waxed paper or a plastic sandwich bag
- Scissors
- A casserole dish or a large bowl

Instructions:

1. Put the sewing needle on the magnet. Set them aside with the needle attached to the magnet.
2. Add water to the casserole dish until it is mostly full.

3. Use the scissors to cut a small rectangle of waxed paper or plastic from the plastic bag. It needs to be about as long as the needle and wide enough so that it can act as a "raft," allowing the needle to float on water (see the picture below).

4. Set the rectangle of waxed paper on the surface of the water in the casserole dish so it floats.
5. Pull the needle off the magnet, and carefully place it on the waxed paper rectangle so that it floats as shown in the picture on the left. Note that I used colored plastic wrap so it is easier to see.
6. Allow the water to settle down, and watch how the needle behaves. If the needle floats on its "raft" to the side of the dish, gently push it more towards the middle. Eventually, it should stop moving.
7. Once it has stopped moving, note the direction in which the pointed end of the needle is pointing.
8. Bring the magnet close (but not too close) to the needle. If you bring it too close, the needle will jump off the "raft" and onto the magnet, which means you have to go back to step 5. As you bring the magnet close, the needle should start to move in response. Play around with the magnet and needle so you have a good idea of how the needle moves in response to the magnet.
9. When you have figured it out, use the magnet to make the needle spin so that its pointed end is facing in a different direction from the direction it faced in step 7.
10. Move the magnet far from the needle so that it doesn't affect the needle anymore, and watch how the needle moves. Once it stops moving, note the direction in which the pointed end is pointing.
11. Repeat steps 9 and 10.
12. Clean up your mess. Put the needle back on the magnet and leave it there, putting them both someplace handy. You will use the same setup again the next time you do science.

What happened in your experiment? The needle was made of steel, which is ferromagnetic. Since it sat on the magnet for a while, it became a magnet. When you put it on the "raft," it was free to move around, and it started responding to the earth's magnetic field, aligning itself with the field. In other words, you made a crude compass. The needle might not have pointed north like a real compass because of the way it was magnetized. Nevertheless, it did end up pointing in a specific direction, and that direction was determined by how it responded to the earth's magnetic field.

When you brought the magnet close to the needle, its magnetic field was able to overcome the earth's magnetic field, and you were able to use it to point the needle in a different direction. However, what happened when you moved the magnet away? The needle should have returned to pointing in its original direction. That's because as long as there aren't any stronger magnets around, a magnet that is free to move will align with the earth's magnetic field. Had the needle not been floating on the "raft," it wouldn't have been free to move. However, since it was, the needle ended up moving so that it pointed in the same direction once you moved the magnet away. You will learn how this relates to the motion of the continents the next time you do science.

Comprehension Check

5.2 Suppose you had a compass and brought the magnet from your kit close to the compass. What would happen to the compass needle? What would happen when you pulled the magnet far from the compass? If you have a compass, feel free to test whether or not your answer is correct.

The Earth's Magnetic Field

You probably already knew that the earth has a magnetic field, because you have probably played with a compass or perhaps even used one to help you determine what direction you were traveling. Your experiment demonstrated that magnetic materials move around in response to that field, as long as there aren't other, stronger magnets in the area. I will tell you how that influenced our understanding of the way the continents move in a moment. However, since I am talking about the earth's magnetic field, I want to discuss it in detail so that you learn how important it is.

The drawing below is one way to represent the earth's magnetic field. Before I discuss the lines you see in the picture, I want you to understand a couple of things about the drawing. The big bar running down the center of the earth is one way to depict a magnet. When something is magnetic (like the needle in your experiment), it has two poles: a north pole and a south pole. The bar in the illustration, then, represents a substance that is magnetic. This is *not* the reason the earth has a magnetic field. It is not made of magnetic material. You will learn why the earth has a magnetic field later. Nevertheless, it is useful to visualize the earth's magnetic field this way, since the earth's magnetic field behaves as if it were made by a huge magnetic bar that runs from near the North Pole to near the South Pole.

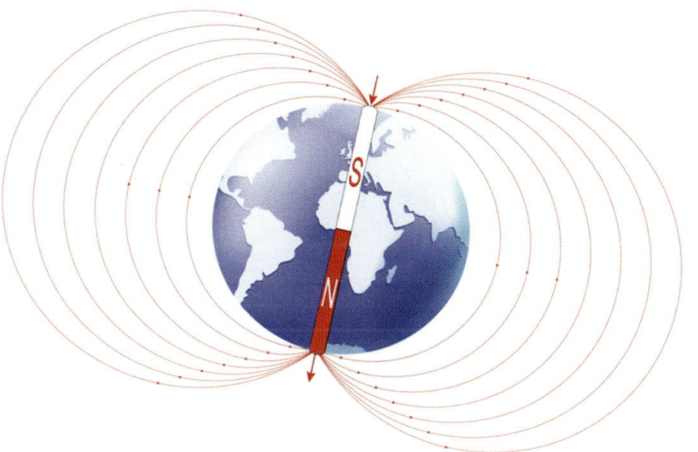

This is one way to illustrate the earth's magnetic field.

Notice also that the magnet is oriented so that its south pole (the big "S") is at the earth's North Pole. Why? The Basic Law of Magnetism says that opposite poles of magnets attract one another, while like poles repel. So which way will the north pole of a compass's magnet point? It will point to the south pole of the earth's magnetic field. In other words, the north pole of a compass needle points north because it is attracted to the earth's south magnetic pole! From a geographic standpoint, however, it's the north end of the earth, so we call it the North Pole, even though it is the south pole of the earth's magnet.

The lines in the illustration represent the earth's magnetic field. You might have seen or done an experiment where you sprinkled bits of iron on a piece of paper that had a magnetic bar underneath. If you have, you saw that the iron tended to form lines like you see in the illustration. Those lines illustrate the magnetic field's influence. The arrows tell you the direction that a magnet's north pole will point if placed on any line, as long as the magnet is free to move. Notice that while the lines curve, they all point north, indicating that any magnet placed in the earth's magnetic field will point its north pole north, as long as it is free to move.

The earth's magnetic field is very useful, because it allows us to navigate. Before the Global Positioning System (GPS), a compass was considered an essential tool when it came to figuring out where you were and what direction you were heading. Indeed, compasses are still very important for people who are traveling and cannot access the GPS. Interestingly enough, there are a lot of animals that use the earth's magnetic field in the same way. Birds and butterflies that must make long journeys, for example, have organs that act like a compass, sensing the earth's magnetic field and telling them which way is north. Salmon, which leave the freshwater stream where they are born, travel to the ocean to mature, and then return to the same stream to spawn, sense the earth's magnetic field in order to help them in their navigation.

More importantly, however, without the earth's magnetic field, there would be no plants, animals, or people on earth. That's because the sun, whose energy is essential for supporting life, also bombards the earth with high-energy charged particles that would destroy life if a sizeable percentage of them were allowed to hit the planet. The earth's magnetic field deflects those charged particles, keeping them from hitting the surface of the earth. The sun's light is not affected by the earth's magnetic field, so it travels unhindered to the surface of the earth, providing life with the energy it needs.

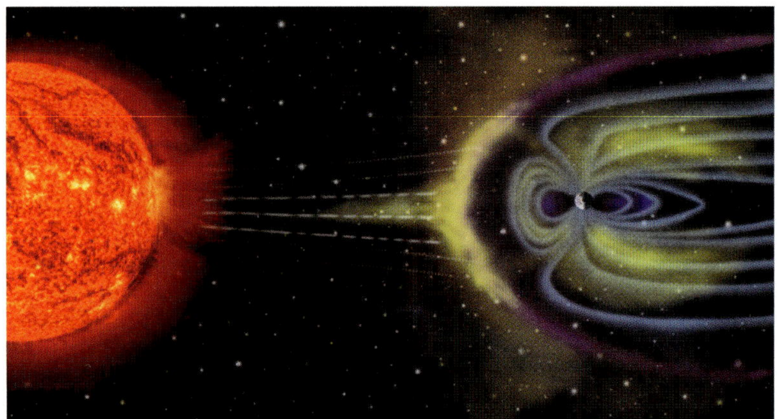

The earth's magnetic field shields us from high-energy charged particles coming from the sun. This distorts the field.

The drawing on the left illustrates this process. The dashed lines represent the charged particles that would hit earth, and the blue lines represent earth's magnetic field. The glowing regions show the interaction between those charged particles and the magnetic field. Notice how the magnetic field in this illustration is not the same as the magnetic field shown in the one on the previous page. That's because the charged particles distort the field. Thus, this illustration shows a more accurate depiction of the earth's magnetic field.

As your experiment demonstrated, the earth's magnetic field isn't very strong. After all, a small magnet was able to overpower its influence on the floating needle when it got close to the needle. This is actually a good thing. Our nervous system, for example, uses moving charged particles to send important signals throughout the body. If the earth's magnetic field were too strong, it could interfere with those moving charged particles the same way it interferes with the ones coming from the sun. That, of course, would be bad.

In order to support life, then, the earth needs a magnetic field that is strong enough to deflect the high-energy charged particles that come from the sun, but it cannot be so strong that it interferes with the motion of charged particles in human and animal bodies. While some consider this just a "fortunate coincidence," I consider it just one of the many design features that demonstrate that the earth was created by God as a haven for life.

What Causes the Earth's Magnetic Field?

Now as I said, even though the illustration on the previous page represents the earth's magnetic field as a bar magnet, that's not how the earth's magnetic field is generated. To learn the actual reason the earth has a magnetic field, perform the following experiment.

Experiment 5.3: An Electromagnet

Supplies:

- All the materials from the previous experiment

- A pair of tongs that are wooden, plastic, or have plastic covers on its ends
- Aluminum foil
- Cellophane tape
- A standard battery (AA or C work best)
- A metal paper clip (It cannot be covered in plastic. It must be bare metal.)
- Someone to help you

Instructions:

1. Wrap the grips of the tongs with aluminum foil, making sure the foil doesn't touch any metal on the tongs (if there is any). Also, make sure there is enough foil so that part of it can be covered in tape with plenty of bare aluminum foil below the tape.
2. Unbend the paper clip, making it mostly straight, but with the ends bent.
3. Have your helper hold the tongs while you tape each bent end to the aluminum foil. Make sure the tape is wrapped all the way around the tongs so that it holds the ends of the paper clip securely, but make sure there is plenty of bare aluminum foil below the tape. Your helper can adjust the distance between the tongs to make this job easier, especially when you are taping the second bent end.
4. Use the needle, waxed paper, scissors, water, and casserole dish to make the same setup you did in the previous experiment.
5. Once the needle has stopped moving on the water, grab the tongs (without the battery shown in the picture) and bring the paper clip very close to the needle without actually touching the needle. Does the needle move in response? It might move a little because you moved the air around it, but it shouldn't move much at all.
6. Now fit the battery in between the tongs and squeeze them to hold it so that each end of the battery is pushing against the aluminum foil below where the tape is. The setup should look something like the picture on the right.

7. Repeat step 5, keeping the battery gripped in the tongs. You should notice a difference. If you don't, make sure the battery is good, each end of the paper clip is firmly in contact with the aluminum foil, and the battery is touching the foil below where the tape is. You need electricity to flow through the paper clip.
8. Once you have seen how you can move the needle when electricity is flowing, clean up your mess. You will not be using the materials anymore, so put everything away.

Why did the needle move when there was electricity flowing through the stretched-out paper clip? Because when electricity flows, it generates a magnetic field. This turned the stretched-out paper clip into an electromagnet, and it's how the earth generates its magnetic field. There are electrical currents that run in the earth's outer core, and those electrical currents produce the earth's magnetic field. In other words, the earth is one big electromagnet!

Comprehension Check

5.3 The electromagnet you made was pretty weak. It didn't move the needle nearly as strongly as the magnet from your kit. What could you do to increase the strength of your electromagnet? Don't try it, because depending on what you do, it could be dangerous!

How Does This Relate to the Motion of the Continents?

Remember that I started this whole discussion about the earth's magnetic field because it gave more evidence for the idea that the continents move. Think about Experiment 5.2, where you first constructed a crude compass. The needle pointed in a specific direction because it responded to the earth's magnetic field. Even when you used a magnet to change where it was pointing, once that magnet was taken away, the needle went back to pointing in the original direction. Now, of course, you used a waxed paper "raft" and water so that the needle could move easily. Thus, the experiment showed that when magnetic objects are free to move easily, they tend to point in a specific direction, which is given by the earth's magnetic field.

Now think about igneous rock forming. It starts out as molten rock, right? As the molten rock cools, minerals precipitate out of it. Suppose there was a sample of magma that started cooling, and the mineral magnetite precipitated out. The rest of the magma is still liquid, right? The magnetite is a solid that has come out of the liquid. While it is in the liquid magma, it is free to move. Magnetite is a ferromagnetic mineral, so what will happen? It will line up with the earth's magnetic field.

Once the magma cools to the point where it becomes solid, the magnetite can't move anymore, so it is "locked in" to its position, pointing in the direction of the earth's magnetic field as it was when the magma froze. Think about what that means. The direction magnetite points in an igneous rock is a record of the earth's magnetic field in the location that the rock formed at the time the magma froze. If you could study igneous rock that was formed in the past, then, you could determine what the earth's magnetic field looked like in the past. This study is called **paleomagnetism** (pay' lee oh mag' nuh tiz uhm).

Paleomagnetism – The study of the magnetic properties of rocks that were determined by the earth's magnetic field at the time they were formed

Paleomagnetism was a difficult endeavor until the late 1950s, when P.M.S. Blackett (a British scientist) developed a device that could measure the magnetic properties of rocks with great sensitivity.

Because of that device, geologists were able to study paleomagnetism in detail. They looked at the magnetite in igneous rocks from around the world, trying to understand what they indicated about the earth's magnetic field over time. After all, based on the Principle of Superposition, the deeper igneous rocks should be older. True, the Principle of Superposition isn't always correct, but it can be used as a general rule of thumb in many cases. So, by looking at deep igneous rocks, we can assume that the magnetite inside was influenced by the earth's magnetic field a long time ago. The igneous rocks nearer to the surface can be assumed to have formed under the influence of the earth's magnetic field more recently.

What they found was astounding! The magnetite showed that the earth's magnetic field seemed to change dramatically over time. In deep igneous rocks, the magnetite indicated that the poles of the earth's magnetic field were in one place, but igneous rocks nearer to the surface indicated that the poles were in a different place! The igneous rocks in between usually showed a transition between the two. Even more disturbing, the igneous rocks on different continents didn't agree with one another! One continent showed the poles moving one way, while another continent showed the poles moving the other way!

How could scientists understand this? Well, it's possible that the earth's magnetic field was changing so much that the position of the poles changed dramatically over the course of time. However, if that were the case, the igneous rocks on one continent should show the same motion as the igneous rocks on another continent. The most obvious explanation was that the continents moved! Indeed, if scientists constructed a supercontinent like Pangaea and then allowed the continents to drift to their current location over time, they could explain the changes seen in the igneous rocks.

Look, for example, at the illustration below. It's a repeat of what you saw when you learned about continental drift. The drawing on the left shows the continents fitting together to form Pangaea. If you were around back then and looked at the magnetite in igneous rock at a location in what would eventually be North America, you would see the magnetite pointing north, which is shown by the white arrow near the words "NORTH AMERICA." Similarly, if you were in the location that would eventually become Africa, you would see the magnetite pointing in the direction given by the other white arrow.

On the right, the white arrows show where the magnetite in deep igneous rocks points, and the yellow arrows show where the magnetite in igneous rocks closer to the surface point. By comparing the left and right drawings, you can see how this would make sense if today's continents were formed by the breakup of Pangaea.

However, if you were to then "fast forward" to today and look at magnetite in recently-formed igneous rock in the same two locations, you would see them pointing to the north again, but because the continents moved, that direction is now given by the yellow arrows. That's what the igneous rocks show. In those two present-day locations, the deeper igneous rocks have their magnetite pointing in the direction of the white arrows, while the igneous rock nearer to the surface is pointing in the direction of the yellow arrows. This provides very strong evidence that the continents themselves actually moved over time!

It Is Not Constant

The discussion you just read assumes that the poles of the earth's magnetic field don't move. It turns out that's not true. The poles themselves do change their positions. However, in most cases, that change is small compared to the way the continents have changed their positions over time. Nevertheless, the fact that the poles change position is important. In fact, it is so important that we have to distinguish between the **geographic North Pole** and the **magnetic north pole**. The compass needle points to the magnetic north pole. Now remember, that's actually the south pole of the earth's magnetic field, but it is on the north part of the globe, so it is the magnetic north pole. The geographic North Pole is the northernmost point on the earth.

While the geographic North Pole doesn't change, the magnetic north pole does. In the 1900s, for example, the magnetic north pole was in Canada, at a latitude of 70° N. It is now very close to the geographic North Pole, which is at a latitude of 90° N. Now don't get confused here. This change in the earth's magnetic north pole cannot explain why the magnetite in igneous rock changed over time.

Those changes are more dramatic and are different in different locations on the earth. Thus, this is *another* change that also must be taken into account when trying to understand the earth's magnetic field in the past.

And, in fact, there is still another change that occurs from time to time. If you look at the igneous rock record carefully, you will find that there is sometimes an abrupt shift, and the magnetite is pointing in the opposite direction! In those cases, the magnetite indicates that the magnetic north pole is near the geographic *South Pole*. In other words, the igneous rocks indicate that at certain times in the earth's past, the compass needle would have pointed south, not north. This is called a **magnetic reversal**.

Magnetic reversal – An event in which the earth's magnetic field switched its orientation to the opposite of what it is today

In addition to the changes that occur to the poles of earth's magnetic field, scientists have been directly measuring the strength of the field since about 1850. Over the entire 170-year period of measurements, its strength has been declining. It is now about 10% weaker than it was in 1850. That might concern you, since you learned that the earth's magnetic field protects us from charged particles that come from the sun. However, the rate of decrease is small, so even if it continues at the pace we have been measuring, the magnetic field will still be strong enough to protect us for more than 2,000 years.

So, the earth's magnetic field is anything but constant. Its poles move, usually by small amounts. However, every now and again, they actually reverse. Finally, at least since we have been making direct measurements, the strength of the magnetic field is getting weaker. Nevertheless, paleomagnetism has allowed us to use this changing magnetic field to develop strong evidence that the continents are not stationary.

Learning From the Ocean

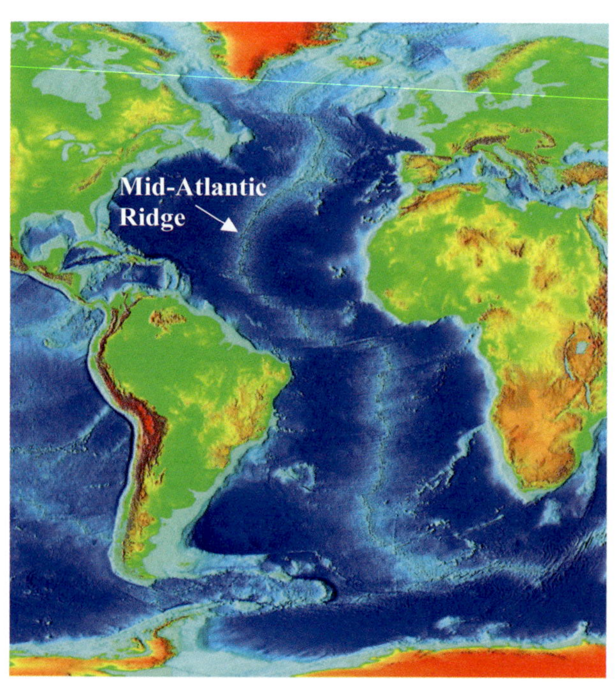

This map shows the Mid-Atlantic Ridge.

Now remember, the problem scientists had with Wegener's idea of continental drift wasn't the evidence. Wegener had geologic and fossil evidence that the continents used to be connected. The problem was explaining what could cause the continents to move. Even with the evidence gained from paleomagnetism, it was still hard to understand how such movement could be possible. However, that changed thanks to Harry Hess, a geology professor who served as the Commander of a ship in World War II. He was very interested in everything related to the earth, so when his ship traveled from battle to battle, he would use his ship's equipment to map the bottom of the ocean. After the war was over, the U.S. Office of Naval Research extended his studies, and the result is a map like the one shown on the left.

Notice the line snaking down the center of the Atlantic Ocean. It's an underwater mountain range called the **Mid-Atlantic Ridge**. Notice how it seems to follow the shapes of the continents. It was surprising enough to find a vast mountain range at the bottom of the ocean, but even more surprising was that in the center of that range had a valley running through it! We now call that valley the Great Global Rift, and it is one of many **ocean rifts** that can be found on the seafloor.

The Mid-Atlantic Ridge and the rift at its center aren't the only surprising features found in the ocean. Once you get away from the mountains of the ridge, the depth of the ocean levels off. There are still mountains and valleys, but compared to the ridge, it is pretty flat. In fact, it is often called the **abyssal** (uh bis' uhl) **plain**. The word "abyssal" refers to the deep ocean, so it's like the plains you find on land, but it's deep in the ocean. However, once you come close to the continents, you can find deep trenches, like the Puerto Rico Trench shown on the right. Its deepest point is several kilometers below the abyssal plain. In fact, the bottom of that trench is the deepest part of the Atlantic Ocean, and it is near the continent of North America. Why is the deepest part of the Atlantic Ocean so close to a continent? Similarly, why is the deepest part of the Pacific Ocean in the Mariana Trench, which is near the continent of Asia?

This map shows the Puerto Rico Trench, which holds the deepest part of the Atlantic Ocean.

There is one more puzzling feature found at the ocean's bottom. The parts of the abyssal plain near the Mid-Atlantic Ridge do not have much sediment on them. However, as you get farther away from the ridge, you find more sediment. The most sediment is found near the continents. We know that sediment accumulates at the bottom of any body of water over time. Why hasn't much sediment accumulated near the ridge, even though a lot has accumulated near the continents?

Harry Hess looked at all these features and came up with an amazing suggestion. In 1962, he published a book entitled, *The History of Ocean Basins*. In that book, he suggested that magma slowly rises from rifts like the Great Global Rift, forming new igneous rock that forms the mountains found in ridges like the Mid-Atlantic Ridge. Since this is a continuous process, igneous rock would constantly form, pushing against the old igneous rock. This would cause the old igneous rock to move away from the ridges where it was formed, making new oceanic crust. That new oceanic crust would "spread out," pushing on anything in its way, including the continents!

So, in Hess's view, the continents were moving apart from each other because the seafloor was spreading out. Now remember, the big problem everyone had with the idea that the continents had moved in the past was that no one had come up with a mechanism which could cause such motion. Hess's theory provided that mechanism, so it was taken more seriously than Pelligrini's ideas or Wegener's continental drift. It was eventually referred to as **seafloor spreading**, and it explained all the observations that existed at the time.

In fact, it even explained the existence of ocean trenches. After all, imagine what will happen as the seafloor starts pushing against the continents because it is spreading. The continents will move, but there will be some resistance. After all, imagine trying to get a car to move by pushing it. It's not easy, because there is a force called **friction** that resists the motion. So as the seafloor spreads, it pushes against the continents, but the continents resist. They move a bit, but the seafloor needs to move more. What happens? The oceanic crust slips under the continental crust, forming a trench!

Hess's theory explained the observations that were made at the time, but one of its most important confirmations came the year after he published his book. Two British geologists, Frederick Vine and Drummond Matthews, studied the orientation of magnetite in igneous rock taken from oceanic crust. They found many examples of magnetic reversals in the crust samples, and they noticed a pattern. The magnetic reversals were found to be the same on both sides of a ridge. If there was a magnetic reversal found in oceanic crust a certain distance on one side of the ridge, the same reversal would be found the same distance away on the other side of the ridge.

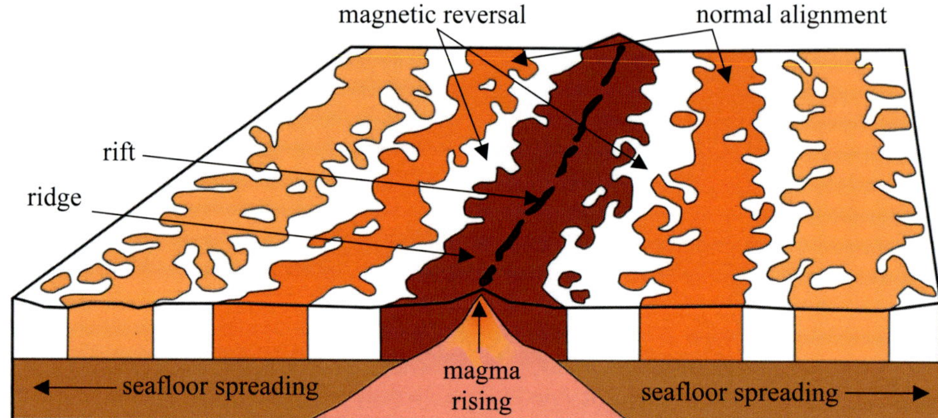

In this illustration, the white patches represent ocean crust that was formed when the earth's magnetic poles were reversed.

As shown in the illustration on the left, this is easy to explain if Hess's theory is correct. After all, if magma coming up from the rift has magnetite in it, the magnetite will precipitate out as the magma becomes lava and cools. As this happens, the magnetite will align itself with the earth's magnetic field, just like the needle did in Experiment 5.2. Once the lava freezes into igneous rock, however, those magnetite samples become locked in place. As those magnetite-containing rocks spread out with the seafloor, they will spread evenly in both directions. Crust found at equal distances from the ridge will have formed at the same time, and the magnetite will be oriented in the same way. That's why the pattern of magnetic reversals is the same in either direction from the ridge.

This record of magnetic reversals was seen as strong evidence that the seafloor did, indeed, spread the way Hess envisioned. As a result, scientists began to look at the seafloor as a giant "conveyor" in which new oceanic crust is made at the rift in the middle of the ocean, and old oceanic crust is "recycled" into the mantle when it slips under continental crust.

Comprehension Check

5.4 Seafloor spreading also explains why there is not much sediment on the seafloor near an ocean ridge, but there is a lot of it on the seafloor next to a continent. How?

5.5 In 2012, film director James Cameron spent three hours exploring the ocean at the deepest point humans have ever explored. Was he near an ocean ridge, in the abyssal plain, or near a continent?

5.6 Suppose you could accurately measure the date at which the igneous rock in the ocean crust formed. If you measured those dates near a continent and then near a ridge, which crust would be the youngest?

Why Does the Oceanic Crust Slip Under Continental Crust?

In Hess's view of seafloor spreading, oceanic crust is made at the rift of a ridge (like the Mid-Atlantic Ridge) and then slips under the crust of a continent, forming ocean trenches. But why does it slip under continental crust? Perform the following experiment to find out.

Experiment 5.4: When Crusts Collide

Supplies:

- The graduated cylinder from the laboratory kit made for this course
- The mass scale from the laboratory kit made for this course
- Corn syrup or another kind of thick syrup
- Water
- Vegetable oil (olive oil, canola oil, etc. will also work.)
- Scissors
- Three small glasses, like juice glasses
- A medicine dropper from the laboratory kit made for this course
- Cardboard that can be cut to fit inside one of the juice glasses

There are sample calculations for this lab at the end of the chapter, right before the answers to the "Comprehension Check" questions.

Instructions:

1. Cut the cardboard so it fits into a juice glass, vertically separating one side of the glass from the other (see the picture below). The fit doesn't have to be great; it just has to be tight enough so the cardboard stands up on its own and presses against the inside of the glass on both sides.
2. Turn on the mass scale, make sure the unit is grams, and hit the "tare" button so it reads 0 g.
3. Put the graduated cylinder on the scale and record its mass, then remove it from the scale.
4. Fill the graduated cylinder to the 50-mL mark with water. If you fill it past the 50-mL mark, use the medicine dropper to remove the excess.
5. Hit the "tare" button again so the mass reads 0 g.
6. Put the graduated cylinder with 50-mL of water on the scale and record its mass.
7. To get the mass of just the water, take the mass you just measured and subtract the mass of the cylinder by itself, which you measured in step 3.
8. Divide that mass by 50 mL to get the density of the water.
9. Dump the water into one of two glasses that doesn't have the cardboard fit to it. Shake the graduated cylinder to get as much water out as possible.
10. Hit the "tare" on your scale again so it reads 0 g.
11. Repeat steps 3-8, but this time for the vegetable oil. Yes, you do need to record the mass of the graduated cylinder again before you put in the vegetable oil.
12. Dump the vegetable oil into the only other glass you haven't used yet.
13. Pull the cardboard out of the glass you fit it to and add corn syrup to that glass until it is about ¼ of the way full.
14. Put the cardboard back in the glass.
15. Hold the other two glasses, one in each hand.
16. At the same time, pour the water on one side of the cardboard and the oil on the other, so that you have something like what is pictured on the right.

17. Look at the glass. Note that the oil and water are both floating on the corn syrup.
18. While you are watching the oil and water, pull the cardboard out of the glass so the water and oil can crash into each other.
19. Observe where the oil and water end up. Look at the densities you measured. Do their final positions make sense to you?
20. Clean up your mess. Everything can go down the drain. You will need lots of soap and warm water to clean the graduated cylinder and get everything down the drain.

What happened in the experiment? Most likely, some of the oil and water crossed the cardboard divider before you pulled it out, but that's okay. When you pulled the divider out, the water and oil crashed into each other. The water slid under the oil, so that in the end, there were three layers: the syrup layer on the bottom, the water layer on top of that, and the oil layer on top of the water. You measured the density of the water and oil, and with that information, it should make sense why the water ended up under the oil: It is denser than the oil. The less dense material (oil) floated on top of it.

Remember what you learned at the beginning of this chapter. Oceanic crust is more dense than continental crust. So in your experiment, the corn syrup represented the asthenosphere. The water represented oceanic crust, and the oil represented continental crust. Just like the water slid under the oil when they collided, oceanic crust slides under continental crust when the two collide.

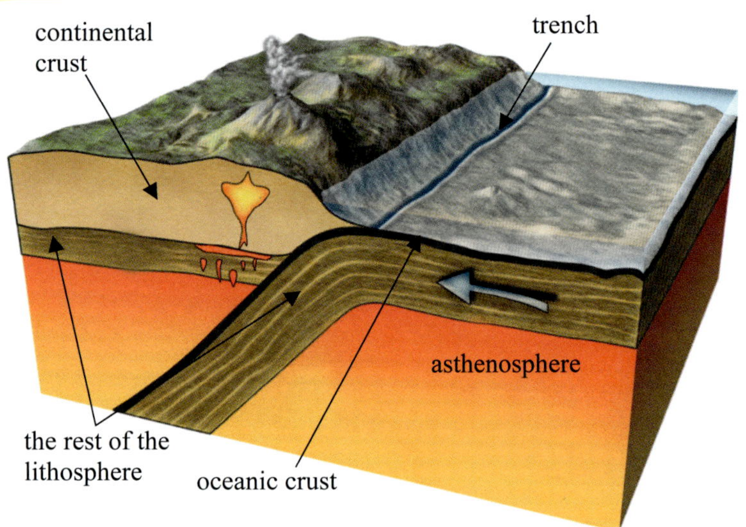

This illustration shows what happens when oceanic crust slips under continental crust.

Now, of course, the experiment wasn't a completely accurate depiction of what goes on, and the illustration on the left shows you why. When the oceanic crust slips under the continental crust, it doesn't just slide under. It actually goes down into the asthenosphere and then into the upper part of the mantle. Eventually, it melts and becomes a part of the mantle, replacing what is being pushed up through the ocean rifts.

Plate Tectonics

Now remember what you learned in the introduction to this chapter. There is rock directly below the Moho, and that rock is part of the lithosphere. That's what "rest of the lithosphere" refers to in the illustration above. It's the part of the lithosphere that is moving, because it is floating on the asthenosphere below.

In the illustration above, there are clearly two different sections of lithosphere. The lithosphere that is sliding down and to the left must be different from the lithosphere of the continent, since they are moving differently. This indicates that the lithosphere is broken up into several units, which are called **tectonic plates**.

Tectonic plate – A massive, irregularly-shaped slab of lithosphere

The more general theory that deals with how these plates move and change is called **plate tectonics**, and it is an important theory that helps us understand many aspects of the geosphere, including the motion of the GPS devices illustrated on page 128.

There are different ways that geologists can list the plates of the lithosphere, but in general, there are 15 recognized major plates. They are shown in the illustration below. You don't need to remember the names of the plates or where they are, but I do want you to study the illustration to see some basic things. First, notice that the Pacific Ocean mostly covers a single plate, the Pacific plate. However, the Atlantic Ocean doesn't have its own plate. Instead, the Atlantic Ocean sits on top of several plates, including the North American, South American, African, and Eurasian plates. Thus, when seafloor spreading from the Mid-Atlantic Ridge occurs, it changes several plates.

This illustration shows the 15 major plates that are recognized by most geologists.

Now look specifically at the Caribbean plate, which is shown as the yellow region between North America and South America. The Puerto Rico Trench illustrated on page 137 forms at the boundary between the Caribbean plate and the North American plate. So, when seafloor spreading happens, it pushes the Caribbean plate into the southwestern edge of the North American plate. Since that portion of the North American plate is made of continental crust and the Caribbean plate is made of oceanic crust, the Caribbean plate is pushed under the North American plate, and the Puerto Rico Trench is formed.

Comprehension Check

5.7 In the experiment, I told you that you had to measure the mass of the graduated cylinder again after you dumped the water out of it. You had already measured the mass of the empty graduated cylinder. Why did I make you measure it again?

5.8 You didn't measure the density of the corn syrup in your experiment, because it would have been really messy. Had you measured it, however, how would it have compared to the densities of water and vegetable oil?

Plate Movements

As you can see, then, geologists figured out that the lithosphere is made up of moving plates well before we could use the GPS to directly measure their motion. After all, when massive slabs of rock move around, they leave behind evidence of their motion. By examining that evidence, geologists can determine how the plates are moving without directly measuring their motion. Even without the aid of the GPS, geologists worked out how the lithosphere's plates were moving and what happened as a result of that motion.

There are three basic ways plates can move in relation to one another. Consider, for example, seafloor spreading. As Hess envisioned, the seafloor spreads away from an ocean rift in both directions. Well, that rift is a boundary between two plates, which means the two plates are moving away from each other. This is called a **divergent boundary**.

Divergent boundary – A boundary between two plates that are moving away from each other

While an ocean rift is a good example of a divergent boundary, there are **continental rifts** that are found at divergent boundaries on continents as well.

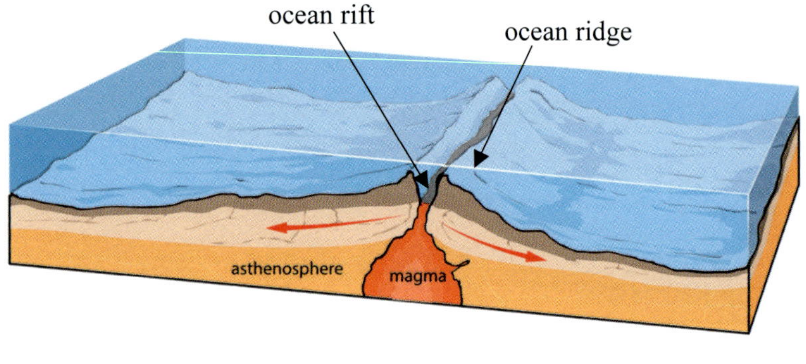

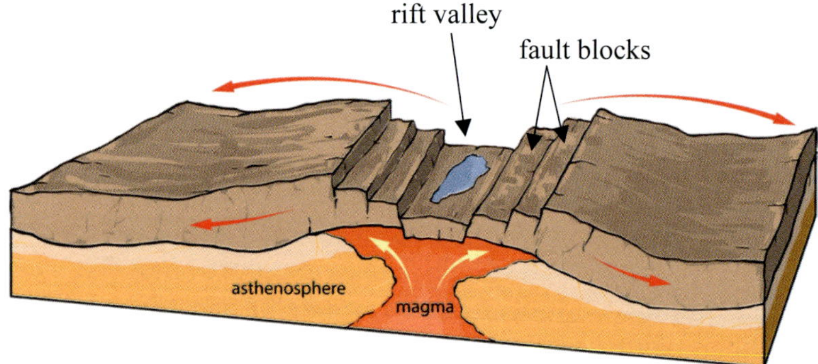

At a divergent plate boundary, plates move away from each other, resulting in different consequences depending on whether it occurs on a continent or in the ocean.

The illustrations on the left compare what happens at an ocean rift and a continental rift. In both cases, the plates move away from one another, which causes magma to push upward; however, the consequences are different. At an ocean rift, mountains form as the lava cools quickly in the water, producing an ocean ridge. At a continental rift, the crust tends to break as the plates move apart from one another. This causes large blocks, called **fault blocks**, to form. The fault blocks end up defining a valley, which is usually called a **rift valley**. There is often volcanic activity in the area, since the magma can escape and flow to the surface. Earthquakes can also happen as the fault blocks form. Also, water can collect in the rift valley, forming rivers and lakes. The results can produce spectacular views, as shown by the continental rift in Iceland, shown below.

A view of the continental rift in Iceland, where the North American and Eurasian plates are diverging.

In my discussion of seafloor spreading, I have also discussed the second type of plate boundary: a **convergent boundary**.

Convergent boundary – A boundary between two plates that are moving toward each other

When an oceanic plate and a continental plate have a convergent boundary, you already learned that the oceanic plate gets pushed down under the continental plate, because it is denser. When one plate goes under another, the process is called **subduction**.

Subduction – The downward motion of a plate into the mantle as it moves under another plate

The subduction forms a trench, as shown in the drawing below. Because the oceanic plate starts to melt as it dives into the mantle, the resulting magma can rise up through the continental crust and form **continental arc volcanoes**. They are called "arc volcanoes" because when multiple volcanoes are made, they usually fall along a curved path (an arc).

Of course, that's not the only kind of convergent boundary that can occur. Two plates can have a convergent boundary in the ocean. In that case, you have two sets of oceanic crust colliding. What happens then? One of the two crusts will be farther from the ridge where it formed. That crust will be more compressed, since it is farther from the hot lava that formed it and has been subjected to pressure for a longer amount of time. As a result, it will be denser than the other oceanic crust, and it will undergo subduction. When that happens, a trench is formed.

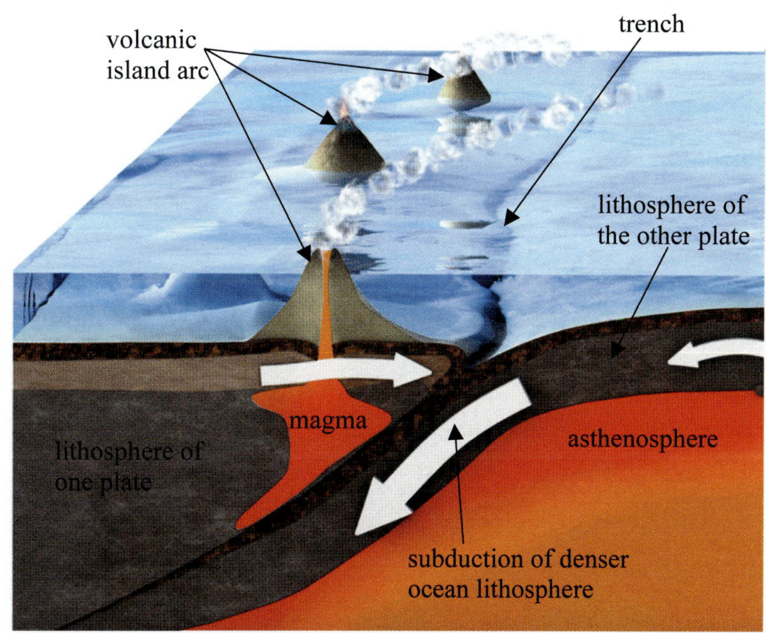

When two oceanic plates converge, the denser one will subduct.

Like what happens when oceanic and continental plates converge, the subducting plate can melt, forming magma that can rise to make volcanoes under the water. If they grow tall enough, the volcanoes can break the surface, forming islands that are not surprisingly called **volcanic islands**. When volcanic islands form like this, they follow an arc just like continental volcanoes formed by converging plates. For example, the Mariana Trench is formed by two oceanic plates converging. As a result, volcanoes formed the Mariana Islands, and those volcanoes follow an arc. Please note, however, that there are other ways volcanoes can form in the ocean (you will learn about another way soon), and they can also reach the surface to form islands. So, not all volcanic islands follow an arc. The Hawaiian Islands, for example, are volcanic islands, but they were not made from subduction at a plate boundary. As a result, they form a line, not an arc.

Of course, the other possibility is having a convergent plate boundary between two continental plates. When that happens, things are very different, because neither crust is significantly denser than the other or the underlying rock of the mantle. Thus, subduction can't occur. Instead, the results of the collision are similar to what happens when two automobiles collide. The plates crumple up, forming mountains. Because there is no subduction going on at this kind of plate boundary, there is no melting of the lithosphere, no rising up of magma, and therefore no volcanoes. However, as you might imagine, there can be a lot of earthquake activity due to the crumpling of the crust. For example, the Himalayas are mountains formed at the convergence of two continental plates, and there are a lot of earthquakes going on there. In 2015, for example, Nepal (which is near the plate boundary) experienced a deadly earthquake that killed nearly 9,000 people and injured almost 22,000 others.

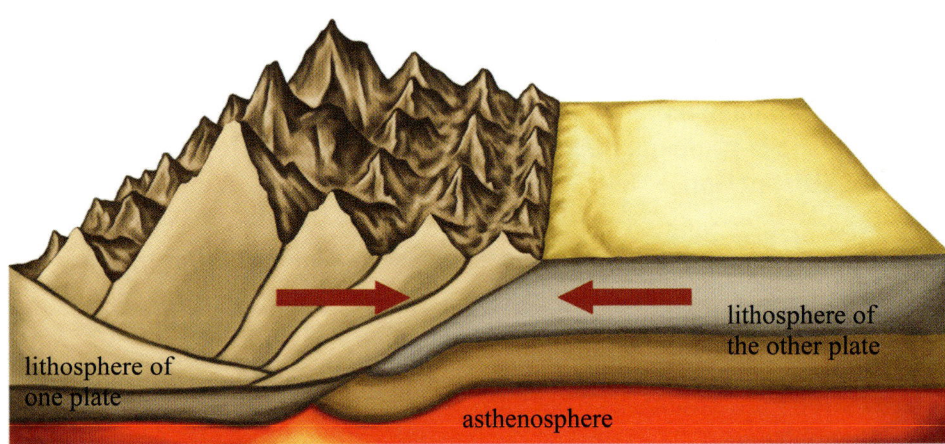

When two continental plates converge, subduction doesn't happen.

Divergent and convergent boundaries are opposites of one another. The last type, **transform boundary** is really nothing like the other two.

Transform boundary – A boundary between two plates that are moving alongside each other

Put your hands together so that your palms are flat against one another, then move one palm forward and the other one backward. That's the kind of motion happening at a transform boundary.

Most transform boundaries are found in the ocean, but some are found on continents, and they can have a big effect on people. Because they aren't diverging or converging, there are no mountains or volcanoes formed at such boundaries. However, the sides of the plates rub together with a lot of friction, and that can produce earthquakes. You have probably heard about the San Andreas Fault in California, for example. What is it best known for? Earthquakes. That's because it is part of a continental transform boundary.

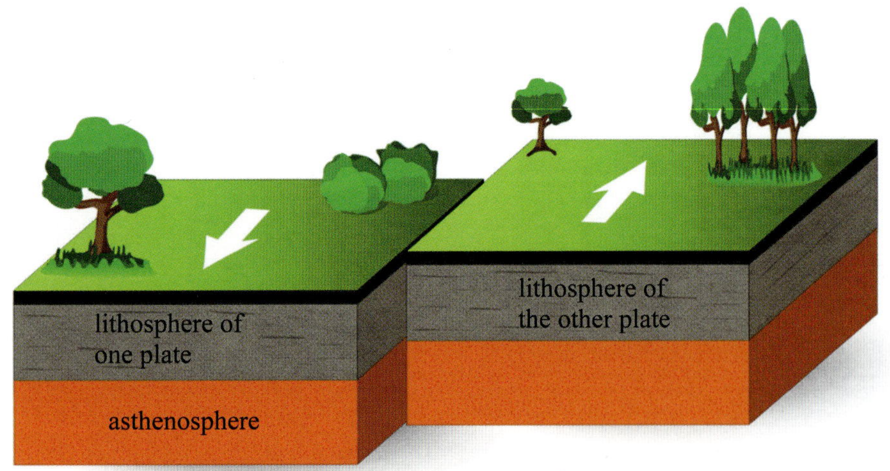

In a transform boundary, plates slide alongside one another.

Speaking of the San Andreas Fault, I need to define a term that is a bit more general than a plate boundary:

Fault – A fracture between two blocks of rock that allows the rocks to move relative to each other

Obviously, plate boundaries are faults, since plates can move relative to each other. The San Andreas fault is a plate boundary, but there are many other faults that are not at a plate boundary. For example, there is a series of faults near New Madrid, Missouri where large blocks of rocks can move relative to one another. It has been a major source of earthquakes in that part of the United States, but it is on the North American Plate, far from any boundary. Fault, then, is a more general term. All plate boundaries are faults, but not all faults are plate boundaries.

Hot Spots and Volcanic Islands

When I was telling you about volcanic islands, I told you that not all volcanic islands form arcs, because not all volcanic islands form at the boundary between convergent oceanic plates. The Hawaiian Islands are a great example. They were formed by a **hotspot** in the mantle.

Mantle hotspot – A relatively stationary plume of magma rising up from the mantle into the crust

From its definition, it should be obvious that a hotspot can form a volcano. If the hotspot is under oceanic crust and forms a tall enough volcano, the result will be a volcanic island. Each Hawaiian island is thought to have formed from the same hotspot.

But wait a minute. If a hotspot is relatively stationary, how can it form a chain of different islands? Well, the hotspot doesn't move, but the oceanic plate does. As a result, multiple volcanoes are formed, making multiple volcanic islands. Look at the illustration below:

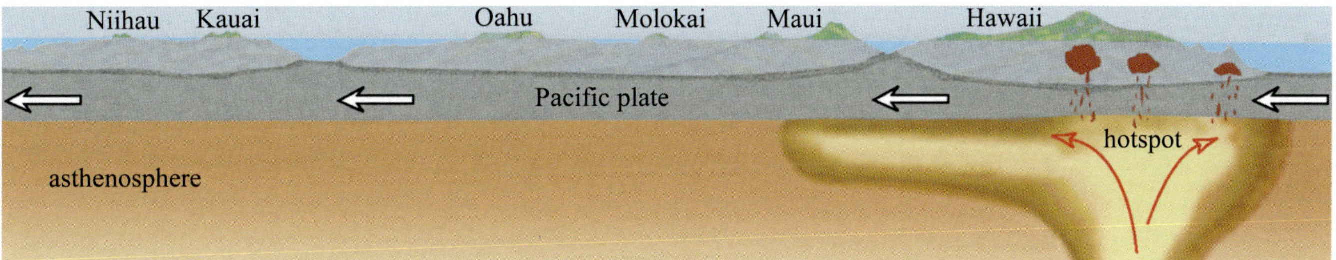

It shows a side view of the part of the Pacific Plate that holds the Hawaiian Islands. In the drawing, the plate is moving to the left.

When the island chain first started forming, the hotspot produced a volcano that became tall enough to break the surface of the ocean and form the island of Niihau (the small island on the far left of the illustration). However, the plate was moving, so eventually, the hotspot was no longer under the island of Niihau. As a result, there was no longer lava being formed there, but the island made by the lava that had formed there remained. As the plate moved, lava still rose out of the hotspot, but not enough to make a tall volcano, so no island was made. Eventually, however, enough lava was produced to make a new volcanic island, but by that time, the plate had moved enough so this new island, Kauai, was completely separate from the older island. This scenario continued as the plate moved, forming Oahu, then Molokai, then Maui, and then the "Big Island" of Hawaii.

When islands are made this way, they tend to form in a line that follows the motion of the plate. If you look at an aerial view of the Hawaiian Islands, you will see that they form a line. In addition, we know that the island farthest from the hotspot formed first, so it is the oldest island. Also, while they are all volcanic islands, most cannot have lava form on them. There is only one island that has an

active volcano (a volcano that can form lava), and that's the newest island, which is over the hotspot. For the Hawaiian chain, then, the island of Hawaii is the only one that holds an active volcano.

Plate tectonics is another excellent example of how scientists can learn a lot by studying things indirectly. The plates are currently moving so slowly that there is no way we could sense their motion until the GPS was put in place. Nevertheless, most of the details of plates and their movements were determined before the GPS was used to study them, because scientists were able to learn about them by studying things like ocean ridges, ocean trenches, magnetic reversals, and volcanic islands.

Comprehension Check

5.9 Rift valleys at divergent plate boundaries can have rivers in them, making them look a bit like canyons. If you are looking at something with high walls of rock and a river flowing through the bottom, how could you tell if it is a canyon or a rift valley?

5.10 You are studying a boundary between two oceanic plates. How could you decide whether it was a convergent plate boundary or a divergent plate boundary by just observing the seafloor at the boundary?

5.11 A mountain range has some volcanoes in it. Were the mountains formed at a plate boundary between two continental plates or a plate boundary between an oceanic plate and a continental plate?

5.12 You are studying a line of volcanic islands. What would you look for to determine which island was formed most recently?

What's Actually Moving the Plates?

As I told you before, scientists had problems with Wegener's idea of continental drift because they couldn't understand what moved the continents. Hess's view of seafloor spreading gave scientists some insight into that, but it wasn't complete. Now, however, we think we have a pretty good idea of what causes the plates to move. It is best if I start the explanation with an experiment.

Experiment 5.5: Convection

Supplies:
- The blue-topped dropper bottle from the laboratory kit made for this course
- The denser rock in the "density rocks" bag from the kit (the one that is not pumice)
- A tall, transparent glass
- Dark food coloring (Blue works best.)
- Aluminum foil
- A rubber band
- A single ice cube
- A sink with a faucet that can deliver both hot and cold water

Instructions:
1. Unscrew the lid from the dropper bottle, pull off the dropper top, empty the bottle's contents into the sink, and rinse it out.
2. Hold the rock underneath the bottle and wrap both it and the bottle in aluminum foil so that the foil holds the rock to the bottom of the bottle (see the picture on the next page). You are using the rock to weigh down the bottle, since you want it to sink in water.

3. Use the rubber band to hold the foil to the bottle (see the picture below).
4. Let cold water come out of the faucet for a little while so that the water is as cold as it can get.
5. Fill the large tall glass nearly full of that cold water. Set it on a level surface.
6. Put five drops of dark food coloring in the dropper bottle.
7. Gently fill the dropper bottle with cold water. *Do not* replace the dropper top or lid.
8. Carefully lower the bottle/rock/foil contraption into the glass, allowing it to sink to the bottom. Your experiment should look a bit like the picture on the right.
9. Observe what happens for a couple of minutes.
10. Carefully pour the water out of the glass and into the sink, catching the bottle/rock/foil contraption as it falls out.
11. Rinse out both the glass and the dropper bottle.
12. Repeat steps 4-6.
13. Repeat steps 6&7, but this time, let the hot water run so it gets as hot as possible and use that to gently fill the dropper bottle. **Be careful with the hot water!**
14. Make sure you have the ice cube handy, but don't do anything with it yet.
15. Repeat step 8. There should be a dramatic difference in what you see this time.
16. Once you notice the difference, put the ice cube in the glass so it floats on the water.
17. Look carefully at the part of the ice cube that is underwater. What do you see?
18. Rinse out the dropper bottle, put the dropper top and lid on, and clean up your mess.

rock/foil/bottle contraption with a rubber band holding it together

What happened in the experiment? When both the glass and bottle had cold water in it, the blue water from the bottle mixed with the water in the glass a bit, but not much. As seen in the picture above, tendrils of blue came out of the bottle, but they were thin tendrils and didn't move in any particular direction. When the dropper bottle had hot, blue water in it, however, the results were dramatically different. The blue water should have risen up to the top of the water in a column, like smoke from a chimney. Why? Well, when water gets hot, it expands. When something expands, its volume gets bigger, so its density gets lower. Hot water, then, is less dense than cold water. Thus, when the water was hot, it rose out of the bottle, since its density allowed it to float in the water of the glass.

When you put the ice cube in the water, you should have noticed that where the blue water touched the ice cube, thin tendrils of blue water started sinking towards the bottom. Why? When the blue water hit the ice cube, it cooled down. The water touching the ice cube became colder, which lowered its volume and increased its density. It became colder than the water in the glass, so it became denser than the water in the glass. Thus, it sunk.

But think about the energy involved in this process. As the blue water rose, it gave energy to the water (and the ice cube) at the top of the glass. That means the blue water itself cooled down. When it hit the ice cube, it cooled down enough to become denser than the surrounding water, and it began to sink. So the motion of the blue water caused energy to be transferred from the blue water to the water at the top of the glass and to the ice cube. When energy is transferred in this way, we call it **convection** (kahn vek' shun).

Convection – The process of transferring energy through the movement of heated or cooled fluid

If you think about it, something similar happens in the mantle. After all, the bottom of the mantle is in contact with the outer core, which is hotter than the mantle. That means the rock near the bottom of the mantle is hotter than the rock farther up. Since it is hotter, it is less dense, so it rises up in the mantle. However, the rock in the mantle that is touching the lithosphere gets cooled down, since the lithosphere is cooler. That makes it denser, so that rock begins to sink. Thus, the rock of the mantle is not moving randomly. Currents of rock are rising in the mantle, while other currents are falling.

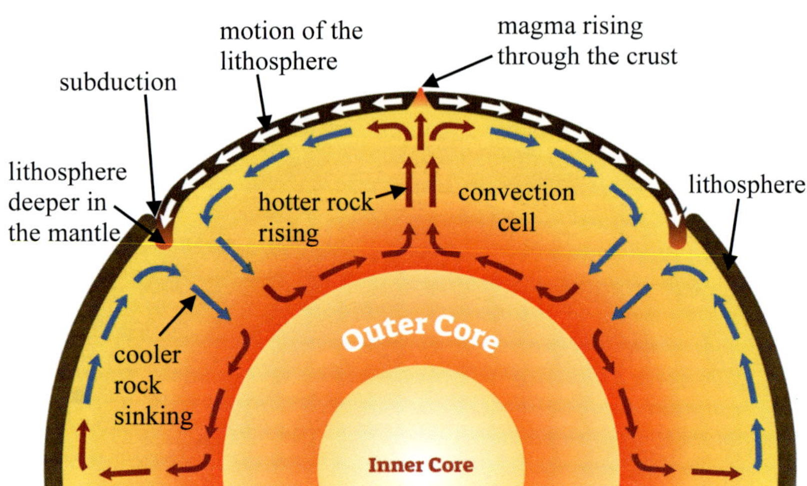

Convection determines the direction in which the plates move.

It turns out that this follows a pattern, illustrated in the drawing on the left. Notice where magma is rising through the crust. Since magma is coming out there, a lot of the less dense, hot mantle rock is pulled towards it. So, there is a current of less dense rock rising in the general area of an ocean ridge. Next, notice the two places where subduction is happening. The cooler rock of the lithosphere reaches deeper into the mantle there, so there is more cooling going on in those regions. As a result, there is a current of denser rock going down in those regions.

Thus, there is a general tendency for the rock of the mantle to rise near an ocean ridge and sink near where there is subduction. This sets up a **convection cell**, where the mantle rock rises mostly in one location and sinks mostly in another. In the illustration, there are four convection cells. In between, the rock of the mantle flows to replace what has been moving. The hotter rock goes to where the rock is mostly rising, and the cooler rock goes to where the rock is mostly sinking.

In other words, the changing density of the molten rock in the mantle (which is caused by the changing temperature of the rock) produces currents of molten rock, as illustrated by the red and blue arrows above. Those currents are called **convection currents**. The plates are floating on the asthenosphere, so they end up moving in the same direction as the convection currents that are found there. The motion of the plates, then, is driven by the temperature difference between the outer core and the lithosphere, and the direction is determined by the convection currents produced from those temperature differences.

Comprehension Check

5.13 In the right portion of the illustration above, the lithosphere is pointed out. If a plate were sitting where the arrow is pointing, in what direction would it be moving?

5.14 As time goes on, what will happen to the temperature difference between the asthenosphere and the lower parts of the mantle?

Sample Data and Calculations for Experiment 5.1

Rock A: mass = 8.6 g
Rock B: mass = 4.2 g

Rock A: new volume = 34.2 mL
volume of rock A = 34.2 mL - 30 mL = 4.2 mL
Rock B: new volume = 35.1 mL
volume of rock B = 35.1 mL - 30 mL = 5.1 mL

Rock A: Density = 8.6 g ÷ 4.2 mL = 2.05 $\frac{g}{mL}$ (Remember, we keep only three digits)

Rock B: Density = 4.2 g ÷ 5.1 mL = 0.82 $\frac{g}{mL}$ (Remember, we keep only three digits)

Sample Data and Calculations for Experiment 5.4

Graduated cylinder: mass = 21.4 g
Graduated cylinder and 50 mL water: mass = 70.9 g

Mass of 50 mL water: 70.9 g - 21.4 g = 49.5 g
Density of water: 49.5 g ÷ 50 mL = 0.99 $\frac{g}{mL}$

Graduated cylinder: mass = 21.5 g
Graduated cylinder and 50 mL oil: mass = 67.0 g

Mass of 50 mL oil: 67.0 g - 21.5 g = 45.5 g
Density of oil: 45.5 g ÷ 50 mL = 0.91 $\frac{g}{mL}$

Answers to the Comprehension Check Questions

5.1 On average, sedimentary rock has the lowest density, while metamorphic rock has the highest density. Just think about how they are formed. Sedimentary rock is made of grains that are "cemented" together. That's not going to lead to tightly-packed matter. Metamorphic rock is formed by heat and/or pressure. If you apply a lot of pressure to a rock, you push the matter closer together, making it more tightly packed. Once again, rocks have a variety of densities. For example, pumice (an igneous rock) is much less dense than limestone (a sedimentary rock). On average, however, sedimentary rocks have densities ranging from 2.3-3.5 $\frac{g}{mL}$, igneous rocks range from 2.5-3.6 $\frac{g}{mL}$, and metamorphic rocks range from 2.3-7.6$\frac{g}{mL}$.

5.2 The compass needle would move in response to the magnet, but when you pulled the magnet far away, it would return to the way it was pointing. That's because the compass needle behaves just like the needle in your experiment. It aligns with the earth's magnetic field unless there is a stronger magnetic field in the area.

5.3 You could use a stronger battery. Since the magnetic field is produced by electricity, the more electricity that flowed, the stronger the magnetic field that could be made. However, in this setup, a lot more electricity could make things dangerously hot, which is why I told you not to test it!

5.4 The nearer you are to the ridge, the newer the seafloor is. As a result, there hasn't been as much time for sediment to collect on it. The oldest parts of the seafloor are nearer to the continents, so there has been more time for sediment to collect on those parts of the seafloor.

5.5 He was near a continent. In fact, he was in the Mariana Trench. Trenches are the deepest parts of the ocean, and they are far from ocean ridges and not part of the abyssal plain.

5.6 The crust near the ridge would be the youngest. Oceanic crust forms at the ridges. As the seafloor spreads, it moves away from the ridges, so the farther it is from the ridges, the older it is. There are methods that measure the age of igneous rocks, and while there is some evidence that they aren't very reliable (as you will learn later), they do show this pattern.

5.7 The mass might have changed because there might have still been some water in the cylinder. Depending on how well you shook the water out of the graduated cylinder, the mass might have been the same the second time you measured it. However, when I did it, the new mass was 0.1 g larger because of the water still clinging to the inside of the graduated cylinder.

5.8 It would have been larger than both. Since the water and oil both floated on the corn syrup, the corn syrup must have been denser than both of them.

5.9 If it is a canyon, the rocks should be reasonably smooth, because erosion tends to smooth things out. A rift valley isn't made by erosion, so the rocks shouldn't be very smooth. Canyons also tend to be a lot wider, since more erosion happens at the top than the bottom. This was a hard one, so don't worry if you got it wrong.

5.10 Look for the presence of a ridge or a trench. Divergent oceanic plates have a rift and a ridge, while convergent oceanic plates have a trench.

5.11 <u>The were formed by the convergence of an oceanic plate and a continental plate</u>. When two continental plates converge, there is no subduction, so there are no volcanoes.

5.12 <u>Look for an island with an active volcano. That will be the one formed most recently</u>. Remember, a line of volcanic islands is made by a hotspot. Motion of the plate makes the line of islands, and only the one that was most recently formed will be over the hotspot.

5.13 <u>It would move up and to the left</u>. It would follow the convection current in the asthenosphere, which is shown by the blue arrows right under the lithosphere.

5.14 <u>It will decrease</u>. As the hot rock rises, it warms up the upper parts of the mantle. As the cold rock sinks, it cools down the lower parts of the mantle. Whenever there is exchange of energy, it flows from hot to cold. As a result, the hot parts get cooler, and the cool parts get warmer. Eventually, the temperature evens out. Now, of course, the mantle is huge, and the temperature differences are enormous, so this would take a very long time. However, it would happen eventually.

Chapter Review

1. Define the following terms:

a. Lithosphere
b. Paleomagnetism
c. Magnetic reversal
d. Tectonic plate
e. Divergent boundary
f. Convergent boundary
g. Subduction
h. Transform boundary
i. Fault
j. Mantle hotspot
k. Convection

2. List the following from least to most dense: oceanic crust, asthenosphere, continental crust.

3. List the following in order of how deep they are found in the earth, starting with the deepest: transition zone, outer core, inner core, lower mantle, lithosphere, asthenosphere.

4. What was Pangaea?

5. What causes the earth's magnetic field, and in what two ways does it help us?

6. What geographic location is the south pole of earth's magnetic field near?

7. Explain how seafloor spreading works, making sure to use the terms "rift," "ridge," "abyssal plain," "subduction," and "trench."

8. How did magnetic reversals and paleomagnetism help confirm the idea of seafloor spreading?

9. What are the major features found at a divergent boundary between oceanic plates? What about divergent boundaries between continental plates?

10. What kinds of plates will produce subduction at a convergent boundary? Which plate will go underneath?

11. Which kinds of plates at what kind of boundary produce continental arc volcanoes? Which kinds of plates at what kind of boundary produce volcanic island arcs? Which kinds of plates at what kind of boundary produce no volcanoes?

12. Distinguish between faults and plate boundaries.

13. If volcanic islands form in a line, what structure in the mantle produced the volcano? If there are several islands, how can you identify the one over that structure?

14. Where are the convection currents that drive the motion of the plates? Is that the only place under the lithosphere where convection currents exist?

15. If molten rock in the mantle is sinking, how does its temperature compare to the surrounding mantle rock? Is it more likely near a place where subduction is occurring or where magma is breaking through the crust and becoming lava?

Chapter 6: More About Motion in the Lithosphere

Introduction

The tectonic plates are constantly moving because of the convection currents in the asthenosphere and mantle, but there are other kinds of movement in the lithosphere. After all, the definition of a fault is a fracture between two blocks of rock that allow the rocks to *move* relative to one another. Just as the motion of the plates can produce things like mountains, volcanoes, and earthquakes, the motion of those blocks can do the same. In this chapter, we will explore the many different ways in which the lithosphere moves.

Different Types of Faults

Look at the wall of rock pictured below. It is composed of different strata of sandstone. Can you figure out what happened to make it appear that way? Look at the block of rock to the left of the fault that is pointed out. Compare that block to the one on the right side. Can you see that the block on the right has essentially the same strata as the block on the left, but they are all higher? Clearly, the sandstone originally formed in continuous, horizontal layers. However, something happened to break the rock into blocks, and those blocks then moved along the fault. The block on the left could have moved down, the block on the right could have moved up, or both of those things could have happened.

The strata in this wall show the movement that must have occurred along the fault. There is also a second fault to the right of the one being discussed in the text.

In order to be able to talk about a specific block of rock along a fault, we call one block the **hanging wall** and the other the **foot wall**. The way we distinguish between them is to imagine a point on the line of the fault, and then imagine that point moving straight up. Whichever block it moves into is the hanging wall. Look, for example, at the yellow line in the picture above. It starts on the fault line and goes directly up, ending in a circle. Thus, the yellow circle is on the hanging wall. If you imagine a point moving down from the fault, it moves into the foot wall, as illustrated by the green line and circle.

Faults can be categorized into one of three broad types: **dip-slip fault**, **strike-slip fault**, and **oblique** (oh bleek')-**slip fault**. Each shows a different kind of slipping motion along the fault. The one pictured above is an example of a dip-slip fault. As the name implies, the blocks slid so that one is dipped below the other. Which is which? Notice that the strata of the hanging wall are lower than the strata of the foot wall. In this case, then, the hanging wall is dipped below the foot wall. This specific kind of dip-slip fault is called a **normal fault**. If the foot wall is dipped below the hanging wall, the dip-slip fault is called a **reverse fault**.

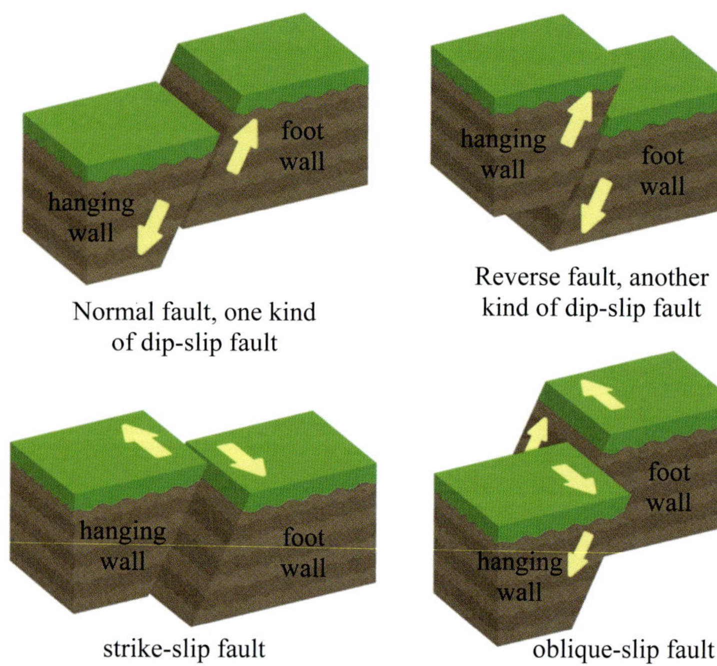

Normal fault, one kind of dip-slip fault

Reverse fault, another kind of dip-slip fault

strike-slip fault

oblique-slip fault

The three types of faults are shown in the illustration above. The top two are both dip-slip faults, but because the vertical motion of the hanging and foot wall is different, we give them different names.

Of course, it's possible for the rocks to move horizontally instead of vertically. When that happens, you have a strike-slip fault. If both horizontal and vertical motion happens along a fault line, you have an oblique-slip fault. Now don't get confused with all of these new terms. As illustrated on the left, there are three basic types of faults: dip-slip faults, strike-slip faults, and oblique-slip faults. However, there are two kinds of dip-slip faults I want you to know, and they are distinguished by which way the hanging wall dips.

I need to emphasize that when we are talking about these three types of faults, we are not talking about boundaries between plates. Plates are huge slabs of rock that float on the asthenosphere. The boundaries between them are typically many, many kilometers long, and the movement at a plate boundary goes all the way down to the asthenosphere. The kinds of faults I am talking about here can be large, but not nearly as large as plate boundaries. Also, the movement doesn't extend down to the asthenosphere. It typically involves just sections of the crust. Now, of course, these kinds of faults are often *caused* by movement at a plate boundary. After all, the motion of the plates involves huge forces, which can easily break blocks of rock. Thus, while these faults are not plate boundaries, they can be made by movements at those boundaries.

Even though they aren't plate boundaries, rocks can still move along these faults. Perform the following experiment to see how that usually happens.

Experiment 6.1: Friction in Faults

Supplies:

- 20 sheets of paper (They need to be 8.5x11 inches, and they can be blank or already used.)
- A transparent jar with a lid (If you don't have one, use a transparent glass and cover it with plastic wrap that is secured by a rubber band. You just want to limit any water that might spill out.)
- Water

Instructions:

1. Fill the jar halfway with water and put the lid on.

2. Lay one sheet of paper on a flat surface.
3. Lay the next sheet of paper on top of the first sheet, but offset a bit along the length. See the picture on the left. Note that you don't have to use different colors of paper. I did that to make a noticeable difference between the first sheet (white in the picture) and the second sheet (green).

4. Repeat steps 2 and 3 nine more times, so that you have 20 sheets of paper stacked on top of one another in two groups that are slightly offset from each other.
5. Place the jar on top of the stack of papers, positioned so that it is on the stack that is offset to the right, but it is on the left end of that stack, as shown in the picture on the right.

6. Hold the edge of the left stack of papers (white in the picture) in your left hand and the edge of the right stack of papers (green in the picture) in your right hand.
7. Continuing to firmly hold the left stack in your left hand, try using your right hand to pull the right stack of papers to the right. If you can do that, watch the water in the jar. How does it react to the motion?
8. You might not be able to pull the right stack. If so, take the jar off the paper, remove the top two pieces of paper (one from each stack), replace the jar, and repeat step 7. If you still can't do it, continue removing the jar, removing two sheets of paper at a time, replacing the jar, and repeating step 7 until you can.
9. Once you can get the motion going and the right stack is no longer in contact with the left stack, wait a moment for the water in the jar to settle down.
10. Continue to pull the right stack to the right, watching the water in the jar. How has the water's behavior changed?
11. Clean up your mess.

Were you surprised at how hard it was to move the right stack? It was hard because of friction, and as I told you in the previous chapter, friction is a force that opposes motion. There is some friction between two sheets of paper, and when you add more sheets, the force of friction becomes stronger. In addition, when two surfaces are in contact with one another, you can increase the strength of the frictional force by pushing them close together. That's one reason I had you put the jar on the stack of paper. Its weight pressed down on all the sheets, pushing them closer together. That's why they were so hard to pull apart from one another.

Regardless of whether it happened the first time you tried or after you removed some sheets, you were eventually able to overcome the force of friction and get the pages to move. But how did they move? Was it smooth, like the motion in step 10? No. It was jerky, and that caused the water in the jar to move around. The same thing happens when blocks of rock move along a fault. A strong force can overcome the friction, the rocks will move a bit, and then they will stop again until friction can be overcome again. That jerky motion causes the crust to shake like the water in the jar, causing an earthquake. You will learn more about earthquakes the next time you do science.

Comprehension Check

6.1 There are two types of friction: static friction and kinetic friction. Static friction must be overcome to get an object moving. Kinetic friction needs to be overcome to keep the object moving once it has started. Based on the experiment (and your general experience), which is stronger?

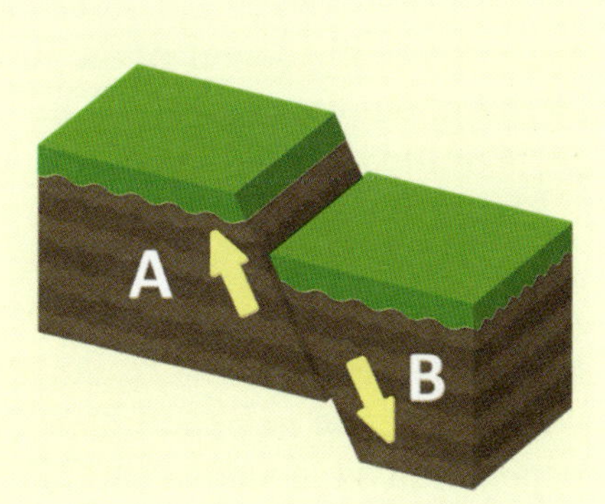

6.2 In the drawing on the right, which block (A or B) is the hanging wall? What kind of fault is this?

Earthquakes

I have talked about earthquakes a couple of times so far, but now I want to concentrate on them for a while. When you think of earthquakes, what place comes to mind? For most people, it's the state of California. While there are a lot of earthquakes in California, there are more in the state of Alaska. You don't hear about them as much, however, because not very many people live in Alaska, so they don't cause nearly as much misery as the ones in California. The deadliest earthquake in the United States happened in San Francisco, California in 1906, killing more than 3,000 people. However, there have been 11 earthquakes in Alaska that were stronger.

As Experiment 6.1 demonstrated, earthquakes are caused by blocks of rock moving along a fault. Sometimes, that fault is a plate boundary, and sometimes, the fault is one of the types discussed in the previous section. Regardless of the type of fault, they are explained with the **elastic rebound theory**, which is illustrated below:

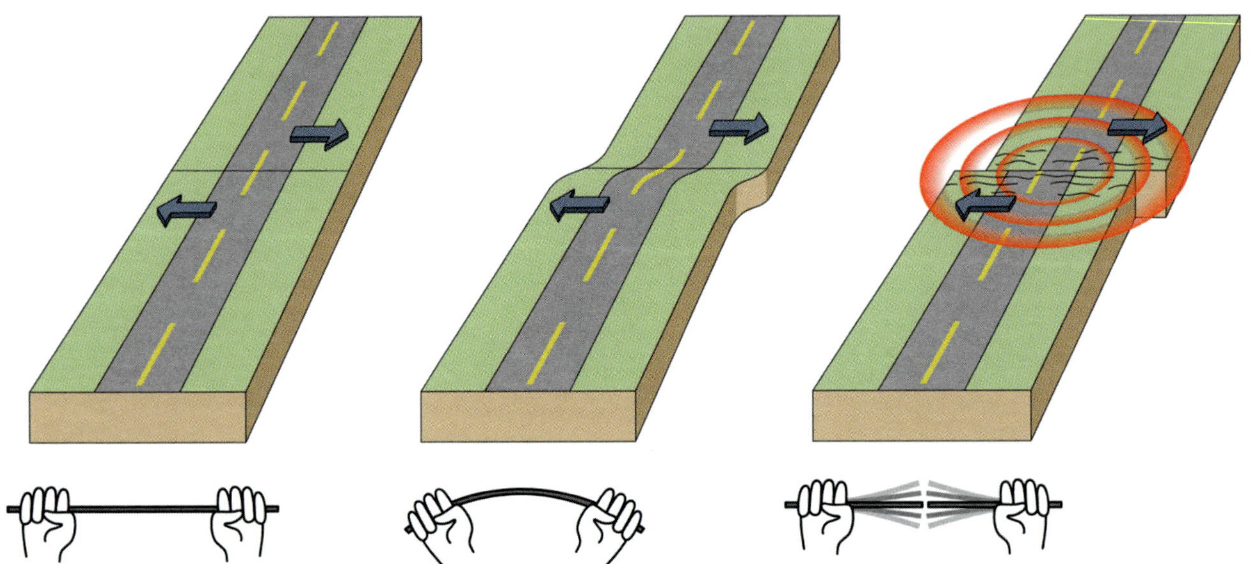

The motion of rock blocks along a fault can be compared to bending and breaking a stick.

On the left side of the illustration, you have two blocks of rock that have a fault between them. The arrows indicate the forces that are acting on them. Notice that the forces are in opposite directions. If the blocks were free to move, they would move in the directions given by the arrows. However, just like what initially happened in Experiment 6.1, they aren't free to move, because friction is resisting the motion.

Initially, there is very little strain, but as the forces keep acting, the rocks have to respond. If the friction between the blocks is still too strong to allow motion to occur, the rocks will start to bend, as shown in the center of the illustration. You don't normally think of rocks bending, because rocks are hard, but they can bend, to a limited degree. Think of it like bending a stick, as shown in the lower parts of the illustration. If you apply enough force, you can get the stick to bend, right?

But what happens if you increase the force you are using? The stick will eventually break, right? Something similar happens to the blocks of rock. Eventually, enough force has built up that friction at the fault can no longer resist the motion, and the blocks move along the fault. Think about what happened in Experiment 6.1. You kept pulling harder and harder, and eventually, the stack of papers moved, because the force you used eventually overcame the friction between the pages.

What happened in the jar when the papers finally moved? The water sloshed around, because the motion of the papers was sudden. In the same way, imagine what would initially happen to the two halves of the stick when it breaks. They would vibrate back and forth. The same thing happens to the blocks of rock. When they move suddenly, they vibrate back and forth as they "rebound" back to their original shape. That vibration, of course, is an earthquake. In other words, earthquakes happen because friction makes the motion of the rocks along a fault jerky. The rocks bend before friction is overcome, and when friction finally is overcome, they rebound to their original shape, causing vibrations to occur.

You can see how the rocks moved along the fault by looking at the yellow stripes.

You can see the results of elastic rebound in the picture on the right. It shows the middle of a road in Utah after an earthquake. When the road was built, of course, the yellow stripes in the picture were continuous. After the earthquake, the rocks had moved, and the road moved with them. The block in the foreground moved to the left, while the block on the other side of the fault moved to the right. As a result, the yellow stripes are no longer continuous.

Now, of course, earthquakes can cause a lot more damage than what is shown in the picture. They can also cause no damage at all. It depends on how the rocks vibrate and shake as they rebound, so it's important to be able to measure those vibrations. The seismologists you learned about in Chapter 2 study this using a **seismograph** (size muh graf'). While you might think of the seismograph as a modern invention, it is not. After all, earthquakes have been happening throughout history, and people have tried to understand them by studying them.

The first seismograph of which historians are aware was invented by Chinese mathematician Zhang Heng in AD 132. He hoped to better understand earthquakes by detecting them and seeing if he could see any pattern in when and where they happened. His seismograph successfully detected its first earthquake six years after he made it. It couldn't tell him how strong the earthquake was, but it could tell him the direction in which the earthquake happened.

Nowadays, seismographs are much more advanced than Zhang Heng's, and they can give us a very detailed analysis of an earthquake. A more modern version of a seismograph is illustrated on the right. The device itself is anchored down so that it will only move when the rocks below it move. A weight is suspended from a wire, and a pen is attached to the weight. The pen touches paper on a rotating drum. If the rocks underneath it shake, the weight will swing back and forth, and the pen will record the swings as peaks and valleys. The larger the peaks and valleys, the

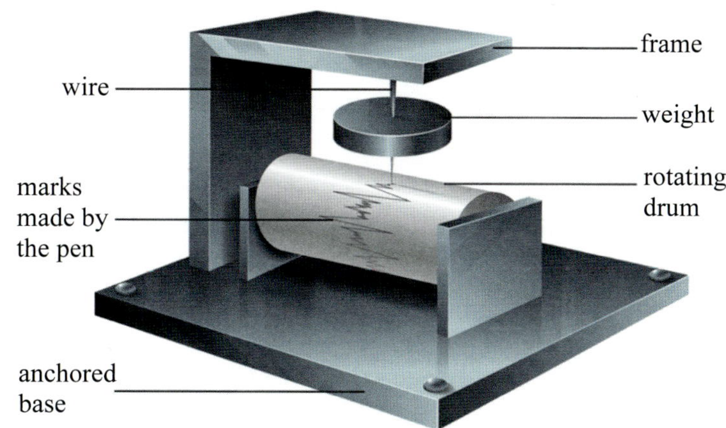

This kind of seismograph was used until digital electronic ones were developed.

stronger the earthquake. Today's seismographs use electronic sensors and record the vibrations digitally so that they can be analyzed by computers and communicated to other seismologists quickly.

Since the size of the peaks and valleys recorded by a seismograph indicate how much the rocks vibrate, we can use a seismograph to determine the strength of an earthquake. In 1935, a geologist in California, Charles Richter, developed a scale that we now call the **Richter scale** so that you can describe how strong an earthquake is with a number. The larger the number, the more energy the earthquake released.

There is a wide range of energy that an earthquake can release. Some earthquakes are so gentle that people don't even notice them. Others are so violent that they can cause a lot of death and destruction. As a result, the Richter scale has to cover a wide range. The drawing on the left gives you a general view of how to interpret the Richter scale. When earthquakes are 2.0 or lower on the Richter scale, they are so gentle that they usually go unnoticed, except by seismographs. From 2 to 4, they get more noticeable. Around 4, earthquakes can release enough energy to cause light damage, and as the numbers increase, the earthquake becomes stronger and stronger. Most earthquakes above 7 can cause major damage, and the damage increases as the number gets larger.

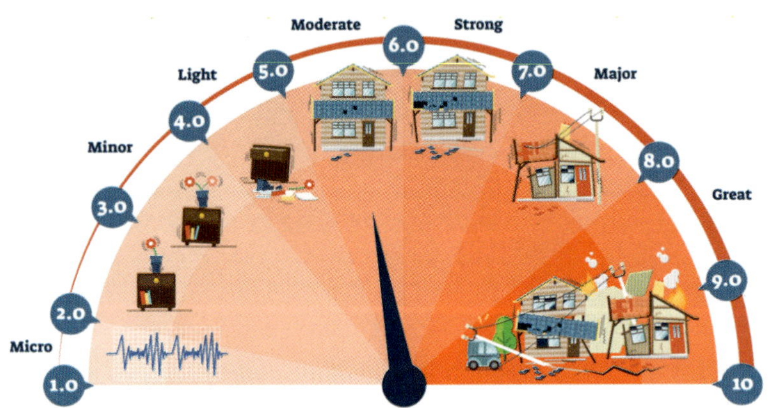

This illustration shows (in a general way) how the Richter Scale relates to the damage an earthquake can cause.

If you think about it, the Richter scale has to cover a wide range, but its numbers go from 1 to 10. How can it do that? Because the Richter scale is not a **linear scale**.

Linear scale – A scale that uses equal divisions to represent equal values

Most (if not all) of the scales with which you are familiar are linear scales. Think, for example, about how we measure mass. If something has a mass of 200 grams, it has twice the amount of matter as something that has a mass of 100 grams. Twice as much mass means twice as much matter, because a mass scale is a linear scale. That's not the case with the Richter scale. It is a **logarithmic** (la guh rith' mik) **scale**

Logarithmic scale – A scale that uses equal divisions to represent multiplying by a specific number

In this kind of scale, going from one number to another represents multiplying by something that is not related to the numbers themselves. I know that sounds strange, but bear with me.

On the Richter scale, each number represents multiplying by 32. So, an earthquake that measures 2 on the Richter scale releases 32 times the energy as an earthquake that measures 1 on the Richter scale. Each time you increase the number by 1, then, you multiply by 32. Think about an earthquake that measures 3 on the Richter scale. How does it compare to an earthquake that measures 1? Well, to go from 1 to 2, you must multiply by 32. To go from 2 to 3, you must multiply again by

32. Thus, an earthquake that measures 3 on the Richter scale releases 32·32 = 1024 times more energy than an earthquake that measures 1. (In case you have forgotten, that dot in between the 32's stands for multiplication.) Rather than saying "measures," most people who use the Richter scale say "magnitude," so I could also say that an earthquake that is magnitude 3 on the Richter scale releases 1024 times more energy than an earthquake that is magnitude 1 on the Richter scale. I want you to make sure you understand this kind of scale by studying the following example problem.

Example 6.1

An earthquake that is magnitude 7 on the Richter scale happens, and then later on, one that has a magnitude of 3 happens. How much more energy was released in the first earthquake?

Each number on the Richter scale represents multiplying by 32. To get from 3 to 7, I must first go from 3 to 4. That means multiplying once by 32. But then I have to go from 4 to 5, which means multiplying by 32 again. Then I must go from 5 to 6, which means multiplying again by 32. Finally, I get to 7 by multiplying one more time. Thus, I have to multiply by 32 a total of four times:

$$32 \cdot 32 \cdot 32 \cdot 32 = 1{,}048{,}576$$

So a magnitude 7 earthquake releases <u>1,048,576 times as much energy</u> as a magnitude 3 earthquake! Notice that I didn't drop any digits like I have done in the past. That's because I can't do that without changing the actual meaning of the number, so I have to report all the digits. For this course, you should only drop digits to the right of the decimal if the number gets too long. When you take chemistry, you will learn how to deal with big numbers like this one.

The Richter scale isn't the only logarithmic scale you will learn in this course. Later on, we will discuss acidity, which is measured with the pH scale. It is also logarithmic, but it is based on multiplying by 10, not 32.

Interestingly enough, while you will often hear news outlets report how an earthquake measures on the Richter scale, most scientists use a different scale, called the "moment magnitude scale." It was developed because scientists have figured out additional methods to measure the details of an earthquake, and the moment magnitude scale includes those other methods. Thus, it is a more accurate depiction of the energy released by an earthquake. It is very similar to the Richter scale, but it is a lot more complicated, so a detailed discussion is beyond the scope of this course.

The Richter scale and the moment magnitude scale are objective measures of how much energy an earthquake releases. However, that doesn't necessarily tell you the destruction and devastation caused by an earthquake. An earthquake in an uninhabited region of the country might be magnitude 9 on the Richter scale, but a magnitude 8 earthquake that happens in a city would cause more destruction and devastation. As a result, the **Mercalli** (mur kah' lee) **scale** was developed to indicate how much an earthquake is felt by people in the area and how much damage was done. It measures earthquakes according to Roman numerals I through XII. Earthquakes that measure I on the scale are not felt at all, and earthquakes that measure XII are catastrophic, destroying nearly everything in the area. You won't be required to know the numerals on the scale; I just wanted you to know that it is an alternate way of measuring an earthquake.

Don't Lose Your Focus

The devastation an earthquake produces at the surface of the earth is important to understand, but please realize that earthquakes don't actually happen at the surface of the earth. Instead, they happen at a fault that is in the crust, under the surface. The vibrations then travel to the surface, and if the earthquake is strong enough, we feel their effects at that point. As a result, we need to define two more terms:

Focus of an earthquake – The place where the earthquake began

Epicenter – The point on the earth's surface directly above the focus

So, the **focus** is the place where the rocks actually moved along a fault, but the **epicenter** is the point on the surface where the earthquake is first felt. If you don't recognize it, the prefix "epi" means "on top of," so the epicenter is the center of the earthquake as seen from on top of the earth's crust.

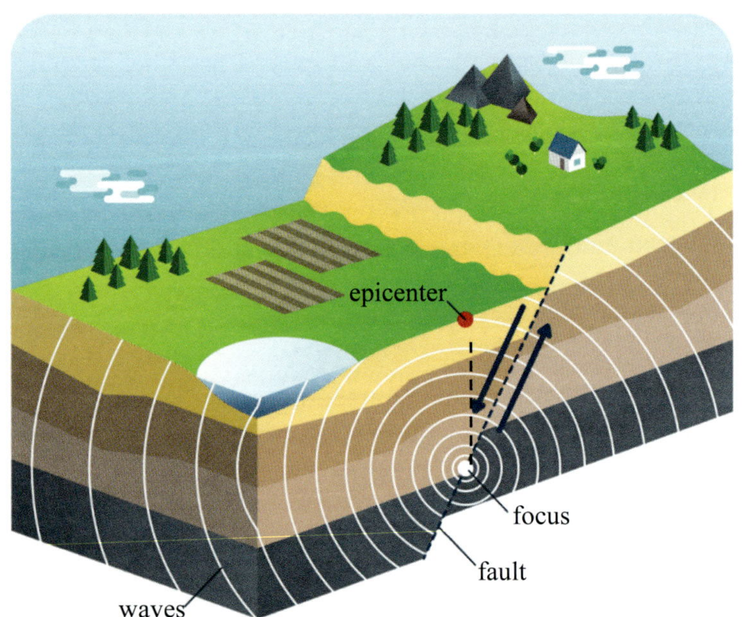

The epicenter of an earthquake is the point on the surface of the earth that is directly above the focus.

When the vibrations form at the focus, they will travel in all directions, as shown in the illustration on the left. If you think about it, the vibrations that travel straight up will reach the surface before any others do, since they travel the shortest distance. As a result, from the surface of the earth, it looks like the earthquake started at the epicenter. The distance the vibrations must travel to get to the surface will affect how strong the earthquake is at its epicenter. If you have two earthquakes and each has the same strength at its focus, the one whose focus is deeper in the earth will not be as strong at its epicenter, since the vibrations will spread out and lose energy as they travel. In the same way, the farther you are from the epicenter of an earthquake, the weaker the vibrations become.

Comprehension Check

6.3 Suppose you could measure the force that is required to overcome friction in a fault and produce an earthquake. If so, how would the strength of an earthquake along a fault where the required force is low compare to the strength of an earthquake along a fault where the required force is high?

6.4 An earthquake is magnitude 5 on the Richter scale, while another one is magnitude 8. Which is the stronger one? How much energy does it release compared to the other?

6.5 Suppose the focus of an earthquake is in the crust below a large lake. A helicopter is hovering above the lake the moment it happens, and the pilot sees waves produced by the earthquake on the surface of the lake. How could the pilot tell where the epicenter of the earthquake is?

Focusing in on Waves

In the previous section, I mentioned the vibrations made by an earthquake. In Chapter 2, you learned that they produce seismic waves, which can tell us what's underneath the earth's crust. I want to spend some more time discussing seismic waves, but to do that, I need to make sure you understand some details about waves themselves. The best way to start is with an experiment.

Experiment 6.2: Waves

Supplies:

- An empty metal can, like the kind vegetables come in
- A can opener
- Plastic wrap
- Tape
- Pepper
- A sink that can be plugged and filled with water
- Water
- A medicine dropper from the laboratory kit made for this course
- Someone to help you

Instructions:

1. Use the can opener to remove the top of the can (if it isn't already removed) and the bottom of the can. That way, the can is open on both ends.
2. If necessary, clean out the can.
3. Use plastic wrap to cover one end of the can. Use tape to hold the wrap to the can, and stretch the wrap tight so there are no wrinkles in the part that is covering the end of the can.
4. Shake pepper onto the plastic wrap so it is spread out over the wrap.
5. Hold the can so that the plastic wrap is on top (as shown in the picture on the right). You should hold it as high as you can while still being able to clearly see the pepper on top.
6. Have your helper get under the can so that his or her mouth is right under the opening of the can.
7. Have your helper sing "ahhh" with the lowest pitch he or she can sing. While your helper is singing, notice how the pepper behaves.

8. Have your helper sing "ahhh" at a much higher pitch, but at about the same volume he or she sang at the lower pitch.
9. Once again, notice how the pepper behaves as your helper holds the note.
10. Did you notice a difference in the pepper's behavior during the high note compared to its behavior during the low note?
11. Repeat steps 7-9 to make sure you see the difference.
12. Have your helper sing "ahhh" at any pitch he or she wants. Your helper should start singing the note with very little volume, but then he should increase the volume as time goes on. Notice what happens to the pepper as your helper increases the volume.
13. Put the can aside and plug the sink.
14. Add water to the sink until it is a bit less than half full.
15. Pull water from the sink up into the dropper.
16. Allow the water to settle so that the surface of the water is smooth.

17. Hold the dropper high above the surface of the water.
18. While you watch the surface of the water, allow a single drop to fall out of the dropper and onto the surface of the water. If you mess up and add several, that's fine. Just allow the water to settle and try again. Once you release a single drop, concentrate on what's happening on the surface of the water.
19. Repeat step 18 a couple of times, concentrating on how the ripples move from where the drop hits the water.
20. Allow the water to settle, and hold the medicine dropper as high as you can while still being able to release a drop onto the water.
21. Repeat step 18.
22. Hold the medicine dropper so that it is about level with your eyes but will still release a drop into the surface of the water.
23. Repeat step 18. How is this different from when you held the medicine dropper really high? If you can't tell, repeat steps 20-22 until you do.
24. Clean up your mess.

Let's start with what happened in the second part of the experiment. You saw ripples on the surface of the water, and they spread out in a circle from the point where the drop hit the water. The ripple was a wave, and it was the result of water rising to form a high point called a **crest**. There was also a low point, called a **trough**, but you might not have noticed that. The crest and trough were vertical, but the motion of the wave was horizontal. When a wave vibrates like that, it is called a **transverse wave**.

Transverse wave – A wave that vibrates perpendicular to the direction it travels

Remember, the crest and trough are made by the vibrations, and they were vertical. The waves moved away from the center horizontally, which is perpendicular to the vertical direction.

What happened when you changed the height of the medicine dropper? You should have noticed that the ripples were easier to see when you held the dropper high, and they were harder to see when you held it lower. That's because the size of the crests and troughs increased the harder the drop hit the water. We call the size of the crests and troughs the **amplitude** of the wave. Now imagine what would have happened if you had a huge sink and released several drops one right after the other. There would be multiple crests and multiple troughs spreading out, right? If you have a situation like that, the distance between the crests is called the **wavelength**. The drawing on the left illustrates all of this.

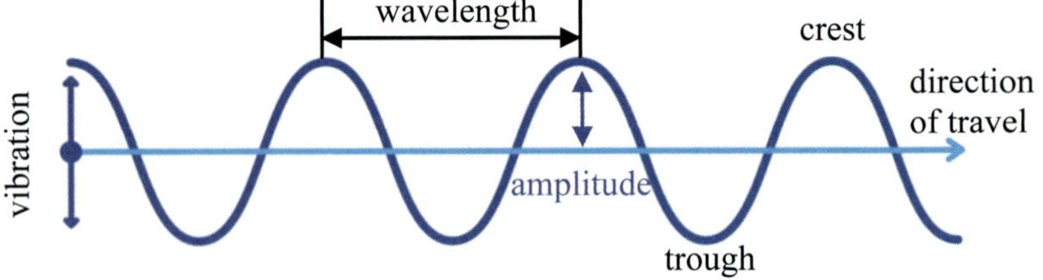

In a transverse wave, the vibration is perpendicular to the direction of travel.

Now think about what happened in the first part of the experiment. When your helper sang, you saw the pepper bounce up and down on the plastic, didn't you? That's because sound is also a wave. In this case, it is a wave built out of air molecules. The pepper bouncing up and down showed you the vibration of the wave. The wave was vibrating up and down. Which way was it traveling? It was traveling up, from your helper's mouth to the plastic. In this case, then, both the vibration *and* the direction of motion were both vertical. That makes a **longitudinal wave**.

Longitudinal wave – A wave that vibrates parallel to the direction it travels

The illustration below gives you an idea of what a longitudinal wave looks like. The crests represent "clumps" of whatever the wave is vibrating. In the case of a sound wave, it is vibrating air, so the crests are places where air molecules have been pushed together. The troughs are the places where the air molecules are spread out. Just like a transverse wave, the distance between the crests is called the wavelength. But what about the amplitude? In a transverse wave, the amplitude is the height of the crests. In a longitudinal wave, it is how thick the "clumps" are. In the case of a sound wave, the more air molecules crammed into a crest, the larger the amplitude.

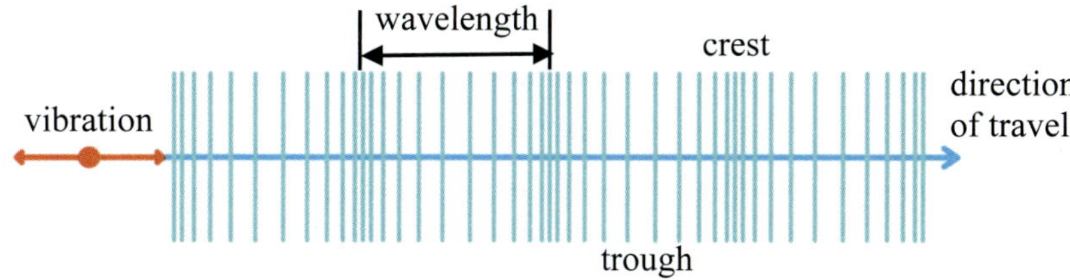

In a longitudinal wave, the vibration is parallel to the direction of travel.

Think about what happened when your helper changed his or her pitch. The higher he or she sang, the faster the pepper bounced up and down, right? That's because the crests were hitting the plastic more often. Why did the crests hit the plastic more often? Because they were closer together. When your helper sang at a low pitch, the pepper didn't bounce as quickly, because the crests were farther apart. In other words, when the wavelength was small (as was the case for the higher pitch), the crests hit more often, and when the wavelength was large (as was the case for the lower pitch), the crests didn't hit as often. We have a word for that. It's **frequency**. The more often the crests of a wave reach a specific point, the higher the frequency. The pitch of a sound, then, is determine by the frequency of the wave. The higher the frequency, the higher the pitch.

Now if you think about it, there is a relationship between wavelength and frequency. After all, if the wavelength is small, the crests are close together. As a result, the crests will reach a given point more often. That means the frequency is high. If the wavelength is long, the crests won't reach a given point as often, so the frequency is low. Wavelength and frequency, then, are opposites. When one is high, the other is low, and vice-versa.

What happened when your helper increased his or her volume? The pepper bounced higher. That's because a sound's volume is determined by the amplitude of the wave. The more air molecules crammed into a crest, the harder it hit the plastic, which means the higher the pepper bounced.

You don't really need to remember the details about sound waves right now, because we are going to focus on seismic waves. Thus, you don't need to remember how volume and pitch relate to sound waves. However, you do need to understand the difference between the two types of waves, because some seismic waves are longitudinal, while others are transverse. You also need to know how amplitude is defined in each. Also, you need to know how wavelength is defined as well as the relationship between wavelength and frequency.

Comprehension Check

6.6 Light is a transverse wave, and the frequency of the wave determines the energy of the light. The higher the frequency, the more energy the light has. You are studying two light waves. The first has a long distance between its crests (a long wavelength). The second has a short distance between its crests (a short wavelength). Which light wave has more energy?

More About Seismic Waves

Now that you've learned about waves in general, I want to discuss seismic waves in more detail. There are two basic kinds of seismic waves: **body waves** and **surface waves**. The body waves move through the interior of the earth. Remember, the focus of an earthquake occurs underground, so the waves must first travel through the interior of the earth before they reach the surface. Once those waves reach the surface, they can cause surface waves, which travel along the surface of the earth away from the epicenter.

There are two kinds of body waves: **primary waves** and **secondary waves**. Primary waves are longitudinal, and they cause the rocks through which they travel to contract and expand. When the rocks contract, they form the crest of the longitudinal wave, and when they expand, they form the trough. The top part of the diagram on the left illustrates a primary wave, which is often abbreviated as "P-wave." Secondary waves (abbreviated as "S-waves"), on the other hand, are transverse waves, as shown in the lower part of the illustration. Now don't be fooled by the limits of the drawing. While there is a surface in the drawing, there is no surface for an S-wave, because it's a body wave. Thus, the entire section of rock drawn there is under the surface of the earth.

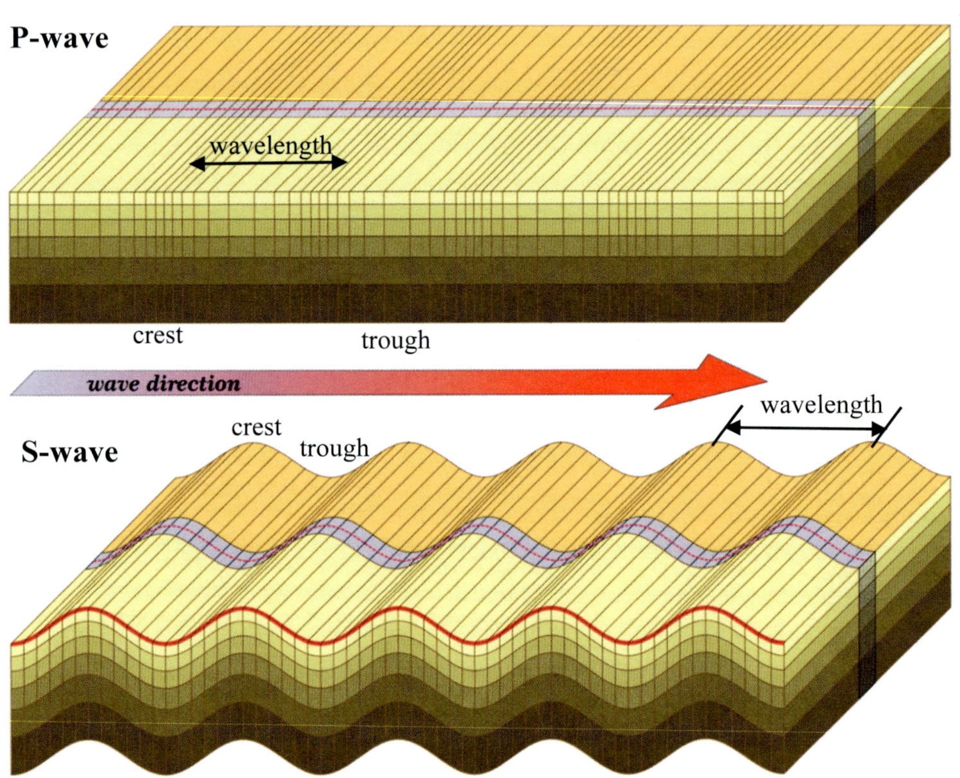

Body waves can be longitudinal (P-waves) or transverse (S-waves)

P-waves travel more quickly than S-waves, so P-waves are the first to reach the surface. Now remember, the waves travel away from the focus in all directions. They hit the epicenter first, because it's directly above the focus. However, they also hit the surface of the earth away from the epicenter. Thus, no matter where you are on the surface of the earth, the first waves to hit you are the body waves. But because P-waves travel faster, the first body waves that hit you are the P-waves. The S-waves come later.

Once the body waves reach the surface of the earth, they start producing surface waves. Don't get confused. S-waves *are not* surface waves, despite the fact that they have an "S" in their name. S-waves are a type of body wave. Surface waves are seismic waves that run along the surface of the earth. These waves end up doing the most damage, because they cause the surface of the earth to move. They are also the slowest of the waves, so they come after the P-waves and S-waves.

Amazingly enough, you can often see all of this by looking at the readout from a seismograph, which is often called a **seismogram**. If you look at the top part of the illustration below, you will see what a seismogram might look like when there is an earthquake. Remember, a seismograph measures the amount that the earth is shaking. If there is no shaking, the seismogram reads as a horizontal line, which is called the **baseline**. Any shaking that happens will cause the seismograph to make a mark either above or below the baseline. The higher the marks are above the baseline and the lower they are below the baseline, the more the earth was shaking at that moment.

The seismogram at the top of the illustration shows that the earth started shaking a bit, but then later on, it started shaking more, and then later on, it started shaking even more. The first shakes that the seismograph records are caused by the P-waves. Remember, they travel fastest, so they arrive at the seismograph first. The S-waves arrive later, since they travel more slowly. Surface waves cannot be formed until the body waves hit the surface, so they come even later. As the illustration shows, P-waves usually cause the smallest response from the seismograph, while S-waves usually produce a larger response, and surface waves an even larger response.

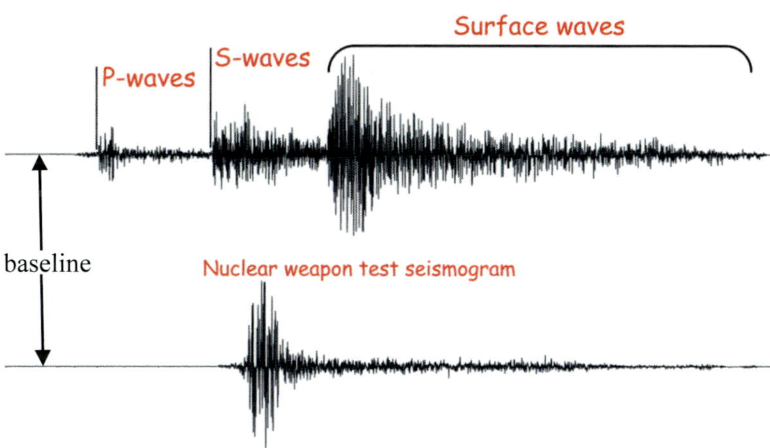

Because of the different waves produced, an earthquake's seismogram has a basic structure that is different from a nuclear weapon test.

Compare the seismogram of an earthquake to that of a nuclear weapon test, which is shown in the lower part of the illustration. In order to contain the explosion that happens, nuclear bombs are usually tested underground. They release a lot of energy really quickly, so they shake the rocks as well. However, they shake the rocks differently from an earthquake. The force of the explosion pushes out in all directions, compressing the rock. That produces a P-wave, the main kind of wave that an underground nuclear bomb produces. Thus, they travel to the surface quickly, but they are the only body waves that happen. Surface waves can follow, but they are usually small in comparison to the surface waves caused by earthquakes.

As you can see, then, the seismogram of an underground nuclear weapon test is quite different from the seismogram of an earthquake. This is how governments can figure out if there was an underground test of a nuclear bomb. The bomb produces seismic waves, and seismographs all over the world can detect them. Furthermore, the seismogram it produces is so different from an earthquake's seismogram that in most cases, it is easy to distinguish between the two.

It turns out that the difference in speed between P-waves and S-waves can tell us something else about the earthquake. Have you ever been watching a thunderstorm and noticed that you hear thunder after you see lightning? As you will study in detail later on, lightning causes thunder. Thus, when you see lightning, you should hear thunder. However, you usually hear the thunder *after* you see the lightning. That's because light travels much more quickly than sound, so the light waves hit your eyes before the sound waves hit your ears.

Has anyone ever told you that you can actually estimate how far away the lightning strike happened by timing how long it takes for you to hear the thunder after the lightning? After all, they both have to travel the same distance, but the light makes the trip very quickly. The sound requires more time to make the same trip. Well, the longer the trip is, the farther behind the sound will get. As a rough estimate, every second that goes by between seeing the lightning and hearing the thunder means that the lightning struck about 340 meters (0.2 miles) away. Thus, if you see lightning and then hear the thunder 2 seconds later, the lightning struck about 680 meters (0.4 miles) away.

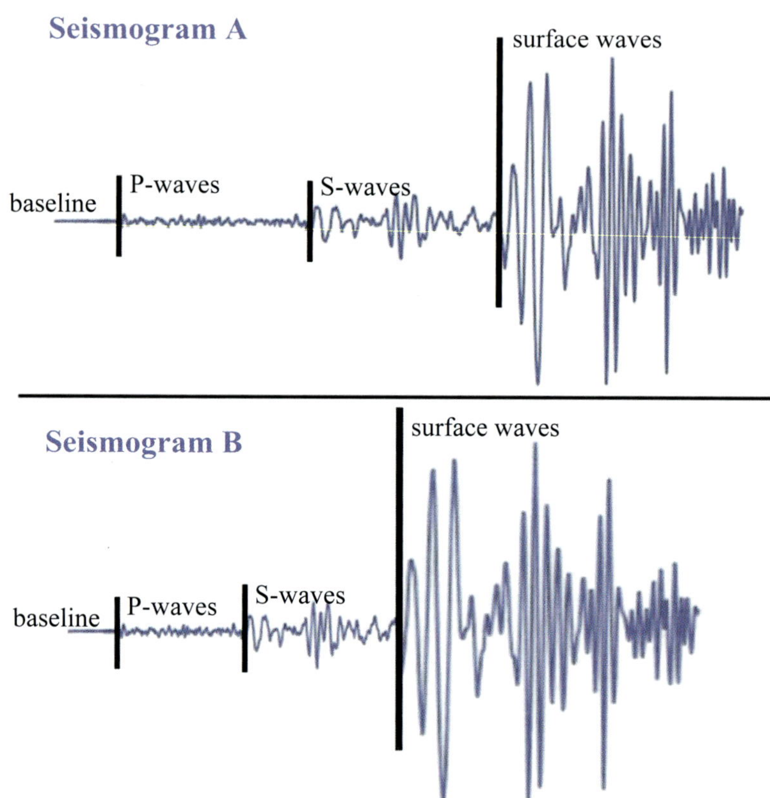

The time interval between P-waves and S-waves tells us that Seismograph B was closer to the earthquake than seismograph A.

The difference in speed between P- and S-waves tells you something similar about earthquakes. Look, for example, at the two seismograms on the left. They both record the same earthquake, but they are from two different locations. As marked in the illustration, you can see the P-waves, S-waves, and surface waves at each location. The P-waves arrive first to both seismographs, as expected, but the time interval between the P-waves and S-waves is different. The S-waves take longer to arrive at seismograph A. What does that tell you? Just like the lightning/thunder situation, the longer it takes for the S-waves to arrive after the P-waves, the farther the waves had to travel. Thus, we can conclude that seismograph B is closer to the earthquake than seismograph A. So, a seismograph doesn't just tell you that an earthquake happened, it allows you to estimate how far away the earthquake was!

While P-waves are always faster than S-waves, their speeds change depending on the type of rock they are traveling through. This is one way we have been able to investigate what is under the crust without actually drilling through it. Remember from Chapter 5 that the Moho separates the crust from the rest of the lithosphere. Andrija Mohorovičić discovered it because he compared different seismograms for the same earthquake. Some of the seismograms indicated that the P-waves were traveling much faster than the other seismograms. By comparing them, he saw that the increase in speed was seen only for P-waves that had to travel about 100 kilometers or more below the earth's surface in order to reach the seismograph. Thus, he reasoned that there must be a different kind of material at that depth. As a result, he decided there must be a boundary between the crust and that different material, and we call it the Moho.

Before I go on any further, I want to let you know that the seismograms I am showing you here are *very easy* to interpret compared to most real seismograms. That's because there are a lot of things that can happen, and they complicate how a seismograph responds to an earthquake. For example, if the seismograph is very close to the earthquake, the P and S-waves might be very close together and hard to distinguish from one another. Also, the relative sizes of the waves aren't always what is shown in the illustrations. Sometimes, the S-waves and surface waves, for example, produce similar responses in a seismograph.

Think back to Experiment 6.1 for a moment. Do you remember how uneven the movement was as you tried to pull the right stack of paper? Something similar can happen at a fault. The blocks of rock can make several slips, each producing an earthquake. In addition, when the rocks rebound, that can cause other disturbances that cause other earthquakes. A seismograph reads all of the shaking that results, and they all add together. This can make the seismogram much more difficult to read.

How do seismologists make sense of all this? They typically look for the largest marks above and below the baseline, and they call that the **main shock** of the earthquake. If there are smaller marks above and below the baseline before the main shock, they represent less violent earthquakes that occurred before the main shock and are called **foreshocks**. If smaller P- and S-waves come after the P- and S-waves of the main shock, those are smaller earthquakes that occurred after the main shock and are called **aftershocks**. If an earthquake is very strong, it typically produces many aftershocks. The aftershocks get weaker and weaker as time goes on, so they eventually aren't noticeable, but seismographs still record them. Foreshocks, mainshocks, and aftershocks can make it more difficult to read seismographs.

There is another issue as well. Depending on where the earthquake happened and where the seismograph is, there might not be any S-waves in the seismogram. Why? Remember that S-waves are transverse waves. They require the material in which the wave moves to vibrate perpendicular to the direction of the wave's motion. This can happen at the surface of a liquid, because the surface is free to move. However, when the liquid is fully enclosed, there is no way to produce a transverse wave. A longitudinal wave can be produced, because the water molecules can move closer together or pull farther apart, but a transverse wave cannot.

Thus, S-waves cannot move through liquid that is enclosed in the earth. This, in fact, is how we know that the outer core is liquid even though we haven't sampled it directly. Look at the illustration on the right. It has an earthquake occurring at the North Pole and follows the paths of the S-waves. As long as the S-waves have solid rock to travel through, they will make it to a seismograph at the surface of the earth. However, if they encounter liquid, they will no longer be able to travel and will simply stop.

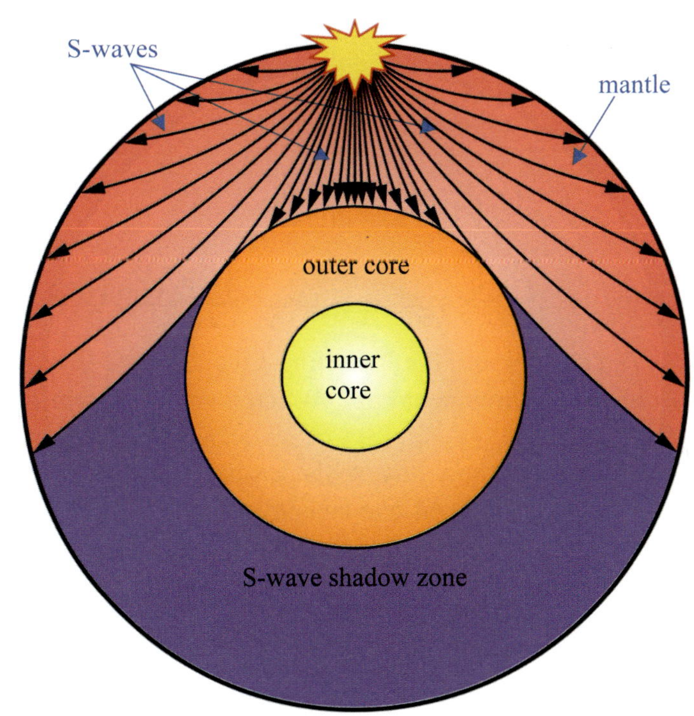

We know that there is a liquid outer core in the earth because of the S-wave shadow zone.

In the illustration, then, any seismograph that is at the surface of the earth in the purple zone of the drawing will not see any S-waves. They might still see P-waves, however, since P-waves are longitudinal. Those P-waves will then make surface waves, so the seismographs might see surface waves as well. However, they will not see S-waves. This is called the **S-wave shadow zone**, and it is caused by the fact that the outer core of the earth is liquid. Seismologists can map the S-wave shadow zone for any earthquake. Based on the results of several earthquakes around the world, they can tell how big the liquid outer core is.

But wait a minute. How do we know that there is a solid inner core? The liquid outer core stops the S-waves. What allows us to determine that there is something different inside? We learn that from the P-waves. However, to understand how the P-waves help us understand the inner core, you need to learn something else about waves, which you will learn the next time you do science.

Comprehension Check

6.7 Imagine that you are inside the earth, watching a seismic wave. The rocks vibrate up and down, but the wave moves from your left to your right. Were you watching a P-wave, an S-wave, or a surface wave?

6.8 For the seismograms shown below:

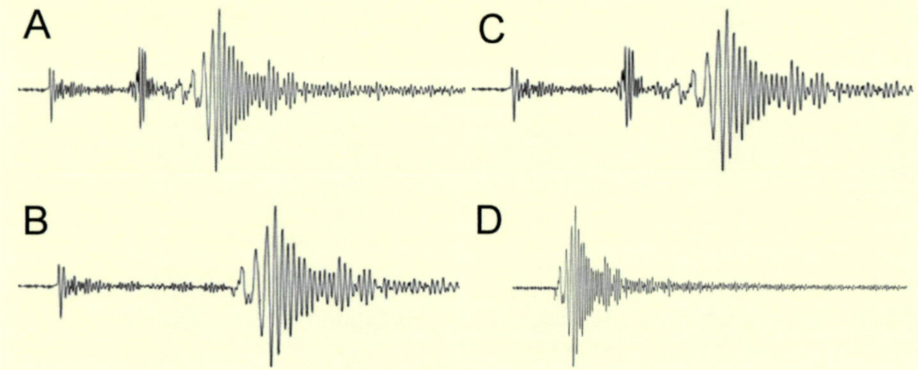

a. Which is not from an earthquake?

b. Of the ones that are from an earthquake, which one can you tell is the closest to it?

c. Of the ones that are from an earthquake, which is in the S-wave shadow zone?

Reflection and Refraction

To understand the last thing I want to cover with respect to seismic waves, you need to learn a bit more about wave behavior. When a wave encounters a new material, there are three things that can happen. First, the wave might be absorbed, which stops it altogether. For example, have you ever noticed that on a sunny day, a black shirt makes you hotter than a white shirt? That's because a black shirt absorbs all the light waves coming to it from the sun. As a result, it absorbs the energy that the waves are carrying. Because the shirt gains energy, it gets hotter. Well, seismic waves can be absorbed as they enter new material as well. This is part of the reason a seismic wave eventually dies out. As it goes from one material to another, some of the energy gets absorbed, reducing the total energy being carried by the wave.

You have a lot of experience with the next thing that can happen – **reflection**. When you look at a mirror, you see yourself. That's because light waves hit you, reflect off you, hit the mirror, and then reflect off the mirror and back into your eyes. Indeed, the very fact that you can see anything is because of light waves being reflected. When light waves hit any object, they can bounce off the object and enter your eyes. Without any light, you can't see anything, because there is nothing to reflect off objects and enter your eyes. Seismic waves can reflect when they encounter new material as well.

Indeed, some of the surface waves in an earthquake are caused by seismic waves reflecting off of new material and traveling back up to the surface. This leads to a lot of complexity when it comes to interpreting seismic waves, so I don't want you to worry about that. However, seismologists must.

What you need to worry about is the last thing that can happen when a wave hits a new material: **refraction** (rih frak' shun).

Refraction – Deflection of a wave due to a change in the material through which it is traveling

The best way to see what this definition means is to do an experiment.

Experiment 6.3: Refraction of Light Waves

Supplies:
- A quarter or other large coin
- A bowl that is not transparent and doesn't have much of a rim at the top
- Water
- A container that will hold enough water to fill the bowl but can be held in one hand (like a pitcher)
- A chair and table

Instructions:
1. Fill the container with enough water to fill the bowl.
2. Take the empty bowl, quarter, and container of water to the table and sit down in the chair.
3. Set the quarter at the bottom of the bowl, near the center.
4. Put the bowl on the table near you so that you can see the quarter while you are sitting up straight.
5. Slowly push the bowl away from you without changing your posture. Continue to sit up straight, and stop pushing the bowl the instant you can no longer see the quarter sitting at the bottom.
6. Gently add water from the container to the bowl. Do it slowly so that it doesn't disturb the quarter, and continue to sit with the same posture as before.
7. Watch the surface of the water as you fill the bowl. You should eventually start to see the quarter again.
8. Stop pouring water into the bowl once you see all but the very edge of the quarter.
9. Hold the bowl and shake it so that water sloshes back and forth in the bowl. You want it sloshing away from and toward you. Don't change the average position of the bowl. Just get the water sloshing back and forth. Don't do it too much, because you don't want the water to spill out of the bowl. Just get the surface of the water sloshing back and forth a bit. As the water sloshes, concentrate on how the appearance of the quarter changes.
10. Clean up your mess.

As the instructions indicated, when you added water to the bowl, you should have started seeing the quarter again, even though you couldn't see it when the bowl was empty. If you got the water sloshing properly, you should have seen that as the water sloshed away from you, more of the quarter was visible. As it sloshed toward you, less of the quarter was visible. Most likely, the quarter also changed shape in weird ways as the water sloshed around.

What explains these results? To see the quarter, light had to hit the quarter, reflect off the quarter, and then enter your eyes, as shown on the left frame of the illustration on the next page. Once you moved the bowl farther away, the light rays that bounced off the quarter and were traveling towards your eyes got blocked by the edge of the bowl. As a result, they didn't enter your eyes, and you could no longer see the quarter, as shown in the middle frame of the illustration below.

When you added water, light traveled through the water, bounced off the quarter, traveled back up through the water, and then left the water to travel to your eyes. Once the water level got high enough, you could see the quarter again. But it wasn't just because the water was high. It was also because of refraction. When the light entered the water, its path was deflected. In fact, the path it traveled became steeper, as shown in the right frame of the illustration below. When the light reflected off the quarter, traveled to the water's surface, and entered the air, its path was deflected again, as shown in the right side of the illustration below. Because of those two deflections, there was now light bouncing off the quarter and entering your eyes again, so you once again saw the quarter.

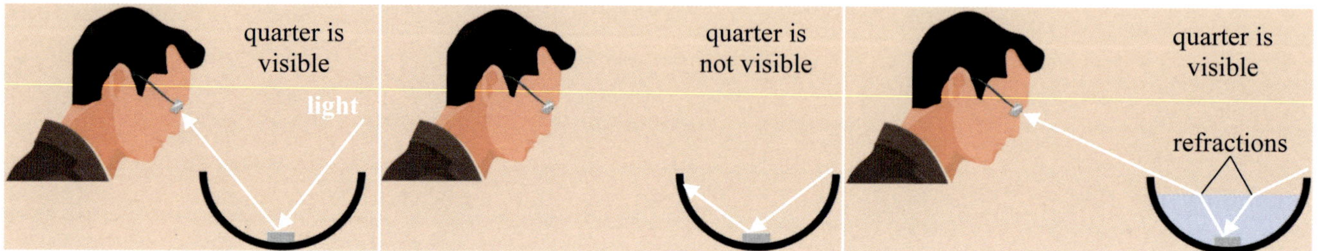

In your experiment, refraction in the water is what caused the quarter to reappear.

Why did I have you slosh the water around at the end of the experiment? I wanted to show you that what you were seeing was a result of how light was affected by the surface of the water. When the water tilted the right way, it changed the deflection, making more or less of the quarter visible. In addition, it distorted the shape of the quarter, all because it was changing the refraction of the light waves.

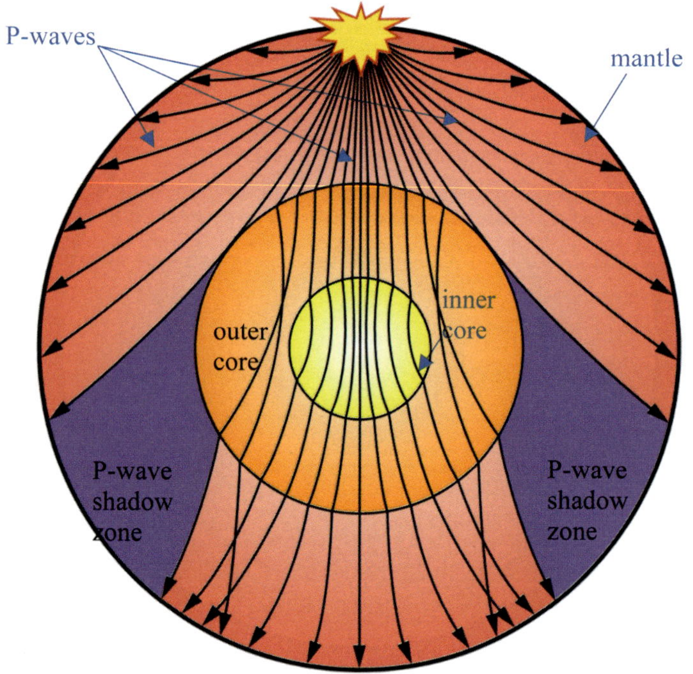

We know there is a solid inner core is because of the way P-waves behave when they travel through the earth.

When you take high school physics, you will learn a lot more about refraction. For the purpose of this course, you just need to realize that refraction deflects seismic waves, like it deflects light waves. As a result, the way seismic waves travel through the earth is sensitive to the changes that occur in the materials through which they move. When P-waves move through the earth, for example, they experience refraction when they encounter the liquid outer core. This changes the path of those waves, producing a P-wave shadow zone, which is shown on the left. Of course, that illustration is flat, but the earth is really a sphere. Thus, while it looks like there are two separate P-wave shadow zones, there is really just one. It forms a three-dimensional "donut" shape that wraps around the entire globe.

As I told you way back in Chapter 2, seismic waves have allowed earth scientists to map out the details of the interior of the earth, even though we have never been below the crust. Now, hopefully, you have a better idea of how that has been accomplished.

Comprehension Check

6.9 Suppose you are standing in a shallow pool and looking down into the water. Off to one side, you see a rock sitting underwater, and you decide to lean down and pick it up. Will the rock be where it appears to be? Why or why not?

Tsunamis

While we normally think about earthquakes destroying things by shaking them violently, there is another way they can wreak havoc. If an earthquake happens along a large dip-slip fault deep in the ocean, the water above the fault is pushed up, forming waves that are initially pretty small. Once they are formed, however, they start moving away from the fault, usually at a speed of about 700-800 km/hr (400-500 mph). As they encounter shallower waters, they get taller and slow down. The closer they get to shore, the taller they get. When they finally reach land, they are huge waves that can cause significant amounts of damage. This series of waves is called a **tsunami** (soo nah' mee). People often call a tsunami a "tidal wave," but that is not correct. A tidal wave is not caused by an earthquake. It is caused by the tides, which you will learn about later on in this course. Also, tidal waves are small and do not cause large amounts of damage unless amplified by very strong winds from a storm.

A tsunami starts as a small wave caused by an earthquake deep in the ocean.

There are two specific tsunamis that I would like to mention because of the damage they caused. In 2004, an earthquake that measured 9.1 on the Richter scale occurred deep in the Indian Ocean. It produced a tsunami that hit many islands, killing about 228,000 people without warning. The earthquake was far enough away and so deep in the ocean that the people on the islands didn't even feel the vibrations. Had it not happened in the ocean, the islands wouldn't have suffered nearly as much damage. Because it happened in the ocean, however, the tsunami was able to carry the earthquake's energy a long way and do a lot of damage.

In 2011, another tsunami was caused by another earthquake that measured 9.1 on the Richter scale. This one was in the Pacific Ocean, off the coast of Japan. Because Japan monitors earthquakes, there was some warning before this tsunami struck the coast. Still, it killed nearly 20,000 people. Also, it destroyed some of the safety systems at the Fukushima Daiichi nuclear power plant on Japan's east coast. This caused several reactors to overheat, which resulted in the release of radioactivity into the surroundings. People living within 30 kilometers of the power plant were forced to evacuate to avoid being exposed to too much radiation. As of the year this book was published (2021), that area is still off-limits to people because of the danger from the radioactivity. Surprisingly, the Onagawa nuclear power plant was *closer* to the epicenter of the earthquake on Japan's east coast, but it did not experience any major damage because it was surrounded by a high wall that was designed to withstand tsunamis. The Fukushima Daiichi nuclear power plant's wall was less than half as high and could not protect the plant against the tsunami.

Earth Has Its Ups and Downs

Of course, motion along a fault has consequences beyond the earthquakes it can produce. It can also form mountains. After all, at a dip-slip fault, one block of rock moves down, while the other moves up. If there is enough movement like that, a mountain and a valley can form. Lots of dip-slip faults close together can produce a chain of mountains, such as the one shown below:

These are some of the Teton Mountains in Wyoming, which are fault-block mountains.

Mountains formed this way are called **fault-block mountains**, because they form as blocks of rocks move along a fault.

Fault-block mountain – A mountain formed as a result of the motion of rocks along a dip-slip fault

Now please understand that while motion along a fault is what initially produced the mountains, weathering and erosion also happen, shaping the mountains into what you see today. Also, please understand that the motion which produced them is also still going on. Thus, the heights of the Teton mountains change. Those changes are not noticeable to you, because they are currently small compared to the heights of the mountains themselves. Nevertheless, they can be measured. Data indicate that on average, the Teton mountains are growing taller at a rate of about 0.8 millimeters per year.

Before I go on, I need to make a point about the picture above. Do you see that a couple of the mountains appear to have smoke coming from them? As a result, you might think that that they are volcanoes. They aren't. In fact, that's not even smoke; it's clouds. As you will learn later on in the course, mountains have an effect on cloud formation, so sometimes pictures of mountains will look like they have smoke coming out of them, but they really don't.

There is another kind of motion that can occur along a fault, which leads to a completely different kind of mountain. Do you remember what happens when two plates containing continental crust collide? Subduction doesn't occur, because neither crust is denser. As a result, the crust just buckles and folds, like cars in a head-on collision. As the illustration at the top of page 144 shows, that forms mountains, which we call **folded mountains**.

Folded mountain – A mountain formed by the folding of the earth's crust caused by the collision of tectonic plates

Interestingly enough, the tallest mountain ranges on the surface of the earth are formed in this way. You already learned that the Himalayas were formed by the collision of two continental plates, so they are folded mountains. The world's highest mountain, Mount Everest, is part of the Himalayas, along with thirty other mountains that are among the highest in the world.

These are some of the Himalaya Mountains in Nepal. Mount Everest is at the center. Notice once again that the clouds make it look like some of the mountains are smoking.

Of course, because the tectonic plates are moving, these mountains are changing as well. As you learned in Chapter 5, there are a lot of earthquakes at this convergent boundary. In addition, the mountains are growing at a rate of about 6.1 centimeters (2.4 inches) each year. Once again, while folding produced these mountains, they are also shaped by weathering and erosion.

Motion along faults isn't the only way that mountains are produced. Remember that hot rocks are always rising in the mantle due to convection currents. If they rise far enough, they can push up against the rocks of the lithosphere. Especially if the rocks are sedimentary, they can bend, forming a dome that rises on the surface. The magma cools, forming intrusive igneous rock, and the resulting rise is called a **dome mountain**.

Dome mountain – A mountain formed by magma pushing up on the lithosphere without breaking through it.

Navajo Mountain in Utah is an example of a domed mountain, and as shown on the next page, the dome shape is very apparent.

This is Navaho Mountain in Utah, as seen from Lake Powell. Its dome shape is very obvious.

Notice that while the dome shape is apparent, it isn't a smooth dome. That's because weathering and erosion have occurred, producing hills and valleys on the dome itself.

Sometimes, the weathering and erosion makes it very hard to see the dome shape. Consider, for example, the Black Hills of South Dakota. A view of those hills is shown in the image below on the left. Does that look like a dome? No. Nevertheless, the Black Hills are formed out of a single dome structure. You can see that if you view it from space, as shown in the satellite image on the right.

The Black Hills of South Dakota (left) are part of a single dome structure, which can be seen in a satellite image (right).

The dome was formed by granite that was lifted vertically, and the sedimentary rocks above were weathered and eroded, forming hills and valleys. In some places, weathering and erosion removed all the sedimentary rock, exposing the granite that formed the dome to begin with. Mount Rushmore, the famous mountain that has the faces of four important U.S. presidents carved into it, is actually part of that exposed granite.

As you have already learned, magma doesn't have to turn into intrusive igneous rock like granite. It might rise all the way to the surface and become lava. At that point, a volcano is formed. The resulting mountain is, not surprisingly, called a **volcanic mountain**.

<u>Volcanic mountain</u> – A mountain formed by a volcano

Volcanic mountains start out as a crack in the crust where magma escapes and becomes lava. The lava cools into igneous rock, surrounding the crack. As more lava is formed, it must rise above and then flow down the newly-formed igneous rock. Once the lava stops flowing, a cone-shaped mountain is produced, such as Mount Rainier (ray near') pictured below. Once again, weathering and erosion can change the shape of the mountain.

Now remember, there are lots of volcanoes under the ocean. If they grow large enough, they can reach the surface of the ocean and form an island. As you learned in the previous chapter, that's how the Hawaiian Islands formed. I told you earlier that the highest mountain in the world is a folded mountain – Mount Everest. While that's true, it isn't the tallest mountain. The peak of Mount Everest is 8.8 km above the surface of the earth, and that is higher than any other mountain. However, the volcanic mountain Mauna Kea (mow' nuh kay' uh), which is on the island of Hawaii, is 10.2 km above the spot from which it formed. Thus, it's the *tallest* mountain in the world. Most of it is underwater, however, so its peak is "only" 4.2 km above the surface of the earth. While it is not the highest mountain in the world, it is the tallest!

Mount Rainier in the state of Washington is a volcanic mountain.

Comprehension Check

6.10 If they have time, boaters are told to quickly move their boats out from the shore and into deep water when a tsunami is approaching. Why?

6.11 Careful measurements show that a fault-block mountain is actually getting shorter, even though it is a part of the block that is rising along the fault. What is happening to cause it to get shorter?

6.12 Geologists are drilling down a dome mountain. They are currently drilling through sedimentary rock. Will they continue to drill through sedimentary rock all the way to the bottom of the mountain? If not, what type of rock (be as specific as you can be) will they eventually hit?

More About Volcanoes

Since the previous section ended with volcanic mountains, I wanted to give you a bit more detail about volcanoes themselves. When magma escapes the crust to become lava, we say that the magma has **erupted**. Thus, when a volcano releases lava, we say that the volcano is **erupting**. You might be surprised to learn that this is a fairly common event. About 50 or 60 volcanoes erupt each year somewhere on the earth. In other words, a volcano erupts somewhere on the earth about once each week! Also, as you might expect, volcanoes are not randomly scattered across the earth. The map on the next page shows you where the major volcanoes on the earth are found.

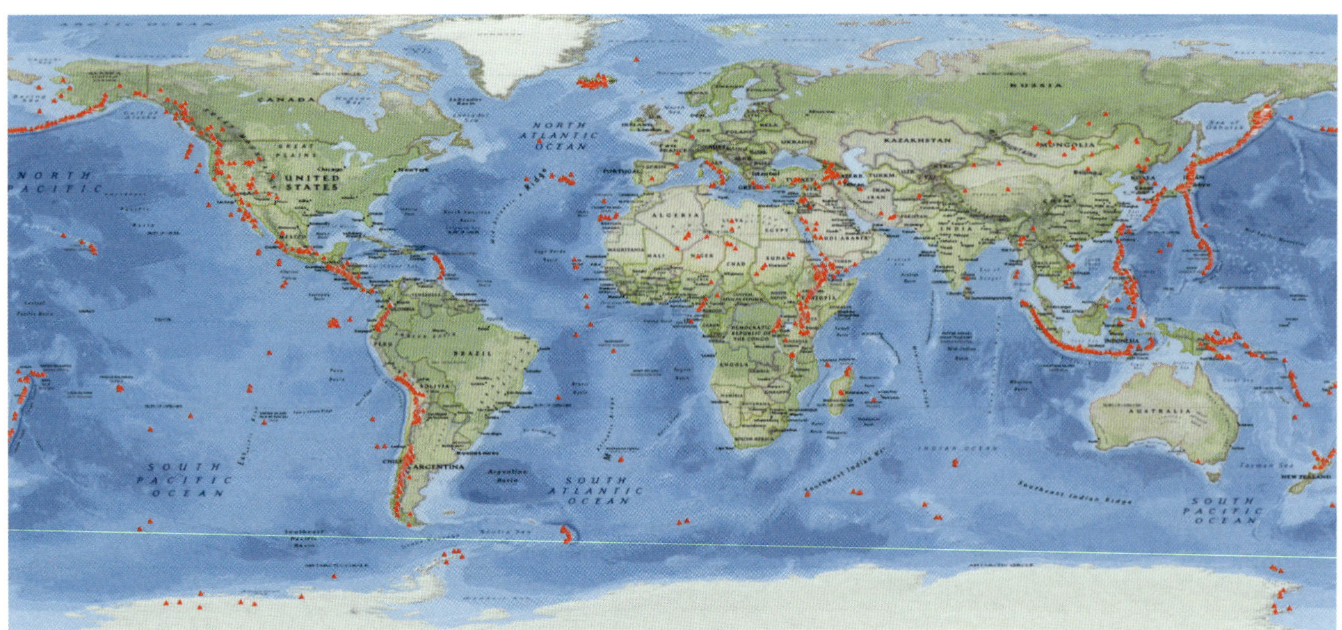

Each red dot on this map indicates a major volcano.

Notice all the volcanoes that line the west coast of the North and South America, the oceans east of Asia, and the oceans north of Australia. They form a ring around the Pacific Ocean, so these volcanoes are said to be a part of the **Ring of Fire**. Can you think of why the ring of fire exists? Remember that the Pacific Ocean sits mostly on one plate. The Ring of Fire is formed by that plate colliding with the other plates. That makes arc volcanoes, as you learned in the previous chapter. Because of all this plate activity, the Ring of Fire has the most volcanic eruptions and the most earthquakes in the world.

While erupting volcanoes emit lava, they can also emit hot pieces of solid rock, which are collectively referred to as **pyroclastic** (pie' roh klas' tik) **material**.

Pyroclastic material – Hot pieces of solid rock emitted during many volcanic eruptions

Pyroclastic material comes in various sizes. When the rock pieces are roughly the size of sand, they are called **volcanic ash**. However, because of the energy involved, an erupting volcano can emit large pieces of rock as well. Larger rocks emitted by an erupting volcano are often called "volcanic bombs."

In fact, the pyroclastic materials emitted by a volcano are the chief danger to people who live close to one. While lava is very dangerous because of its temperature, it moves very slowly. Thus, it is usually quite easy for people to avoid the lava produced by a volcano. However, pyroclastic material gets shot out of the volcano at high speeds, rises in the air, falls back down, and starts to avalanche down the mountain, often at speeds of several hundred kilometers per hour! You might have heard about the ancient city of Pompeii that was destroyed by the eruption of Mount Vesuvius. The people were not killed by the lava that flowed from the eruption. They were killed by the pyroclastic materials that covered the city quickly and violently!

All volcanoes also emit gases when they erupt. The most common gas is water vapor, because there is a lot of water in the mantle. In fact, there is more water in the mantle than there is in all the oceans! There are many other gases that can be emitted as well, including sulfur dioxide, hydrogen sulfide, and some acidic gases. These can be very dangerous. In addition, carbon dioxide is also emitted by volcanoes. While it is a normal component of the air you breathe, its concentration is low. When its concentration increases, it can become dangerous as well. On the island of Hawaii, for example, the Kilauea (key' lu way' uh) Volcano often emits gases that can be dangerous. The state of Hawaii has a warning system that lets its citizens know when they should avoid being outside in certain areas when the gases get too concentrated.

The images below are of two different volcanic eruptions in Indonesia, which is part of the Ring of Fire. On the left, you can clearly see lava flowing down the sides of the volcano as well as being thrown up in the air and falling back down in "arcs of fire." However, notice the black "smoke." That's the pyroclastic material. On the right, you see a closeup view of an eruption that has only a little lava (the red glows in the picture) but a lot of pyroclastic material. Both of the eruptions are also emitting gases, but you don't see them in the pictures.

A volcano emitting mostly lava and some pyroclastic material (left) and one emitting mostly pyroclastic material (right)

One other thing that most people don't think about when it comes to volcanoes is that their tops are often covered in snow and ice, as shown in the picture of Mount Rainier on page 175. When such a volcano erupts, the heat melts the snow and ice, and the pyroclastic materials mix with the resulting water to make mud. That mud then flows down the volcano, making a **mudflow**.

If you live near a volcanic mountain, should you be concerned about it erupting again? That depends on what's going on underneath. Generally speaking, volcanoes can be classified as either **active**, **dormant**, or **extinct**. An active volcano is one that has erupted recently and is expected to erupt at some time in the future. What do I mean by "recently?" That depends. Some earth scientists say that if it erupted within recorded history, it is an active volcano. Others extend the date back into the more distant past.

A dormant volcano, on the other hand, is one that hasn't erupted recently. Once again, the exact definition of "recently" changes depending on which scientists you ask, but a dormant volcano is one that hasn't erupted for quite some time. However, there is still the chance for it to erupt again. An extinct volcano, on the other hand, is one that is never expected to erupt again.

There are actually three different types of volcanoes, and you can usually tell the difference by looking at their shapes. However, they all have one thing in common: a **vent** from which magma escapes and becomes lava. First, there are **cinder cones**, the smallest type of volcano. They mostly

These pictures show a cinder cone (top), a composite cone (middle), and a shield volcano (bottom).

erupt pyroclastic materials, which fall as cinders around the vent. Most cinder cones have a bowl-shaped crater at the top, as shown in the picture on the left.

Composite cones, also called stratovolcanoes, are some of the largest and most breathtaking of the earth's volcanoes. They generally have steep sides, and the rocks that make them up are a composite of igneous rock formed from the lava that flowed forth and pyroclastic materials that were emitted during eruptions. Most composite cones have a crater at the top of the summit and a vent that travels all the way up to the crater. There are often cracks in the sides of the cone as well, so lava can also escape the vent through the sides of the volcano. While you can't see the crater or vent in the composite cone shown on the left, you can see its steep sides.

Shield volcanoes are not nearly as steep as composite cones, and they are formed almost completely from lava that flows gently from the vent. Because of the nature of the flow, the volcanoes are much wider than they are tall, as shown in the picture on the left. They get their name because they look like a round shield that has been laid on the ground.

Now think about the magma that forms a volcano. It has risen up from the mantle, but until the eruption, it is being held underground by the surrounding rock. When the magma finally escapes through the volcano's vent, that can leave an empty space where there once was a lot of hot, pressurized magma. What happens to that empty space? Well, it's possible that the surrounding rocks will collapse into the chamber. When that happens, a large crater forms at the top of the volcano. That crater is called a **caldera** (kal der' uh). Depending on the climate, that caldera can become filled with water, forming a lake like Crater Lake in Oregon, which is pictured on the left.

Crater Lake in Oregon is a water-filled caldera.

Volcanoes can give us a glimpse into the mantle, because they are fed from magma that originated there. As I already told you, we know that there is a lot of water in the mantle, and one reason we know that is because water vapor is the most abundant gas emitted by volcanoes. That's just the beginning, however. Look at the picture below. If you remember Chapter 3 well, you might recognize that the green oval in the center is the mineral olivine. It is surrounded by basalt, which you already learned is an extrusive igneous rock.

So, the basalt in the picture was formed from lava that came from a volcano. However, it surrounds the mineral olivine. What does that tell you? The olivine was already formed before the magma escaped, right? After all, if the olivine was liquid when the lava was flowing, it would be spread out and mixed with the lava. The only way the olivine could look like that is if it were already solid while the lava flowed. Since the olivine wasn't part of the lava that formed the basalt, we call it a **xenolith** (zee' nuh lith). Xenoliths give earth scientists great insights into what is happening in the mantle, because the properties of the xenolith depend on the pressure and temperature at which it was formed.

The olivine in this basalt is called a xenolith.

Think about diamonds, which you learned about in Chapter 3. Remember, diamond and graphite are different forms of the same element: carbon. Under most conditions, carbon makes graphite. However, at very high temperatures (like 1200 °C) and very high pressures (50,000 times more than atmospheric pressure), carbon forms diamond. Well, diamonds are found as xenoliths in kimberlite, an intrusive igneous rock. The very presence of diamonds in igneous rock means that there must be a place in the mantle where such temperatures and pressures exist.

Comprehension Check

6.13 Identify the phase (solid, liquid, or gas) of each of the following volcanic emissions: pyroclastic material, water, lava.

6.14 Are the volcanoes in the Ring of Fire more likely to be grouped in straight lines or arcs?

6.15 A volcano is very, very tall. Most likely, what kind of volcano is it?

6.16 A xenolith in some extrusive, igneous rock has properties that indicate it was formed at a temperature of 1,000 °C. Did it form above or below the depth where diamonds form?

Answers to the Comprehension Check Questions

6.1 Static friction is stronger. You should have noticed that once you got the papers moving, it became a bit easier to keep them moving until they stopped. You have probably noticed this when trying to push something along the ground. It takes more force to get something moving than to keep it moving.

6.2. B is the hanging wall, and it is a normal fault. Imagine any line on the fault and then move straight up (not diagonally along the fault – straight up vertically). That's the hanging wall. Since the hanging wall moved down, it is a normal fault.

6.3 The large force of friction produces a stronger earthquake. Remember, the stress that builds up before the rebound is due to friction. The more friction there is, the more stress is needed, so the more violent the rebound.

6.4 The magnitude 8 earthquake is 32,768 times stronger than the magnitude 5 earthquake. Each number on the Richter scale represents multiplying by 32. To get from 5 to 8, you have to increase the number by 3, so you have to multiply by 32 three times. That means a magnitude 8 earthquake is $32 \cdot 32 \cdot 32 = 32{,}768$ times stronger.

6.5 The waves will appear at the epicenter first. They will also appear to be traveling away from the epicenter in circles, but that didn't have to be a part of your answer.

6.6 The second has more energy. Since frequency determines energy, this question is asking which light wave has the higher frequency. Well, frequency and wavelength are opposites, so the higher-frequency wave will be the one with the lower wavelength. Thus, it's the one with the shorter distance between its crests.

6.7 You were watching an S-wave. Since it was vibrating up and down but moving left to right, the vibration was perpendicular to the direction of travel. That means it is transverse. There are transverse surface waves, but they don't happen inside the earth; they happen at the surface. The only transverse waves inside the earth are S-waves.

6.8 a. D. An earthquake seismogram should have at least two different kinds of waves.

b. A. The shorter the time between the P-waves and S-waves, the closer. Of the two that have both S-waves and P-waves, A has the shorter time between them. B has no S-waves, so you can't tell how close it is just from the seismograph.

c. B. The other two earthquake seismograms have S-waves. B has P-waves and surface waves, but the S-waves are missing.

6.9 It will not be where it appears to be, because of refraction. Think about what happened in your experiment. When you added water to the bowl, you could see the quarter, even though it wasn't at a place where you could see it. That's because refraction caused the quarter to appear to be somewhere else. The deflection of light made it appear to be in a different place. If you have ever been taught how to spear fish while you are wading in water, you learned that you have to aim lower and in front of the fish you see underwater, since refraction makes it appear to be somewhere it is not.

6.10 The waves of a tsunami are small when you are in deep water. If you can get into deep enough water, the waves will be small enough that the boat will not be harmed.

6.11 Weathering and erosion are making it shorter. You could have just said erosion, since it's the fact that sediments are being carried away that shortens the mountain. Remember, motion along a fault isn't the only thing shaping the mountains.

6.12 They will eventually hit intrusive igneous rock. Since magma causes the dome to form, there must be igneous rock at the bottom, because frozen magma is igneous. Since it formed underneath the earth, it is intrusive. If you answered "granite," you were a bit too specific, because there are other intrusive igneous rocks, so not all dome mountains have granite in them.

6.13 Pyroclastic material is solid, water is gas, and lava is liquid. Remember, at the temperatures involved in a volcano, water is not a liquid. It is a gas.

6.14 They are more likely to be grouped in arcs. Since the ring of fire involves a convergent plate in which at least one plate is oceanic, there will be subduction, which tends to form arcs of volcanoes, as you learned in Chapter 5.

6.15 It is probably a composite cone. Shield volcanoes are wide and not very tall, and cinder cones are pretty small.

6.16 It formed above where diamonds form. Remember, the deeper you go in the geosphere, the hotter it gets. Diamonds form at about 1200 °C, so that must be deeper than the part of the geosphere that is 1,000 °C.

Chapter Review

1. Define the following terms:

a. Linear scale
b. Logarithmic scale
c. Focus of an earthquake
d. Epicenter
e. Transverse wave
f. Longitudinal wave
g. Refraction
h. Fault-block mountain
i. Folded mountain
j. Dome mountain
k. Volcanic mountain
l. Pyroclastic material

2. Draw and label the four types of faults, indicating the hanging wall and foot wall in each type.

3. Explain elastic rebound theory in your own words.

4. A seismograph measures an earthquake of magnitude 3 on the Richter scale, followed by one of magnitude 7, followed by one of magnitude 4. Identify the foreshock, aftershock, and main earthquake. Compare the strength the main earthquake to the foreshock and the aftershock.

5. You are comparing two waves. The first has a longer wavelength than the second. How do their frequencies compare?

6. Of the two types of seismic body waves, which is longitudinal and which is transverse? Which moves faster? Which cannot travel through liquids underneath the earth's crust?

7. Consider the four seismograms shown below, each of which comes from the same earthquake:

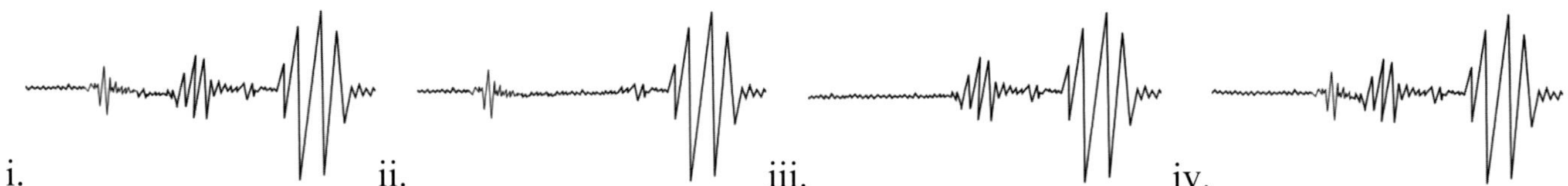

a. Which is from a seismograph in the P-wave shadow zone?
b. Which is from a seismograph in the S-wave shadow zone?
c. Which is closer to the epicenter: The seismograph that made (i) or the one that made (iv)?
d. Which are the surface waves in the sesimograms?

8. What one word can be used to explain why there is a P-wave shadow zone?

9. Where must the epicenter of an earthquake be in order to produce a tsunami? In what depth of water does the tsunami move quickly? In what depth of water are the waves large?

10. Which types of mountains can be formed by magma? What type can be formed by the convergence of plates that both have continental crust?

11. What single plate is responsible for the volcanoes in the Ring of Fire?

12. The tallest volcanoes are what type of volcano? What type is made mostly from pyroclastic material? What type is made mostly from lava and tends to be wide instead of really tall? What type is made from a lot of both lava and pyroclastic material?

13. How do xenoliths help us understand the geosphere below the crust of the earth?

Chapter 7: Fossils in Rocks

Introduction

Since I have covered everything that I want you to know about the geosphere as a whole, I now want to discuss one of the most fascinating aspects of the geosphere: **fossils** found in the crust. Of course, when I use the term "fossil," you probably know what I mean:

Fossil – Evidence of past life preserved within the geological record

However, please understand that when natural philosophers (what scientists used to be called) first started studying fossils, they didn't understand them at all. They had no idea that fossils were from creatures that were once alive! They actually thought that fossils were rocks that just happened to be formed in the shape of an animal or part of an animal. Indeed, the word "fossil" comes from the Greek word *fossus*, which means "to be dug up." Thus, initially the term "fossil" applied to anything that was dug up from the ground.

While you might find it odd that a scientist would think something shaped like an animal was just a rock that happened to look like the animal, you have to understand that the way scientists look at things is strongly influenced by the general beliefs of their culture. At that time, people believed that there was a "shaping force" in nature that determined the shape of everything. A clam, for example, got its shape not to suit its biological needs, but because the shaping force made clams look the way they did. Well, if this shaping force could make clams look the way they do, why couldn't it make rocks that looked like clams?

This stood as the standard explanation for why you could find things in rocks that looked like animals for quite some time. There were always those who disagreed. Leonardo Da Vinci, for example, was convinced that these interestingly-shaped "rocks" were, indeed, the preserved remains of once-living creatures. However, the vast majority of natural philosophers disagreed with him.

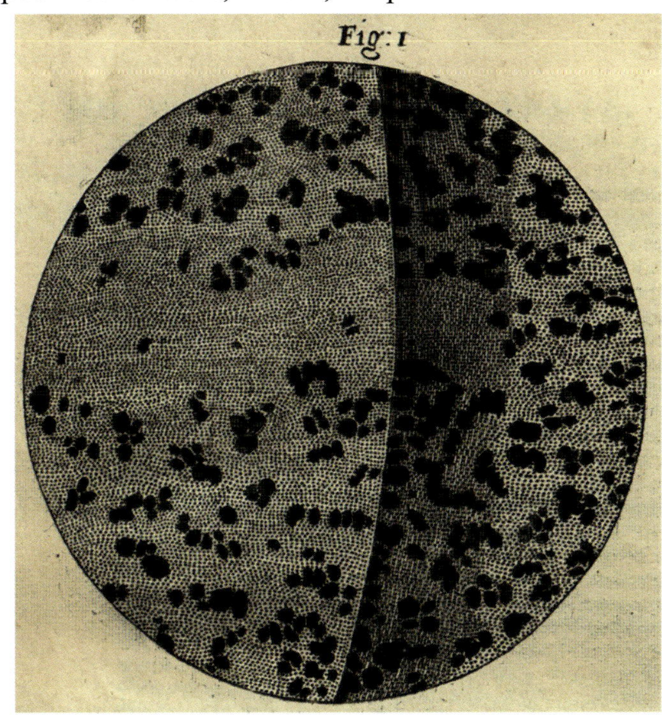

This is Robert Hooke Hooke's drawing of what petrified wood looks like under a microscope.

This started to change in the mid-1600s, when Robert Hooke, an English natural philosopher, started studying things from nature under a microscope that he had made. He looked at cork under the microscope, for example, and saw rectangular structures that looked like the cells that monks live in. In fact, that's what he called them. We now know that those cells are the basic building blocks of life.

In addition to being the first to describe and draw cells, he was also the first to carefully compare recently-dead organisms to the fossils that they resembled. He saw that their microscopic features were very similar. For example, the drawing on the right comes from his book, *Micrographia*, in which he reports on his microscope studies. It shows the structure of a "rock" that is shaped like a piece of wood. He compared that to what rotting oak looked like under a microscope, and concluded that the "rock" was wood at one time, and it hardened into something stonelike because of minerals that were deposited in the wood. Similarly, his microscopic comparison of shell-like "rocks" to actual animal shells led him to conclude that the "rocks" really were animal shells at one time. While his conclusions were controversial at the time, we now know that he (and others like Da Vinci) were correct. Fossils really do come from once-living organisms.

Different Fossils

What happens when an animal dies? Generally, its body decomposes, rotting away entirely. This is good, because the chemicals that make up the animal are useful, and when it decomposes, those chemicals can be used by other living things. However, if decomposition were the only thing that happened, we wouldn't have fossils. What keeps a fossil from decomposing? Actually, there are several answers to that question, but before we start exploring those answers, I want you to take a first look at the fossils in your kit.

Experiment 7.1: Examining Fossils

Supplies:
- The fossils in the Geology Basics Kit from the laboratory kit made for this course
- The magnifying glass from the laboratory kit made for this course
- A well-lit room

Instructions:
1. Remove specimens 16-20 from your Geology Basics Kit and put them on a flat surface.
2. Pick up specimen 16 and examine it with your naked eye, turning it over so you look at all sides.
3. Rub your finger on both sides of the fossil, concentrating on how it feels.
4. Make sure there is a lot of light in the room and examine it with the magnifying glass, once again looking at all sides. Just like you did when examining rocks, look for sparkles that indicate minerals as well as different colors that indicate different chemicals.
5. Repeat steps 2-4 with the other specimens, trying to notice as many differences between the specimens as possible. Is there a difference in the number and size of sparkles that you see? Is there a difference in the uniformity of the specimens? Specimen 18 is a fossil shark tooth, and it might be in a zipped bag. That's fine. Just pull it out and examine it like the others, but return it to the bag when you are done.
6. Keep the specimens out while you continue the reading, because you might want to look at them again.

Hopefully, you noticed differences among the specimens, but you should have noticed something they all shared – they are hard, like rocks. This is one reason natural philosophers confused them with rocks. Also like rocks, they had different textures. Specimens 16, 17, 19, and 20 were rough all over, while a large part of specimen 18 was smooth. The top was rougher, like the other specimens, but most of specimen 18 was pretty smooth.

Specimens 16, 17, and 20 were probably fairly uniform in color. There might have been some different colors in the specimens, but probably not as many different colors as you saw in specimen 19. Also, while the differences in color were distributed throughout all of those specimens, the differences in color for specimen 18 were found in specific parts of the fossil. Most of the fossil (the smooth part) was one color, while the U-shaped top was a different color. Most likely, there was also a slightly-different color in a band between the U-shaped top and the smooth part of the fossil.

Finally, you might have seen some sparkles in all the specimens, but you should have seen the most sparkles in specimen 19. What does that tell you? The mineral crystals in that specimen were bigger than the mineral crystals in the other specimens. It's possible you didn't see sparkles in some or all of the other specimens, because the mineral crystals were pretty small. Also, for specimen 18, the smooth part of the fossil probably had a nice sheen to it, but there probably weren't individual sparkles. You might have seen some small sparkles coming from the rougher parts of the specimen.

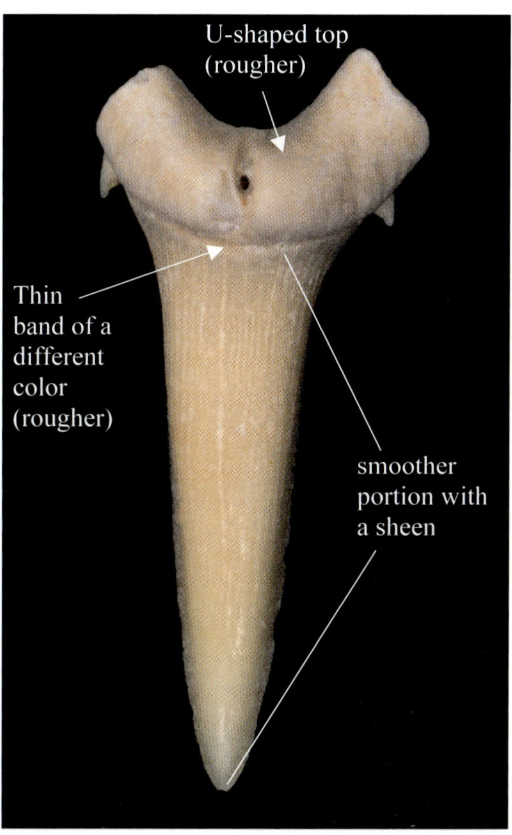

Specimen 18 in your kit might not have looked exactly like this, but it should have had the features pointed out in the picture.

As you will learn in this chapter, fossils form via different processes. Even when they form by the same process, the details of the process depend on the surroundings. As a result, the details of fossils can be very different from one another. However, regardless of the details, all the fossils you examined were as hard as rocks. That's what kept them from decomposing. Instead of rotting away like the remains of most animals, something caused these remains to turn hard and be preserved.

Regardless of the details, there is one thing we can say about all fossils – compared to the number of animals that have lived and died since this planet was created, fossils are incredibly rare. The processes that preserve a dead animal (or parts of a dead animal) are very specific, and they don't happen very often.

7. Put away your fossils and magnifying glass.

Comprehension Check

7.1 When you looked at specimen 16, you probably noticed right away that it was a fossil of a clam. What made you think it was a fossil and not just a rock that simply looked like a clam?

<u>The Key to Fossilization</u>

Fossils can form many different ways; however, those processes have something in common: They keep the remains of the animal from decomposing, at least for a while. After all, as I already mentioned, a dead organism tends to decompose, returning its chemicals to the environment so they can be used by other organisms. If a fossil is going to form, decomposition has to be stopped or delayed in some way. How does that happen? Let's start with an experiment.

Experiment 7.2: Partially Preventing a Reaction

Supplies:
- A medicine dropper from the laboratory kit made for this course
- Two small glasses, like juice glasses
- Vegetable oil (Olive oil or any other liquid cooking oil will work.)
- Active, dry yeast (available in the baking department of any supermarket)
- Hydrogen peroxide from the laboratory kit made for this course
- A ¼-teaspoon measuring spoon
- A spoon for stirring
- A sink

Instructions:
1. Add enough vegetable oil to one of the glasses so that it is about half full.
2. Add ¼ teaspoon of yeast to the oil in the glass.
3. Use the spoon to stir the yeast and oil together so they mix really well. Once you are done, you should see yeast throughout the oil, slowly sinking.
4. Put the glass off to the side.
5. Pour a small amount of hydrogen peroxide from the bottle in your kit into the other glass (the one that is still empty). You only need enough to fill the medicine dropper a few times.
6. Take the glass and dropper close to where the glass with oil is. Tilt the glass that has hydrogen peroxide in it and pull hydrogen peroxide into the dropper.
7. Look at the glass with the oil and yeast. Almost all the yeast should have settled to the bottom of the glass. If there is some yeast still floating, that's fine. You just want a fine covering of yeast at the bottom of the glass. If you don't see that, wait a few minutes until you do.
8. Hold the glass of yeast and oil in one hand at eye level so you are looking through the oil.
9. Using the other hand, hold the dropper over the center of the glass.
10. Give the dropper a squeeze so that a stream of hydrogen peroxide shoots out of the dropper and into the oil.
11. Watch what happens. If hydrogen peroxide didn't reach the bottom of the glass and cause an interesting effect, try again.
12. Once the effect dies away, repeat steps 10 and 11.
13. Clean up your mess. You will need to use a lot of soap to clean the oil out of the glass. Also, you will need to pull water into the medicine dropper and squeeze it out a few times.

If things went well in your experiment, you should have seen the hydrogen peroxide sink to the bottom of the glass, where bubbles started forming. Those bubbles then rose through the oil, forming what looked a bit like a tornado. However, you should have seen that the bubbles were isolated. They may have formed in a few places, but they should have formed a stream in those places, with no bubbles forming around them. It's even possible that the individual streams of bubbles merged together. Regardless, at the bottom of the glass, there should have been definite places where bubbles were streaming up to the surface, and other places where no bubbles formed at all.

Why did this happen? I am sure you know that oil and water don't mix. The hydrogen peroxide that you squirted into the oil is a water-based solution, so it did not mix with the oil. However, it is denser than oil. Thus, it sunk to the bottom of the glass. At the bottom of the glass, it encountered yeast. Yeast caused the hydrogen peroxide to quickly change into water and oxygen, and the oxygen made bubbles. Since the bubbles of oxygen were less dense than water, they floated up to the surface.

Because the hydrogen peroxide solution and oil didn't mix, the hydrogen peroxide couldn't move throughout the glass. The oil prevented that. As a result, the reaction was isolated to the place where it had been squirted to the bottom of the glass. Now, of course, the bubbles caused a lot of movement, and the hydrogen peroxide might have moved with the bubbles, merging with other streams of bubbles. Nevertheless, even while it was moving like that, it was "enveloped" by the oil, which restricted where it could move. To prevent decomposition, something like that has to happen to a dead organism – the dead organism has to be "enveloped" by something to "protect" it.

What must the organism be protected from? You might have already learned that there are other organisms whose main job is to decompose dead organisms. Not surprisingly, these organisms are called **decomposers.** Lots of bacteria are decomposers, as are many fungi. These decomposers have to be kept from the dead organism (at least for a while) so that fossilization can occur.

But that's not the end of the story. While decomposers speed up the process, lots of decomposition happens when the right chemicals are present. Thus, the dead organism must be protected from certain chemicals. That doesn't necessarily mean the chemicals must be completely blocked from the specimen. Remember, a chemical's behavior depends on its concentration. Decomposition can be slowed significantly by just reducing the amount of certain chemicals present. In addition, chemical reactions usually slow down at lower temperatures. Thus, cold temperatures slow down decomposition significantly.

How is all this actually accomplished? Let's start with a couple of straightforward examples. Look at the picture below. It shows you a sample of **amber**, which is a hard, partially transparent substance that looks like a mineral. In fact, it is often used in pieces of jewelry. Amber starts out as a thick, sticky liquid, called **resin**, that is produced mostly by certain trees but also by a few other plants. When such a plant is injured, it produces resin in order to cover the injury. Also, if an organism is burrowing into the plant, resin can sometimes kill the organism or at least discourage it from continuing to harm the plant. When resin is produced, it tends to flow on the surface of the plant, and any small animal that is there can get caught in the thick, sticky liquid. Once it is caught, it might not be able to escape, and as more resin is produced, the insect can become completely covered in resin.

Over time, the resin can harden into amber. Any insect that was covered in the resin will then be entombed in the amber. That's what you are seeing in the picture on the right. The insect you see there got covered by resin, and when the resin hardened, the insect was surrounded by amber. Because lots of insects crawl around on plants that can make resin, many samples of amber have insects entombed in them.

This is a sample of amber that has an insect entombed in it.

Now think about what happens once the insect is trapped by the resin. It dies, of course, but remember that resin is designed to protect the plant that makes it. It does so by keeping chemicals and organisms out. Thus, the resin "protects" the dead organism as well, which slows down decomposition. Depending on the resin and the chemical components that are in the surroundings, it might come close to stopping decomposition altogether. Once the resin hardens into amber, it is even more difficult for chemicals to get to the dead organism, so decomposition slows to a near halt.

However, decomposition never fully stops. The chemicals that make up an organism are very complex, and complex chemicals tend to decay over time, even if there aren't chemicals and organisms to help the process. So even in amber, decomposition does occur. As a result, the organisms in there are not as pristine as when they first died. Nevertheless, as you can see from the picture on the previous page, they can be remarkably well preserved! As a result, many **paleontologists** (pay' lee ahn' tah luh jists – scientists who study fossils), especially the ones who are interested in small animals that live around trees, are always on the lookout for amber.

Another way decomposition can be slowed significantly is when an animal becomes entombed in ice. The picture below is of a fossil mammoth, which is an elephant-like animal that had lots of hair on its body. The fossil was found in Siberia in 2014, where it was surrounded by ice. It was removed from the ice and shipped to a museum, and the shipping container was also filled with ice. The picture shown here was taken shortly after it was removed from the shipping container and placed in a freezer, so the white that you see on the top of the mammoth is ice that was still sticking to it after it was removed from its shipping container.

This mammoth was encased in ice, which is why it didn't decompose much.

The ice it was surrounded by after it died helped keep decomposers and chemicals away from the dead animal, and the low temperature slowed down what decomposition was happening, which is why the fossil was kept in ice while being shipped. As a result, the animal is very well-preserved. For example, it is rare to find skin and hair on a fossil, since they decompose easily. However, you can see both on this fossil.

Being entombed in ice or amber produces such remarkably well-preserved fossils that you can often find some original chemicals from the organism that have not yet decayed away! That allows paleontologists to learn a lot about the organism. However, those aren't the only ways decomposition can be slowed so that fossilization can occur. If a fossil is quickly covered in sediments, especially ones that have a lot of silt and/or clay, the decomposers and chemicals can be blocked well enough to allow fossilization to occur. You will learn more about that the next time you do science.

Comprehension Check

7.2 Suppose you are looking at a gallery of fossils that were entombed in amber. Next door, there is a gallery of fossils that were entombed in ice. Overall, which gallery would house the larger fossils? Why?

Fossil Molds and Casts

In the previous section, you learned about fossils that were made because decomposition had been slowed so much that the remains of the organism did not fully decompose. But that's not required to form a fossil. In fact, most fossils are from organisms that completely or almost completely decomposed. However, decomposition was slowed long enough for something else to happen. Consider, for example, the picture on the right. As you can see, there is a pattern in the center of the rock. That pattern was left by an animal known as an **ammonite** (am' uh night'). The sediments that make up the rock covered the ammonite, and the ammonite was partially protected from the organisms and chemicals that would cause decomposition. As a result, it lasted long enough for the sediments to undergo lithification.

This fossil mold is actually on a castle wall in Poland. The rock was probably split by the builders, exposing half of the mold.

As time went on, decomposition continued, and the ammonite eventually decayed away. However, the rock was already hard around it, so an ammonite-shaped space was left inside the rock. Eventually, the rock was broken (either by weathering or by someone who wanted to use the rock), and one side of that ammonite-shaped space was exposed. If the rock split nicely, there should be another rock that contains the other side. So here we have a fossil of an animal, but the animal is completely decayed away. However, its shape is preserved, and as a result, we can learn something about the animal. These kinds of fossils are called **fossil molds**.

Fossil mold – A fossil formed when sediments lithify around an organism and it decays afterwards, leaving behind a cavity in the rock

Now, of course, a fossil mold preserves only the shape of an organism, but that can still tell us something. If you look at the fossil mold above, you can see that the animal was shaped like a spiral. There are spiral-shaped organisms that are alive today, such as the nautilus that is pictured on the right. Is the fossil mold above from a nautilus? No. While the overall shape is similar, notice that the nautilus's body is smooth. It wouldn't leave the bumpy spiral that is preserved in the mold. In fact, there is no spiral-shaped animal we have seen alive today that has the bumpy spiral shape we see in the mold. As a result, we assume that the fossil is from an **extinct** animal, meaning there were once living versions of it, but now there are none. Even though the animal is extinct, its shape suggests it might be similar to a nautilus, which suggests that perhaps this is an animal that lived underwater, like the nautilus does today.

This spiral-shaped creature is called a nautilus. It is similar to, but not the same as, an ammonite.

This picture shows a mold (right) and a cast (left) of a trilobite.

Now consider the picture on the left. It shows a rock that has been split open. What do you see? On the right side, you see a fossil mold. But what do you see on the left? You see something that looks like it made the mold. This fossil is unlike any known living animal, so once again, we think it is extinct. Notice that its body has three sections. Those sections are called "lobes," so this is a three-lobed animal, which we call a **trilobite** (try' luh byte'). So are you seeing a preserved trilobite on the left side of the picture? Surprisingly, the answer is, "No."

While it looks like the object on the left made the object on the right, that's actually backwards. This fossil started off as a mold. Over time, however, that mold filled with sediments, and those sediments eventually lithified (went through lithification), forming a rock inside another rock. Of course, that rock took the shape of the mold. Fossils formed this way are called **fossil casts**.

Fossil cast – A fossil formed when sediments enter a fossil mold and lithify

So while the object on the left side of the photo above looks like it could be a preserved trilobite, it is really just a trilobite-shaped rock that was formed from the mold the actual trilobite left behind.

Just like a mold, a cast only tells us about the exterior of the organism that left the fossil behind. Thus, these kinds of fossils don't tell us nearly as much as fossils that were entombed in ice or amber. Nevertheless, they can show us enough details about the organism to teach us something. In the case of both fossils shown in this section, for example, we can learn enough to conclude that the animals which made them are not the same as any known animals alive today. Thus, they are remains of extinct animals.

It Looks Like a Mold, But…

This leaf fossilized because it was compressed into sedimentary rock.

Now look at the picture on the left. What does that look like to you? It looks a lot like a mold. However, notice how flat it is. It's so flat that it barely has any thickness at all. So, while the leaf that was once alive got covered in sediment, it didn't really leave a cavity in the rock. That's because the sediment experienced a lot of pressure before and/or after it lithified. This compressed the fossil. A lot of organisms would get crushed or at least distorted by the pressure, but leaves are so flat that they don't get destroyed by the pressure. In the end, they just leave a flat imprint behind. Fossils like these are called **compression fossils**, because they are formed as a result of being compressed.

Compression Fossil – A fossil preserved in sedimentary rock that has undergone compression

Besides being really flat instead of three-dimensional, there are other differences between molds and compression fossils. First, go back and look at the fossil leaf again. See how much detail is preserved. Notice you can see the veins in the leaf as well as the blade. That's more than just the outside shape of the organism.

Second, unlike molds, compression fossils often have some remnant of the chemicals that made up the organism. Most of the molecules that are found in an organism contain carbon, and as a compression fossil forms, those molecules get pressed into the rock. The chemicals either partially or completely break down, producing carbon. Carbon is black, and compression can make it stain the rock, making a film. Once again, that film is mostly made of carbon, but there might be chemicals from the organism that aren't completely broken down. As a result, it is sometimes possible to analyze the stain and learn something about the chemicals that were once in the organism.

This compression fossil of a feather is not the one that paleontologists found the color-producing structures in, but it is similar.

For example, fossils don't retain the colors of the organism. Thus, you might think there is no way we can ever figure out the colors that were present on an organism that we know only from its fossils. However, that's not completely true. For example, paleontologists have studied compression fossils of feathers like the one shown on the right. They have actually found evidence of color-producing structures that are found in cells. From their analysis of those structures, they could tell that the feather they studied had a reddish color. For that particular fossil, then, they can be fairly sure that at least some of the feathers on the animal had a reddish color!

Now even though I have shown you compression fossils of two very flat things (feathers and leaves), there are examples of animals leaving compression fossils as well. Not surprisingly, those compression fossils distort the shape of the animal, but they still allow you to learn a lot about the animal that left each fossil behind. Fish, for example, are often found as compression fossils, because if a fish is covered in sediment, it is usually at the bottom of the lake, river or stream where it lived. Water is pretty heavy, so the sediments at the bottom of a deep lake or the ocean are under a lot of pressure, which is what is needed to form compression fossils. The picture on the right is an example of a fish

This compression fossil of a herring probably formed at the bottom of the ocean.

compression fossil. Obviously, the fish wasn't flat like its fossil indicates. The flatness is a result of the compression that formed the fossil.

One thing compression fossils have in common with a lot of fossil molds is that there should be two halves to them. After all, if the animal is encased in sedimentary rock, you have to break the rock to reveal it. That will result in two parts, and they each should have a compression fossil in them, as shown in the picture on the left. The two halves of the fossil are called the **part** and **counterpart**, but some paleontologists call them the slab and counter slab. As you can see in the picture, the compression fossil in the part should be a mirror image of the fossil in the counterpart. It doesn't matter which half you call the part or counterpart, since they must both exist and fit together.

This compression fossil of a flying reptile (*Pterodactylus*) shows the part and counterpart.

This is a useful fact, because certain fossils can be very valuable. As a result, there are people who spend a lot of time finding fossils and then selling them to collectors or paleontologists. The problem is that some of those people will fake a fossil to make it appear to be more valuable than it really is. If the fossil is a mold or a compression fossil, you can use the part and counterpart to see if something like that happened.

For example, in 1999, the magazine *National Geographic* ran a sensational article about a fossil that was named *Archaeoraptor*. It had been purchased on the black market and studied by paleontologists. Some of them decided that it was a fossil of an animal that was part bird and part dinosaur. There was a lot of press coverage about it, and many people hailed the fossil as a great discovery. However, the article was written based solely on one half of the fossil (let's call it the "part"), and it disregarded the opinions of many experts who questioned the fossil's validity. Eventually, one of the paleontologists who had originally studied the fossil was able to find the counterpart, and when he compared the two, he saw that the part had lots of differences from the counterpart. In other words, it was a deliberate forgery, and *National Geographic* had to publish the fact that they had been fooled by a fake fossil.

Fossil Footprints

So far, I have been discussing the kinds of fossils that occur when the remains of organisms end up leaving their mark on sedimentary rock before the organism decomposes. However, not all fossils are caused by an organism's remains. Sometimes, the organism leaves a mark while it is still alive, and that mark gets preserved by the rocks. For example, organisms can leave footprints that end up being preserved in the rocks. We call such fossils **trace fossils**, because they are produced by the organism passing through an area (i.e., leaving a trace).

Trace fossils – A fossil that is formed by a trace of the organism rather than the organism itself

While a footprint is an example of a trace fossil, anything an organism leaves behind (like a nest or a burrow) that ends up being preserved is a trace fossil.

It should be obvious that footprints are rarely preserved. After all, if you walk through mud or sand, your footprints don't last very long. They are generally erased by rain, ocean waves, etc. However, imagine what would happen if you left a footprint in mud and then there was no rain for a while. The mud would harden, making it easier for the footprint to withstand being wiped away. Then, imagine that before the mud could get really wet again, it got covered in sediment. The sediment would fill the depression made by the footprint. At that point, it would be hard for the footprint to be wiped away.

In this picture, you see the photographer's foot next to the fossil footprint so you have an idea of how big the footprint is.

If the mud and sediment eventually lithified, you would have one layer of rock with a footprint in it, and the other layer of rock filling in and covering the footprint. Over time, suppose enough weathering and erosion occurred to wipe away that top layer of sedimentary rock. What would you see? You would see the footprint in the rock formed by the mud. That's most likely how the fossilized footprint shown on the right formed.

Even though a trace fossil like a footprint doesn't include the remains of the organism itself, it can still tell you a lot about the organism. For example, you know that whatever left the footprint pictured above was bigger than a person, since its footprint is so much bigger than a person's foot. In addition, unlike a person, the organism that left the footprint clearly has only three toes. If you have several footprints to examine, you might even be able to tell a bit about how the organism walked.

Comprehension Check

7.3 Classify each of the fossils below as being a fossil mold, fossil cast, or compression fossil.

a. 

b. 

c. 

7.4 A paleontologist says that she has found chemicals from an extinct animal in a fossil. Of the fossils you have learned about so far, she was probably analyzing one of two kinds. Which kinds?

7.5 Gastroliths (gas' truh liths) are rocks that some reptiles swallow to aid in digestion. If a reptile vomits a gastrolith that ends up preserved as a fossil, which of the definitions you learned best fits it?

NOTE: The next day's experiment requires you to wait for 45 minutes. Please plan accordingly.

Petrifaction

None of the fossils that you have in your kit were made by any of the processes you have learned so far. Instead, they were made by **petrifaction** (peh' truh fak' shun), which is also called **petrification**.

Petrifaction – The process by which an organism's remains are replaced by minerals, preserving them

How does this make a fossil? Perform the following experiment to find out.

Experiment 7.3: A Quick Petrifaction

Supplies:

- Epsom salt from the kit that is made for this course
- Water
- A bowl
- An oven
- A baking dish, like a casserole dish
- Toilet paper
- A measuring tablespoon
- A ½-cup measuring cup
- A spoon for stirring
- Oven mitts or pads

Instructions:

1. Set the oven to 300 °F (150 °C) and turn it on.
2. Use the measuring spoon to add a tablespoon of Epsom salt to the bowl.
3. Use the measuring cup to add ½ cup of water to the Epsom salt in the bowl.
4. Stir with the spoon. You won't get all of the Epsom salt to dissolve, so the solution will be cloudy, but stir it well.
5. Tear off a single square of toilet paper.
6. Put the square of toilet paper into the saltwater solution and use your hand to gently move the toilet paper around in the solution. Don't tear the paper, but move it around a lot so that it completely soaks up the solution. It can touch the solid Epsom salt that is left as well. You just want it to be thoroughly soaked with the solution.
7. After you are sure it is thoroughly soaked, roll it up into a ball. Try not to squeeze any of the water out of it while you are doing that.
8. Put the ball of solution-soaked toilet paper in the center of the baking dish.
9. Put the baking dish in the oven. It doesn't have to be at 300 °F (150 °C) yet. As long as it is getting hot, go ahead and put it in the oven.
10. Let the ball of solution-soaked toilet paper "bake" for 45 minutes. Like anything you are cooking, you should check on it from time to time. Do some other schoolwork while you are waiting and checking so that you are being efficient.
11. Use the oven mitts to pull the dish out of the oven. **Careful – it will be hot!**
12. Set the dish on a surface that's safe for hot things and allow it to cool for a few minutes.
13. **Being careful to not touch the dish itself,** touch the top of the baked ball with your finger. What does it feel like?

14. Once again, being careful not to touch the dish itself, tap the baked ball with your index finger. Does it feel the way you expect a ball of toilet paper to feel?
15. Allow the dish to sit for a while until it is completely cool. Then, clean out the dish first with just water and a little scrubbing. Then use soap and water. Clean up the rest of your mess as well.

Did the way the baked ball feel surprise you? It should have been much harder than toilet paper. That's because of the Epsom salt. The solution you made had as much Epsom salt as possible dissolved in it. The toilet paper soaked up that solution, so there was a lot of Epsom salt and water distributed throughout the toilet paper. When you put the wet toilet paper in the oven, the water evaporated. After all, water boils at 100 °C (212 °F), and the oven temperature was higher than that. Thus, once the ball of toilet paper got heated up, the water in it boiled, which means it started to turn into a gas. When a liquid turns into a gas, we say it has **evaporated**.

Epsom salt doesn't melt at 150 °C (300 °F), so nothing happened to it. However, as the water began to turn into a gas, there was less liquid water for the salt to dissolve into. As a result, the Epsom salt started to precipitate out, turning back into a solid. Eventually, all the water turned into gas, so all the salt had to precipitate out. Since the Epsom salt was distributed throughout the toilet paper, that means solid started appearing throughout the toilet paper. This turned the ball of toilet paper hard, like a solid lump of salt.

This is essentially what happens during petrifaction. An organism dies and is covered by sediment, which slows decomposition. If there is water in the sediments and that water has a lot of minerals dissolved in it, the mineral-filled water can soak into the organism's remains. Then, if the water evaporates, the minerals are left behind. As the organism's remains decay, the minerals can be left behind right where the decay occurred. As a result, you have a mineral version of the organism's remains.

A particularly striking example of petrifaction is shown in the picture on the right. If you look at the picture, you can see that it's a log. However, you have probably never seen a log with such interesting colors. It looks like that because it has been petrified. The tree died, and water that had lots of minerals in it soaked into the wood. Over time, those minerals precipitated out and replaced the wood as the wood decayed away. The colors come from those minerals. If you were to touch that log, it would not feel like wood. It would feel like a rock, because it is no longer made of wood. In your kit (which you will use the next time you do science), specimen 19 is a sample of petrified wood. It doesn't feel like wood, because there is little or no wood left in it. It feels like a rock, because it is made of minerals.

This log has undergone petrifaction. The colors are from the minerals that replaced the wood.

Now remember, petrifaction happens because minerals replace the dead organism's remains as those remains start to decay away. Over time, then, the remains get harder and harder, as more and more minerals accumulate. As that happens, the remains that haven't yet decayed away become even better protected from the organisms and chemicals that cause decomposition. Thus, those remains end up decaying even more slowly. As a result, it is always possible that when you have a fossil formed by petrifaction, there are still some of the organism's actual remains in the fossil.

In fact, there are many, many examples of petrified fossils that still have some of the organism's remains still in them. In 2005, for example, a paleontologist named Dr. Mary Schweitzer was examining the petrified remains of a dinosaur called *Tyrannosaurus*. She had to break the fossil, something paleontologists don't like to do. Fossils are precious; they shouldn't be broken. When she did break it, she noticed that there was soft, flexible tissue inside, which she didn't expect. After performing chemical tests on the fossil, she showed that it had exactly the chemicals you would expect if it were made by a reptile-like animal, so she concluded that the fossil had not completely petrified. Instead, some of the dinosaur's original tissue was still in the fossil!

While Schweitzer's results were controversial at the time, many, many examples of soft tissue have since been found in dinosaur fossils. The picture below, for example, is from the fossil of a dinosaur called a diplodocid (duh plah' duh sid). After the mineral content was removed by a chemical process, soft tissue was found. That soft tissue was put under a microscope, and pictures were taken. The left side shows a lower magnification picture, and the box shows a higher magnification one. As is pointed out, the long, thin piece of tissue is one of the dinosaur's blood vessels. In the part that shows higher magnification, you can actually see bone cells from the dinosaur!

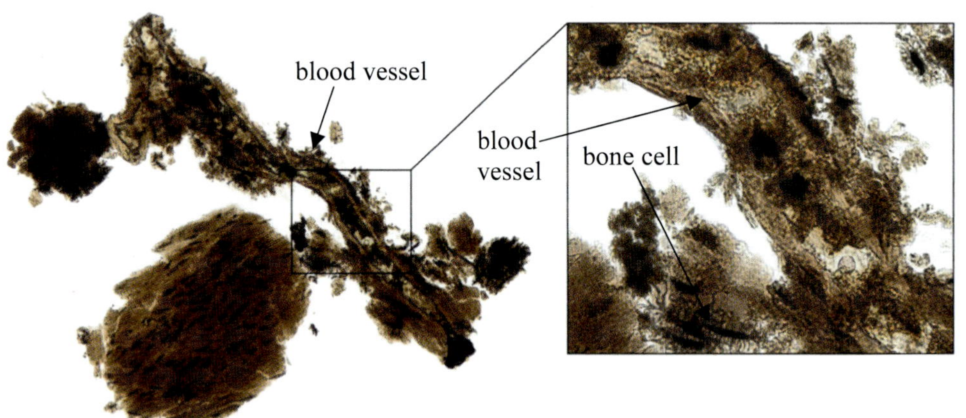

This image, taken with a microscope, shows a blood vessel that was extracted from a dinosaur fossil. The square shows it at a higher magnification, which allows you to see bone cells from the dinosaur.

When I was at university, I was told that we would never have access to soft tissue from dinosaurs, because their fossils were completely petrified. As a result, no one bothered to look for dinosaur soft tissue. However, after Schweitzer's accidental discovery, scientists have been searching petrified dinosaur fossils and have found lots of soft tissue. This is an example of how science can develop a "blind spot." Scientists simply assumed dinosaur fossils are fully petrified, so they never even thought to look for original dinosaur tissue in them. As a result, it went undiscovered until 2005!

Comprehension Check

7.6 Hot springs are produced when water running underground is naturally heated and then rises to the surface. Would water from such a spring make petrifaction more likely or less likely?

What Can We Learn from Petrified Fossils?

While it is wonderful that we can sometimes extract actual remains from a petrified fossil, most of what we know about many extinct organisms comes only from examining the fossils themselves. Your kit contains petrified fossils, so I want you to spend some time examining them to see what they can tell us about the animal that produced them.

Experiment 7.4: Detailed Examination of Petrified Fossils

Supplies:

- The fossils from the Geology Basics Kit from the laboratory kit made for this course
- The magnifying glass from the laboratory kit made for this course
- A well-lit room

Instructions:

1. Make sure the room is well lit so that you can see everything you need to see.
2. Pick up specimen 16 from the kit and examine it on all sides, like you did in the first experiment. This time, instead of concentrating on mineral characteristics, try to understand the nature of the animal that left the fossil. This fossil comes from a clam, which belongs to a group of animals called **bivalves**. They are given that name because their shells are made of two halves (the "bi" means "two") that are roughly mirror images of each other. If you look at the fossil on all sides, you should see the shape is curved on one side and a bit more pointed on the other. That pointed end has a hinge that allows the animal to open its shell. You are probably familiar with the general layout of a clam, but if you had never seen a clam before, would you bc ablc to tell that this fossil is from an animal whose shell comes in two halves which can be opened?

3. Use the magnifying glass to see if you can find the division between the top half of the shell and the bottom half. It should appear as a slit. If you have trouble seeing it, focus on the hinge and see if you can find a slit that originates there.
4. Does your clam specimen have ridges that follow the curve of the shell? Those ridges represent the shell's growth. The more ridges, the older the clam. Some clams form a ridge or band once per year, so you can determine their age by counting the ridges or bands, just like counting the rings in a tree. Other clams form ridges or bands at a different rate. Thus, when it comes to a fossil, all we can really say is that for very similar fossil clams, the more ridges or bands you find, the older the clam was when it died.
5. Put specimen 16 away and pull out specimen 17, which is a **brachiopod** (bray' kee uh pod').
6. Once again, look at the specimen from all sides. You should see that like the clam fossil, it has one side that is more curved and another side that is more pointed. Look at the pointed side. Can you see that it makes a hinge, which is something like the hinge on the clam fossil?
7. Use the magnifying glass to look at the hingc. Once again, you should be able to find a slit that separates the two parts of the shell that can open. If you can identify that slit, try to see if you can now recognize it without using the magnifying glass. This might take a couple of attempts, but you

should eventually be able to see the separation between the two parts. How do these two parts compare to the two halves of the clam fossil? You should see that in this case, the parts aren't mirror images of one another. Instead, one part is much bigger than the other part. That's one characteristic biologists use to distinguish between bivalves and brachiopods. They both have two parts to their shell that can open, but the two parts of the brachiopod's shell are not as similar as the two parts of a bivalve's shell. The smaller part is the bottom of the brachiopod, while the larger part is the top.

8. Look at the ridges on your brachiopod fossil. How are they different from the clam fossil? They don't follow the curve of the slit that separates the two parts, do they? Instead, they run lengthwise down the fossil. They can't be used to get any information about the age of the fossil.
9. Use the magnifying glass to look at the top of the fossil. Can you see lines that go the opposite way, much like the ridges on your clam's fossil? If you can find them, those are the lines that are formed by the shell's growth. They are harder to see, though, so you might not find them.
10. Put away specimen 17 and pull out specimen 18, the shark's tooth.
11. As I pointed out after the first experiment (p. 184), there are three distinct parts of the tooth. The largest part is smooth, and it's the body of the tooth. It is covered in enamel, which is the hardest substance in a shark's body. In fact, petrifaction hasn't affected it much, since it is so hard to begin with. In order to see the difference between the enamel in your fossil and the enamel in a living shark's tooth, you would need some pretty sophisticated equipment.
12. However, the U-shaped top looks very different from the body of the tooth. That's the root of the tooth, which is what anchored it in the shark's mouth. That tissue is not hard, so it had to be petrified to be preserved. You can see that by comparing the root to the enamel. It looks very different, because the root is now made from the minerals that produced petrifaction. It turns out that the characteristics of the root (shape, size, etc.) can be used to help identify the type of shark that the tooth came from. The portion in between the root and the enamel is called the "chevron," and it is a transition between the root and the enamel.

13. Put away specimen 18 and pull out specimen 19, petrified wood.
14. The three specimens you have examined so far come from structures that are pretty hard. However, this was originally wood. It doesn't feel like wood, of course, because the wood has been mostly replaced by minerals, so it feels hard like a rock. Also, unlike the other fossils, it isn't shaped in a familiar way. Nevertheless, if you look at it closely,

you should be able to see that it was originally wood. Turn the fossil over in your hands, and see if you can tell what part was the bark and what part was the wood of the tree (use the picture on the previous page as a guide). How thick is the bark compared to the rest of the fossil?

15. Can you see a layer in between the bark and the inner wood? That's the inner bark. What most people call "bark" is actually the outer bark of a tree, and it is made of dead wood. The inner bark is made of living cells, which reproduce to form the wood inside the tree. Your fossil might not be good enough to distinguish the inner bark, but there should be an obvious difference between the outer bark and the wood inside.
16. Notice the fact that the minerals seem to be laid down in vertical lines. That's because a tree has "tubes" called xylem and phloem that run vertically through the tree.
17. Put away specimen 19 and pull out specimen 20, a crinoid (krih noyd') stem. This fossil is the hardest to recognize, since it is incomplete – just a portion of a stem. A complete fossil crinoid is shown in the picture on the right. It has a "head" that contains "tentacles" and a mouth. The "tentacles" filter out food particles that are floating in water, and they are put in the mouth, which is at the center of the "tentacles." The bottom of the stem is used to anchor the crinoid to one place. Even though you have only a portion of the stem, do you see that it has horizontal lines running around it? That's because the stem is made of individual plates stacked together, like a tall stack of pennies. The plates are made of a hard substance that doesn't decompose rapidly, which is why you are much more likely to find fossils of the stem than you are of the "head."

18. Put everything away.

Before paleontologists found out there is soft tissue remaining in many fossils, all they could learn about an extinct animal came from analyzing fossils much like you just did. As you found out, there is quite a bit you can learn that way. The difference between bivalves and brachiopods was easy to see in the fossils, the differences in tissues in a shark's tooth were easy to see, and depending on the fossil, even petrified wood can teach you a lot about the structure of a tree. The crinoid fossil gave the least information, but even with it, you could see that crinoids have stems made of stacked plates.

Of course, paleontologists not only have better equipment with which to analyze fossils, but they also have access to more fossils than you do. By looking at a lot of fossils, they can better understand even the partial fossils. The picture of the crinoid fossil above, for example, is more informative than your crinoid fossil, but once you have seen the picture, you can understand that your crinoid fossil comes from an individual whose stem was made from smaller, thinner plates than the one in the picture. That can sometimes be enough to distinguish one type of crinoid from another.

Comprehension Check

7.7 You have a fossil that is shaped like a log. One person tells you it's petrified wood. Another tells you it's part of a crinoid stem. How can you tell which person is correct?

Surprising Fossils

It makes sense that things like shells, shark's teeth, and the plates that make up crinoid stems can fossilize. After all, they are hard, so they don't decompose very quickly. Even wood lasts for a long time before it decays away. Thus, it is easy to imagine how such things could decay away slowly enough for fossilization to occur. However, even the softest parts of an organism can fossilize, as long as the conditions are right. Look, for example, at the picture on the left. It shows a sample of limestone from Germany. You can see that there is a pattern in the stone. While it might not be obvious to you, paleontologists have identified this as a fossil jellyfish.

This is a jellyfish fossil in limestone.

Have you ever seen a jellyfish floating in the ocean or in a tank at an aquarium? There isn't anything hard in the entire animal. Indeed, it gets its name from the fact that its entire body has the consistency of jelly. Nevertheless, this jellyfish was preserved in limestone. How? It is thought that the jellyfish died, and its body sank to the bottom of the ocean where there was very fine calcium carbonate sediment. Most likely, the jellyfish body sank into the sediment and quickly got covered. That slowed decomposition down enough for the sediments to lithify into limestone, preserving the shape of the jellyfish. Fossils of soft bodies like this are rare, of course, because hard substances are more likely to be preserved. Nevertheless, they are found from time to time.

Believe it or not, animal feces (solid excrement) can be petrified and become fossils. To make them sound more appealing, they are called **coprolite** (kahp ruh light').

Coprolite – Fossilized animal feces

The picture below gives you an example. It obviously has the shape you would expect for animal feces, but how do we know it's actually coprolite?

This is an example of coprolite.

Let's go back to the early 1800s, when an amateur fossil-hunter, Mary Anning, discovered a fossil skeleton of an extinct swimming reptile that we now call ichthyosaurus (ik' thee uh sor' us). Near the end of the ribcage, she found fossils that were feces-shaped. Since that's where you would expect to find feces that had not yet been released by a reptile, she thought they were fossils of those feces. When she broke them apart and looked inside, she found fossils of scales and small bones. It made sense to her that those were fossils of the things that the ichthyosaurus had eaten. She sent them to one of the most important geologists of that time, William Buckland, who had them chemically analyzed. The analysis found chemicals you would expect to find in an animal's feces, so he concluded that Mary Anning was correct.

When modern paleontologists find fossils that look like they might be coprolite, a similar process is followed. Nowadays we don't necessarily need to break open the fossil to look inside, since we have instruments that can see inside things without harming them. We can also chemically analyze samples using only a small portion of the fossil. That way, it isn't damaged too much. As a result, paleontologists can usually identify coprolite with a lot of confidence. Once they do that, it can tell us a surprising amount about the animal that left it. As is the case with Mary Anning's coprolite, for example, it can often preserve things from what the animal had eaten!

Surprisingly, even some organisms that are so small you must use a microscope to study them can leave fossils behind. Fossils like that are called **microfossils**. The pictures below show you four examples of microfossils. The pictures are made with an electron microscope, which uses beams of electrons to image things instead of light. If you have ever used a microscope, it probably used light as a means of seeing the things you were looking at. However, beams of electrons can be used as well, and they can actually produce higher magnifications. However, since color is based on the energy of light, electron microscopes are not sensitive to color. That's why the images are in black and white.

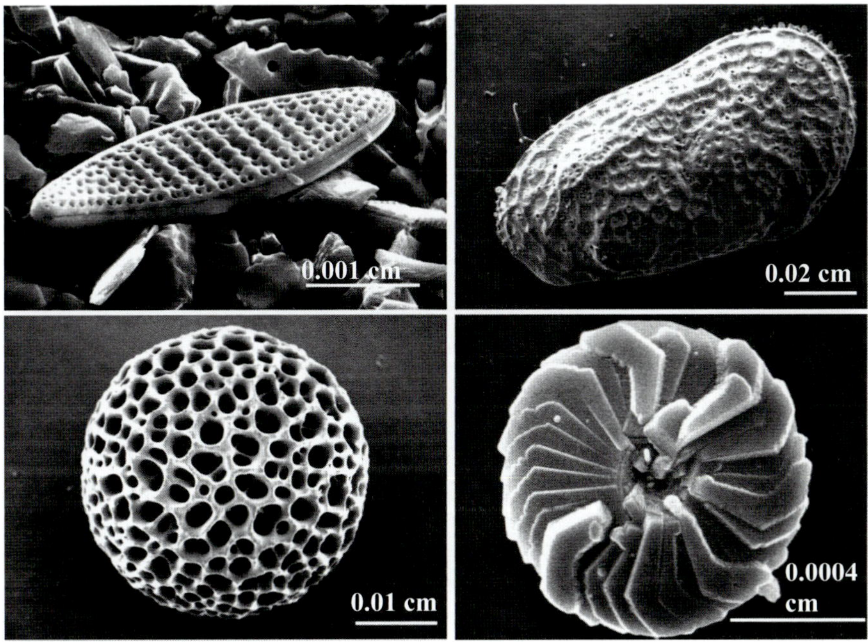

These are microfossils of a diatom (top left), an ostracod (top right), a radiolarian (bottom left), and a plate from a coccolithophore (bottom right).

Since you are seeing an image from a microscope, you have to be told how big the things in the images are. That's what the white bars in the bottom right part of each image tells you. They show you how long a specific measurement is at the magnification used by the microscope. In the top left, for example, the bar is labeled as 0.001 centimeters (0.0004 inches) long. Since you can fit about three of those lines across the widest part of the fossil, that tells you the fossil is roughly 0.003 centimeters (0.0012 inches) across.

The top left image shows a fossil left behind by a single-celled creature called a diatom (dy' uh tom). It surrounds itself with a silicate mineral, which makes it resistant to decomposition. The bottom left is from another single-celled organism called a radiolarian (ray' dee uh lair' ee un). It also surrounds itself with a silicate mineral. The top right fossil comes from an ostracod (ah' struh kod), which is actually a tiny animal that has a shell. The bottom right image isn't of a complete organism. It's a calcium carbonate plate made by a single-celled organism known as a coccolithophore (kah' kuh lih' thu for). A single coccolithophore covers itself with lots of plates, which is why that fossil is so much smaller than the others.

Most people think of fossils as prehistoric, because we study such fossils quite a lot. After all, if something was alive even a few hundred years ago, there is a good chance it was studied by scientists back then. If that's the case, we probably know more about the organism from those studies than what we can learn from a fossil. With organisms that went extinct before they could be studied scientifically, the only information we can learn about them comes from their fossils. As a result, we usually think of fossils as remnants from prehistoric creatures.

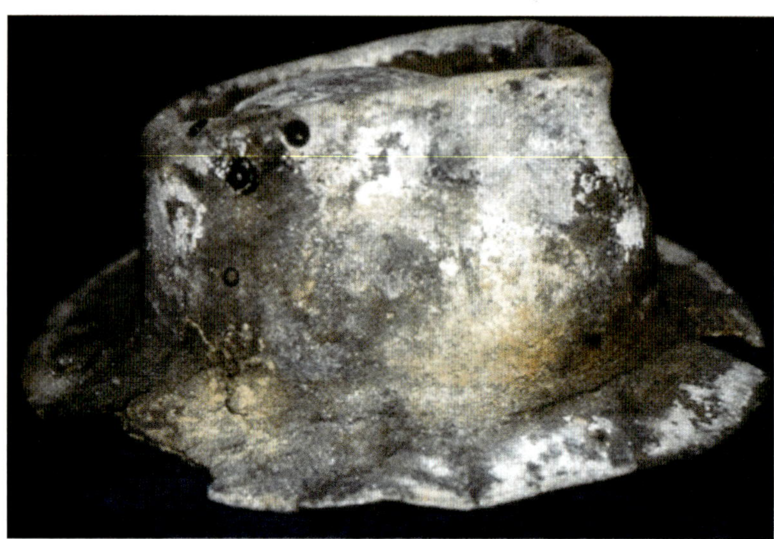
This hat has been completely petrified.

Please understand, however, that anything which has been preserved and was connected to a living organism is a fossil. Consider, for example, the picture on the left. You might think it is just an old hat. However, it is a very special old hat. It is a fossil! If you were to hold the hat, it would be very heavy and hard as a rock, because it has undergone petrifaction. The hat is only about 75 years old, but it was left in a Tasmanian mine by someone who was working there. Water that had a lot of dissolved minerals flooded the mine, and that mineral-rich water covered the hat. As a result, it petrified! Remember the spark plug I showed you back on page 117 in Chapter 4? It's another example of something relatively recent that fossilized. So remember, not all fossils are prehistoric. Some can be of rather recent origin!

Reading too Much Into Fossils

In the previous section, you spent time examining your fossils and trying to learn what you could about the organisms that left them behind. While there was much you could learn, there was more that you simply couldn't learn. You had no idea what was inside the clam or the brachiopod, because the shells were petrified closed. Even if you had an X-ray machine or something else that would allow you to see inside the shell, it's not clear how much you would have learned. You might have been able to notice some differences between certain parts of the interior (like the differences between the different parts of the shark's tooth or the petrified wood), but that would be about it. In the end, then, the information that comes from the fossils is limited.

Nevertheless, paleontologists want to learn as much as possible from the fossil, so they often try to "fill in the gaps." Consider, for example, your clam fossil. Since you could open up a live clam and see what's inside, you might think it is reasonable to assume that the clam which left your fossil had roughly the same insides. That's probably a reasonable guess. Paleontologists do that a lot. They look at a fossil and compare it to the same parts in a living organism. If the parts of the fossil match the same parts on a living organism, it is often reasonable to think that the same basic organism left the fossil. Sometimes, however, the results of such reasoning are disastrous!

In 1917, for example, a rancher who also studied fossils found a fossil tooth that was unlike any tooth with which he was familiar. Five years later, he sent the fossil to a paleontologist, Dr. Henry Osborn, who decided that it matched an ape's tooth pretty well, but it had some characteristics that made it more like a person's tooth. At the time, many paleontologists were eager to find evidence that humans evolved from apelike animals, so Osborn published a paper in 1922 that said the tooth was from an "ape man" – an animal that represented an evolutionary link between apelike animals and people. Since the rancher was from Nebraska, this "ape man" became known as Nebraska Man. That same year, a magazine named the *Illustrated London News* ran an article about this amazing find, and the article included Nebraska Man's illustration, as shown on the right.

This illustration of "Nebraska Man" was published five years before it was officially determined that the fossil upon which it was based came from a specific type of pig.

Many paleontologists were skeptical of Osborn's claim, while some embraced it enthusiastically. Because of the skepticism, some paleontologists went to where the tooth fossil had been found and tried to find other fossils that might help them determine whether or not Osborn was correct. Eventually, they found several fossils of a peccary, which is a kind of pig. With that in mind, they were able to show that the fossil tooth was, in fact, a peccary tooth that had been weathered in a way that fooled Dr. Osborn.

Now, of course, that happened well over 100 years ago. Surely mistakes like that aren't made in modern times, right? Wrong! Once again, fossils have incomplete information, and because paleontologists want to learn as much as they can, they often read too much into the fossil. In 1983, for example, one of the most important scientific journals in the world, *Science*, published an article about portions of a petrified skeleton found in Pakistan. Based on the fact that some of the bones were similar to the bones of whales, the article described this animal as a small whale-like creature that was named *Pakicetus*, which combines "Pakistan" with "*cetus*," which refers to a whale. The journal put an illustration of *Pakicetus* on the issue's cover. While the drawing on the right is not directly from the magazine, it shows the same features that are shown in the journal's illustration. This was the accepted view of what *Pakicetus* looked like from 1983 until 2001.

This was the accepted view of the animal known as *Pakicetus* as determined by the initial fossil find. It took 18 years for paleontologists to find out that it is wrong.

What happened in 2001? More bones were found so that a more complete skeleton could be analyzed. As a result of having more bones, it was clear that the original interpretation of the fossil was very wrong. The picture below is now considered the accurate view of what *Pakicetus* looked like. Of course, we don't really know whether or not that is completely correct, since all we have is portions of the petrified skeleton. The pieces we have do a good job of telling us its general shape, but we have no hair, skin, eyes, etc. Thus, all of that is just artistic license. Now please understand, I am not telling you about these mistakes to insult the scientists who made them. Even the greatest scientists make mistakes. I am just illustrating the kind of mistakes that come from people trying to read too much into a fossil.

This is currently considered the correct view of *Pakicetus*.

Now remember, there is also the problem of fake fossils. As I mentioned previously, *National Geographic* was tricked into thinking that a fake fossil was real. The only reason we now know it is a fake is because one of the paleontologists went looking for the counterpart of the fossil, which showed that the part they had been studying had been tampered with. Fake fossils are becoming more common today, because fossils can sell for a lot of money, and the more interesting the fossil, the more money it sells for.

Does all this mean that we shouldn't trust the conclusions of paleontology? Of course not! However, it does tell us that we should be cautious, especially when there are few fossils available. After all, Nebraska Man was a mistake because it was based on a single tooth. The incorrect view of *Pakicetus* was the result of not having enough fossil bones to get a clear view of what the animal looked like. In addition, it is easy to fake a few fossils, but hard to fake a lot of them. Thus, the more fossil evidence there is, the more confidence you can have in the conclusions that paleontologists draw from them.

Comprehension Check

7.8 Of all the fossils pictured in this section, which one or ones would be considered trace fossils?

7.9 You find what you think is coprolite from an extinct animal that was a vegetarian. What would you expect to find inside the coprolite?

7.10 In 2015, paleontologists published a paper saying that they had discovered a new kind of prehistoric human based on finding a single fossilized pinky finger. How confident should they be of their conclusion?

<u>Extinct or Not?</u>

I have mentioned extinct organisms many times already. That's because there are a great many fossils of plants and animals that we cannot find living examples of today. Since we can find fossils of them, we know that they lived at some point in the earth's history. However, we assume something happened to cause them all to die out, so they are now extinct. There is a problem, however. It's difficult to know whether or not an organism is actually extinct, because the earth is a big place, and there are portions of it that are largely unexplored, at least by people who are trying to observe it scientifically. As a result, there are times that scientists say an animal is extinct, but they find out later that it is not!

To make the discussion a bit easier, I want to define a word you might not have heard before. I didn't define "extinct," since it is a familiar word. But what's the opposite of extinct? In science, we say it is **extant**.

Extant – Something that exists presently

When something is extinct, it no longer exists. If it exists today, we say it is extant. All fossils must represent organisms that are either extinct or extant.

The **coelacanth** (see' luh kanth') is a famous example of an animal that was thought to be extinct but is extant. A fossil is shown below. In 1839, one like it was described by Louis Agassiz, a famous Swiss biologist and geologist. At the time, no living fish like it was known, so it was considered extinct. Nearly 100 years later, however, a fisherman brought a dead fish that had been caught in his nets to the East London museum. The woman who examined it, Marjorie Courtney-Latimer, had never seen a fish like it before, so she made a detailed drawing of it and sent it to a well-known fish expert in South Africa, James Leonard Brierly Smith. He said that based on the drawing, the fisherman had actually caught a coelacanth! Since then, many live specimens have been studied, so we know that the coelacanth is extant.

This is one of many coelacanth fossils that have been found.

I ended the previous section with examples of paleontologists reading too much into fossils. Well, that can happen even when scientists have more than just the fossils. James Leonard Brierly Smith not only had fossil coelacanths to study, but he also had the sketch that Marjorie Courtney-Latimer had made. The sketch made it clear that the coelacanth was not extinct, so Smith decided it was only a matter of time before people found coelacanths living in their natural habitats. As a result, he wrote a book entitled *Old Four Legs: The Story of the Coelacanth.* In that book, he said that when scientists found the coelacanth in the ocean, they would find that it used its fins like legs to crawl along the bottom of the ocean.

Why did he think that? Well, the fossils of coelacanths indicated that their fins were stronger than those of most fish. Latimer's sketch confirmed this and indicated that the fins were muscular and contained more bones than the fins of most fish. Thus, Smith thought that the coelacanth used its fins as legs. That's where the title of his book comes from. However, we now know that coelacanths don't do that at all. They use their muscular fins to swim quickly and make precise turns in the water. Thus, while the fins might look different from the fins of most fish, they are used in the same way other fish use their fins.

While the coelacanth is probably the most famous example of an organism that was thought to be extinct but is now known to be extant, there are many others. In 1895, for example, a mouse-sized animal called the mountain pygmy possum was described based on a fossil. No such creature was known, so it was declared to be extinct. Several living specimens were found in 1966. This happens for plants as well. The Wollemi pine was described based only on fossils and was declared extinct, but living specimens were found in Australia in 1994. Thus, while it is common to hear that plants and animals known only from their fossils are extinct, there is no way to be completely certain.

At the same time, however, we have explored a lot of the earth and have described a lot of different organisms. So, when we find fossils of organisms for which we have no living specimens, we can at least say that it is very possible that those organisms are extinct. In fact, the vast majority of fossils that we find today are from organisms that are now thought to be extinct. Thus, extinction seems to be the rule when it comes to the earth's history.

There Is a Pattern

So far, I have been concentrating on how fossils form and what they can tell us about the organisms that left them behind. Now I want to step back and concentrate not on the individuals, but on the entirety of the fossil record. To do that, you have to remember that fossils are usually found in sedimentary rock, and sedimentary rock usually forms nice strata that are relatively easy to distinguish from one another. However, those strata are mostly underground. If we see them, it is either because weathering and erosion have exposed them, or because we dug down into the crust to investigate them. Well, weathering and erosion expose only certain areas, and we can only dig in certain areas. As a result, we can't see the big picture. We only get a glimpse here and a glimpse there.

To see what I mean, look at the drawing below. It shows cliffs that can be found in three national parks in the United States: Bryce Canyon National Park, Zion Canyon National Park, and Grand Canyon National Park. These parks are within about 230 kilometers (140 miles) of each other.

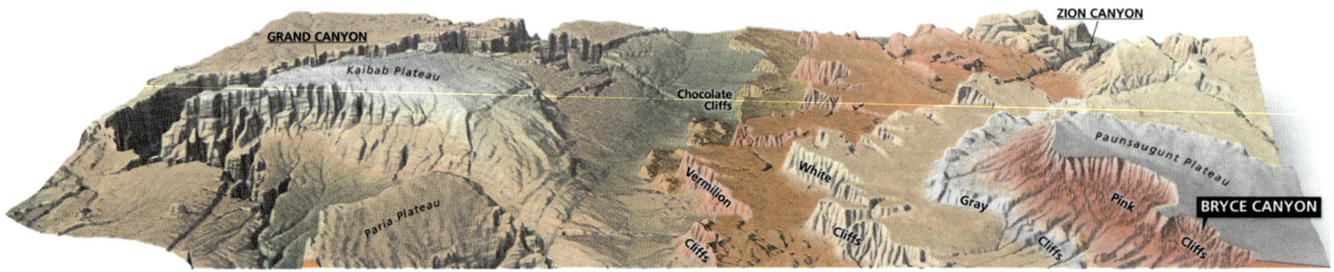

Looking from right to left, you can see that the cliffs known as "Pink Cliffs" are above the "Gray Cliffs," which are above the "White Cliffs," which are above the "Vermillion Cliffs." Each set of cliffs has strata of sedimentary rock exposed. Why are the Gray Cliffs below the Pink Cliffs? The most reasonable explanation is that the sedimentary rocks that make up the Pink Cliffs were originally on top of the Gray Cliffs, but they eroded away. Similarly, the sedimentary rocks that make up the Gray Cliffs were originally above the White Cliffs, but they eroded away. Thus, the different levels of cliffs resulted from sedimentary layers above them being removed by weathering and erosion.

As you move farther to the left in the drawing, you see that the land rises between the Vermillion Cliffs and the Grand Canyon. Does that mean less erosion took place there? Surprisingly, the answer is "No!" Even though the land rises above the Vermillion Cliffs, we know that the upper strata of sedimentary rock have eroded away. How? Because the sedimentary rocks that make up the Pink Cliffs have a specific composition of minerals, and that composition isn't found anywhere near the Grand Canyon. The rocks that make up places like the Paria Plateau and the Kaibab Plateau in the drawing are nothing like the rocks that make up the Pink Cliffs. Instead, they are very similar in composition to the rocks that make up the very bottom of the Vermillion Cliffs.

In the end, if we match the strata of rocks according to their composition, we find something like what is seen below.

The surface that is drawn is exactly the same, but the stripes of color represent strata of matching composition. If you look at all the stripes on the right side of the drawing, you see that they form the Pink Cliffs, Gray Cliffs, White Cliffs, and Vermillion Cliffs. If you were studying just one set of cliffs, you would see only one or two sets of strata. However, if you studied all the cliffs and then correlated your results, you would understand that all those strata exist under the Paunsaugunt Plateau, which is the gray area on the right of the drawing.

You can also see that the reason the area between the Paria Plateau and Kaibab Plateau near the Grand Canyon is high because other strata of sedimentary rock have formed a dome underneath. How do we know? Because the Grand Canyon exposes those strata. Since they are different in mineral composition from the strata under the Paunsaugunt Plateau, we know they are not the same rocks that form the Pink, Gray, White, and Vermillion Cliffs. This kind of reasoning is called **physical correlation**

Physical correlation – The matching of sedimentary strata based on their physical characteristics

Physical correlation allows us to figure out what strata exist underground, even though we can't see them all in a single place.

While physical correlation allows us to learn a great deal about rock strata in a region, it can't be used to make sense of all strata. Fortunately, there is another way we can try to develop relationships between different strata of rock: **temporal correlation**.

Temporal correlation – The matching of sedimentary strata based on the time they were formed

This is where fossils come in. One of the striking features of the fossil record is that certain fossils are found in a specific order when it comes to the depth of the strata in which they are found. For example, remember the trilobite I showed when I talked about molds and casts (page 190)? As far as we know, all trilobites are extinct, so there are limits to what we know about them. However, we do know that when their fossils are found in the same general area as the fossils of birds and mammals, the trilobite fossils are always found in strata that are *underneath* the strata that hold the mammal and bird fossils. Now, of course, there are a lot of places on earth where no trilobite fossils are found at all. When they are found, however, they are found in the lower fossil-bearing strata. This is an example of the **Principle of Faunal** (faw' nul) **Succession**:

Principle of Faunal Succession – The strata in which certain fossils are found can have a predictable position relative to the strata that contain other fossils

This principle can be used to relate the time at which sedimentary strata were laid down, as shown in the illustration below, which shows three sets of sedimentary strata taken from three different parts of the earth. There is no way to use physical correlation to determine how the strata relate to one another, since they have different mineral contents. However, look at the strata on the right. The lowest layer holds a trilobite fossil, and the next one up holds a brachiopod that is sort of butterfly-shaped. Above that is a fossil reptile known as dimetrodon, and above that is a fossil dinosaur known as stegosaurus. Finally, there is a horse-like mammal called *Hyracotherium* in the top layer. Using the Principle of Superposition, we can say that the lower the fossil, the older the layer in which it is found. Thus, the layer with the trilobite is the oldest, and the layer with the horse-like mammal is the youngest.

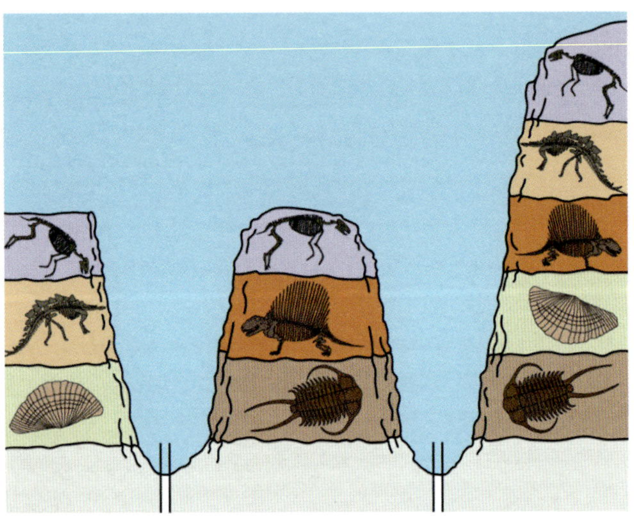

The Principle of Faunal Succession can be used to provide temporal correlation to the strata illustrated here.

Now look at the center set of strata. Notice that the trilobite is still at the bottom. There are no butterfly-shaped brachiopods, but there is a dimetrodon below a horse-like mammal. Assuming the Principle of Faunal Succession is correct, we can say that the bottom layer of rock in the center was laid down at roughly the same time as the bottom layer on the right. But the next layer up in the center was laid down roughly at the same time as the third layer from the bottom on the right, because it has the dimetrodon in it. Finally, the top layer in the center was laid down roughly when the top layer on the right was laid down, because they both have the horse-like fossil. Similarly, the bottom layer on the left was probably laid down when the second deepest layer on the right was laid down, and the top two layers on the left were laid down when the top two layers on the right were laid down.

Now please understand that I chose the fossils in the illustration very carefully. You can't do this with all fossils. However, there are certain fossils that seem to follow the Principle of Faunal Succession well, so they can be used to temporally correlate strata. They are called **index fossils**.

Index fossil – A fossil that can be used to temporally correlate strata of sedimentary rock

The fossils I included in the illustration are index fossils. They seem to be characteristic of specific strata of sedimentary rock.

The illustration above shows one of the challenges of understanding the fossil record on a global scale. You can't go to any one place on earth and find all index fossils in one place. The set on the left has three strata that each has an index fossil, the one in the middle also has three, but two of them are different from the ones on the left. The one on the right has five, but that's not all the strata on earth which have index fossils. Thus, in any given place on earth, there are only some strata that contain index fossils. Nevertheless, paleontologists would like to use fossils to construct a history of the entire earth. How can they do that if each geological structure has only some of the strata and fossils?

This is where the **geological column** (illustrated on the right) comes in. Since index fossils can be used to do temporal correlation among strata, we can use reasoning like I used on the previous page to make a single, hypothetical structure that represents all the fossil-bearing strata on the earth. There is nowhere on earth that the geological column illustrated here actually exists, but if the Principle of Faunal Succession is correct, we can say that the geological column is a representation of how different strata of fossil-bearing rocks correlate to one another around the world.

In the geological column, each layer of rock is given a name, and the lower the layer is, the earlier it was formed. Precambrian rock, then, represents the earliest sedimentary rock, while Quaternary rock is the sedimentary rock that is most recent. Notice that there is a pattern to the fossils. Fossils in the lowest layers are ocean dwellers. In low layers, the fossils are of the "simpler" organisms, but as you go up the column, you find more complex organisms. Plants and animals that live on land are much higher in the column, while mammals and most of what we see today are at the top.

What does the geological column mean? That depends on how you interpret it. Uniformitarians see the geological column in one way, while most catastrophists see it in a different way. Each view has its strengths and weaknesses, and we will explore those together in the upcoming chapters.

The geological column attempts to correlate sedimentary strata found around the world.

Comprehension Check

7.11 Suppose you are studying two dinosaurs: diplodocus (26 meters long) and microraptor (77 centimeters long). For which can you be more confident that it is actually extinct?

7.12 The Principle of Superposition is also used to correlate strata. Does it help in physical correlation or temporal correlation?

7.13 Clam fossils have been found in rocks that contain trilobite fossils and rocks that contain mammoth fossils. Are they good index fossils? If so, which layer of rock would they be used for?

Answers to the Comprehension Check Questions

7.1 It had many of the same characteristics as clams, and we don't normally see that kind of detail in rocks. If you looked from the side, you should have been able to see where the two halves of the clamshell came together. If you looked at the back, you should have seen the "hinge" that opens the two halves up. It also had the ridges that you normally see on a clam's shell. Robert Hooke used this reasoning (including similarities on the microscopic level) to conclude that fossils are not just interestingly-shaped rocks. Of course, your answer can be different from mine, since I asked what made *you* think it wasn't a rock. However, if your answer is similar to mine, it means you are thinking like a scientist.

7.2 The gallery containing fossils entombed in ice would house the larger fossils, because you can find very large samples of ice on the earth, such as in glaciers. Amber is emitted by trees, so you don't find large samples of it in any one place. As a result, it isn't plentiful enough to entomb large animals. Surprisingly enough, however, some medium-sized animals have been found in amber, such as frogs and scorpions!

7.3 a. Compression fossil. It is very flat, which generally indicates that it is from compression.

b. Fossil mold. It is a depression in the rock. Thus, it's part of a cavity.

c. Fossil cast. It isn't flat, so it can't be a compression fossil. Also, it isn't a cavity of any kind. From those three choices, then, it must be a cast.

7.4 A fossil entombed in amber or ice or a compression fossil. The first two slow decomposition so much that original chemicals can be found. The text also gave an example of chemicals from the organism being found in a compression fossil.

7.5 It would be a trace fossil. The gastrolith isn't from the organism's remains. It's something the organism left behind as a trace of its passing.

7.6 It would make petrifaction more likely. First, as you learned in Chapter 3, the hotter a liquid is, the more minerals it will have. Thus, there would be more minerals to precipitate out. Second, hot water evaporates faster, so the speed at which minerals precipitate would be greater.

7.7 Look for horizontal stripes around the "log." Petrified wood might have vertical stripes, and depending on the fossil, it might even show the rings of wood inside. However, a crinoid fossil has horizontal lines running around the fossil, because the stem is made of stacked plates. Please note that's just the most obvious thing to look for. There are others. For example, you expect a petrified tree to have outer bark, inner bark, and wood, which can all be distinguished from one another. A crinoid stem would not have those.

7.8 The coprolite and the hat would be trace fossils. They are both associated with living things, but they are not parts of the living things themselves. Now, of course, if the hat had been made from leather, you could say that it wasn't a trace fossil of the animal out of which the leather was made. However, it was a trace fossil of the man who wore the hat.

7.9 You would expect to find the remains of plants in it. Remember, coprolite can contain remnants from the animal's food. Seeds, for example, often survive digestion and are expelled from the body.

7.10 <u>They should not be at all confident</u>. When there is only one bone, you could be faced with a situation like Nebraska man, which was clearly a mistake.

7.11 <u>You can have more confidence that diplodocus is extinct</u>. After all, we determine whether or not an organism is extinct by seeing if we can find a living one today. If we can't, we assume it is extinct. Well, it would be hard to miss a living version of diplodocus, since it is so big! The smaller the animal, the more likely it would be to have escaped our notice.

7.12 <u>It helps with temporal correlation</u>. The Principle of Superposition is used to determine which layer is laid down first. That's a correlation in time.

7.13 <u>They would not be good index fossils</u>. An index fossil has to be characteristic of a particular layer. Trilobites are found much lower in the geological column than mammoths. Thus, there is no way to determine which layer of rock you have if the only fossils you can find are clams.

Chapter Review

1. Define the following terms:

a. Fossil	d. Compression Fossil	g. Coprolite	j. Temporal correlation
b. Fossil mold	e. Trace fossil	h. Extant	k. Principle of Faunal Succession
c. Fossil cast	f. Petrifaction	i. Physical correlation	l. Index fossil

2. What must be slowed down in order for fossilization to occur?

3. By what two things can the remains of an organism be entombed in order to preserve them? Which can entomb the remains of larger organisms? Which comes from trees? Explain how each helps to preserve the organism's remains.

4. If you see a fossil mold and cast of the same organism, which formed first?

5. For the two statuses "extinct" and "extant," which can you be certain of?

6. Explain the part and counterpart of a fossil. How can they help to determine if it is fake?

7. You see fossils of a bone, a tooth, feces, a shell, and footprints. Which are trace fossils?

8. Of the fossils you learned about, which ones can contain soft tissue from the organism?

9. What kind of fossil could have remains of chemicals from the organism but not soft tissue?

10. How does soft tissue in dinosaur fossils demonstrate that science can develop blind spots?

11. Which is more likely to fossilize: a clam or a jellyfish. Why?

12. Are all fossils prehistoric?

13. What do Nebraska Man and *Pakicetus* tell us about interpreting fossils?

14. Why do people sometimes make fake fossils?

15. James Leonard Brierly Smith had seen coelacanth fossils and a sketch of a recently-dead coelacanth. Nevertheless, he made a big mistake in his interpretation. What was it?

16. A paleontologist is showing you fossils she recently discovered. She shows you a dinosaur fossil, an elephant fossil, and a trilobite fossil. Referring to the geological column on page 209, which was found in the lowest layer of rock? Which was found in the highest layer?

17. What kind of correlation is used to construct the geological column? On which principle is it based?

18. Is there somewhere on earth that you can go and see all the layers of the geological column and the fossils upon which it is based?

NOTE: The first section in the next chapter has an experiment that must sit for 30 minutes. Please plan your day accordingly.

Chapter 8: Interpreting the Geological Record

Introduction

In the past several chapters, I have discussed the data regarding the geosphere. Aside from a few places where I discussed differences in the way scientists interpret the data, I have focused on what we know based on observations that have been made. Thus, nearly every scientist who studies the earth would agree with the basics of what has been discussed so far. The problem arises when we start to interpret the data in order to learn more.

Consider, for example, the fossil record. We know there are fossils of organisms that look nothing like what we see today. Those fossils tell us something about the earth's past. At minimum, they tell us that at some time in the past, there were organisms alive unlike any living thing we have discovered. While that's interesting, as scientists, we want to learn much more than that. As a result, we want to use the fossil record to find out more about what happened long ago.

The problem is that the fossil record must be interpreted, and sometimes, that interpretation can be horribly wrong. Nebraska Man and the original interpretation of *Pakicetus* are two examples of that. Both of them were the result of trying to interpret incomplete fossils. Well, the entire fossil record is incomplete. There are no fossils that preserve everything in the organism, and fossilization is pretty rare, so the vast majority of organisms that lived in the past don't appear in the fossil record. Thus, any interpretation we make based on the fossil record might be completely wrong.

Because of these and other factors, different scientists interpret the fossil record and geology as a whole quite differently. In this chapter, I want to spend time discussing how uniformitarians and most catastrophists interpret the geological column. As you will see, their interpretations are so radically different that both cannot be correct!

Making Interpretations Based on Limited Data

To give you an idea of how difficult it can be to make interpretations when you have limited data, perform the following experiment.

Experiment 8.1: Did it Change?

Supplies:

- The pH Universal Indicator Paper package from the laboratory kit made for this course
- Baking soda
- Water
- Two small glasses, like juice glasses
- A measuring tablespoon
- A measuring teaspoon
- A 1-cup measuring cup
- Two spoons for stirring
- An oven
- Oven mitts
- A glass baking dish
- A napkin or paper towel

Instructions:

1. Set the oven to bake at 450 degrees (200 in Celsius) and turn it on.
2. Use the measuring tablespoon to get a tablespoon of baking soda and then add it to the glass baking dish so that it forms a thin layer covering the bottom of the dish.
3. Look at the solid for a moment to get an idea of its appearance.
4. Put the baking dish in the oven, even though it has not reached 450 degrees yet. It will warm to the proper temperature soon enough.
5. Let the dish sit in the oven for 30 minutes.
6. While you are waiting, put the two small glasses next to each other on a counter.
7. Use the measuring teaspoon to add one teaspoon of baking soda to one of the glasses.
8. You will have a lot more time to wait, so do some other schoolwork or chores until the dish has been in the oven for 30 minutes.
9. Use the oven mitts to remove the dish from the oven. **Be careful! It is very hot!**
10. Look for a moment at the solid in the dish. Based on what you see, has any change (other than it now being very hot) occurred?
11. Tilt and shake the dish gently so that most of the solid accumulates in one part of the dish. You can even gently tap the dish against a surface that is safe for hot things to get the solid to accumulate.
12. Set the dish on a surface that is safe for hot things.
13. **Being careful to not touch the hot dish**, use the measuring teaspoon to collect one teaspoon of the solid in the dish.
14. Add the white solid to the small glass that is still empty.
15. Pick up both glasses, one in each hand, and compare the solids. Shake the glasses gently so the solids move around in the bottom of the glass.
16. Based on your observations, how much difference is there between the two solids?
17. Put the glasses down.
18. Use the measuring cup to add one cup of warm water to each glass.
19. Using a different spoon for each glass, stir the solid and water so that most of the solid dissolves. The solutions will probably be cloudy, but that's okay.
20. Compare the two solutions. Based on your new observations, how much difference is there between the two solids?
21. Open the package of pH Universal Indicator Paper. It looks a lot like a matchbook, with thin strips rising from the bottom of the container.
22. Each section is covered with a protective strip that looks noticeably different from the strips of paper underneath it. Tear away the protective strip nearest to one of the edges.
23. Tear away the first strip of paper that was underneath the protective strip.
24. Dip the strip of paper into one of the solutions you made so that about half of it gets wet with the solution.
25. Pull the strip out and set it on the napkin or paper towel. Compare the color of the dry part of the paper to the color of the wet part.
26. Tear away the strip of paper that was underneath the strip you just pulled out, and repeat steps 24 and 25 with the other solution.
27. Set that strip of paper on the napkin or paper towel.
28. Compare the two strips of paper.
29. Based on your new observations, how much difference is there between the two solids?
30. Clean up your mess, and be sure to turn the oven off if you haven't already.

When you pulled the baking dish out of the oven, did you notice anything that had changed? Probably not. It was a white solid when you put it in the oven, and it was a white solid when you pulled it out of the oven. In your experience, baking often changes the color and appearance of things, but in this case, you probably didn't see a change. Thus, based on that observation alone, heating up the baking soda to 450 degrees didn't produce any change.

When you put the baked solid into the second glass and then compared the two glasses, did you notice any difference? You might have. The solid that had been baked was not quite as "clumpy" as the solid in the other glass. If you didn't see it, that's okay. The difference is subtle, and it takes a good eye to see it. So, based on your observations at this point, it's still likely that you concluded that heating up the baking soda didn't cause any change to it.

When you added water and stirred, you might have seen a difference there as well. If you observed things carefully, you might have seen that the solution made with the heated baking soda was cloudier than the solution made with the unheated baking soda. That tells you the heated baking soda didn't dissolve as easily in water as the unheated baking soda.

The last part of the experiment is where there should have been a very noticeable difference between the two. The paper turned from yellowish to green in the solution made with the unheated baking soda, but it turned from yellowish to blue in the solution made with the heated baking soda. Why? The paper turns colors based on the level of acid to which it is exposed. You will learn more about this in a later chapter, so don't worry about that now. Just understand that the different colors on the paper indicated that the solids were actually different chemicals!

How could heating the baking soda turn it into a different chemical? Heat is a form of energy, and when you give energy to chemicals, they can react to it in many ways. One way is to change their chemical structure. In this case, the following chemical reaction happened:

$$2NaHCO_3 \rightarrow Na_2CO_3 + CO_2 + H_2O$$

The chemical on the reactants side of the equation ($NaHCO_3$) is baking soda. The first chemical on the products side (Na_2CO_3) is often called "washing soda." In your experiment, then, you heated baking soda, turning it into washing soda. That's a very important chemical change, but it probably wasn't noticeable without making solutions and testing them with the paper.

But wait a minute. What happened to the CO_2 and H_2O? After all, H_2O is the chemical formula for water, and the solid wasn't wet! Well, CO_2 is a gas, so it floated away. You are used to water being a liquid, but that's only at room temperature. At 450 degrees Fahrenheit, it is also a gas, so it also floated away. As a result, the only thing left in the baking dish was Na_2CO_3.

I wanted you to do the experiment to specifically see how your interpretations of a situation change based on new data. You probably concluded that the solids were the same (or at least very, very similar) throughout most of the experiment. However, new observations eventually led you to the conclusion that a significant change had occurred. In geology, we have limited observations. Based on those limited observations, our interpretations of the data can be wrong. As a result, we have to be very careful about putting too much faith in our interpretations.

Comprehension Check

8.1 If you heat washing soda (Na_2CO_3) to about 850 °C, it will melt and form liquid washing soda. Is it possible to melt baking soda ($NaHCO_3$)?

Uniformitarianism's Approach to the Geological Record

In chapter 2, you learned the definition of uniformitarianism and how it is used by most geologists to interpret geological processes. It can be best summed up by the phrase, "The present is the key to the past." In other words, uniformitarians look at processes that happen now, and they try to envision how those processes could have produced the geological features we see today. Consider, for example, plate tectonics. As you have already learned, we can use the global positioning system to measure the speed at which the plates are moving. Not surprisingly, they are moving very slowly. Uniformitarians assume that while there have been some changes in the speed at which the plates moved in the past, those changes were small. In the end, then, they assume that throughout the course of history, the plates have always moved pretty much the way they are moving right now.

Obviously, if the plates have always been moving slowly, it took a long, long time for them to separate from Pangaea and move to where we see them today. Indeed, in the uniformitarian view, Pangaea existed about 250 million years ago. The continents slowly moved with the plates, and after another 50 million years they formed two separate continents. What are now North America, Greenland, and the Eurasian continent formed a big continent called Laurasia (lah' ray shuh), and the other continents formed a big continent called Gondwanaland (gahnd wah' nuh land'). It took another 200 million years for them to reach the positions in which we see them today. Now remember, the plates are continually moving, so the continents didn't start out arranged as Pangaea. They started out as a different supercontinent called "Rodinia," but that's beyond the scope of this course.

250 million years ago

200 million years ago

Present Day

In the uniformitarian view, Pangaea existed millions of years ago, and the continents slowly moved apart to reach their present-day positions.

Depending on the science books you have read in the past, you may or may not be used to seeing references to things that happened on the earth millions of years ago. It is common in most science books, because most of them are written from a uniformitarian perspective. While early uniformitarians simply speculated about how old the earth was, once radioactivity was discovered and studied in detail, the age of the earth and the rocks that make up the geosphere became things that could be measured, as long as you made certain assumptions. To learn how uniformitarians use radioactivity to measure the age of the things they are studying, you first have to understand a bit about what it is and how it works.

Atoms and Radioactivity

Radioactivity was discovered in the late 1800s, and as scientists started learning about it, they were taken by surprise. One of the most important theories guiding science at the time was Dalton's Atomic Theory, and it stated that atoms cannot change. They can become part of a molecule, which would result in something with different properties than the atom, but the atom itself would not change. If you "plucked" the atom out of the molecule, it would be exactly the same as it was before it went into the molecule. Radioactivity showed that Dalton's Atomic Theory was wrong on that point. In radioactivity, an atom changes into a completely different atom. For example, there are radioactive carbon atoms. Over time, they change into nitrogen atoms.

How can such a thing happen? To understand the process, you need to understand some basics about atoms. Hopefully, this part will be review for you. If not, I will go over everything you need to know, so don't worry. If this is review for you, please read it anyway, just to make sure you remember what you need to know in order to understand radioactivity.

As shown in the illustration on the right, an atom is made up of three components: **protons, neutrons**, and **electrons**. Protons have positive electrical charge, and neutrons have no electrical charge. The protons and neutrons are crammed together in the center of the atom, called the **nucleus**. Electrons are negatively charged, and they travel around the nucleus. If you have studied the atom in detail, you know that this illustration isn't really correct. First, the electrons are drawn way too large. They are so small that they would not even show up in the illustration if they were drawn to scale. Also, the illustration has them circling the nucleus in orbits. That's not really how they travel around the nucleus, but their actual paths make things so complex that for the purposes of most scientific discussions, we just show them as traveling in circular orbits.

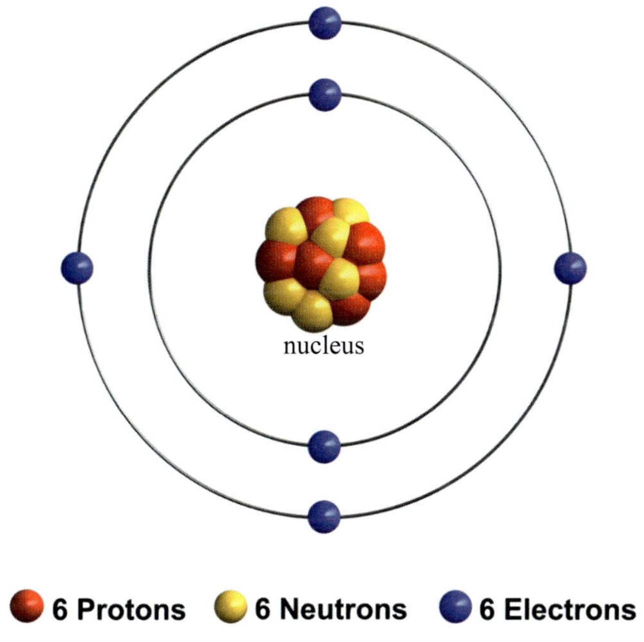

This is one way to visualize a carbon-12 atom.

The specific atom shown in the illustration above is a carbon-12 atom, which is often labeled ^{12}C. All carbon atoms must have 6 protons, so because the atom in the illustration has 6 protons, it is a carbon atom. Atoms must have equal numbers of protons and electrons, because atoms cannot have any overall electrical charge. Because protons are positive and electrons are negative, their charges cancel each other. An equal number of protons and electrons ensures that atoms have no overall electrical charge. Since all carbon atoms have 6 protons, they must all have 6 electrons. Why are there two electrons in one orbit and four in another? Don't worry about that. It represents the fact that only a few electrons can be close to the nucleus. The more electrons an atom has, the farther away some of its electrons have to be from the nucleus. When it comes to radioactivity, the electrons are not important, and we will start ignoring them in a moment.

In radioactivity, the nucleus is what's important. After all, the number of protons determines the type of atom. Since radioactivity involves one atom changing into another type of atom, the changes that are produced by radioactivity happen in the nucleus. What's in the nucleus of the atom shown above? There are 6 protons and 6 neutrons. That's where the "12" in the atom's name comes from. It represents the total number of protons and neutrons in the nucleus. Since this atom has 6 protons and 6 neutrons in its nucleus, and 6 + 6 = 12, this is a carbon-12 (^{12}C) atom.

It turns out that this is not the only kind of carbon atom found in creation. There are atoms that have 6 protons and 7 neutrons in their nucleus. Now remember, any atom that has 6 protons in its nucleus is a carbon atom, so these atoms are carbon atoms. But they are different from carbon-12 atoms. They have one extra neutron. Thus, they are called carbon-13 (^{13}C) atoms, since they have a total of 13 protons and neutrons (6 protons + 7 neutrons) in their nucleus.

When atoms have the same number of protons but different numbers of neutrons, they are called **isotopes** (eye' suh tohps).

Isotopes – Atoms that have the same number of protons but different numbers of neutrons

It turns out that there are isotopes for pretty much every type of atom. For carbon atoms, there are actually three different isotopes found in nature. They are shown below:

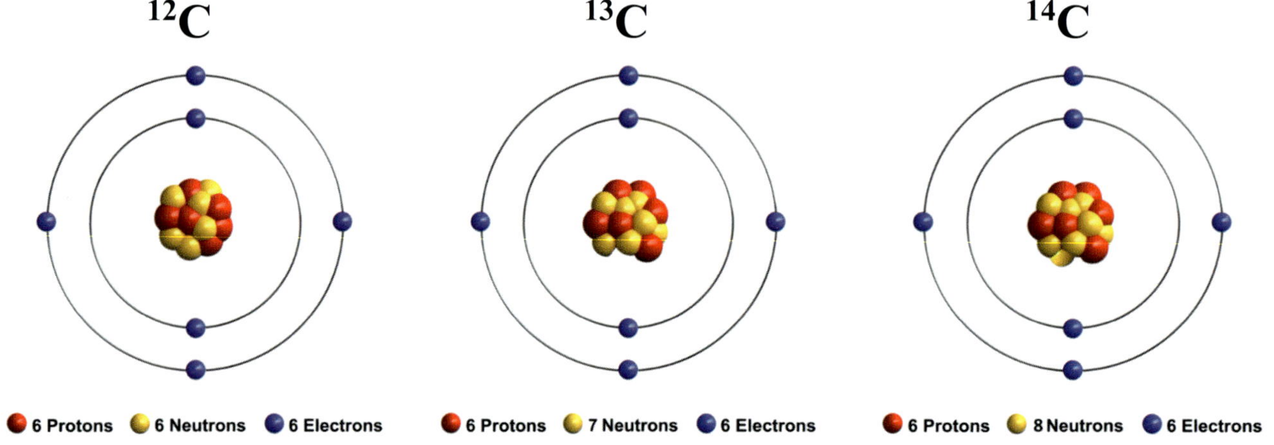

These are all isotopes of carbon. They are all carbon atoms, because they all have six protons. However, they are different atoms, because they have different numbers of neutrons. When atoms have the same number of protons but different numbers of neutrons, they are isotopes.

The carbon-12 pictured on the previous page is once again shown on the left. It's the atom with 6 protons and 6 neutrons. The middle atom is carbon-13. It is called that because it has 6 protons and 7 neutrons. The carbon on the right is carbon-14, because it has 6 protons and 8 neutrons. I can make other isotopes of carbon in the lab, but these three are the only ones found naturally in creation.

Now it turns out that the vast majority of carbon atoms in creation are carbon-12 atoms. Indeed, if I gathered a trillion (1,000,000,000,000) atoms, 989,000,000,000 of them would be carbon-12 atoms. 10,999,999,999 of them would be carbon-13 atoms, and *only one* of them would be a carbon-14 atom. In other words, carbon-14 is very rare. There is a reason for that. Carbon-12 and carbon-13 are stable. That means they don't want to change into a different kind of atom. However, carbon-14 is radioactive. It's unstable, and over time, it will change into a nitrogen-14 atom.

Because carbon-14 changes into a different atom, we say that it is a **radioactive isotope** of carbon. Carbon-12 and carbon-13 are stable. They won't change into another atom unless something really unusual happens. However, as long as you wait long enough, carbon-14 will change into nitrogen-14. When that happens, we say that the carbon-14 has **decayed** into nitrogen-14.

Radioactive isotope – An isotope that is unstable and eventually decays to become a different atom

But how does carbon-14 decay and become nitrogen-14? Well, all nitrogen atoms have 7 protons. That means nitrogen-14 must have 7 protons and 7 neutrons in its nucleus. How is that different from carbon-14, which has 6 protons and 8 neutrons? There is one more proton and one less neutron. Can you think about what might have happened to make the change? A neutron turned into a proton! Think about it. Carbon-14 has 6 protons and 8 neutrons. If one neutron changes into a proton, it will now have 7 protons and 7 neutrons. That's nitrogen-14!

How does a neutron decay into a proton? It actually emits an electron! Think about it. A neutron has no charge. A proton has a positive charge. For a neutron to turn into a proton, it needs to get rid of a negative charge. That way, it will become positive. The illustration on the right shows the process. I have removed the electrons that are orbiting the nucleus just so you can concentrate on the nucleus. The nucleus of a carbon-14 atom has 6 protons and 8 neutrons. When one of those neutrons spits out an electron, it turns into a proton. That makes a nucleus with 7 protons and 7 neutrons, which is nitrogen-14.

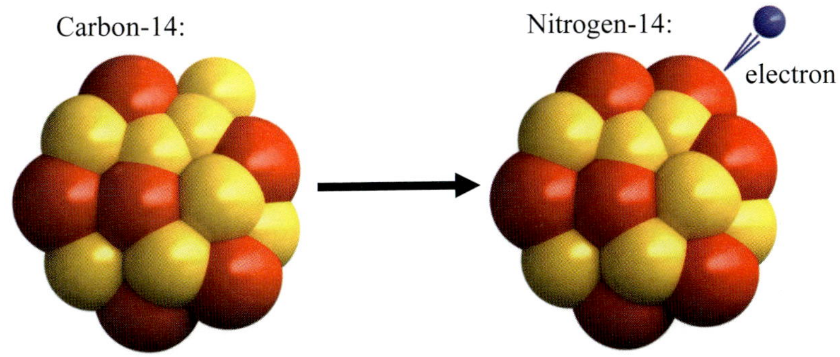

In beta decay, a neutron emits an electron to become a proton.

This process is called **beta decay**, and it is a very common form of natural radioactive decay found in creation.

Beta decay – Radioactive decay in which a neutron emits an electron to become a proton

Why is it called beta decay? Before scientists really understood radioactivity, they noticed that radioactive isotopes like carbon-14 emitted negative particles. They didn't identify those negative particles as electrons at first, so they called them "beta particles." Even though we now know that beta particles are electrons, we still call this process beta decay, and scientists still call the electrons emitted in the process beta particles.

It's important to note that I am simplifying this quite a bit. There is actually another particle emitted during beta decay, but it isn't necessary for the purpose of this course, and it really complicates the discussion. In addition, there are actually two different kinds of beta decay, but this is the only kind that relates to geology, so it's all I want to talk about.

There are two other kinds of radioactive decay that are important to geology, and one of them is really the reverse of this one. If a neutron emits an electron, it becomes a proton. What do you think might happen if a proton absorbs an electron? It will become a neutron. That's what happens in **electron capture**.

Electron capture – Radioactive decay in which a proton absorbs an electron to become a neutron

Carbon-11, for example can be made in the lab. It has 6 protons and 5 neutrons in its nucleus. One of those protons can actually absorb an electron that gets close to the nucleus. When that happens, the proton becomes a neutron. That means there will now be one less proton and one more neutron, so the new atom has 5 protons and 6 neutrons in the nucleus, which makes it an boron-11 atom.

Both beta decay and electron capture involve something inside an atom's nucleus changing its identity, which changes the identity of the atom as well. However, the last type of radioactive decay that we need to discuss is very different. It involves a nucleus getting rid of protons and neutrons by spitting out parts of itself. This is called **alpha decay**.

Alpha decay – Radioactive decay in which a nucleus ejects two protons and two neutrons as a single particle

Just like beta particles were named before scientists knew they were electrons, alpha particles were named before we knew they were made of two protons and two neutrons. We now know that alpha particles are actually the nucleus of a helium-4 atom, but we still call them alpha particles.

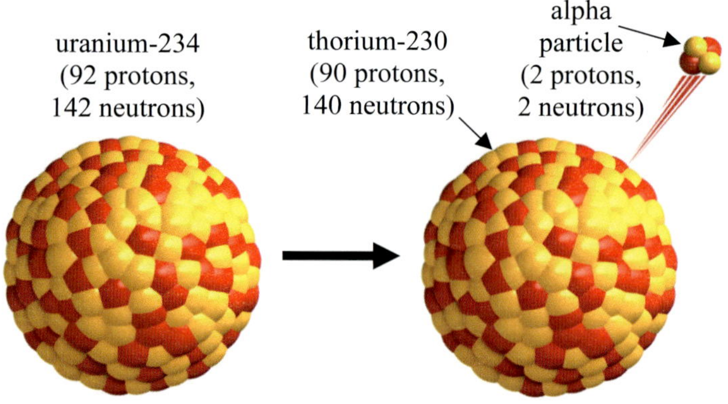

In alpha decay, a nucleus ejects an alpha particle to lose two neutrons and two protons.

One example of alpha decay that happens in the geosphere is illustrated on the left. The isotope shown, uranium-234, has 92 protons and 142 neutrons in its nucleus. All uranium atoms have 92 protons, and adding the neutrons and protons tells us this is uranium-234. It can emit an alpha particle, which means it will lose 2 protons and 2 neutrons. That means it will now have 90 protons in its nucleus. That makes it a thorium atom, since all thorium atoms have 90 protons. Since it also loses 2 neutrons, it has only 140 neutrons in its nucleus, so it is thorium-230.

Everyone seems to know that radioactivity is dangerous, but few people really know why. Radioactivity isn't dangerous because of the chemical reactions it can do. Indeed, radioactivity isn't a chemical process at all. Thus, radioactive isotopes aren't like poisons. The danger of radioactivity is caused by the particles that are emitted. Each time a radioactive isotope decays, a particle (like an alpha or beta particle) speeds away from the isotope. That particle is like a tiny bullet. If it hits you, it can deposit energy in you. That energy might kill a cell, or it might mutate the DNA in a cell.

Believe it or not, having a few cells killed or having a few cells' DNA mutated isn't a big deal. In fact, more than two million cells die in your body every second! You even have a few cells with mutated DNA. However, your body has been wonderfully designed to replace lots of dead cells and eliminate cells with mutated DNA. The problem with radioactivity is that it adds to the dead cells and mutated DNA. If it adds too many dead cells, your body can't replace them fast enough, and you get a burn, which is typically called a "radiation burn." Severe radiation burns destroy organs. If it makes too many cells with mutated DNA, your body can't eliminate them, and you end up having cancer.

So you can think of radioactivity as a tiny "machine gun" that is shooting tiny "bullets" at you. If you can block those bullets, you won't be harmed. That's why people who work with radioactivity usually have some sort of shielding between them and the radioactivity. The shielding stops the bullets. You can also move away from the radioactivity. The farther you are away, the fewer "bullets" that hit you.

Comprehension Check

8.2 Lithium-8 has three protons and 5 neutrons in its nucleus. If it undergoes beta decay, how many protons and neutrons will it have?

8.3 Potassium-40 is an isotope found in nature. It has 19 protons and 21 neutrons in its nucleus. If it decays by electron capture, how many protons and neutrons will it have?

8.4 Americium-241 has 95 protons and 146 neutrons in its nucleus. If it undergoes alpha decay, how many protons and neutrons will it have?

The Speed at Which Radioactive Decay Occurs

In the previous section, you learned about what radioactivity is. While this is important, the more important issue in earth science is how fast it happens. Let's explore that with an experiment.

Experiment 8.2: Half Life

Supplies:

- 60 coins (You can actually use 60 of any small objects that have two easily-distinguished sides, like M&M candies. When the experiment and text refer to "heads" and "tails," you can just substitute the two sides of whatever object you are using.)
- A plastic container with a lid that is wider than it is tall and would easily hold hundreds of the items you are using.

Instructions:

1. Place the 60 items in the container, all with their "heads" side up.
2. Put the lid on the container.
3. Shake the container up and down multiple times so that the items move around and flip over several times.
4. Put the container down on a level surface.
5. Take off the lid.
6. Look at the coins. Some will be lying with their "heads" side up, and others will be lying with their "tails" side up. Remove any coins with the "tails" side up. If coins are stacked on top of one another, that's not a problem. Carefully move the coin on top to reveal the coin underneath, making sure you do not change which side of the coin is facing up.
7. When all of the coins with the "tails" side up have been removed, count the number of coins that remain and record it.
8. Put the lid back on the container.
9. Repeat steps 3-7.
10. Continue to repeat steps 2-7 until there are two or fewer items left in the container.
11. Clean up your mess.

The experiment you just did is a reasonably accurate model of how radioactive atoms decay. As I told you before, carbon-14 atoms turn into nitrogen-14 atoms via beta decay, but they don't do so all at once. They change slowly over time. Your experiment simulated this by representing the radioactive carbon isotopes with coins. When a coin was in the container with its "heads" side up, it was still a carbon-14 atom. Any coin with the "tails" side up represented a carbon-14 atom that had changed into a nitrogen-14 atom.

Each time you shook the container, you were allowing a certain "amount of time" to pass. As that time passed, some of the carbon-14 atoms changed into nitrogen-14 atoms, and some did not. When you removed the coins with the "tails" side up, you were removing the carbon-14 atoms that had changed. When you got done removing them, the coins left in the container represented carbon-14 atoms. Each time you repeated this process, you were allowing the same "amount of time" to pass.

Now look at the number of coins that remained after every shake. While it might not have worked out exactly right, the number of coins that remained after a shake should have been roughly half of the number of coins that you had before the shake. This should make sense. After all, you were using "heads" and "tails" to decide which coins to keep and which to remove. If you flip a coin around many times, you will find that it will have its "heads" side up half the time and its "tails" side up the other half. Thus, each time you shook the container, you ended up removing roughly half the coins and leaving the other half. You didn't end up removing exactly half, because there is randomness in the process. As a result, one time you might have removed slightly more than half, and the next time slightly less than half. On average, however, you removed about half the coins each time.

I want you to actually make a graph of this, because it is instructive. On the vertical (y) axis of the graph, record the number of coins remaining in the box after you removed the ones that were "tails" side up. On the horizonal (x) axis of the graph, record how many times you shook the container. In the graph below, for example, the blue marks show my results. Before I did any shaking, there were 60 coins in the container, so the first blue mark is at zero times shaken and 60 coins remaining. The first time I shook the container, I removed 28 coins, so there were 32 remaining. The blue mark at one time shaken is at a value of 32 on the vertical axis, which is a little above 30. The second time, there were 15 coins remaining, so the blue mark at two times shaken is halfway between 10 and 20. The third time, 8 remained, the fourth time, 4 remained, and the fifth time 2 remained, which ended the experiment. Your graph won't be the same, but it should show that overall, the number of coins remaining got smaller each time the container was shaken, and it didn't take many shakes before you reached the point where you were told to stop.

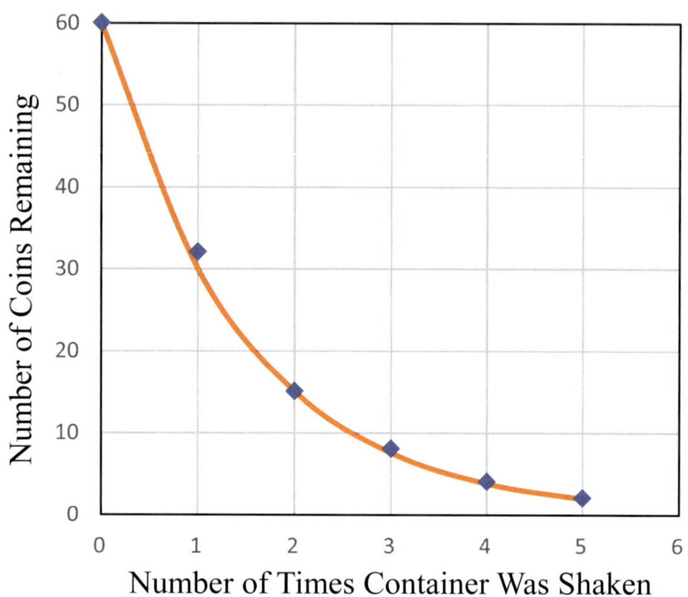

Carbon-14 atoms turn into nitrogen-14 atoms in a very similar way. Had your experiment been a perfect simulation of the way it happens, the marks should have fallen on the orange curve that is also on the graph. Notice that the orange curve starts at 60 and drops to 30 at the first time shaken. If your experiment had gone perfectly, you would have removed 30 coins, which is half of the 60 you started with. The next time you shook the container, you would have removed 15 coins, which is half of the number that existed before shaking. In other words, if the experiment had happened perfectly, you would have always removed half of the coins that were in the container.

If you have a sample of carbon-14 atoms and wait a certain amount of time, half of them will decay into nitrogen-14 atoms. If you wait the same amount of time again, half of the *remaining* carbon-14 atoms will decay. This length of time is called the half-life of the radioactive isotope.

Half-life – The time it takes for half a sample of radioactive atoms to decay

In your experiment, then, shaking the container represented one half-life of carbon-14 atoms. When you removed coins, you were removing the atoms that had decayed. No matter how many atoms you started with, each half-life should have resulted in about half of them decaying.

In your experiment, I had you start with 60 pennies. That probably seemed like a lot when you were counting them out, but in the end, you probably shook the container 5 times or less before you had to stop. Suppose you started with 960 pennies instead. That would have been really annoying, but surprisingly, it wouldn't have given you a lot more shakes. If the experiment worked perfectly, the first shake would reduce it to 480 pennies, the next to 240 pennies, the next to 120 pennies, and the next to 60 pennies. Then you would be back to what you did in your experiment. So, starting with 16 *times* more pennies resulted in only four more shakes! That's the nature of radioactive decay. When something cuts itself in half in a set amount of time, it takes *a lot* more of it to last just a few more half-lives.

But how long is a half-life? That depends on the isotope. The half-life of carbon-14 is 5,730 years. That means if you started with 1,000 carbon-14 atoms, it would take 5,730 years for half of them to decay. At that point, you would have 500 carbon-14 atoms and 500 nitrogen-14 atoms. The nitrogen-14 atoms are not radioactive, so they won't change. However, if you waited *another* 5,730 years, 250 of the remaining carbon-14 atoms would decay, leaving you with only 250 carbon-14 atoms left and a total of 750 nitrogen-14 atoms.

While 5,730 years sounds like a long time, there are radioactive half-lives that are a lot longer. Potassium-40, for example, decays with a half-life of 1.3 *billion* years. You might wonder how we can measure half-lives of thousands of years or billions of years, given that we have only been studying radioactivity for a bit more than 100 years. Well, go back and look at the graph on the previous page. Radioactive isotopes always produce a graph like that when they decay. Thus, the data always fall along a curve like the orange curve, but the specific equation of the curve depends on the half-life. So I can study the decay of carbon-14 or potassium-40 and make a graph like the one on the previous page. Then, I can find the equation of the curve that fits that data, and the equation will tell me the half-life. Even just a few years' worth of data will allow me to fit an equation that tells me the half-life.

Not all radioactive isotopes have long half-lives. Iodine-131, for example, undergoes beta decay with a half-life of 8 days. It can be made in a lab and used for short-term experiments, and it is even used to treat a specific form of cancer. Polonium-214 undergoes alpha decay with a half-life of 0.00016 seconds. Obviously, it exists for only a short time after it has been formed. While radioactive decay is important regardless of the half-life involved, for the uniformitarian interpretation of geology, only radioactive isotopes with very long half-lives are of interest. Indeed, carbon-14 has the shortest half-life of any radioactive isotope used in the study of geology.

Comprehension Check

8.5 A radioactive atom has a half-life of 15 minutes. If you have 1,000 atoms and wait 15 minutes, how many will have decayed? If you wait 15 more minutes, how many more will have decayed?

Using Radioactive Decay to Determine the Age of Rocks

If you think about it, radioactive decay can act as a "clock" that can be used to measure how old a rock is. After all, some rocks will contain radioactive isotopes. The longer the rocks exist, the more half-lives will pass. As a result, the fewer radioactive isotopes there will be and the more products of the decay there will be. If we can analyze a rock and find how much of each isotope exists, we should be able to figure out how many half-lives have passed, and that should tell us how long ago the rock was formed.

Of course, the problem is that rocks are formed and destroyed through the rock cycle, which you learned about in Chapter 4. A sedimentary rock, for example, might contain radioactive isotopes and the products of their decay, but those might have originally been part of a completely different rock that was weathered. The sediments formed as a part of that weathering could have then formed the rock you are studying now. Thus, while you can study the radioactive isotopes in sedimentary rock, they won't tell you much about when that specific rock formed.

Think, however, about igneous rock. It forms from magma that was originally in the mantle. The mantle has lots of radioactive isotopes in it. In fact, that's the main reason the mantle is so hot. Radioactive decay also produces a lot of energy, which heats up the interior of the earth. When the magma rises and freezes, or when it forms lava and freezes, it will have those radioactive isotopes in it. However, it might not have the products of their decay in it. What do I mean? Perform the following experiment to find out.

Experiment 8.3: Gases in Liquids

Supplies:
- Any kind of carbonated beverage like soda pop, seltzer water, etc.
- A small pot or pan for boiling
- A stove
- Two small glasses, like a juice glasses
- A ¼-cup measuring cup
- A two spoons for stirring

Instructions:
1. Use the measuring cup to add ¼-cup of carbonated beverage to one of the small glasses.
2. Use the measuring cup to add ¼-cup of carbonated beverage to the small pot or pan.
3. Turn a stove burner on "high" and put the pan or pot on it.
4. Watch the carbonated beverage inside the pan. Once you see lots of large bubbles forming and popping, wait one minute and then remove it from the burner and turn the burner off.
5. Carefully pour the boiled beverage from the pot or pan into the empty glass.
6. Put a spoon in each glass, and at the same time, agitate the liquid in both glasses vigorously for a few seconds.
7. Pull the spoons out of the glasses and compare the liquids. Which has more bubbles? If neither has many bubbles, you didn't agitate the liquids enough. Move the spoon back and forth vigorously to really agitate the liquids. One liquid should be bubbling a lot more than the other.
8. Use the spoons to gently stir the liquids for one minute. In this case, you aren't trying to make bubbles. You are just trying to keep each liquid moving. This will allow the warm liquid to cool a bit.
9. Carefully taste the liquid that you had boiled. It is probably still warm, but it shouldn't be any warmer than tea or hot chocolate.
10. Now taste the liquid from the other glass. What's the difference?
11. Please note that you should **NEVER** taste anything in an experiment unless a knowledgeable source tells you to do so, since experiments can produce some nasty things!
12. Clean up your mess.

What differences did you notice in the experiment? First, the recently-boiled beverage should have produced a lot fewer bubbles when it was agitated than the beverage that had not been heated. In addition, the recently-heated beverage should have tasted "flatter" than the one that wasn't heated. Why? Because there was a big difference in the amount of gas dissolved in each liquid.

Carbonated beverages cause a "tingling" sensation in your mouth because carbon dioxide gas has been dissolved in them. That's where the term "carbonated" comes from. When you open a carbonated beverage and pour it into a glass, you see a lot of bubbles. Those bubbles are made by the carbon dioxide gas that had been dissolved into the beverage and is escaping. The same thing happens when you agitate the beverage.

The foam on top of this root beer is made of carbon dioxide bubbles.

When you agitated the two beverages in your experiment, the heated beverage didn't make many bubbles because not much carbon dioxide was dissolved in it. That's also why it tasted "flatter" than the non-heated beverage. Since the dissolved carbon dioxide causes the tingle in your mouth, less dissolved carbon dioxide means less of a tingle.

But where did that carbon dioxide go? It boiled away. When you boil water, the bubbles you see are bubbles of water vapor that are rising and escaping the liquid water. When you boiled the carbonated beverage, the initial bubbles were not from water vapor. They were from carbon dioxide rising and escaping the liquid. Eventually, the water in the beverage got hot enough to form water vapor bubbles, so by the time you were done, there were also water vapor bubbles. However, most of the bubbles you saw in the short time you were boiling the beverage came from carbon dioxide, not water vapor.

So what does this tell you about how well gases dissolve into liquids? At low temperatures, gases dissolve into liquids really well. The higher the temperature, however, the harder it is to dissolve a gas in a liquid. As you heated the carbonated beverage, then, it became less carbonated because carbon dioxide gas just couldn't stay dissolved in the hot liquid.

What does this have to do with igneous rock? Since it starts out as magma, it is really, really hot. What does that tell you about how well gases can dissolve in it? They don't dissolve very well. So we can assume that when magma is rising up from the mantle, most gases will leave the magma, just like the carbon dioxide left the carbonated beverage in your experiment. As a result, when the magma (or lava) cools and freezes, the rock should have very few (if any) gases in it. Nevertheless, if you actually analyze igneous rock, you will find gases in it. In fact, argon gas was discovered by analyzing igneous rocks!

So if the magma or lava from which igneous rock forms has few (if any) gases in it, where did that argon come from? Well, it turns out that potassium atoms are found in the mantle, and potassium is a solid at room temperature and a liquid at magma temperatures. Thus, when igneous rock forms, it will often have potassium in it. Well, as you learned in Comprehension Check question 8.3, potassium-40 is radioactive. What does it decay into? Argon-40, which is a gas!

In igneous rock, then, there should be very little (if any) argon in the magma or lava from which it comes. However, after the magma or lava freezes, any potassium-40 that is in the rock will start decaying and making argon-40. The longer it has been since the rock formed, the less potassium-40 there should be, and the more argon-40 there should be. Thus, if I analyze an igneous rock, the amounts of potassium-40 and argon-40 inside should be an indicator of how long ago the rock formed.

This is the general principle behind using radioactive decay to determine the age of a rock. A radioactive isotope decays into another atom. If I can figure out how much of either the radioactive isotope or the atom into which it decays was there when the rock was formed, I can measure how much is there now. That should tell me how many half-lives passed since the rock formed. That can tell me how old the rock is.

There are a lot of different radioactive isotopes that are used for this purpose. The decay of potassium-40 into argon-40 is (not surprisingly) called the **potassium-argon** dating method. Another method for dating igneous rocks is called the **rubidium-strontium** dating method, and it uses the beta decay of rubidium-87 into strontium-87. There is also the **carbon-14** dating method, but it is not used on rocks. I will discuss that method later on. Collectively, these methods are called **radiometric dating methods**.

Radiometric dating – The process by which radioactive decay is used to determine the age of a specimen

Now remember, sedimentary rocks cannot be dated using radiometric methods, because the sediments that make them up might have come from much older rocks that were weathered. Radiometric dating can work on metamorphic rocks, but it is not very straightforward. Thus, most radiometric dating is done on igneous rocks.

Comprehension Check

8.6 Uranium-234 can alpha decay into thorium-230 with a half-life of 80,000 years. If the rock is analyzed for those two specific atoms and it is determined that uranium-234 has been reduced to ¼ of its original content since the rock formed, how old would radiometric dating say it is?

Radiometric Dating and the Geological Column

Now that you know the basics of radiometric dating, you are ready to learn how uniformitarians use it to reconstruct the earth's past as represented by the geological column. Remember, the geological column doesn't exist anywhere in the world, but it uses the Principle of Faunal Succession to relate the various layers of fossil-bearing rocks found throughout the world. The fossils are mostly found in sedimentary rock, and radiometric dating can't be used on sedimentary rock. How, then, can it be related to the geological column?

This rock formation has strata of fossil-bearing sedimentary rock, and some of those strata have rock made from volcanic ash above and below them.

Look at the beautiful formation on the left. It is called Cathedral Rock, and it contains sedimentary strata that hold lots of fossils. In addition, those sedimentary strata often have layers of rock formed by volcanic ash. While the sedimentary rock cannot be dated with radiometric methods, the rock made of volcanic ash can be, since it is igneous. It is difficult to see in the photo, but there are actually strata of rock made from volcanic ash both above and below layers of sedimentary rock.

Think about what that tells you from a uniformitarian point of view. The Principle of Superposition says that the lower layers are older than the upper layers. So, suppose you find a layer of rock formed by volcanic ash, and it is radiometrically dated as 55 million years old. Right above that, you find a layer of fossil-bearing sedimentary rock, and right above that, there is another layer of rock formed by volcanic ash, which is radiometrically dated to 45 million years old. How old is the layer of sedimentary rock? It is between 55 and 45 million years old. Any fossils that are found in that layer, then, must be somewhere between 45 million and 55 million years old. This process is called **bracketing**, because the volcanic ash can be used to bracket the age of the sedimentary rock.

Now, of course, not all fossil-bearing strata can be bracketed, because lots of fossil-bearing rock is found in places with no volcanic activity at all. However, there are a lot of volcanoes on the earth, and many of them were quite active for quite a while. As a result, in certain areas of the earth, there are lots of fossil-bearing sedimentary rock layers that are bracketed, so uniformitarians think they have a good idea of how old those fossils are.

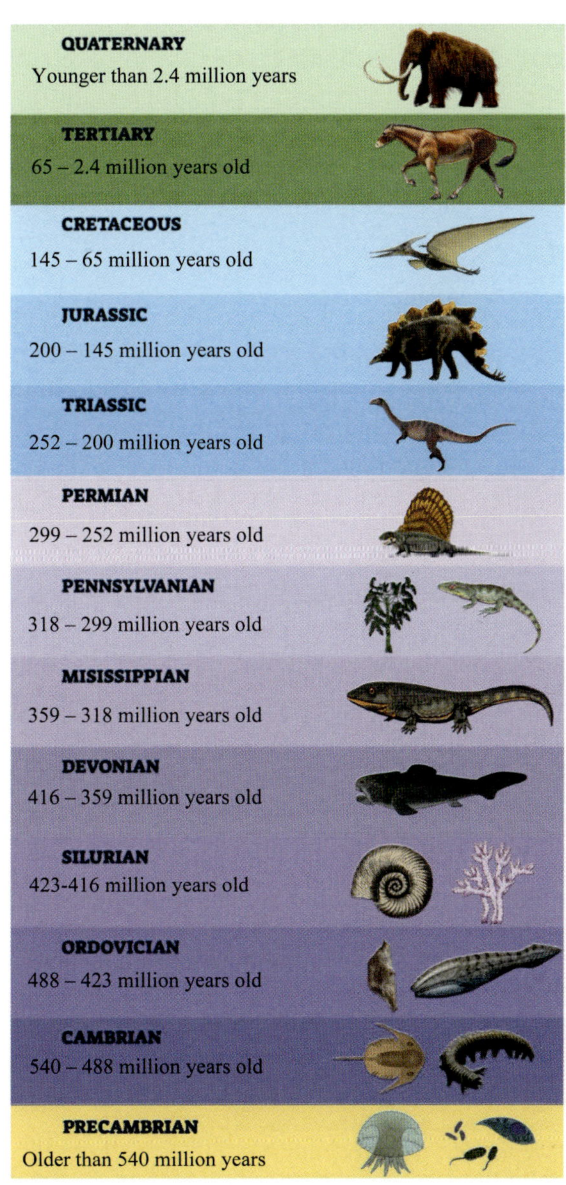

This is the geological column as it is interpreted by uniformitarians.

How does that help with the geological column, which tries to relate fossil-bearing rock layers around the entire world? Well, remember that there are some fossils that are used as index fossils. They are thought to be characteristic of specific layers of sedimentary rock. Suppose, then, paleontologists are lucky enough to find an index fossil in a layer of rock that is bracketed by igneous rock. The bracketing tells them the age of that layer of sedimentary rock, and the index fossil is characteristic of that layer. Thus, if I find that same index fossil anywhere else on the earth, whether or not it's bracketed, I know the age of the rock it is in.

Using that kind of reasoning, uniformitarians have assigned ages to the layers of rocks in the geological column, which also gives ages to the fossils. The result is shown on the right. Notice that each layer took *a long* time to form. In the uniformitarian view, then, each layer of the geological column represents a particular time period in the earth's history. The fossils found in those layers represent the organisms that lived during that time period. As a result, uniformitarians often refer to the layers in the geological column as eras in the earth's history. The fossils in Cambrian rock, then, represent the animals that lived during the Cambrian era.

I am sure you have heard that most scientists think dinosaurs and people never lived on the earth at the same time. The uniformitarian interpretation of the geological column shows you why they think that. After all, look at where the dinosaur fossils are found. They are found in Cretaceous (kruh tay' she' us) rock and some of the layers below. Thus, they lived in the Cretaceous era and in earlier eras of the earth's history. However, they are not found in layers above Cretaceous rock. Since the "youngest" Cretaceous rock is 65 million years years old, that means according to this interpretation of the geological column, dinosaurs went extinct 65 million years ago. Human fossils, on the other hand, are found only in Quaternary (kwah' dur nair' ee) rock. That means humans didn't exist until a maximum of 2.4 million years ago, Thus, humans didn't exist until long after the dinosaurs became extinct.

Now remember, this view of the geological column is based on several different things. First, it uses the Principle of Faunal Succession, which says that there is an overall pattern to the fossils found in layers of sedimentary rock found all over the world. Second, it is based on the idea that index fossils really do characterize certain layers in the geological column. Third, it depends on the Principle of Superposition. After all, the ages for the sedimentary layers come from bracketing some of them with igneous rock, and it is assumed the igenous rock underneath was formed first, then the sedimentary rock formed, and then the ingeous rock on top formed. Finally, the actual ages of the layers in the geological column are given by radiometric dating of igneous rock layers that bracket sedimentary rock layers, so they depend on radiometric dating being a reliable way of determining the age of igneous rocks.

We already know that the Principle of Superposition isn't always correct, as I discussed in Chapter 4. Uniformitarians realize this, but they think that rock formations like the one I show on page 118 are rare. Thus, while there are specific cases in which the Principle of Superposition is wrong, they are few and far between. As a result, it doesn't affect the overall interpretation of the geological column. In the same way, some of the other things upon which their interpretation depends may have isolated cases in which they aren't true, but the Principle of Faunal Succession, index fossils, and radiometric dating are all reliable enough in general to produce a reliable interpretation of the geological column.

The Earth's History According to Uniformitarians

The uniformitarian view of the geological column leads to a specific view of the earth's history. First, the earth is very old. The oldest thing dated reliably by radiometric methods is a crystal from Australia that is made of a mineral called zircon. The result is 4.38 billion years old. However, according to the uniformitarian view, at its youngest, the earth was too hot to make minerals or crystals. However, radiometric dating of rocks in space, along with specific theories about how the earth was formed, lead uniformitarians to think that the earth is 4.5 billion years old.

This is a fossil stromatolite. Uniformitarians say it is one of the oldest forms of life on the earth.

Since the earth was initially too hot to form rocks and minerals, it certainly wasn't initially hospitable to life. But as it cooled, it eventually became hospitable to life. Uniformitarians can't exactly agree on when the first life form appeared on earth, because finding microfossils is tricky, and some things that look like microfossils turn out to be just patterns formed in the rocks. However, fossils like the one shown on the left have been found in rocks that are dated to be about 3.5 billion years old. While it looks to be just a group of circles in the rock, it actually turns out to be something that can by identified as being made by single-celled organisms: a **stromatolite** (stroh mad' uh light').

Believe it or not, scientists can say with some confidence that the fossil on the previous page was made by microscopic organisms, even though the fossil itself is not microscopic. How can they say that? Look at the picture on the right. Those circular objects are stromatolites that are currently being formed. By analyzing them, we know that these structures are produced when microscopic organisms called **cyanobacteria** (sigh' an oh bak tear' ee uh) trap sediments that are floating in the water in which they live. Chemicals emitted by the cyanobacteria cement the sediments together, forming rock structures just like the fossil shown on the previous page.

These are stromatolites that are currently being formed by cyanobacteria.

Cyanobacteria are single-celled organisms, and that single cell is the "simplest" kind of cell that exists today. I put quotation marks around that word because even the "simplest" cell is more mind-bogglingly complex than the most advanced technology that exists today. Nevertheless, there are even more complicated cells in creation, like the ones that make up your body. As a result, cyanobacteria cells are considered "simple" by comparison.

The earliest forms of life, then, were "simple" single-celled organisms that lived in water, because the oldest fossils found were made by such organisms. However, as you move up the geological column (but remain in the Precambrian layer), you eventually find fossils of organisms that were composed of many cells. Based on their fossils, these organisms also lived in water. As a result, uniformitarians think the first living things appeared in the oceans, not on land. They were also much "simpler" than the organisms that are found in the oceans today.

As you move up into the Cambrian era, there is an "explosion" of diversity. The fossils are from organisms that are much more complex than the Precambrian organisms, and there is huge variety in terms of the different kinds of organisms found. They are all thought to be organisms that lived in the ocean, and some of them (like sponges and corals) are very similar to what we find today. Others look very strange, but they have characteristics that we see in animals today. For example, the drawing on the right shows you what some of the animals whose fossils are found in Cambrian rock might have looked like. Even though many of them are very different from the ocean animals we see today, many of them have hard, segmented

This is an artist's conception of what animals whose fossils are found in Cambrian rock might have looked like.

bodies that are very similar to the bodies of lobsters and other organisms that biologists call arthropods (ar' thruh pahdz').

As you go up the geological column, the fossils start looking more and more familiar. The first fossils of plants that lived on land appear in the Ordovician (or' duh vih she' un) era, and shark fossils are found in the Silurian (suh loor' ee uhn) era. Fossils of land animals do not appear until you reach the Devonian (dee vohn' ee uhn) era. What does that tell you? According to the uniformitarian interpretation of the geological column, life resided in the oceans for most of earth's history. Eventually, however, plants appeared on land, followed by animals.

As you move farther up the geological column, you see a lot more fossils of land animals, and eventually, you find dinosaur fossils. Dinosaurs "ruled" the earth for well over a hundred million years, but at the end of the Cretaceous era, they went extinct, because their fossils are not found in rock layers above Cretaceous rock. While some mammals appeared in the Triassic (try as' ik) era, they did not become really plentiful on earth until the Tertiary era, after dinosaurs went extinct. People were a very late arrival, appearing "only" a few hundred thousand years ago.

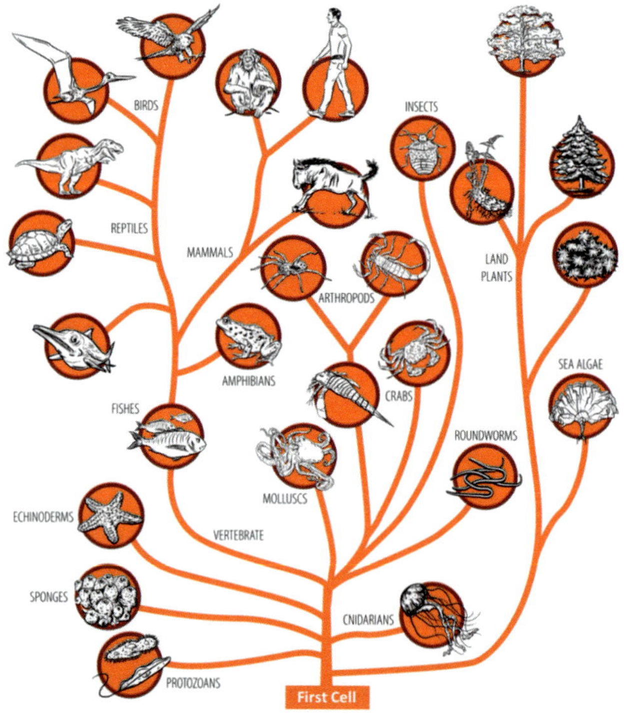

This drawing illustrates the theory of evolution. Life started out simple, and over billions of years, it became what we see today.

In this interpretation of the geological column, then, life on earth starts out as "simple" as possible – single-celled organisms made of the simplest kind of cells. Now once again, even the simplest organism is mind-bogglingly complex. However, single-celled organisms are clearly simpler than organisms that are made of many cells working together. As time went on, organisms got more and more complex. This fits nicely with the **Theory of Evolution**, which says that all life can be traced back to a "simple" common ancestor. That common ancestor began reproducing, and over billions of years, it gave rise to all the organisms we see today. The biological details of how this might have happened are beyond the scope of this course (and are still somewhat of a mystery among those who believe it), but the uniformitarian interpretation of the geological column gives a lot of supporting evidence for it, so we will be discussing it more in the next chapter.

At the same time, however, it is important to note that not all uniformitarians believe in the theory of evolution. While they support the idea that the earth is billions of years old and the strata in the geological column represent eras in earth's history, they don't think it is possible for natural processes to cause organisms to change in the way the Theory of Evolution requires. Some of these scientists are called **old-earth creationists**. They are creationists because they think life is not a product of evolution. Instead, it was specially designed and made by God. Since they believe in the uniformitarian interpretation of the geological column, however, they believe the earth is billions of years old. There is also a group of scientists who follow a view called **intelligent design**. Most of them also believe in the uniformitarian view of the geological column, but they also think that life cannot be solely the product of evolution. It must have been designed at some level. Unlike old-earth creationists, however, those who hold to intelligent design think that scientists should not speculate on who (or what) did the designing.

Now you might be wondering where the "simple" organism that started the process of evolution came from. Uniformitarians who are **atheists** (they do not believe in God) think it arose from random chemical reactions and other processes that happen naturally on the earth. After all, they reason, all the elements in life are found here on earth, and at least hundreds of millions of years passed on earth before the first organisms appear. Hundreds of millions of years is a long time, and lots of unlikely things can happen, including the production of a living organism from non-living things.

This brings us to yet another group of uniformitarians – **theistic evolutionists**. The term "theistic" refers to someone who believes in God and thinks that He interacts with His creation. They are also convinced that evolution happened roughly as the Theory of Evolution suggests. However, they recognize that there is no way a living organism can come from non-living things through natural processes, so they suggest that God either created the first organism or manipulated specific natural processes to produce the first organism. Some theistic evolutionists say that this was God's only role in creating life, but others say that evolution is so unlikely that God had to manipulate natural processes many times throughout history to "nudge" evolution along. Theistic evolutionists often prefer to be called "evolutionary creationists," since they see evolution as the mechanism by which God created things.

Atheists, old-earth creationists, theistic evolutionists, and most of those who follow intelligent design have very different views of how life came to be and why it looks as it does today. However, they all have one thing in common: they are convinced that the uniformitarian view is the best way to understand the earth's crust and the fossils found in it.

Comprehension Check

8.7 An intrusion of igneous rock is found in a layer of sedimentary rock that has an index fossil indicating it is Permian rock. Which of the following is possible for an age of that igneous intrusion: 450 million years, 350 million years, or 250 million years? (Feel free to use the geological column illustration on page 227.)

8.8 A student is studying a rock layer that he says must be Silurian. You find a dinosaur fossil in it. What does that tell you about the student's conclusion? (Feel free to use the geological column illustration on page 227.)

8.9 If you read much about evolution, you will eventually hear about the "Cambrian Explosion." To what does that refer?

Catastrophism and the Worldwide Flood

Most catastrophists today have a completely different approach to the earth's history, because they do not think that the present is the key to the past. They think that both small- and large-scale catastrophes are responsible for the majority of the earth's geological features. The most important of all these catastrophes is a worldwide Flood. Since most catastrophists are Christians, they use the Biblical account (Genesis chapters 6-9) as a guide for understanding some of the basics of how and when the Flood occurred. However, it is important to note that while you and I are probably most familiar with the Biblical account, a worldwide Flood is a common part of the history of most cultures, and there are many similarities among these various accounts.

Consider, for example, the *Epic of Gilgamesh*, which is an ancient story from Mesopotamia. The oldest nearly-complete version is found on stone tablets that are thought to have been made in about 700 BC. However, the story is much older than that, since fragments of it have been found in older artifacts. It is thought that the story might be as much as 5,000 years old, but there is no way to know for sure. We only know that it is ancient. In one part of the story, the hero (Gilgamesh) seeks out immortality, and his search brings him to another man, Utnapishtim, who tells Gilgamesh how he achieved immortality.

According to Utnapishtim, a god who was annoyed with how loud people had become wanted to destroy them with a worldwide flood. A god who liked humans warned Utnapishtim about the flood, so he built a huge ship. He gathered his relatives, as well as pairs of the animals that he wanted to save, and he brought them on his ship. When the flood came, the people and animals on the ship were saved. To see when the floodwaters had receded, he regularly released birds to search for land. Eventually, the ship came to rest at the top of a mountain, and the people and animals left the ship. The god who warned him granted Utnapishtim immortality because he saved the human race and the animals.

This picture shows the clay tablet that contains the flood account in the *Epic of Gilgamesh*. The writing is called "cuneiform," which was used in ancient Mesopotamia.

Does any of that sound familiar? If you have read Genesis 6-9, it should. God warned Noah about a worldwide Flood, so he built a great ship (the ark) and gathered his family as well as two of every kind of animal (and more of some kinds) to save them from the flood. He released a dove to find out when the floodwaters receded, and when they did, his ark eventually came to rest at the top of a mountain. While Noah was not given immortality, he and his sons were blessed by God.

What can we make of the striking similarities between these two stories? It's possible that one was adapted from the other. The Jewish people and the Mesopotamians had many interactions, so it's possible that one group heard the other telling this story and then adapted it to their culture. However, that explanation is hard to believe when you find out that lots of different cultures, many of which never had any real contact with the others in ancient times, also have legends of a worldwide Flood. Several Native American cultures have a worldwide Flood account as a part of their histories, as do the native Hawaiians. There are also worldwide Flood accounts from China, Russia, India, Greece, Egypt, Italy, and many other cultures. While they aren't all as similar to the Biblical account as is the flood story in the *Epic of Gilgamesh*, many of them do contain several of the same elements. According to Dr. Duane Gish, there are more than 250 separate worldwide Flood accounts from various cultures throughout the world.

Since it is hard to imagine how all these ancient cultures could have been in contact with one another, a more reasonable explanation for these flood accounts and their similarities is that all of them are reporting on an actual event that happened, and these cultures remember it because they are descended from people who were saved from the worldwide Flood. Of course, as is typical with any cultural account, the details got a bit garbled over the years, so the accounts don't agree completely. However, their similarities point to the fact that they are all discussing an actual event that really happened.

Of course, most catastrophists today think that the Biblical account is the most accurate, and there are very good reasons for that. Several books such as *Evidence that Demands a Verdict* by Josh McDowell and *On the Reliability of the Old Testament* by Dr. Kenneth A. Kitchen demonstrate that the Old Testament is a very reliable source of ancient history. Because they use the Bible's account of a worldwide Flood as a guide for understanding the geological column, most modern-day catastrophists are Christians who believe that the Bible gives an accurate account of the earth's history. Because a straightforward reading of the Bible's account indicates that the earth is only thousands of years old, they are generally referred to as **young-earth creationists**, which I will abbreviate as "**YECs**."

The Young-Earth Creationist View of the Geological Column

Because of the powerful effects of weathering and erosion, you can imagine that a worldwide Flood would have a profound effect on the geological record. As a result, YECs typically split the geological record into three sections: rocks formed before the Flood, rocks formed during the Flood, and rocks formed after the Flood.

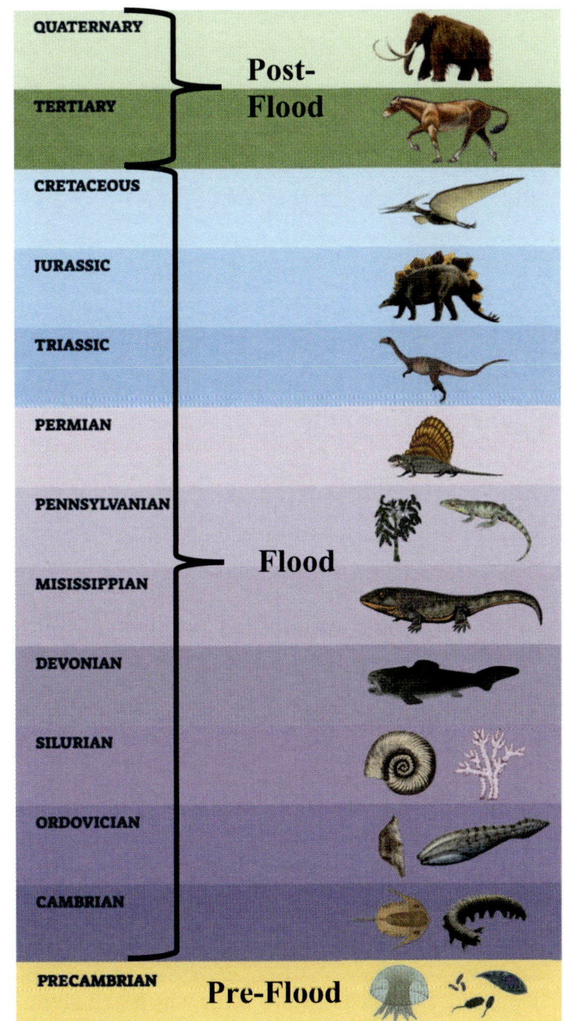

This illustration shows the YEC view of the geological column.

The rocks formed before the Flood would be those formed in the initial creation of the earth, along with any rocks formed by small catastrophes that happened before the Flood. After all, the plates have been moving since they were made, so there was almost certainly some volcanic activity prior to the Flood. That would form igneous rock. There were probably also floods that affected individual regions of the earth, and they could lay down new sediment that could undergo lithification. However, catastrophes like that would affect only specific regions, so they are generally referred to as **local catastrophes**. It is generally thought that Precambrian rocks are from the original creation and local catastrophes that happened prior to the Flood.

The Flood would have an enormous impact on the geological record, so the majority of the geological column is composed of rocks that were laid down during the Flood. While there is still some debate among YECs, for the purpose of this course, we will treat the rock layers from Cambrian through Cretaceous as being formed during the Flood. The floodwaters would have weathered and eroded a large portion of the rock that was already there, so that the Precambrian rocks we see today are just a small remnant of the rocks that existed prior to the Flood. The sediment produced by the weathering and erosion formed the sedimentary rocks found in the Cambrian through Cretaceous layers, but there was a lot of volcanic activity that produced igneous rock. In addition, heat from the volcanic activity and pressure from the water and newly-formed sediment layers formed metamorphic rock.

Once the floodwaters receded, the earth began to "recover" from the devastation. Nevertheless, local catastrophes continued to happen, which would produce additional rocks on top of the rocks formed during the Flood. Those are found in the Tertiary and Quaternary layers of the geological column. The illustration on the previous page summarizes the YEC view of the geological column.

If you think about the implications of this interpretation, the fossils found in Precambrian rock sample the life that lived on the earth before the Flood. However, since so much weathering and erosion would have occurred during a worldwide Flood, the fossils would be sparse. The fossils found in Cambrian through Cretaceous rock would represent the organisms that were killed by the Flood. The fossils in Tertiary and Quaternary rock would represent the organisms that lived after the Flood. So while the YEC interpretation of the geological column does separate it into different periods of the earth's history, those periods are very different from the eras used by uniformitarians.

Now remember, the geological column is theoretical. There is no complete representation of it in any one place on the earth. However, the Principle of Faunal Succession that is used to build it can be verified in most geological formations that contain several different layers of sedimentary rock. The uniformitarian view of the geological formation explains the Principle of Faunal Succession quite easily. What about the YEC interpretation? How does it deal with that principle?

Well, take a look at the artist's interpretation of Cambrian life that appears at the bottom of page 229. What do you see? You see ocean-dwelling organisms that live on or near the bottom of the ocean. In the context of the worldwide Flood discussed in the Bible, this makes perfect sense. Consider, for example, how the Bible describes the beginning of the Flood: "In the six hundredth year of Noah's life, in the second month, on the seventeenth day of the month, on the same day all the fountains of the great deep burst open, and the floodgates of the sky were opened. The rain fell upon the earth for forty days and forty nights." (Genesis 7:11) So while most people think of the worldwide Flood as being caused by rain, the Biblical Flood was caused by water rising from the fountains of the deep *and* water falling from the sky.

This picture of hydrothermal vents was taken at the bottom of a very deep part of the ocean. It is completely dark down there, so the lighting is artificial.

What were the fountains of the great deep? Look at the picture on the left. It was taken at the bottom of a very deep part of the ocean, and it shows something that is found there – **hydrothermal** (hi' droh thur mul) **vents**. The "smoke" that you see there isn't smoke at all; it is a mixture of water and minerals. Hot water below the crust has a lot of minerals dissolved in it, and when a fissure opens in the sea floor, that hot water and its minerals are released. As soon as the mixture hits the cold water of the sea floor, most of the dissolved minerals precipitate, causing what looks like smoke.

Hydrothermal vent – An opening in the seafloor that emits hot, mineral-rich water

Depending on the minerals in the water, a hydrothermal vent might produce a white precipitate, like the one shown in the photo on the previous page. As a result, it's called a "white smoker." Other minerals can make a dark precipitate, and hydrothermal vents that emit water containing those minerals are called "black smokers."

A hydrothermal vent is a small opening in the seafloor. What would you have if there was a large opening? You would have a huge rush of mineral-filled water, producing a huge amount of precipitate. That's what YECs think the fountains of the great deep were. They think large amounts of mineral-rich water came spewing from the seafloor. The precipitates formed would quickly cover things on the seafloor, and the current of the water could carry those precipitates, along with other sediment already present, across the bottom of the ocean, killing and trapping large numbers of animals that live on or near the seafloor. The fossils we find in Cambrian rock are from animals like that, so Cambrian rock represents sediments laid down during the earliest parts of the Flood.

As the fountains of the great deep continued to spew water and precipitates around, organisms that lived in the ocean but higher than the seafloor would eventually get overcome and trapped by sediments, forming the next layers in the geological column. The stronger swimmers could survive until later in the Flood, so they are found in the higher layers. Eventually, the floodwaters rose to the point where sediments began trapping land-based plants and animals, which are in even higher layers of the geological column. Once again, the more mobile the organism, the longer it could survive before being engulfed by the sediments, so the higher it will appear it the geological column. Very mobile (and very smart) organisms might be able to escape to very high land. They would eventually be covered by water and die, but because they weren't rapidly covered in sediments, they would leave no fossils behind. Instead, their bodies would rot away, like most dead bodies do. This is why some organisms (like horses and people) didn't leave any fossils in the rocks laid down during the Flood. Their fossils appear only in the rocks that were laid down by post-Flood catastrophes.

But wait a minute. Does that really make sense? Well, consider the fossil footprints shown in the picture below. They are in Devonian rock found in Poland. The rock is dated to be 395 million years old. However, no fossil of an animal that could leave these tracks has been found in any rock that deep in the geological column. Instead, the first non-trace fossil of an animal that could leave such tracks is in rock that is dated at 380 million years old. So the tracks occur *lower* in the geological column than the body. Now remember, footprints require very specific conditions to fossilize – much more specific conditions than animal remains.

These fossil footprints are found in Devonian rock that is deeper in the geological column than any fossils of an animal that could leave such tracks.

Why are footprints found in rocks *below* where the animals that could have made them are found? Because the animals were fleeing the approaching floodwaters. As a result, the sediments carried by the floodwaters preserved the footprints *before* they were able to preserve the remains of the animals themselves. The fossil tracks I am discussing are not an isolated example. In many geological formations around the world, you find fossil footprints (and other trace fossils) in rocks that are below the fossils of the animals that made them. This indicates that worldwide, animals were fleeing something that ended up forming fossils of them.

Now please understand that there is more to the YEC explanation of the Principle of Faunal Succession. It's not just about animals being able to run away from the floodwaters. Consider, for example, the picture on the left. It was taken at a lake in Finland. You might think you are looking a normal island in the middle of the lake, but that's not what it is. It is actually a "raft" made from vegetation and accumulated sediment. In other words, it is a **floating island**. It floats freely in the water, but it is thick and stable enough to support the trees that are growing on it!

This island is freely floating in the lake.

As the floodwaters rose onto land, lots of vegetation would be ripped out of the ground, and it would probably come together to make similar floating islands. Some animals (especially insects and other small animals) might survive on such an island. Those animals, as well as the plants that make up such an island, could survive until the later stages of the flood, which would put them in higher sedimentary strata.

In the end, then, the YEC interpretation views the succession of fossils in Cambrian through Cretaceous rock as being caused by the combination of different stages of the Flood and the different ecosystems that the sediment-laden floodwaters would engulf as they rose. In addition, temporary ecosystems, like floating islands, would affect the progression of fossils seen in any specific geological formation. Since YECs don't consider the layers of the geological column to represent long eras in the earth's history, they are more concerned about understanding the fossil progression in each specific geological formation than in the theoretical construct that is called the geological column.

Comprehension Check

8.10 According to YECs, when were dinosaur fossils formed? What about mammoth fossils? (Feel free to refer to the illustration on page 233.)

8.11 Suppose someone found a rabbit fossil in Cambrian rock. What would that tell you about the YEC interpretation of the geological column?

8.12 Evergreen tree fossils are usually found in rock layers below the rocks where you find fossils of deciduous trees (trees that lose their leaves in the winter). In the YEC interpretation, which trees would have been found at lower elevations at the time of the Flood?

The Earth's History According to Young-Earth Creationists

Suppose you are examining the ruins of an ancient city and want to learn as much as you can about when it was built, how it was built, and how it fell into ruin. You see some of the remains of buildings, streets, walls, etc., but nothing has been preserved intact. You can learn a lot by investigating the ruins, but your conclusions will be based on your interpretation of what you see. Now suppose you found out that there was a book written shortly after the city was built, and it discusses the politics of the city for several centuries. While the focus of the book is on the government, it does cover many aspects of how and when the city was built.

Would you completely ignore the book and just examine the ruins, relying on your own interpretation to determine the city's history? Of course not! If you wanted to learn the truth about the city's history, you would read the book and let it help you interpret the ruins that you are investigating. This is how YECs study the geological record. They believe they have a book (the Bible) that comes from the Creator Himself. While the book focuses on more important things like salvation, morality, and our duties to God, it does discuss the creation of the universe, the earth, the organisms that lived on earth, etc. Since YECs consider the Bible to be an accurate source of history, they use it as a guide to studying the "ruins" of the geological column and fossil record. There's a lot more to the history of the earth than what is in the Bible, but at least the Bible gives YECs a starting point to help their interpretation of the geological record.

The Bible says that the entire universe, including the earth and the life that inhabits it, was created miraculously over a period of six days. On the seventh day, God rested (Genesis 1:1–2:3). There are many different ways to interpret the word "day" in the original language of the Old Testament (Hebrew), but YECs use the most obvious meaning – a standard day on the earth. They say this is confirmed by the text itself, since the days are marked by evening and morning.

The activities of those six days are illustrated below. God created the entire universe in the first day, but it was in an unformed state. In addition, He separated light from darkness. On the second day, He separated the waters above from the waters below. Most YECs interpret that to mean that the earth's atmosphere was created then. At this point, the earth was still covered in water, but on day 3, God formed land and the plants that live on it. Throughout these three days, any light that illuminated the earth did not come from the sun, because the sun, moon, and stars were not created until the fourth day. On the fifth day, God created water-dwelling animals and the flying animals. On the sixth day, He created the animals that live on land, and his last creative act of the day was to make two people (Adam and Eve). One of the important things to realize about the YEC view is that the originally-created animals were similar

This illustration shows the progression of creation as given in the Bible.

to, but not the same as, the ones we see today. The Bible discusses animals in terms of **kinds**. That's not the same as species. We will discuss this more in the next chapter.

After the six-day period, Adam and Eve lived in a paradise called the Garden of Eden. The Bible doesn't say how long they lived there, but it does say that they were eventually cast out of the garden because they sinned against God. Their act had far-reaching consequences, because the entire creation was cursed as a result (Genesis 3). This event is often referred to as the **Fall of Man**, or just the **Fall**. After the Fall, the Bible records a list of Adam's children, their children, and so on. A list like that is called a **genealogy** (jee' nee ol' uh jee), and since it includes the ages of the fathers when they had children, YECs can determine that there were about 1650 years between the Fall and the Flood.

In the geological column, anything created during the six-day creation or produced between the Fall and the Flood would be found in Precambrian rock. However, the Flood would have caused so much weathering and erosion that most of the rocks and fossils formed during this time period were probably destroyed. Nevertheless, geological processes like plate tectonics and the rock cycle must have been happening, so Precambrian rock can tell us a bit about what went on during that time.

The Flood, of course, is the major focus of geology according to the YEC interpretation, and it seems to have left an obvious mark on the geological record. Look at the photo below. There is an obvious separation between the igneous and metamorphic rocks near the bottom of the formation and the sedimentary rocks (sandstone) near the top. This is called the **Great Unconformity**. Recall that an unconformity is a separation between rock layers caused by a period where there is no deposition. This unconformity is called "great" because it is found all over the world! In the U.S., it can be seen from California to New York. It is seen in rocks on every other continent as well. It separates fossil-bearing sedimentary strata from igneous and metamorphic rock or fossil-poor sedimentary rock. This kind of worldwide presence indicates the cause was global, which leads YECs to say it marks the point at which the Flood started to shape the geological record.

The Great Unconformity can be found around the world, separating fossil-rich sedimentary rock from igneous and metamorphic rock or fossil-poor sedimentary rock.

Now remember the Flood began with the fountains of the great deep and the floodgates of the sky opening. As I said previously, plate tectonics must have been happening prior to the Flood, and it is likely that the plates moved fairly slowly then, like we see them moving today. But the fountains of the great deep opening would change things immensley! YECs describe this process with **catastrophic plate tectonics**, which uses the general principles of the plate tectonics that you have already learned, but it allows the plates to move much more quickly due to rapid changes in the mantle.

In catastrophic plate tectonics, the supercontinent Pangaea did not exist before the Flood. Like the uniformitarians, YECs think that the original continents were arranged in a different way, perhaps forming a large supercontinent like Rodinia, which I mentioned previously. In the most common version of catastrophic plate tectonics, the ocean crust near the continents is broken. Because the ocean crust is so dense, it sinks rapidly into the mantle. This is called **runaway subduction**, and it produces lots of heat. The pressure is also reduced as mantle material rises in the fountains of the great deep. That makes the mantle rock less plastic, allowing faster motion of the plates. The plates start moving at speeds of a few kilometers per hour instead of centimeters per year, like we see today.

At those speeds, the plates slammed the continents together to form Pangaea, and the supercontinent was stable for perhaps a month or two. During that time, the geological features that we use to figure out how the continents fit together into Pangaea were formed. However, the motion of the plates eventually pulled the continents apart again, forming Laurasia and Gondwanaland. The motion of the plates slowed as the flood waned, but during that time the continents continued to separate until they reached positions near what we see today.

But how did water end up covering even the tallest mountains (Genesis 7:19)? First, remember that the mountains we see today are formed by plate movements, movements along faults, and volcanic activity. Thus, the mountains that existed prior to the Flood were different from the ones we see today. Nevertheless, the pre-Flood world had mountains, since they are mentioned in Genesis.

Next, think about what happens when water gets hot under pressure. When it is released, it shoots out violently. We see examples of that today. They are called **geysers**.

Geyser – A vent in the earth's surface that ejects a column of hot water and steam

The picture below shows you a geyser in Iceland. Now imagine what would happen when large vents near the bottom of the ocean opened up. The rising magma would carry hot water up to hit the cold water of the ocean. That cold water would rapidly heat up and steam would rise up through the ocean and into the air, producing geysers that would make the one in the picture look weak and puny. All that water and steam would then turn into rain, which explains how it could rain for 40 days and 40 nights (Genesis 7:12).

The geysers associated with the worldwide Flood would have been enormous compared to this one in Iceland.

But that's not the end of the story. The *new* ocean crust formed by the magma coming up from the mantle would be less dense than normal ocean crust, because it would be formed under such hot conditions. This would cause the bottom of the ocean to rise, which would push water up and over the continents. In addition, remember that the mantle fluid became thinner because of the increased heat and lower pressure, so the continental crust would sink more than normal. In other words, as the ocean floor rose, the continents would sink. Both effects would cause water to cover the continents.

There is still one more issue to consider, and it is important. All of this heat would increase the temperature of the ocean water. What does water (or pretty much anything else) do as it warms up? It expands. So the heated ocean water would expand. However, it can't expand down, so it expands upwards and outwards. Once again, that would cause the water to cover the continents. All of these effects combined would be more than enough to cause water to cover the mountains of the pre-Flood world.

But what caused the floodwaters to drain away? Well, remember that the old oceanic crust (which is dense) started sinking rapidly into the mantle. At the same time, however, new oceanic crust was being made at the vents where magma rose into the ocean, and that crust was less dense. Eventually, the less dense oceanic crust would reach the continents, but it wouldn't sink into the mantle as rapidly. In other words, the less-dense oceanic crust would not participate in runaway subduction.

Because the runaway subduction ended, the mantle would start to "firm up" again, which would cause the continental crust to rise. In addition, the oceanic crust would cool, causing it to become more dense and sink a bit. Finally, the water would begin to cool, and as it cooled, it would contract. All of these effects together would cause the floodwaters to recede.

While everything I have just discussed is consistent with what we know about plate tectonics and the properties of oceanic and continental crust, it would be helpful to have some evidence related to whether or not such things actually happened. It turns out that there is. In the early 1980s, YECs used catastrophic plate tectonics to predict that we should find slabs of oceanic crust that had sunk into the mantle but had not become hot enough to completely merge with the mantle. After all, the crust sank fast, and in the YEC interpretation, this all happened only a few thousand years ago. While that sounds like a long time, it would take longer for large slabs of crust to heat up like that.

Starting in 1987, detailed seismic wave studies began showing large slabs of material near the very bottom of the mantle that were *significantly* cooler than the surrounding mantle rock. While the mantle has temperatures of about 3,500 °C down there, the slabs had temperatures of "only" about 2,500 °C. This is what was predicted using catastrophic plate tectonics. If the plates always moved as slowly as they do today, there would have been plenty of time for the subducted oceanic crust to come to the same temperature as the mantle. Thus, this indicates that at one time, there was very rapid subduction.

This painting by Edward Hicks depicts animals boarding Noah's ark.

While all of this was going on, most living things would have been killed, and some of their remains and traces would be preserved as fossils. As discussed, then, the YEC interpretation says that all the fossils in Cambrian through Cretaceous rock are from organisms that died in the Flood. However, in the YEC view, the Flood didn't kill every living thing. That's because God warned Noah, who built an ark that took two of every kind of animal aboard as well as more of some kinds (Genesis 7:2). Of course, Hebrew is not the same as English, so what we call "animals" aren't necessarily what the Bible calls "animals" in that passage. For example, Noah didn't take water-dwelling organisms, because they don't fit the Hebrew word that is translated "animal" in that passage of the Bible. Insects aren't necessarily referenced by that word, either, so it is possible that there were no insect passengers on the ark. They could have easily survived on the floating islands that I already discussed. In the same way, as I mentioned previously, "kind" does not mean "species." It is a much broader term. Thus, Noah didn't bring two lions and two tigers on the ark. He brought two members of the cat kind. I will discuss this in more detail in the next chapter.

Once the ark landed, the animals could leave the ark and spread out as the floodwaters continued to recede. Even though the continents were arranged much like they are today, they might have been more connected than they are today. For example, look at the satellite image on the right. It shows where the Asian continent almost touches the North American continent. The body of water in between is called the Bering Strait. Most likely, there used to be land connecting the two continents there, but over time, rising sea levels covered it. During the time it wasn't under water, however, people and animals could travel between the continents. It is possible that other continents were connected by land bridges as well.

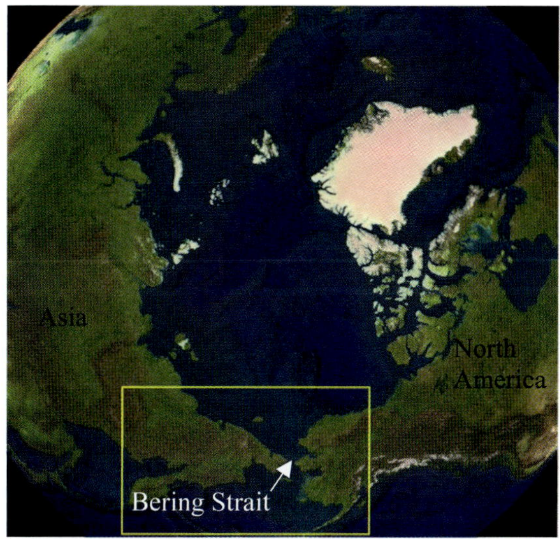

A land bridge probably once connected Asia to North America across the Bering Strait.

Another way animals could have dispersed throughout the earth is by riding on floating islands. A floating island (sometimes called a floating mat of vegetation) might have been on the shore, and animals might have wandered onto it. If the currents then took them out to sea, they would be forced to "ride" the floating island to the next land mass. While this sounds a bit far-fetched, both uniformitarians and YECs invoke this idea to explain certain details of how animals are distributed throughout the earth. It is referred to as "rafting."

As the animals moved across the earth, they would adapt to their new environment. As a result, they would change. For example, imagine the two cats that were on the ark. They probably didn't look like any of the cats we see today. However, as they reproduced and their descendants moved to different environments, they would become specialized to those environments. YECs think that all cats today, from lions to tigers to domesticated cats, came from two cats that walked off the ark. As the animals and people dispersed throughout the earth, local floods, vocanic eruptions, etc. produced Tertiary and Quaternary rock. Thus, the fossils found in those layers represent the animals and people that lived after the Flood.

Now that you know the basics of the uniformitarian and catastrophist views, we will spend the next chapter comparing them to the scientific evidence.

Comprehension Check

8.13 If a Cretaceous index fossil were found below the Great Unconformity, what would that mean about the YEC interpretation of the Great Unconformity?

8.14 Why can't runaway subduction occur at a boundary between two continental plates?

8.15 Compare the depth of the oceans during the Flood to after the Flood.

Answers to the Comprehension Check Questions

8.1 No, it is not. As your experiment demonstrated, when you heat up baking soda, it never has the chance to melt. Instead, it turns into washing soda, carbon dioxide, and water.

8.2 It will have 4 protons and 4 neutrons. Beta decay turns a neutron into a proton. So the number of neutrons decreases by one, and the number of protons increases by one. All atoms with 4 protons are beryllium atoms, so this is beryllium-8.

8.3 It will have 18 protons and 22 neutrons. Electron capture turns a proton into a neutron. So there will be one fewer proton and one more neutron. All atoms with 18 protons are argon atoms, so this is argon-40. This specific example of electron capture is important in geology, as you will soon see.

8.4 It will have 93 protons and 144 neutrons. Alpha decay gets rid of 2 protons and 2 neutrons. That means two less of each. All atoms with 93 protons are neptunium atoms, so this is neptunium-237.

8.5 500 will decay after 15 minutes, and another 250 will decay in the next 15 minutes. The half-life tells you how long it takes for half of them to decay. If you start with 1,000, half of them will decay in the first 15 minutes. That means 500 will decay, which means there will still be 500 left. In the next 15 minutes, half of them will decay, which means 250 more will decay.

8.6 It would say the rock is 160,000 years old. After one half-life, the uranium-234 would be reduced by half, and half would remain. The remaining half would be cut in half in another half-life, leaving you with ¼, which is what the analysis says. Thus, two half-lives have passed. Since each half-life is 80,000 years, a total of 160,000 years passed.

8.7 250 million years. According to the illustration, the Permian Era lasted from 299 million years ago to 253 million years ago. Remember that an intrusion is made when the rock cracks and magma fills the crack. That can happen only *after* the rock as formed, so the intrusion must be younger than the rock in which it is found.

8.8 The student is wrong. Dinosaurs did not exist until much more recently than the Silurian Era, according to the uniformitarian view.

8.9 It refers to the huge increase in complexity and diversity that appears in Cambrian rock. This is a problem for the Theory of Evolution, since in evolutionary terms, the Cambrian Era was too "short" to produce so much diversity. You will learn more about that later.

8.10 Dinosaur fossils formed during the Flood, since they are found in rocks that were laid down during the Flood. Mammoth fossils formed after the Flood, since they are found in the post-Flood rocks.

8.11 It would tell you the YEC interpretation is wrong. Cambrian rock represents the first stages of the Flood. Rabbits would have been fossilized at a much later stage, if at all.

8.12 Evergreen trees were found at lower elevations. After all, the floodwaters would have reached the lower elevations first, so the evergreens would be the first to fossilize. Since trees can't flee the flood, they would have mostly been fossilized where they grew.

8.13 It would mean there is something wrong with equating the Great Unconformity to the beginning of the Flood. Since Cretaceous rock was laid down in the very late stages of the Flood, it could not have been laid down before the Flood, which would be necessary in order to find a Cretaceous index fossil below the Great Unconformity.

8.14 There is no subduction at the boundary of two continental plates. You could also say because neither plate is denser than the other. Remember that subduction happens because one plate is denser than the other. Runaway subduction would require that as well.

8.15 The oceans were shallower during the Flood than after. Remember, the new oceanic crust was less dense than the old oceanic crust, so it would rise. At the same time, the mantle rock was thinner, allowing the continental crust to sink. A higher ocean bottom and lower continents would produce a shallower ocean.

Chapter Review

1. Define the following terms:

a. Isotopes
b. Radioactive isotope
c. Beta decay
d. Electron capture
e. Alpha decay
f. Half-life
g. Radiometric dating
h. Hydrothermal vent
i. Geyser

2. Why must we be careful about putting too much faith in our geological interpretations?

3. Does "The present is the key to the past" describe uniformitarianism or catastrophism?

4. What's the main difference between the uniformitarian's approach to plate tectonics and the catastrophist approach?

5. You have 100 grams of a stable isotope and 100 grams of a radioactive isotope, each in a sealed jar. If you wait a long time, will anything change in either jar? If so, what will change in which?

6. A radioactive isotope has a half-life of 1 year. If you start with 1,000 atoms and wait 3 years, how many atoms of the radioactive isotope will be left?

7. You have 100 atoms each of two different radioactive isotopes. After 24 hours, you have 75 atoms of the first isotope and only 12 atoms of the second. Which has the shorter half-life?

8. Of the three kinds of rocks, which is best suited for radiometric dating?

9. What does the potassium-argon dating method assume about the amount of argon in an igneous rock when it is first formed?

10. Explain what bracketing is and how it is used by uniformitarians to determine how old sedimentary rock is.

11. In the uniformitarian interpretation, what do the layers in the geological column represent? How is the Principle of Faunal Succession explained in that interpretation?

12. What are the differences between atheists, theistic evolutionists, those who believe in intelligent design, old-earth creationists, and YECs when it comes to the earth's history?

13. What historical evidence (outside of the Bible) points to a worldwide Flood?

14. In the YEC interpretation, what do the layers in the geological column represent? How is the Principle of Faunal Succession explained in that interpretation?

15. In the YEC interpretation, what does the Great Unconformity represent?

16. In the YEC interpretation, what is runaway subduction? Is there evidence that it happened?

17. What three processes work together in the YEC interpretation to explain how the entire surface of the earth was covered in water? What happened to cause the water to recede?

Chapter 9: Uniformitarianism Versus Catastrophism

Introduction

Now that you have seen the ways uniformitarians and YECs view the geological record, it is time to compare those views to the evidence. However, I need you to know two things up front. First, *neither* view is 100% consistent with the evidence. There are some observations that fit very well in the uniformitarian framework, but there are others that seem unexplainable in that same framework. In the same way, some data make perfect sense in terms of the YEC view, while other data do not. Thus, there is no clear-cut scientific way to say that one is better than the other.

That leads me to the second thing you need to know: *Every scientist* (including me) is biased on this issue. Because some data fit better in one view than the other, as scientists, we must make judgement calls on which data are the most important. Because of the way I was trained as a scientist, I will weight certain data as more important than other data. Another scientist might weigh the data in a completely different way. Thus, the discussion you read will be biased, despite the fact that I am trying my level best to be as even-handed as possible!

Using Present-Day Observations to Interpret the Past

While I have touched on this before, it is important for you to realize just how difficult it can be to interpret the past using present-day observations. This is especially true for trying to understand how old a rock or a fossil is. To give you one more illustration of this, I want you to do the following experiment.

Experiment 9.1: How Long Has It Been Sitting There?

Supplies:

- Vinegar (White is best, because you can see through it, but any will work.)
- TUMS antacid tablets from the laboratory kit made for this course (The colored tablets in the bag that holds the pH paper and teas.)
- Two small glasses, like juice glasses
- The graduated cylinder from the laboratory kit made for this course
- The mass scale from the laboratory kit made for this course
- A paper towel
- A small pot for boiling (One that has a pour spout is best.)
- A stove
- A freezer
- A teaspoon (not a measuring teaspoon)

Instructions:

1. Fill the graduated cylinder to the 50 mL mark with vinegar.
2. Pour the 50 mL of vinegar into one of the glasses.
3. Put that glass into the freezer.
4. Fill the graduated cylinder to the 50 mL mark with vinegar.
5. Pour the 50 mL of vinegar into the pot.
6. Put the pot on one of the stove's burners and heat it until the vinegar starts to boil vigorously.
7. While you are waiting for the vinegar to boil, use the mass scale like you did in Experiment 5.1 to measure the mass of one TUMS tablet. Record the color with the mass so you can identify it later.

8. Repeat step 7 with a TUMS tablet of a different color.
9. Once the vinegar is boiling vigorously, remove the pot from the stove and carefully pour the vinegar into the remaining glass. **Be careful! It is hot!**
10. Turn off the burner on the stove.
11. Get the glass out of the freezer and place it near the glass that has the hot vinegar in it.
12. Add one TUMS tablet to one glass and the other to the other glass, noting which color went into the hot vinegar and which went into the cold vinegar.
13. Observe what is going on in both glasses.
14. Wait 20 minutes. Do other schoolwork or chores to be more efficient.
15. After 20 minutes, touch the sides of the two glasses. Even though one started out cold and the other hot, they should feel like they are at roughly the same temperature now.
16. Once again, observe what is going on in both glasses.
17. There is probably some solid floating on top of the liquids. Use the back of the spoon to move it aside and scoop out what is left of one tablet. You want to get just the tablet; try to avoid collecting the floating solid as you lift each tablet out of the liquid.
18. Once the tablet is out of the liquid, put the tip of the spoon against the side of the glass and tilt the spoon to drain any liquid in the spoon without losing the tablet.
19. Carefully place the tablet on the paper towel.
20. Repeat steps 17-19 for the other tablet, placing it on the same paper towel but away from the first tablet.
21. Compare the tablets. Is one noticeably smaller than the other?
22. Use your fingers to gently pick up one of the tablets and measure its mass again, remembering to make sure the scale is reading grams and hit "tare" so it reads zero before you put the tablet on it. If the tablet breaks, try to collect the parts and put them on the scale as well.
23. Remove the tablet from the scale and gently wipe away any residue that might be on the scale.
24. Repeat steps 22 and 23 for the other tablet.
25. For each tablet, subtract the mass you just measured from the mass you measured at the beginning of the experiment. This tells you (roughly) the mass of the tablet that was used up in the process.
26. Clean up your mess.

What happened in your experiment? The active ingredient in vinegar ($C_2H_4O_2$) is a weak acid, and TUMS ($CaCO_3$) is an antacid. The two chemicals react according to the following equation:

$$2C_2H_4O_2 + CaCO_3 \rightarrow CaC_4H_6O_4 + H_2O + CO_2$$

Since carbon dioxide is a gas at these temperatures, it formed bubbles and bubbled out of the liquid. What did you notice about the number of bubbles being formed? You should have seen that the tablet in the hot vinegar was producing more bubbles than the tablet in the cold vinegar. That should have made sense. Remember, cold temperatures slow down chemical reactions; hot temperatures speed them up. Thus, the reaction happened more quickly in the hot vinegar than in the cold vinegar.

As time went on, however, the hot vinegar cooled down, and the cold vinegar warmed up. Of course, the reaction kept going in both glasses. As a result, the reaction in the hot vinegar started to slow down as the vinegar cooled, and the reaction in the cold vinegar started to speed up as the vinegar warmed. After 20 minutes, the temperatures changed enough that there shouldn't have been much of a difference in the temperatures or the speeds of the reactions.

Now let's think about the masses you measured. You measured the mass of the tablets before the experiment started, and they should have been similar, since the tablets contain roughly the same amount of matter. When you measured the masses at the end of the experiment, you measured the masses that remained *plus* the mass of whatever liquid was soaked into the tablet. However, you did try to drain excess liquid from the tablet, and the paper towel soaked up some more of the liquid, so while the measurement wasn't perfect, it gave you a rough idea of each tablet's mass after the experiment was over. When you subtracted that mass from the initial mass you measured, you got a rough idea of how much of the tablet was used up during the reaction.

What did you find? The mass that was used up from the tablet in the initially cold vinegar should have been lower than the mass that was used up from the tablet in the initially hot vinegar. That should make sense. The tablet gets used up in the reaction, so the faster the reaction occurred, the larger the amount of the tablet that was used up. Since the reaction was faster in the initially hot vinegar, the amount of mass used up in that vinegar should have been greater.

What does this have to do with uniformitarianism and catastrophism? Suppose someone else had done the first part of the experiment, and you were told to finish it after the 20-minute waiting period. Suppose further that no one told you anything about how the experiment was set up. You were just told the initial masses of the two tablets. As you started step 15, you would conclude that both glasses were at roughly the same temperature. You would also conclude that the reaction is happening at about the same speed in both. You would then finish the experiment, measuring the masses of both tablets and determining how much of each tablet was consumed.

What conclusion would you reach about how long the reactions had been happening in the glasses? Remember, you don't know anything about the initial difference in temperatures. From your investigations, the temperatures of the glasses were the same. However, more of one tablet was consumed than the other tablet. The most reasonable conclusion, then, is that the one with more mass consumed had been in the glass longer than the one with less mass consumed. In other words, the reaction in one glass was "older" than the reaction in the other glass.

Assuming you came to that conclusion, you were thinking like a uniformitarian. You saw the reactions happening at the same temperature and at the same rate, and you decided they must have always been at the same temperature and the same rate. Thus, the speeds of the reactions were the same the entire time, and therefore, whatever TUMS tablet lost more mass had been reacting longer. That would have been a valid conclusion, if your uniformitarian assumption had been valid. It wasn't, however, so your conclusion wasn't valid.

Of course, had you looked at the situation as a catastrophist, you probably couldn't have come to any conclusion at all about how "old" each reaction was. After all, if you are willing to believe that the reactions had different speeds in the past, you would have to know something about the past to determine what was going on. Without asking whoever started the experiment, you would have no way of knowing that, so you couldn't have drawn any conclusion at all. In this case, then, uniformitarianism would have given us an incorrect conclusion, while catastrophism would have given us no conclusion at all.

Comprehension Check

9.1 The reaction in the hot vinegar was faster than the reaction in the cold vinegar, but that's not the only reason more bubbles were produced in the hot vinegar. What's the other reason? (HINT: Think about an experiment you did in the previous chapter.)

Potassium-Argon Dating

The experiment you did previously was an analogy for radiometric dating. We have been studying radioactive decay for just over 100 years, and during that time, it has been operating consistently. Each radioactive isotope decays based on a half-life, and while each isotope's half-life is different, each individual value stays the same. Since we have been measuring it, for example, potassium-40 decays via electron capture into argon-40 with a half-life of 1.3 billion years. The half-life has never gone up or down during the several decades we have been studying it. Uniformitarians assume that this has always been the case, so no matter what igneous rock they are studying, they can determine the age of the rock by assuming potassium-40 always decays with that half-life.

Catastrophists say that you don't really know that the half-life of an isotope has always remained the same. It's possible that in the past, half-lives were shorter or longer. Thus, the radiometric dating rests on an assumption whose reliability is not known. In addition, radiometric dating requires that we know something about the isotopes that existed in the object when it was formed. As you learned in the previous chapter, for potassium-argon dating, it is assumed that when the rock first cooled and froze, there was no argon-40 in the rock, since gases don't dissolve well in hot liquids. Thus, when the rock was magma or lava, there was no argon-40 in it. Any argon-40 that is found in it now must be from the decay of potassium-40. Once again, catastrophists will say that there is no way to confirm that.

Who is correct? The short answer is that we do not know. Based on our understanding of radioactivity, there is no reason to expect the half-life of an isotope to change over the course of time. In addition, radioactive processes can only be altered when an enormous amount of energy is used. Chemical reactions like you used in the previous experiment can be altered by changing the temperature by just a few degrees. Even if you change the temperature by thousands of degrees, radioactive half-lives remain the same. Nevertheless, we don't understand radiation nearly as well as we understand chemistry, so there may be something we are missing.

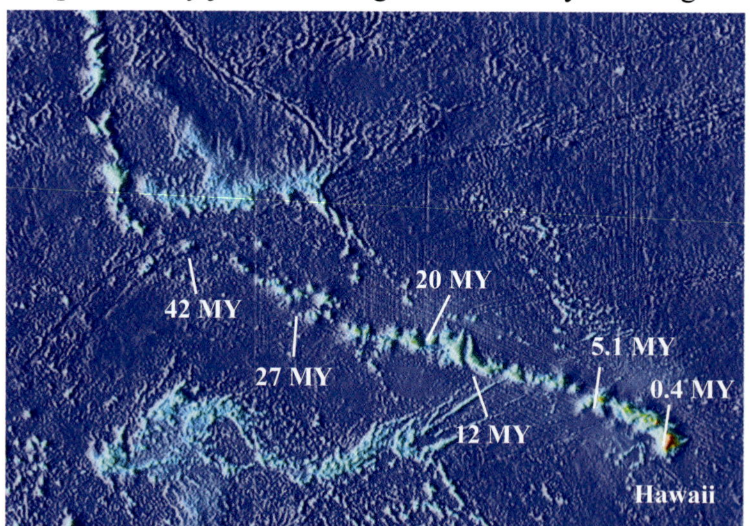

This map shows the Hawaiian Mountain Chain at the bottom of the Pacific Ocean. The numbers are the ages of the mountains as determined by potassium-argon dating. "MY" stands for million years.

However, there is a way we can "test" radiometric dating. We can look at some of its results and see if they make sense based on what we know. Consider, for example, the map on the left. It shows the Pacific Ocean floor in the vicinity of the Hawaiian Islands. As you learned previously, the islands are actually volcanic mountains that rise from the floor of the ocean to above its surface. If you look underneath the surface, however, you see a long line of volcanic mountains, which are usually referred to as the "Hawaiian chain." Notice that until the bend in the upper left of the map, the chain is straight. As you learned previously, this is what you expect from a mantle hotspot and a moving plate. The numbers indicate how old the mountains are according to potassium-argon dating.

Now remember, according to plate tectonics, the Pacific plate is moving over a stationary hotspot in the mantle. As the magma escapes, it forms volcanic mountains, and the motion of the plate causes those mountains to form in a straight line. The line bends at the top left, but that's the result of something else you don't have to worry about. Focusing on the straight line, we know that the island of Hawaii is the most recent volcanic mountain formed, since it still holds an active volcano. The other mountains in the chain must be older than Hawaii, and the farther they are from the island of Hawaii, the older they should be.

Now look at the ages of the igneous rocks (given by MY, which means "million years") in the mountains as determined by potassium-argon dating. Notice that they follow exactly the pattern that you expect. The island of Hawaii has the youngest igneous rocks (0.4 million years), and as you move down the line from Hawaii, the ages get older and older. In addition, we can measure the distance between Hawaii and the last mountain I have pointed out and then divide by the 42 million years that potassium-argon dating says it took to produce all those volcanic mountains. If you do that, you find that the plate had to be moving at an average speed of about 7 centimeters per year in order to make the chain. Guess what? The GPS says that right now, that part of the plate is moving at about 6 centimeters per year. That's remarkably similar to what you get from the potassium-argon dates!

The fact that the dates get older exactly the way they should for this kind of mountain chain, and the fact that the speed of the plate as determined by those dates is very similar to the currently measured speed of the plates provide evidence that potassium-argon dating really does work for igneous rock. And, of course, the Hawaiian chain isn't the only example. Remember the seafloor spreading at the Mid-Atlantic ridge? If you do potassium-argon dating on the igneous rocks on either side of the ridge, you will see that the dates get older the farther you get from the ridge, and they have roughly the same values on either side of the ridge. This is exactly what you expect for seafloor spreading. From a uniformitarian perspective, then, the fact that all these dates are consistent with what we expect provides strong evidence that potassium-argon dating really does work.

This map shows just the Hawaiian Islands and the potassium-argon ages of certain igneous rocks found on them.

YECs, of course, look at this issue in a completely different way. For example, they point out that while many radiometric dates are consistent with what one would expect, there are several examples that are not, and these examples give us strong reasons to doubt the reliability of potassium-argon dating. Consider, for example, the map on the right. It shows the Hawaiian Islands and several potassium-argon dates of igneous rocks found there. Notice, for example, there are igneous rocks on the island of Hawaii that have been dated as 1.4 million years old. That's a problem, since according to the dates presented on the previous page, the island of Hawaii didn't form until 0.4 million years ago. Thus, the potassium-argon dating method says the rocks on the island are older than the island itself. Worse yet, history records the volcanic eruption that produced the lava which formed those rocks. It happened in 1800 and 1801. Thus, we know those rocks formed a bit more than 200 years ago, but the potassium-argon dating method says they formed 1.4 million years ago.

Notice that two other parts of Hawaii have igneous rocks that also date older than the island is supposed to be. In both cases, the ages of those rocks were determined by potassium-argon dating to be about 60 times older than the island. On the island of Oahu, there are also igneous rocks that the potassium-argon method indicates are older than the island. To make matters even more confusing, there are igneous rocks on the island of Kauai that are much younger than the island. That's also a problem, because once the plate moved the island away from the hotspot, no more igneous rock can form. Thus, there should be no igneous rock that is significantly younger than the island. So while there are potassium-argon dates that make perfect sense in the uniformitarian view, there are other potassium-argon dates that conflict with that view.

The brown colors you see inside this diamond are xenoliths. Some diamond xenoliths are igneous and can be dated with the potassium-argon dating method.

The Hawaiian Islands aren't the only place where potassium-argon dates conflict with what we know. A volcanic mountain in the state of Washington, Mount Saint Helens, experienced a series of eruptions from 1980 to 1986. The igneous rocks that formed during those eruptions were dated using the potassium-argon method, and depending on where the rock samples were taken and the specific contents of the rocks, it was determined that they formed between 340,000 years and 2.8 million years ago. However, we know that they formed in the 1980s. In Central Africa, igneous xenoliths in diamonds were potassium-argon dated to be 6 billion years old. This is a problem, of course, since uniformitarians think that the earth is "only" 4.5 billion years old.

YECs say that all these examples of potassium-argon dates that conflict with other potassium-argon dates indicate that the dating method is not reliable. Of course, uniformitarians disagree. They would argue that these examples represent only a small percentage of potassium-argon dates, and so they don't cast doubt on the method itself, just specific ways in which it might be used.

Carbon-14 Dating

Now remember, potassium-argon dating is only one of several different radiometric dating techniques. However, it is the one most responsible for the dates seen on the uniformitarian geological column, which is why I have concentrated on it. Before I move on to another subject, however, I do want to mention one other radiometric dating method, because you have probably already heard about it: **carbon-14 dating**, which is also called "radiocarbon dating" or just "carbon dating." While it is never used by uniformitarians to date rocks, it is used to date certain fossils and archaeological artifacts.

As you learned previously, there are three naturally-occurring isotopes of carbon: carbon-12, carbon-13, and carbon-14. The last one undergoes beta decay to become nitrogen-14 with a half-life of

5,730 years. Because of this, anything that contains carbon has some carbon-14 in it and is therefore a candidate to be dated with the carbon-14 dating method. This includes fossils, since fossils are the remains of once-living organisms, and all living organisms contain *a lot* of carbon. As a side note, that means all living organisms (including you) are radioactive! In fact, you also have potassium-40 in your body, so you have two radioactive isotopes in you, as do most living organisms! Obviously, the concentration of these isotopes is low, so the amount of radioactivity you emit is low. However, you are constantly emitting particles that are being shot out of radioactive isotopes that are in your body.

Even though fossils can be dated with the carbon-14 dating method, uniformitarians rarely use it. That's because it has a short half-life, at least within the uniformitarian view. Remember that because the amount of radioactive isotope gets cut in half every half-life, the amount remaining gets pretty low after just a few half-lives. As a result, no matter how much radioactive isotope you start with, after about 10 half-lives, the amount has dwindled to the point where it is simply not detectable with the scientific equipment we have today. Thus, as a general rule of thumb, the limit for any radiometric dating method is about 10 times the half-life of the isotope being used. This means that the limit for carbon-14 dating is about 57,300 years. That sounds like a long time, but in the uniformitarian view of the geological column, that doesn't get us past the first layer of fossil-bearing rock (Quaternary rock)! As a result, carbon-14 dating is mostly used by archaeologists who are studying human-made or human-used artifacts.

How is carbon-14 dating done? First, we assume that the amount of carbon-14 in any living organism is the same, regardless of when it was alive. We assume this because even though an organism's body loses carbon-14 because it is decaying away, it also replenishes its carbon-14 supply by either eating or making its own food. However, when the organism dies, it stops replenishing its carbon-14 supply, so the amount of carbon-14 starts to decrease. Thus, if we measure the amount of carbon-14 in a fossil and subtract it from the amount that it was assumed to have, we know how much carbon-14 decayed since the organism died, so we know how old the fossil is.

The assumption that the amount of carbon-14 in any living organism is the same no matter when it was alive is not very good, but it turns out that doesn't matter in some cases, because we can correct the faulty assumption. How? Most people know that trees usually form one ring each year. Thus, if you examine the rings on a tree, you can count backwards and determine how old each ring is. What most people don't know is that when the tree forms a new ring, the old one dies! At that point, it acts like a fossil that died that year. As a result, we can measure the amount of carbon-14 in a dead ring and use that measurement to determine the age that carbon-14 dating says it died. However, we know the actual date that the ring died, so we know how old the ring "fossil" actually is. If the carbon-14 age is wrong, we can correct the carbon-14 age. After that, any object that ends up having the same carbon-14 age can get the same correction, and that way, we can arrive at the correct age for anything dated with carbon-14 dating.

When a new ring forms in a tree, the old ring dies. As a result, each ring is like a fossil that died the year the next ring formed.

This process is called **calibration** (kal uh bray' shun), and it is a common technique used in science.

Calibration – The process by which a measurement is compared to results with values that are known to be correct and adjusted so that the measurement is also correct

In other words, the actual ages of the tree rings are known to be correct, because counting tree-rings is known to be an accurate way of measuring the age of the rings. The carbon-14 ages are not necessarily correct, but we can compare each carbon-14 age to the known tree-ring measurements and change the carbon-14 age so that it is correct.

This works really well, until we run out of tree rings from trees that are alive. The oldest living tree with countable rings is about 4,850 years old, but very old trees like that are protected, so they can't be used for calibration. The oldest living tree that scientists could kill so they could use its rings was about 3,500 years old, so the reliable calibration only extends back about 3,500 years. However, there are older things we can estimate the age of and add them to the calibration as well, so those who do carbon-14 dating have a calibration that stretches back 50,000 years. However, after 3,500 years, the calibration is based on estimates, and in general, the older the estimate, the less reliable it is. So carbon-14 dating is very accurate for things that are roughly 3,500 years old or less. However, its accuracy becomes more and more questionable the farther back in time the measurement goes.

Uniformitarians are willing to rely on carbon-14 dating all the way back to 50,000 years ago, because they think the entire calibration is good. However, YECs think that there is a real problem with carbon-14 dating, because they have shown some serious conflicts between carbon-14 dating and other forms of dating. For example, YECs have used carbon-14 dating on six separate dinosaur fossils that were discovered in five different parts of the United States. The carbon-14 ages range from 22,000 years old to 38,000 years old. All of those ages are within the calibration range of carbon-14 dating, but they are clearly at odds with the geological column (whose dates are determined by the potassium-argon dating system), which says the dinosaurs went extinct 65 million years ago. YECs have also used the carbon-14 dating system on other fossils (and even diamonds) that are supposed to be millions of years old, and the results were mostly less than 50,000 years.

Uniformitarians would say that there is obviously contamination in these things, so they don't invalidate carbon-14 dating. However, catastrophists would say that an enormous amount of carbon-14 ages fall within the ranges they measured for dinosaur and other fossils, which means all those carbon-14 ages could also be the result of contamination and therefore also be incorrect.

Comprehension Check

9.2 In the illustration on page 248, the 0.4 MY points to Hawaii, and the 5.1 MY points to Kauai. According to uniformitarians, did Oahu (which is in between) formed 11 MY, 6 MY, or 3 MY ago?

9.3 If an igneous intrusion in Quaternary rock (the top layer of the geological column) was determined to have a potassium-argon age of 32 MY, would a uniformitarian think it was accurate?

9.4 A bird fossil found in Quaternary rock has a carbon-14 age of 26,000 years. Would a uniformitarian believe it was an accurate age? What evidence could a YEC use to show that the uniformitarian must at least be a bit skeptical about the date?

Evidence for the Uniformitarian Interpretation of the Geological Column

You already know that uniformitarians and YECs look at the geological column very differently. Although there is no way to directly confirm which (if either) interpretation is correct, we can evaluate each interpretation based on how consistent it is with all the observable data. Now, of course, it would take several books to cover all the data and whether or not they are consistent with each interpretation. Nevertheless, I want to give you an idea of how it would be done by highlighting just a few examples. Let's start with evidence that supports the uniformitarian interpretation.

Consider, for example, what happens when you compare the fossils found in Cretaceous rock with those found in Tertiary rock. There are a lot of different types of animals preserved in Cretaceous rock, but many of those same animals are not found in Tertiary rock or any layers above. In the uniformitarian view, that means they went extinct at the end of the Cretaceous era. The thing is that most of these animals are found throughout Cretaceous rock, even near the top. Then, they just don't appear in Tertiary rock, even in the very lowest parts.

From a uniformitarian point of view, this means that the animals went extinct in a short amount of time, at least when compared to the millions of years that each layer of the geological column represents. This is particularly striking, because estimates indicate that it applies to roughly 70% of the different types of animals living in the Cretaceous era. This is often called a **mass extinction**.

Mass extinction – The extinction of a large percentage of animals over a short geological time

Because it was so quick and devastating, uniformitarians think there must have been a single event (or a series of events that happened nearly at the same time) to cause this mass extinction.

In 1980, several geologists reported discovering a thin layer of sedimentary rock in between Cretaceous rock and Tertiary rock. This particular layer of rock was very different from the other sedimentary layers in the geological column, because it was very rich in iridium (ih rid' ee um), an element that is rarely found in the earth's crust. However, it is commonly found in meteors. Thus, they concluded that the mass extinction seen between the Cretaceous and Tertiary layers must have been caused by a large meteor impacting earth, which caused not only widespread destruction but also caused so much dust to be blown into the air that it darkened the earth, cooling it for many years. Other geologists began searching for this same layer, and it was found in many places on the earth. Thus, it was concluded that the meteor's impact did, indeed have global consequences, such as the extinction of many different types of animals, including the dinosaurs.

This sample shows the layer of iridium-rich sedimentary rock between Cretaceous and Tertiary rock.

Of course, if there were such a meteor impact, it should have left its mark on the earth. While geologists at the time knew about some craters that were thought to be the result of meteors hitting the earth, none of them were big enough to produce the devastation required to cause a mass extinction. Ten years later, however, some data that had been gathered by a group of scientists looking for oil in the Yucatan region of Mexico were published. The data indicated that there was a ring embedded in the crust, and its diameter was nearly 180 kilometers! Part of the ring was on the Yucatan Peninsula, and part was on the Gulf of Mexico's seafloor. A meteor impact like that could produce the devastation needed to cause a mass extinction.

When the ring structure, called **Chicxulub** (cheech' oo loob), was radiometrically dated, it was found to be 65 million years old, which is exactly how old it would have to be to have caused the mass extinction. From a uniformitarian point of view, this kind of consistency provides a lot of evidence that their interpretation is correct. After all, their interpretation required a global event that would kill off 70% of the types of animals that were alive 65 million years ago. First, a thin layer of sedimentary rock that was consistent with a meteor impact was found exactly where it should be if a meteor impact had been that event. Second, a huge crater was then found, and is the age it needs to be if it was made by that meteor.

In the YEC interpretation of the geological column, Chicxulub was probably the result of a meteor hitting the earth, but it would have taken place near the end or after the Flood. Thus, it didn't cause any mass extinction. It could also be responsible for the thick layer of iridium-rich sedimentary rock between Cretaceous and Tertiary rock, since Cretaceous rock is supposed to be the last of the rocks laid down by the Flood, and Tertiary rock is supposed to be the first layer of rock laid down after the Flood. However, it's a bit hard to understand how the animals that were spreading out across the world after the floodwaters receded could have survived such a devastating event.

This Japanese cedar tree is producing pollen that must travel to another tree so they can reproduce.

Another example of consistency in the uniformitarian interpretation of the geological column comes from fossils of plants. Plants reproduce differently, depending on how they are designed. Some plants, like ferns, reproduce by making spores. Other plants, like trees, reproduce by transferring pollen from one individual to another. However, different kinds of trees produce different kinds of pollen. Evergreens, for example, are part of a group of plants called **gymnosperms** (jim' nuh spurmz).

Gymnosperms – Plants like evergreens that make seeds which are open to the air

Other trees, like the ones that lose their leaves in the winter, are part of a group of plants called **angiosperms** (an' jee uh spurmz).

Angiosperms – Plants that make enclosed seeds using flowers

Because of the different ways gymnosperms and angiosperms make their seeds, the pollen they make is different as well. Thus, it is easy to distinguish between gymnosperm pollen and angiosperm pollen.

In the geological column, you find plants like ferns in Silurian rock. You also find spores that are produced by such plants in Silurian rock, but you don't find any pollen there. You don't find gymnosperms until you get up to Carboniferous rock. At that point, you start seeing gymnosperm pollen as well as spores. You don't find angiosperms until you get to Cretaceous rock, at which point you start seeing angiosperm pollen in addition to gymnosperm pollen and spores. Now remember, spores and pollen can separate from the plants that make them, traveling great distances on the wind, on insects, etc. The fact that you don't find the spores or pollen associated with the plants until you find fossils of those plants is another point of consistency that adds evidence to the uniformitarian interpretation of the geological column.

Now think about how the YEC interpretation would view this situation. According to the YEC interpretation, ferns appear lowest in the geological column because they existed in regions that were taken over by the Floodwaters before the regions occupied by gymnosperms. The gymnosperms appear higher than the ferns because they occupied regions taken over later, and the angiosperms appear highest because they lived in regions that were taken over last. However, as shown in the picture on the previous page, pollen travels through the air. It isn't isolated in different regions, so it should appear everywhere, even in Precambrian rock.

Now it turns out that there have been reports of pollen in Precambrian and Cambrian rock. However, these reports are more than 40 years old, and others who have searched for pollen in such rock have not been able to find any. Thus, most paleontologists think that the old reports are in error. However, YECs often point to those old reports as evidence of an inconsistency in the uniformitarian interpretation. They also point out that there are other plant-related inconsistencies in the uniformitarian interpretation of the geological column.

For example, when you see drawings of dinosaurs in what was thought to be their natural habitat, you never see them walking on grass. They are always walking on ferns and mosses, because the earliest identifiable grass fossils appear in Tertiary rock. However, dinosaurs went extinct at the end of the Cretaceous era, so in the uniformitarian interpretation, grasses didn't exist until dinosaurs went extinct. However, several different samples of coprolite (fossilized feces) found in Cretaceous rock and thought to be from dinosaurs contain the remains of grasses. This indicates that dinosaurs (or some other animals that lived in the Cretaceous era) ate grasses. That would be impossible, of course, if the uniformitarian view of when grasses existed is true.

This is typical of how dinosaurs are depicted. They do not walk on grass, because grass wasn't supposed to exist with the dinosaurs.

In addition, amber formed from the resins of gymnosperms is easy to distinguish from the amber formed from the resin of angiosperms. However, amber that has all the characteristics of being formed from angiosperm resin has been found in Carboniferous rock, which was supposed to have been formed before angiosperms existed. Uniformitarians counter that both the grass remains in coprolite and the angiosperm-like amber are isolated cases. There are so many more examples of fossils that fit with the uniformitarian interpretation that the best way to address such isolated cases is to do more investigation to better understand them. Perhaps, for example, some grasses did exist before the dinosaurs went extinct. However, they were rare, so the grasses themselves didn't fossilize – only their remains in coprolite did.

If you could look at the strata of this formation very closely, you would see pairs of very thin layers, called varves.

The uniformitarian interpretation of the geological column is also supported by the fact that many features in it are easiest to understand in the context of long timespans. Consider, for example, the picture on the left. It comes from the Green River formation in Wyoming. You can clearly see that the rock (slate) is stratified, but notice how many strata there are. It turns out that if you could look at the formation up close, you would see a *huge* number of very thin strata. In some parts of the formation, there seem to be *several million* thin layers of slate. If you looked very carefully, you would actually see that the layers mostly come in pairs: a dark layer and a light layer.

How does such a slate formation occur? If we look at sediments accumulating in certain freshwater lakes today, we can see them being laid down in thin layers that come in pairs. A darker layer accumulates during the growing season, when there is a lot of activity in the surrounding soil. A lighter layer accumulates during the part of the year where there isn't a lot of plant growth. Thus, each pair takes an entire year to accumulate. Since the present is the key to the past (in the uniformitarian view), that means each pair of thin strata in the Green River formation was laid down in one year. Uniformitarians call this a **varve**.

Varve – A pair of thin sedimentary strata of contrasting color that were deposited in one year

Those who study the Green River formation say that there are 20 million varves in rocks, which indicates that it took at least 20 million years to form.

If you look throughout the geological column, you can find other examples of sedimentary strata that come in varves. While the Green River formation currently holds the record for the total number of varves in a given formation, there are others worth mentioning. There is a formation in Siberia near Lake Baikal that has 5 million varves. Below Lake Suigetsu in Japan, there is a rock formation with 60,000 varves. All of these formations speak of long spans of time, which is what one expects using the uniformitarian view.

YECs, of course, disagree with this interpretation. For example, they point to the fact that there are abundant fossils throughout the Green River formation. If the thin layers of slate were formed one year at a time, it would take many, many years for a dead organism to be covered in the sediment. During that time, the organism's remains would decay away. Fossilization occurs when something slows down or stops decomposition, and slowly-accumulating sediments in a lake are not going to do that. Thus, the large number of fossils indicate that the varves were not laid down two at a time over many, many years. They also point to formations like the one I showed on page 118, where at least hundreds of very thin layers of sediment were laid down all at once. However, unlike the slate layers in the Green River, there are no alternating light and dark pairs in those formations.

I will end this section with one more example of a formation that uniformitarians use to provide evidence for the idea that geology is best understood by thinking about very long timespans. The picture on the right shows the White Cliffs of Dover. They are an amazing sight. I can tell you that firsthand, because I have viewed them from a ship on the ocean and from various places on the shore. In fact, my wife took that picture from the ship we were on.

Based on the processes we see happening today, it would have taken more than a million years to form these white cliffs.

The white rock is limestone, and unlike many examples of limestone, it is almost pure calcite. If you look at a sample with a microscope, you will find that the calcite crystals are very, very small. They are much smaller than the grains that are usually formed by weathering and erosion. Generally, the only way scientists have seen such small crystals form is when calcium carbonate precipitates out of water or when microscopic organisms called coccolithophores shed parts of their skeletons, called coccoliths. You were shown a picture of a coccolith on page 201. Since it is very rare to find calcium carbonate precipitating out of water in nature, it is thought that the tiny crystals of calcium carbonate in the White Cliffs of Dover are the remains of billions upon billions of coccoliths that were shed.

Based on what we know about coccolithophores, it is estimated to have taken more than a million years for enough coccolithophores to live, die, and shed their coccoliths in order to produce the White Cliffs of Dover. YECs would counter that the Flood probably produced a lot of calcium carbonate when minerals from the mantle interacted with the water in the ocean, and most likely, the calcium carbonate in the cliffs was produced by precipitation, not coccolithophores.

Comprehension Check

9.5 If Chicxulub had been dated to be 100 million years old, how would that have changed the way uniformitarians explained the mass extinction at the end of the Cretaceous era?

9.6 There are certain insects that depend on nectar from flowers in order to survive. Would a uniformitarian expect to find them in Silurian rock, which contains only fossils of gymnosperms?

9.7 If you are examining a series of very thin strata, what would you look for to see if they are varves?

9.8 A geologist says a bed of limestone was deposited quickly by erosion caused by a catastrophe. How would the size of the mineral crystals compare to those in the White Cliffs of Dover?

Evidence for the YEC Interpretation of the Geological Column

Of course, YECs also point to observations that indicate their interpretation of the geological column is correct. For example, consider the picture below. It shows a fossil of a **marine** creature. The term "marine" means the creature lived in the ocean. However, this fossil was found on Mount Fisht, the highest mountain in the Caucasus range, which is in Russia. What does that tell you about the rocks that make up the mountain? At one time, they had to be under the ocean. Similarly, fossil ammonites have been found at the top of the highest mountain on the planet, Mount Everest. Once again, since ammonites lived in the ocean, that means the rocks that make up the very peak of Mount Everest must have been under the ocean when the fossils were formed.

This fossil of a marine animal was found on a very tall mountain in Russia.

Surprisingly, fossils of marine creatures can be found at the tops of very high mountains on every continent. Thus, we can confidently state that at one time, even some of the highest elevations of each continent were under the ocean. YECs say that the easiest way to understand this is if all the continents were under the ocean at one time, and that's exactly what would be expected if there was once a worldwide Flood.

Now remember, YECs think that the world changed significantly during and after the Flood. Thus, Mount Everest might not have been nearly as tall before the Flood as it is today. After all, YECs think that the new oceanic crust rose and the continental crust sank in the early stages of the Flood. Towards the end, the reverse happened. All of that motion, along with lots of weathering and erosion, would have probably resulted in much taller mountains after the Flood as compared to before the Flood. Nevertheless, prevalence of marine fossils at high elevations on every continent is very easily explained by a worldwide Flood.

Uniformitarians counter that it isn't odd for marine fossils to be found on the top of a mountain. After all, as you already learned, we know that some mountains are currently getting higher because of the way the crust below them is moving. If this went on for millions and millions of years, something that was under the ocean at one time could easily rise above the ocean and form a very tall mountain. However, for this explanation to work, it must apply to every mountain upon which marine fossils are found. YECs contend that a worldwide Flood is a better explanation, because it accounts for the fact that this is a worldwide phenomenon.

Another really interesting feature of the geological column is the scale of some of its sedimentary deposits. Remember that geologists can use physical correlation to connect a sedimentary rock deposit in one region of the earth to a deposit in another region. Well, in the Grand Canyon, there is layer of sedimentary rock called the Tapeats Sandstone. It sits right on top of the Great Unconformity, and it is considered to be Cambrian rock. Geologists have found very similar sandstone deposits running throughout the United States, Canada, and Greenland, indicating by physical correlation that they are all part of the same overall sandstone formation.

To give you an idea of just how large this formation of sandstone is, look at the map on the right. The yellow shape superimposed on the map of North America shows all the areas in which the same sandstone exists. Since this is the same sandstone in the same geological sequence (it's all Cambrian rock), it was laid down at the same time. Incredibly, this is just one of many examples of enormous sedimentary rock deposits. A single deposit of limestone chalk extends throughout England, into Northern Ireland, across many parts of Europe, into Turkey, Israel and even parts of Central Asia. In fact, some coal beds in the United States can be physically correlated to coal beds in Europe and the former Soviet Union!

This map traces a single sandstone deposit across North America.

We don't see sediments accumulating slowly over such huge areas today. Instead, YECs say that large deposits like this are evidence of a global catastrophe that produced huge amounts of sediment which got deposited quickly. Once again, of course, uniformitarians have explanations for how each of these deposits might have formed, but they each involve using processes that only seem to happen today on much smaller scales. YECs think that the large scales of these deposits are best explained using a large-scale event.

Now remember, sedimentary rock requires sediments. One of the ways sediments are formed is through weathering of existing rocks. If those sediments are then carried away by erosion, they can accumulate elsewhere and form new rocks. That's all part of the rock cycle that you learned about previously. In some cases, geologists think that they can examine existing sedimentary rock and determine where the sediments that formed it came from.

Consider, for example, the Navajo Sandstone, which is found in Southern Utah. It is made of sediments that are quite different from what is found in other sandstone in and around the area. Those sediments contain zircon, a mineral that often contains uranium, which has radioactive isotopes that decay to lead through a series of alpha and beta emissions. If the amounts of uranium and lead in those zircons are measured, they match the zircons found in the parts of the Appalachian mountain range found in New York and Pennsylvania. As a result, most uniformitarians and catastrophists agree that the sediments which made up the Navajo Sandstone come from weathering and erosion that occurred in the mountains of what are now New York and Pennsylvania.

Think about what this means. In order to form the Navajo Sandstone, the mountains in New York and Pennsylvania had to experience some weathering so that sediments would be produced. Erosion then had to carry those sediments to southern Utah, which is roughly 1,250 miles away! How does that happen? YECs say it is easy to understand that in the context of a global Flood, as there would be times in the Flood when large quantities of water were flowing all in one direction. The large quantities of water would do large amounts of weathering, and then those same waters would carry the sediments all in one direction, eventually depositing them where the current died out or was shifted to a different direction.

Uniformitarians would say that sediments can be carried long distances by rivers. However, the amount of sediment required would take millions of years at the rates it is transported today. YECs would counter that it's hard to imagine a river running in the same basic direction for millions of years, depositing its sediments in the same basic way over all that time. Lots of sediment transported over vast distances is better understood in the context of a large catastrophe rather than slow, steady processes.

Another striking aspect of the geological column that YECs point to as evidence to support their interpretation are the places where you find huge numbers of fossils all buried together in a jumbled mess. Many YECs refer to such deposits as "fossil graveyards," and others refer to them as "bone beds." Consider, for example, the picture below. It shows a small portion of sedimentary rock from Morocco. The fossils you see there are of marine creatures called nautiloids. Many nautiloids are spiral, but there are straight ones as well, and those are the kind that left the fossils in the picture.

This collection of nautiloids must have been buried rapidly by a large amount of sediment.

Obviously, this fossil graveyard was not produced by individual nautiloids dying, sinking to the bottom of the ocean, and then being fossilized. When the sediments that made this rock were laid down, they must have trapped a huge number of nautiloids all at once. This indicates a catastrophic process that moved a lot of sediment very quickly. Uniformitarians have no problem with this, of course, because they understand that catastrophes do happen in specific areas and can lead to rapid burial of a large number of fossils. For example, an underwater earthquake could have caused an underwater landslide, which caused lots of sediment to fall on a population of nautiloids, killing and then preserving them.

While the fossil graveyard pictured above might be explained by a local catastrophe that happened in a specific area, there are fossil graveyards around the world that are truly massive. For example, there is a fossil graveyard in the Grand Canyon that contains nautiloid fossils similar to the ones shown above. They are found in a 2-meter (6 feet) thick section of a sedimentary rock layer called the Redwall Limestone. This fossil graveyard, which is taller than me, stretches across 290 kilometers (180 miles), covering a total area of 30,000 square kilometers (12,000 square miles). To give you some perspective, that's roughly the area of the entire state of Maryland! What could have trapped enough animals to produce a fossil graveyard that would cover the state of Maryland to a depth of 2 meters? YECs say only a massive catastrophe could do that!

There are many fossil graveyards found around the world, and amazingly, some of them contain an incredible mix of fossils from very different animals. For example, there is a large coal formation near Montceau-les-Mines, France that contains hundreds of thousands of fossils. The most fascinating aspect of this fossil graveyard is that it contains fossils from marine animals, animals that lived in freshwater, and animals that lived on land. How would animals from such different ecosystems end up buried together so that they could all fossilize in one place?

YECs say that fossil graveyards like this one (and many others found throughout the world) are very easy to understand in the context of a global Flood, which would carry sediments long distances, trapping organisms from many different ecosystems along the way. Those trapped organisms would be carried over long distances with the sediments. Think, for example, about how far the sediments of the Navajo sandstone traveled. If they traveled rapidly, they would carry along anything that was trapped in them. When they finally came to rest, they would deposit a huge number of fossils, all from different ecosystems. Uniformitarians, on the other hand, must resort to assuming that there was an area in which marine, freshwater, and land animals all lived in close enough proximity to be buried and fossilized together.

YECs also point to the odd nature of many unconformities between different layers of sedimentary rock. Consider, for example, the photograph below. It shows some hikers on a trail in the Grand Canyon. The entire wall of rock above and below them is made of limestone, but you can see that the limestone above them is noticeably different from the limestone below them. The rocks above them are part of what is called the "Redwall Limestone" formation, for obvious reasons. The limestone below them is called "Muav Limestone" formation. Notice how gradually the Muav Limestone turns into the Redwall Limestone. From catastrophist point of view, this is easy to understand. If sediment is deposited by water and more sediment is laid down quickly on top of it, the sediments will probably mix a bit, making it very hard to see exactly where the top of one sediment layer ends and the bottom begins. That's what you see in the picture.

The limestone above these hikers is very different from the limestone below them, but the two types of limestone blend together.

YECs are quick to point out that this is a real problem for uniformitarians. The Muav Limestone formation is Cambrian, so in the uniformitarian view, it should have finished being laid down about 500 million years ago. The Redwall Limestone formation, however, is Mississippian rock and is thought by uniformitarians to have been laid down about 330 million years ago. Thus, there should have been roughly 170 million years between when the last of the Muav Limestone formation was laid down and the first of the Redwall Limestone formation was being deposited. During that time, there should have been lots of erosion, so when the Redwall Limestone sediment started being laid down, there should have been a clear separation between the old rock and the new sediments. However, that's not the case.

In Chapter 4, you learned that this situation was called a paraconformity – a bedding plane that is thought to be an unconformity, but it doesn't look like an unconformity because there is no clear separation between the two sedimentary strata. Paraconformities are common in the geological column, and they are difficult to understand in the uniformitarian framework. However, they are very easy to understand in the YEC framework.

In Chapter 4, you also learned about folded sedimentary strata, such as those pictured below. As you learned back then, rocks tend to crack when they are bent. In the absence of cracks, then, these formations are easiest to understand if the strata were bent before they were fully lithified. In other words, they bent when they were at least somewhat soft. YECs say that this makes perfect sense in their interpretation of the geological column, because they think that lots of strata were laid down in a short amount of time. As a result, they were laid down while they were soft, and they then took time to undergo lithification. Thus, folded sedimentary rock like that is best understood as sedimentary strata that were laid down during the Flood and then folded before they hardened. They then finished lithification after they were folded.

YECs contend that large-scale folding in sedimentary rock, as shown in this limestone, is best understood if the folding occurred before the rock underwent lithification.

Uniformitarians say that while rocks do crack and break when they are folded at cooler temperatures, the warmer the rock gets, the easier it becomes to bend. Since the geosphere gets hotter the deeper you go, you can understand folded sedimentary rocks as rocks that were heated until they were soft, then bent, and then cooled again. YECs counter that while that might happen on a small scale, there are large-scale formations of folded sedimentary rock that are best understood as being folded while the sediments were still soft.

Comprehension Check

9.9 Do the marine fossils at the top of Mount Everest tell YECs that the Flood waters were as deep as Mount Everest is tall?

9.10 Why wouldn't a YEC think that the sandstone found throughout North America could be explained by local catastrophes that covered specific regions of the continent?

9.11 Suppose you are looking at a series of sedimentary strata. You find lots of saltwater fish fossils in one layer, lots of freshwater fish in the next layer above, and lots of land animal fossils in the layer above that. You then look at a second series of sedimentary strata. In that series, you find lots of saltwater fish fossils, freshwater fish fossils, and land animal fossils all mixed together in one layer. Which gives evidence for the YEC view?

Evolution and the Geological Column

For most (but not all) uniformitarians, their interpretation of the geological column g support for the theory of evolution. After all, if you look at fossils found in Precambrian rock, you only simple, marine organisms. As you move into Cambrian rock, the organisms get a lot more complex, but they are still all marine organisms. The first fossils of land-based organisms don't appear until the Devonian era. This implies that life on earth started out in the oceans and then eventually spread to the continents. In addition, the fossils get more and more complex the higher you go on the geological column, which implies that life got more and more complex as time went on.

According to the theory of evolution, this is exactly what you expect the fossil record to indicate. The vast majority of evolutionists think that life started out as a single cell that lived in the ocean. As that cell reproduced, some of its offspring were slightly different. Sometimes, the slight difference made the offspring less likely to survive, and as a result, it was less likely to produce its own offspring. However, some of the slight differences would make the organism more likely to survive. As a result, it would be more likely to produce offspring. Any slight differences that made an organism more likely to survive and reproduce would be passed on through the generations, while slight differences that made an organism less likely to survive and reproduce would tend to disappear over the generations.

As you may already know, this process is called **natural selection**, because it says that nature "selects" the traits that make an organism more likely to survive and reproduce, while it discards the ones that make an organism less likely to survive and reproduce.

<u>Natural selection</u> – The process by which traits that make an organism more likely to survive and reproduce are preserved in subsequent generations, while traits that do the opposite are not preserved

Natural selection is a well-understood feature of nature, and many examples of it have been documented.

Consider, for example, the finch pictured on the right. Its beak is short and stout because it eats seeds, and it needs a stout beak to crack them open. A population of these finches that lived on an island were studied for many years, and it was found that when the island experienced a long-term drought, the average beak size of the population got larger. How? The birds didn't grow larger beaks during long-term droughts. Instead, each time new finches hatched, some had larger beaks, and some had smaller beaks. During long-term drought conditions, seeds became scarce, so the birds had to eat seeds they normally wouldn't eat. Birds with larger beaks were able to crack open harder seeds, so they were less likely to starve to death. After several generations of new finches, then, the average size of the beaks was larger. That's natural selection – the finches that had better beaks for the drought conditions were "selected" to survive.

This finch has a short, stout beak so it can break open seeds and eat them.

, of course, a bird with a larger beak isn't much different from the same type of bird with a k, so when scientists have seen natural selection in action, they haven't seen it produce anges. However, since uniformitarians believe the present is the key to the past, they suggest over hundreds of thousands or millions of generations, the small changes we see happening today can "pile up," producing large changes over a long period of time. However, since we haven't actually seen that happen, evolution is often separated into two different categories: **macroevolution** and **microevolution**.

Microevolution – The theory that natural processes can, over time, transform an organism into a more specialized version of that organism

Macroevolution – The idea that natural processes can, over a long period of time, transform an organism into a completely different kind of organism

Obviously, what was observed in the study of finches that I just discussed is microevolution, because a finch with a larger beak is just a more specialized finch. If, over many generations, the finch beaks got longer and sharper, their bodies got bigger, and their feet developed talons, they might become like hawks. That would be macroevolution. For the sake of brevity, I will use the term "evolutionist" to refer to uniformitarians who think macroevolution is responsible for the organisms we see today.

Pakicetus: 41-50 MYA

Ambulocetus: 47-48 MYA

Rodhocetus: 47 MYA

Dorudon: 41-37 MYA

Brygmophyseter: 15 MYA

Evolutionists say these fossils demonstrate macroevolution from land mammals to whales.

Evolutionists would say that we cannot hope to document macroevolution happening because it would take thousands to millions of years. However, if we look at the fossils in the geological column, we can see evidence of how it has happened over and over again in the past. Do you remember *Pakicetus*? While paleontologists originally got its appearance wrong, they now think they have a good idea of what it looked like. Its fossils are found in rocks that are supposed to have been laid down 41-50 million years ago (MYA). While their fossils indicate they lived on land, their ear bones are similar to those of whales, so it is thought that they spent at least some of their time in the water.

In rocks that are dated as being laid down 47-48 MYA, a bit later than the rocks in which *Pakicetus* is found, we find fossils of *Ambulocetus* (am' byuh luh' see dus). This animal looks a lot like a land animal, but because its eyes are near the top of its head and it has a long, powerful snout, it is thought to have behaved much like a crocodile. Thus, it spent more time in water than *Pakicetus*. In rocks that were laid down 47 MYA, we find fossils of *Rodhocetus* (rohd' huh see dus). It had shorter limbs than *Ambulocetus*, and its feet were large. It has many skeletal features that are similar to whales, so evolutionists think it lived most of its life in the water. Because of this, they think its feet were webbed, even though there is no fossil evidence to confirm that.

In rocks dated to have been laid down several million years more recently, there are fossils of an animal called *Dorudon* (dor' oo don). It has more in common with whales than *Rodhocetus*, and based on its skeleton, it is thought to have lived its entire life in water. One big difference between *Dorudon* and *Rodhocetus* is that *Dorudon's* legs are smaller, especially the back legs. The back legs are so small that evolutionists think that they couldn't have been very useful to the animal. In rocks dated to have been laid down about 20 million years more recently, we find fossils of *Brygmophyseter* (brig' moh fiz' ih tur), which is very, very similar to modern sperm whales, which have no rear legs at all. This is all summarized by the illustration on the previous page.

To an evolutionist, this is excellent evidence for macroevolution. After all, we know that about 50 million years ago, there was a land animal (*Pakicetus*) whose ear structure was similar to the ear structure of a whale, which indicates it probably spent at least some time in the water. A couple of million years later, an animal existed (*Ambulocetus*) that had many features in common with *Pakicetus* but more in common with whales. It makes sense that over a couple of million years, animals like *Pakicetus* would be born with traits that made them better at hunting in the water, so it makes sense that over time, natural selection could turn animals like *Pakicetus* into animals like *Ambulocetus*.

As animals like *Ambulocetus* reproduced, they would eventually have offspring with even more traits that made them better at hunting in the water, so in another million years or so, *Rodhocetus* would exist. As *Rodhocetus* spent more time in the water and reproduced, some of its offspring would develop traits that would end up making them most comfortable in water, so that would lead to *Dorudon*. Over time, *Dorudon* would produce offspring with even better traits for water, making something very similar to the whales we see today. So while we didn't directly observe land animals like *Pakicetus* undergoing macroevolution to eventually produce whales, the fossil record has preserved the process.

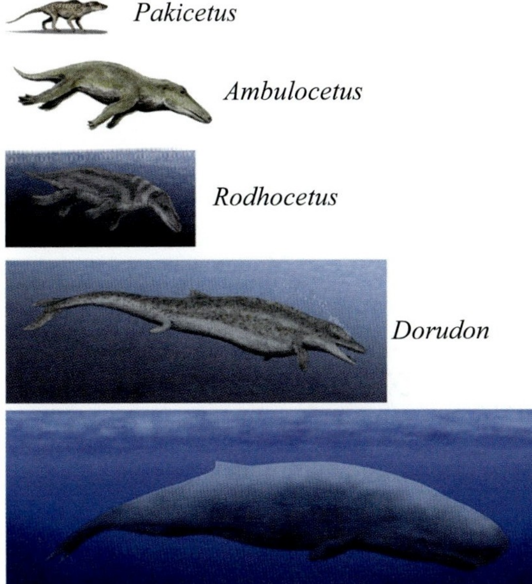

This illustration shows the animals from the previous page properly in terms of their sizes.

Not surprisingly, YECs disagree with the [illegible]sts' interpretation of these fossils. For example, [illegible]nt out a diagram like the one presented on the [illegible] page isn't being honest when it comes to what we [illegible]out these animals based on their fossils. For [illegible]e, the fossils give us a good idea of how big each [illegible]al was, and their relative sizes are shown in the illustration on the right. When you view them like this, they suddenly don't seem as similar. Also, the change in size doesn't make sense, since *Ambulocetus* is a lot bigger than *Pakicetus*, but then *Rodhocetus* is smaller. If macroevolution is making a small animal like *Pakicetus* into huge animals like whales, why isn't the change consistent?

More importantly, YECs would say that there is an enormous amount of overinterpretation going on when it comes to the fossils themselves. Yes, the ear bones in *Pakicetus* have some things in common with the ear bones in whales, but that doesn't mean *Pakicetus* spent any time in the water. Its skeleton looks completely like a land animal. In the same way, the fossils of *Ambulocetus* make it look like a land animal. Yes, it has characteristics that are similar to that of a crocodile, but its ear bones are actually *less* like those of a whale than the ear bones of *Pakicetus*, which would be a step backwards in the macroevolutionary sequence, at least when it comes to the ears. Similarly, *Rodhocetus* has all the structures you expect for a land animal. While evolutionists think it had webbed feet, there is no actual evidence that it did. In addition, while *Dorudon*'s rear legs are small, they are not necessarily useless. Many snakes have small structures in roughly that same position, and they use them when mating. In addition, YECs would say that the ages overlap significantly. *Pakicetus* was alive when *Ambulocetus*, *Rodhocetus*, and *Dorudon* were also alive. That doesn't look much like a progression through time.

Finally, YECs are quick to point out that as is the case with the formation of the Hawaiian Islands, this kind of analysis leaves out some rather important data. For example, a fossil jaw that looks nearly identical to a modern whale jaw was discovered in Antarctica, and uniformitarians say it is 49 million years old. The jaw indicates the animal was in between *Dorudon* and *Brygmophyseter* in terms of size. Of course, 49 million years ago is when *Pakicetus* was supposed to be alive, which means that *Pakicetus* couldn't have given rise to whales. According to this fossil jaw, whales already existed by then! In addition, another fossil that includes the skull and some other bones is very similar to *Dorudon*, but it is supposed to be 45 million years old, which means it existed at roughly the same time as *Ambulocetus*. When you start looking at the details, then, YECs would say that the evolutionists' view of whale macroevolution just doesn't make sense.

Evolutionists counter that there is a lot of variety in modern whales, so you would expect a lot of variety to spring up during their macroevolution. Thus, while the illustration on page 264 gives a general idea of how macroevolution produced whales, what actually happened wasn't exactly as shown in the illustration. Also, there are many other fossils of whale-like animals that fit within that time period, but their fossils aren't nearly as complete, so they aren't typically presented in a book like this. However, they do tend to fit in the pattern.

Do Similarities Imply Macroevolution?

Remember the reasoning employed in my discussion of whale macroevolution. The ear bones of *Pakicetus* are similar to the ear bones of whales, which is why evolutionists think that *Pakicetus* started the macroevolutionary process that produced whales. As the animals being discussed became more and more like whales, the macroevolutionary trend became more apparent. But that brings up an important question: When two animals are very similar, are they related by some macroevolutionary sequence? Consider the two pictures below. One is of a sugar glider, and the other is of a flying squirrel. Despite the squirrel's name, neither of these animals actually flies. They both use their unique body design to glide through the air as they jump from tree to tree. The two animals look very similar, don't they? They look *a lot* more similar than any of the animals that were part of the whale macroevolution discussion. Nevertheless, while they look incredibly similar, evolutionists say that they must have evolved separately, and the time

The animal on the left is a sugar glider, while the animal on the right is a flying squirrel. They are not closely related, according to evolutionists.

scale over which that evolution took place had to be longer than the time scale over which whale macroevolution took place!

Why do evolutionists say this? Sugar gliders are part of a group of animals called marsupials (mar soo' pee uhlz), which includes kangaroos. When they give birth, their offspring crawl into a pouch on the mother's body and develop there until they are ready to start venturing out into the world. Squirrels, on the other hand, are placental (plu sent' uhl) animals. Their offspring develop inside the mother's body, and they are not born until they are ready to live in the outside world. According to macroevolution, these two types of animals are separated by more than 100 million years' worth of evolution! So even though they look similar, they are not thought by evolutionists to be part of the same macroevolutionary sequence. Evolutionists have a name for this. They call it **convergent evolution**.

<u>Convergent evolution</u> – The process by which animals that are not closely related develop similar traits

In the end, then, when we see fossils that are similar, they might not be related to one another in a macrevolutionary sequence. Instead, their similarities might be a result of convergent evolution.

As you might imagine, convergent evolution can make it difficult for an evolutionist to figure out how macroevolution happened. When the evolutionist finds fossils that are very similar to one another, can he or she really figure out whether those similarities are a result of them being related to one another through macroevolution or just the result of convergent evolution? In the case of sugar gliders and flying squirrels, we know enough about their reproductive processes to know that they are actually very different animals, despite their similar appearance. Also, evolutionists think they understand how placental animals evolved and how marsupial animals evolved, so they know that sugar gliders and flying squirrels are not closely related.

YECs say that convergent evolution produces a real problem for macroevolution. After all, if similarities imply evolutionary relatedness, then the more similar organisms are, the closer they should be related. However, there are many, many examples in which this simply doesn't work, so there are many, many times evolutionists must "explain away" such similarities in order to preserve macroevolution. YECs say when proponents of an interpretation spend so much time "explaining away" the data, that indicates there is something wrong with the interpretation.

Comprehension Check

9.12 Cane toads were brought to Australia in an attempt to get rid of a pest in the sugar cane fields. The toads are very large, and they are poisonous to snakes. Snakes in the parts of Australia where cane toads exist have, on average, smaller heads than snakes in parts of Australia where cane toads are not found. Explain this in terms of natural selection.

9.13 Suppose a fossil that is part of the whale macroevolution sequence was found in rocks that are supposed to be 40 million years old. How would an evolutionist expect its rear legs to compare to those of *Rodhocetus*? (Look at the illustration on page 264.)

9.14 The eyes of land reptiles are nearly identical to those of marine animals like the octopus, despite the fact that the octopus is not a part of the macroevolutionary sequence that is supposed to have led to land-based reptiles. How does an evolutionist explain this?

Data That Are Hard To Reconcile With Macroevolution

YECs say that overall, the fossil record speaks against macroevolution. Think, for example, about the fossils related to whale evolution. YECs say that the fossils are overinterpreted, and when you include all of them, the timeline for how it happens doesn't make much sense. Evolutionists, on the other hand, consider the fossil record of whales to be one of the best sets of evidence for macroevolution. In fact, a scientific journal published an article with the rather long title, "Whale Origins as a Poster Child for Macroevolution: Fossils collected in the last decade document the ways in which Cetacea (whales, dolphins, and porpoises) became aquatic, a transition that is one of the best documented examples of macroevolution in mammals." (*BioScience*, Volume 51, Issue 12, December 2001, Pages 1037–1049)

Indeed, despite the fact that there are so many holes in the evidence, the journal article is right. If you look through the fossil record, you find lots of organisms that have similarities, but you see very few examples of fossils that can be put into a consistent macroevolutionary sequence. In the end, despite its difficulties, the fossil record for the evolution of whales really is one of the best-documented sequences for macroevolution. YECs say that if this is the best fossil evidence evolutionists have, there is a serious problem with macroevolution.

Evolutionists, on the other hand, say that the process of macroevolution is complex, and fossilization is rare. As a result, you don't expect to find fossils of all the animals that make up a macroevolutionary sequence. Thus, all macroevolutionary sequences determined from the fossil record will be fragmentary. In addition, evolutionists say that you don't have to find detailed sequences in the fossil record to understand that macroevolution occurred. The overall features of the fossil record, showing a gradual change from "simple" organisms to complex organisms shows that it happened.

The top fossil is of a trilobite found in Devonian rock. The bottom one is of a typical housefly.

YECs are quick to point out, however, that the fossil record really doesn't show a lot of change, even if you believe the uniformitarian interpretation of the geological column. Consider, for example, the two images on the left. The top one shows a trilobite fossil that is dated to be about 400 million years old. Look at its eyes. Do you see all the bumps on them? They tell us trilobites have **compound eyes**, which are eyes that have multiple lenses in them. Your eyes, and the eyes of most of the animals with which you are familiar, are not compound eyes. They have one lens each. For the ways you use your sense of sight, a single lens is best. However, for some animals, compound eyes work better. Now look at the picture on the bottom. That's a close up of a typical housefly. Notice that it has compound eyes as well.

Think, for a moment, about what this tells us from the evolutionist point of view. Compound eyes evolved very early in the history of animals. In fact, there are trilobite eyes in the fossil record that are dated to be 521 million years old. Over the past 521 million years, what would an evolutionist expect? He would expect them to change at least a bit, right? After all, the earliest compound eyes would have to be primitive, and with 521 million years of time, natural selection should have "refined" them to make them better, right? However, that's not what we see. When you look at the number of lenses in the earliest fossil compound eye, the relative sizes of each lens, and how they are arranged in the eye, they are nearly identical to those of insects that are alive today. There doesn't seem to be any change, even over 521 million years.

The top picture shows a horseshoe crab fossil that is supposed to be 400 million years old, and the bottom one shows a living version of the animal.

This seems to be a standard property of the fossil record. Using the uniformitarian interpretation, the earliest version of an animal or biological structure is often nearly identical to the living version of that animal. As one more example, the picture on the top right is of a horseshoe crab fossil. These fossils have been found in rock dated to be as old as 445 million years. Nevertheless, compare that fossil to the living horseshoe crab shown below it. Notice that they are nearly identical. Once again, shouldn't there be at least some change, if natural selection had 445 million years to make the horseshoe crab more survivable?

Evolutionists are quick to say that not all organisms need to experience change throughout the fossil record. Remember, evolution happens when traits appear that make the organisms *better* able to survive. Perhaps the horseshoe crab is already ideally survivable for its surroundings. As a result, there is nothing for natural selection to do except get rid of any traits that make the animal less likely to survive. In the same way, perhaps the way compound eyes formed in trilobites was already ideal. If that's true, there is nothing natural selection can do to improve on them.

In addition, uniformitarians say we know there must be change in the fossil record. After all, we know that the oldest fossils are of single-celled organisms, and the most recent fossils are of really complicated things like oak trees, horses, and people. There's definitely a lot of change in the fossil record, but YECs say that it's not the kind of change that macroevolution would produce. After all, macroevolution is gradual. Small changes "pile up" over the generations, leading to new kinds of organisms. With the uniformitarian interpretation, that's not what you see. Instead, you see short "bursts" of change followed by long periods of little to no change.

According to the uniformitarian view, life started out as single-celled organisms at least 3.5 billion years ago. The oldest known fossil that paleontologists think is an animal has been found in Precambrian rock dated at 558 million years ago. Called *Dickinsonia*, its fossils indicate it was a flat, oval animal that lived at the bottom of the ocean and ate single-celled creatures. Some fossils indicate that it had a digestive system that might have also served as a means of distributing food through the animal. As animals go, however, it was not very complex. Other animal fossils in Precambrian rock are few and far between, and none of them are significantly more complex.

When you get to Cambrian rock, however, the situation becomes completely different. In rocks dated from 541 million years ago to about 521 million years ago, you find lots of different fossils of lots of different kinds of animals. Even though all the animals are marine animals, they represent every basic animal body that exists today. There are animals without backbones, animals with backbones, animals that make their own shells, animals that swim well, animals that float with the current, etc., etc. The fact that the number of different kinds of animals increases dramatically during this time is called the **Cambrian Explosion**.

The organisms before the Cambrian Explosion (top) were simple and not diverse. The Cambrian Explosion produced a huge diversity of complex organisms (bottom).

Cambrian Explosion – The abrupt appearance of many different kinds of animals in a short period of time during the Cambrian era

Remember, from the uniformitarian view, animals didn't appear on earth for nearly 3 billion years after life appeared. Then, a few types of animals evolved, but they were "simple," and there wasn't much variety. But then over a "mere" 20 million years (from 540 million to 521 million years ago), life suddenly got very complex and very diverse. In the uniformitarian view, this is difficult to understand. After all, macroevolution is supposed to happen slowly over the generations, with tiny changes "piling up" to eventually form new kinds of animals. Based on what we see happening today, it is very difficult to understand how macroevolution could produce all the different animals of the Cambrian era in only 20 million years or so.

Data That Are Hard To Reconcile With Young-Earth Creationism

While YECs are quick to discuss the Cambrian Explosion as a problem for macroevolution, they have their own explosion of evolution that also seems hard to understand. YECs don't think macroevolution is possible or ever happened on earth, but they recognize that microevolution is real. After all, it has been observed. More importantly, YECs require a huge amount of microevolution to have occurred in a very short amount of time.

Remember, YECs think that all the land animals (excluding perhaps the insects) were saved from the global Flood because they were taken on Noah's ark. Now even though the ark was incredibly huge, it couldn't hold two of every land animal that exists today! YECs say that when the Bible says Noah took two of every kind of animal on the ark, it's not saying he took two of every species. Instead, he took two of every *category* of animal. He didn't take two lions, two tigers, two bobcats, etc. He took two *cats* on the ark. Most likely, the cats that he took on the ark didn't look like any species of cat we see today. However, they had all the biological information necessary to produce the cat species that we do see today.

Of course, in order for something like that to happen, there had to be *a lot* of microevolution taking place. As these two cats started reproducing, their offspring spread out, reproducing even more. When some of them ended up staying in the grasslands of Africa, microevolution started producing the traits that made them more likely to survive, and eventually, lions were produced. When other cat offspring ended up in the forests of Asia, microevolution started producing the traits that made them more likely to survive, and tigers resulted. As time went on, people started breeding cats, selecting out the traits they liked, and domesticated cats were the result. All of these species of cats, however, can be traced back to the pair of cats that walked off the ark after the Floodwaters subsided.

In the YEC view, a pair of cats were taken on the ark, represented by the center drawing. As those cats reproduced and their offspring started encountering new places, natural selection produced versions of the cats that were more survivable in those places, resulting in all the different species of cat we see today.

Most YECs think the Flood happened 4,000 – 4,500 years ago, so if you think about it, all that cat microevolution had to take place in "only" a few thousand years. Of course, that had to happen over and over again with all the different kinds of animals. Based on the genetic studies that have been done, YECs think that there were 1,500 – 7,000 kinds of animals on the ark. However, scientists recognize at least 30,000 different species of animals that would have come from the animals on the ark. That means in just a few thousand years, microevolution had to increase the diversity of animals by as much as *20 times*! That's a lot of microevolution in a very short amount of time! Based on what we see today, it is hard to understand how microevolution could produce such a diversity of organisms.

Another problem that YECs face is explaining the distribution of animals that we see on earth today, the study of which is called **biogeography**.

<u>Biogeography</u> – The study of the geographical distribution of animals on this planet

As I stated previously, YECs don't have a problem understanding how animals got from the ark to the various places we see them on earth today. There were most likely land bridges between several continents that have since eroded away. Also, both YECs and evolutionists must sometimes assume that animals "rafted" from one continent to another on a floating island. However, YECs do have more difficulty than evolutionists explaining the details of how those animals are distributed.

The fact that potoroos (top), kangaroos (middle), and wallabies (bottom) are only found in Australia can be a problem for the YEC view. They could all be part of the kangaroo kind, but why do we not find traces of their journey from the Ark?

Consider, for example, the fact there is only one continent to which kangaroos are native – Australia. The same can be said of wombats, koalas, platypuses, potoroos, and many other animals. In the evolutionist view, they evolved in Australia from animals that are common in other parts of the world. For example, kangaroos are thought to have evolved from a possum-like animal over a timespan of about 30 million years. You only find them in Australia because only there did natural selection preserve exactly the traits needed to turn the possum-like animal into a kangaroo.

YECs cannot take that approach. They believe that animals like kangaroos couldn't survive the Flood without being on the ark, so there must have been two of some kind of animal that had the ability to become kangaroos through microevolution. Those animals walked or hopped off the ark and then began to spread out. When they got to Australia (via land bridges or rafting), microevolution ended up producing kangaroos and other, similar creatures, like wallabies. The problem, of course, is that we don't find fossils of animals like that except in Australia. Thus, this group of animals that eventually became kangaroos had to travel from the ark to Australia without leaving fossils, but then once in Australia, they left *a lot* of fossils.

YECs will point out that fossilization is a rare event, and usually happens as a result of a catastrophe. If the members of the "kangaroo kind" were constantly migrating, their chance of getting caught in a catastrophe would be low, so they would be much more likely to leave fossils where they ended up taking up long-term residence, which is in Australia. Nevertheless, evolutionists suggest that there ought to be at least some fossils that demonstrate the long journey the kangaroo kind had to make from the Middle East (where the Ark landed) to Australia!

Comprehension Check

9.15 Evolutionists are trying to find Precambrian fossils of animals that are more complex than the ones that are currently found. If they do, how would that help the problem of the Cambrian Explosion?

9.16 Suppose you are looking at pictures of a wolf, a domestic dog, a horse, a fox, and a rat. If a YEC told you that three of those animals all came from a single pair of animals that were on the ark, what three would he or she be referring to?

It's Time to Talk About Time

In this chapter, I have been discussing the uniformitarian and YEC views of geology in light of the evidence. Hopefully, you have seen that each view can explain some observations better than the other view, and each view has serious problems with other observations. As a result, there is no way to definitively state which is better, at least not from a scientific point of view. However, there is one piece of evidence that would provide a definitive answer to the question of which view is better: the age of the earth. The uniformitarian view *requires* the earth to be billions of years old. The YEC view *requires* the earth to be thousands of years old. If we could determine the age of the earth, we would know for certain which view is best.

Unfortunately, when we look at the evidence, we once again find that there is no definitive scientific answer. There are some observations that seem to indicate the earth is billions of years old. However, there are others that seem to indicate that it is thousands of years old. There are even some observations (which I will not discuss in this book) that indicate the age of the earth is somewhere in between billions of years and thousands of years. For the purposes of this course, I would like to give you what I consider the best evidence for each view.

Let's start with the uniformitarian view. Probably the most compelling evidence for the earth being billions of years old comes from radioactivity. However, it's not radiometric dating. Hopefully, you have seen that there are lots of inconsistencies in radiometric dating, so it's not going to give us a definitive answer to how old the earth is. Indeed, while most radiometric dates indicate the earth is 4.5 billion years old, there are radiometric dates (such as those that come from diamond xenoliths I mentioned previously) that indicate the earth is even older!

However, there is one general observation about radiation that, as far as we know, has no exceptions. To understand the observation, you need to remember that different radioactive isotopes have radically different half-lives. Potassium-40, for example, decays into argon-40 with a half-life of 1.3 billion years. However, carbon-22 decays into nitrogen-22 with a half-life of 6 *thousandths* of a second! There are isotopes that decay with longer half-lives than that of potassium-40 and shorter half-lives than that of carbon-22. Depending on the isotope, then, you can find ones that have ridiculously long half-lives and others that have ridiculously short half-lives. There are also many with half-lives in between.

However, if we look only at naturally-occurring radioactive isotopes, we mostly see ones that have very long half-lives. You can find potassium-40, uranium-238, and thorium-232 on the earth, for example, but their half-lives are all longer than a billion years. You can also find uranium-235 on earth, but its half-life is 700 million years. However, you cannot find naturally-occurring isotopes like iodine-129, which has a half-life of 16 million years. People can make such isotopes using nuclear reactions, but they are not found in nature. Why?

If the earth is 4.5 billion years old, this is very easy to understand. Remember, after about 10 half-lives, the amount of any radioactive isotope that remains is so small that it becomes undetectable. Why can't you find isotopes like iodine-129 naturally? Because the earth has been around for much longer than ten half-lives (160 million years for iodine-129). As a result, any iodine-129 that was originally on the earth has decayed to the point that it is simply not detectable. This is true for any radioactive isotope whose half-life is even 400 million years. After all, an isotope with that half-life would have experienced more than 10 half-lives of decay in earth's 4.5 billion years of existence, so you would not expect anyone to find detectable amounts of it. For isotopes with a half-life of 500 million years or more, you would expect to find some naturally-occurring samples, since they haven't yet experienced 10 half-lives yet.

In general, that's what scientists see when they go looking for naturally-occurring radioactive isotopes. It is very easy to find isotopes whose half-lives are greater than 500 million years or so, but it is very difficult to find isotopes whose half-lives are less than 500 million years. But wait a minute. "very difficult" is different from impossible. If the earth really is 4.5 billion years old, it should be *impossible* to find isotopes whose half-lives are less than 500 million years, right? Not necessarily.

There are some radioactive isotopes that have short half-lives but are constantly being made due to natural processes. We have already talked about one of them: carbon-14. Carbon-14 decays into nitrogen-14 with a half-life of 5,730 years. The reason uniformitarians aren't surprised to find it on the earth is because it is constantly being made by the process illustrated below. The sun produces neutrons that constantly slam into the earth. Some of those neutrons end up hitting the nucleus of a nitrogen-14 atom, knocking out a proton. The neutron takes the proton's place, and the result is carbon-14. That carbon-14 will eventually decay back into nitrogen-14, but by the time that happens, another carbon-14 will be made. Thus, even though carbon-14 decays away, it is constantly replenished. As a result, even though its half-life is "short" by uniformitarian standards, it can still be found on the earth.

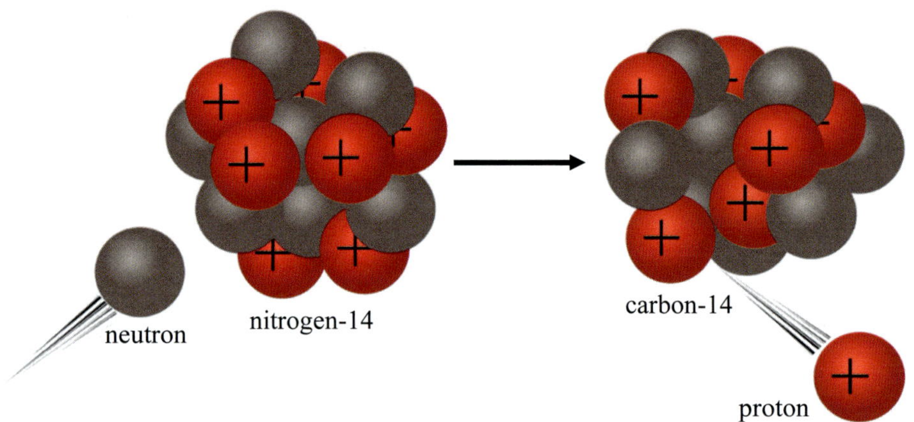

A neutron (brown ball) can hit a nitrogen-14 nucleus, knocking out a proton (red ball) and taking its place. This produces carbon-14. Since the sun is constantly producing neutrons that hit the earth, any carbon-14 that decays is replenished.

It turns out that all naturally-occurring radioactive isotopes on the earth that have half-lives of less than 500 million years are made through a process like the one that makes carbon-14. In other words, there is a natural process that replenishes them. However, there are many other radioactive isotopes with half-lives that are less than 500 million years which can be made in the lab but are not found anywhere on the earth. In fact, in some cases, we find their decay products on the earth. This implies that they existed at one time, but they are not found anymore. The most obvious conclusion is that the earth has been in existence long enough for them to have decayed for more than ten half-lives, indicating that the earth is billions of years old.

Of course, YECs try to explain this observation. They say that such reasoning assumes that radioactive half-lives have always been what we measure them to be now. That's an easy assumption to make if you are a uniformitarian, but catastrophists don't like such assumptions. YECs say that we can understand the lack of such radioactive isotopes if half-lives were shorter in the past. If they were short enough, these isotopes could have all decayed away in only thousands of years.

Uniformitarians are quick to point out that based on how we understand radioactivity right now, there is no reason to expect half-lives to be shorter in the past. More importantly, we know that radioactivity produces heat. Indeed, the main reason the earth's core and mantle are so hot is because of the radioactive decay that is going on down there. If half-lives were shorter in the past, lots more heat would have been produced. In order for half-lives to be short enough to understand what we see today in an earth that is thousands of years old, so much heat would have been produced that the oceans would have been vaporized! In the end, the types of radioactive isotopes you find on earth are best understood if the earth is billions of years old.

On the other hand, there are other things that are best understood if the earth is only thousands of years old. Consider, for example, the earth's magnetic field. You already learned that there are electrical currents in the outer core that produce the planet's magnetic field. However, one thing I have not told you up to this point is that we have been measuring the strength of the field for almost 200 years now, and over time, we have seen it decrease, as shown in the graph on the right. Don't worry about what "amp·m^2" means. Just realize that it is a way of measuring the strength of a magnetic field. The larger the number, the stronger the field. Notice that the more recent the measurement, the smaller the value. This means that the earth's magnetic field is getting weaker over time.

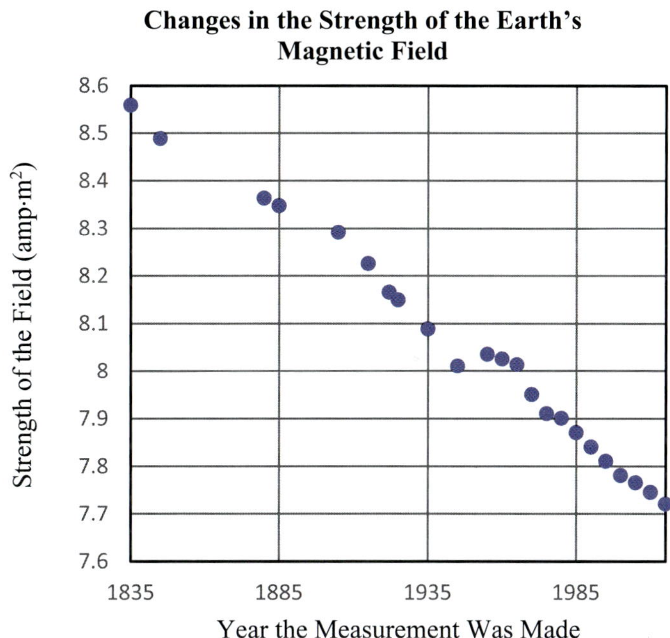

Why? Well, that depends on how you think the electrical currents in the core are being produced. Uniformitarians think that the currents are being produced and maintained by the earth's rotation (which turns night into day) as well as convection currents in the outer core. YECs refer to 2 Peter 3:5, which says, "For when they maintain this, it escapes their notice that by the word of God the heavens existed long ago and the earth was formed out of water and by water." Thus, they suggest that initially, the earth was a big ball of water. Then, God transformed the planet from a ball of water into the structure we see today.

While the details are a bit too advanced for this course, if that process happened in a specific way, physics tells us it would result in a current of electricity flowing in the outer core. However, since that current was started by a process which happened only once at the very beginning, there is nothing keeping the flow of electricity going. Thus, it would slowly decrease, which explains why the earth's magnetic field has been decreasing in strength over the time we have been measuring it.

YECs go much further than that, however. Since we know the factors that affect the flow of electricity, and since we know the basic makeup of the outer core, YECs can estimate how the flow of electricity has decayed over time. They can also estimate how big the flow of electricity was at the very beginning. As a result, using the values that have been measured, they can determine how long the field has been decaying and thus how long ago it was created. According to that analysis, the earth's magnetic field was created 6,020 years ago. Since the earth's magnetic field had to come into existence along with the earth as we know it today, that indicates the earth is 6,020 years old.

Now, of course, this is just a hypothesis, because after all, we don't know that the earth was created the way YECs interpret 2 Peter 3:5. However, the theory can be tested, because we have measured the magnetic fields of other planets. If God created the earth out of water, He probably created the other planets that way as well, so we can use the same hypothesis to determine the magnetic fields of the other planets. The YEC who came up with this hypothesis, Dr. Russell Humphreys, did this back in 1984. At that time, the magnetic fields of all the planets in the solar system except Uranus and Neptune had been measured. The magnetic field of Pluto had also not been measured, but it is no longer considered a planet.

In his 1984 publication, Dr. Humphreys showed that his hypothesis matched the data for all the planets whose magnetic fields were known. However, as a scientist, he knew that to demonstrate the accuracy of a scientific theory, it's not enough to explain what is already known. It is important to predict something that is currently unknown and see whether or not that prediction can be confirmed at a later date. As a result, he predicted what the magnetic fields of Uranus and Neptune should be. In 1986 and 1989, they were measured, and they were both consistent with Dr. Humphreys's prediction. This provides a lot of independent support of the hypothesis, so it is now a scientific theory.

But that's not the end of the story. It was known at the time that Mars did not have a measurable magnetic field, and the theory indicated that as well. However, according to the theory, Mars *had* a magnetic field when it was created. The magnetic field has just decayed away. As a result, Humphreys predicted that at some point, evidence will be produced to indicate that Mars had a magnetic field sometime in the past. In the 1990s, this prediction was also confirmed by analyzing meteorites that originated on Mars.

Mercury provides supporting evidence for the YEC theory of planetary magnetic fields and evidence against the uniformitarian hypothesis.

Probably the most stunning prediction made by the YEC theory has to do with the planet Mercury. When Dr. Humphreys published his theory in 1984, the only measurement of Mercury's magnetic field had been made in 1975 by the robotic spacecraft named "Mariner 10." However, another robotic spacecraft visited the planet in 2010 and measured its magnetic field again. Four years before that second measurement was made, Dr. Humphreys used his theory to predict that Mercury's magnetic field should have weakened by 4-6%. When the second measurement was made, the decrease was measured to be about 8%, which is more than Humphreys predicted. However, it still was a decrease. These confirmed predictions provide evidence for the theory, which in turn provide evidence for the earth being about 6,000 years old.

Of course, uniformitarians have their own theory for how the earth's magnetic field (and those of the other planets) came into being. However, it has not been confirmed by the data. For example, when Mariner 10 went to Mercury, almost all scientists thought it would not measure any magnetic field. Remember, the electrical currents are supposed to be caused by a planet's rotation. Mercury's rotation is too slow to produce such currents, so uniformitarians thought that it would have no magnetic field. However, Mariner measured a very strong magnetic field. In the same way, according to uniformitarians, a planet's magnetic field changes so slowly that there is no way the strength of Mercury's magnetic field would have changed over the 35 years between Mariner 10's measurement and the new one. Once again, the uniformitarian view was contradicted by the data. The uniformitarian view also says that as long as a planet rotates fast enough, it should have electrical currents in its core and therefore a magnetic field. However, Mars rotates quickly enough to produce currents, but it has no magnetic field.

In the end, then, the only theory of planetary magnetic fields that is consistent with our observations is the YEC theory, and it measures the age of the earth to be 6,020 years old. Now, of course, there could be an even better theory out there that hasn't been developed yet, and that theory might end up being consistent with the earth being billions of years old. However, based on science as we know it today, planetary magnetic fields indicate that all the planets in the solar system are only thousands of years old, not billions.

Science Cannot Say Which Is Better

I hope that by now, you have learned that there is no clear scientific answer regarding whether uniformitarianism is better than the YEC view. Some evidence favors the uniformitarian view, while other evidence favors the YEC view. In the same way, some evidence (like radioactivity) indicates that the earth is billions of years old, while other evidence (like planetary magnetic fields) indicates it is thousands of years old. As a scientist, the best you can do is look at all the data and decide what the majority of the evidence favors. That's how I have decided what to believe when it comes to the question of uniformitarianism versus the YEC view.

Before I end this chapter, however, I need to make you aware of one *very* important fact. As a knowledgeable scientist, I could have focused on just the evidence favoring uniformitarianism and against the YEC view. As a result, I could have *easily* convinced you that science clearly demonstrates that the earth is billions of years old and the uniformitarian view of the geosphere is correct. I could also have focused on the evidence for catastrophism and against uniformitarianism, and I could have *easily* convinced you that science clearly demonstrates the earth is only thousands of years old, and the YEC view of the geosphere is correct.

Unfortunately, this is something every student experiences. Because your teachers and the authors of your textbooks know *a lot* more than you do, they can easily convince you of pretty much anything they want. Being a good scientist requires that you respectfully read and listen, but also investigate the issue for yourself. This is probably the most important thing to learn from this chapter. In fact, it is probably the most important thing to learn from all your education:

Regardless of how convincing teachers or textbooks are, do not form an opinion until you have looked at all sides of an issue. Otherwise, you might end up being fooled.

Comprehension Check

9.17 Suppose someone discovers a radioactive isotope in nature that has a half-life of 1 million years. What would uniformitarians most likely do? What about YECs?

9.18 We don't know whether or not Pluto (which rotates very slowly) has a magnetic field. The YEC theory says that its magnetic field should have decayed long ago. Once Pluto's magnetic field is measured, can the result be used to indicate the superiority of one theory over the other?

Answers to the Comprehension Check Questions

9.1 Carbon dioxide doesn't dissolve as well in hot liquids. Remember, carbon dioxide can dissolve in liquids. Thus, some of the carbon dioxide made in the reaction would dissolve in the vinegar. However, as Experiment 8.3 showed you, not as much carbon dioxide could dissolve in a hot liquid. As a result, more would have to bubble out.

9.2 3 MY. Since the line of Hawaiian Islands was caused by plate movement over a mantle hotspot, and since Oahu is in between Kauai and Hawaii, it had to form after Kauai had moved off the hotpot and before Hawaii was formed. 3 MY is the only date in between 5.1 MY and 0.4 MY.

9.3 No, a uniformitarian would not. After all, according to uniformitarians, Quaternary rock started forming about 2 million years ago. An intrusion would have to form after that. As a point of interest, the 32 MY date pointed out in Hawaii on page 249 is in Quaternary rock.

9.4 Yes, a uniformitarian would believe it, since he or she thinks Quaternary rock is younger than 2 million years old. A YEC could say it is within the age range that carbon-14 dating gives for dinosaur bones (22,000 – 38,000 years). Since uniformitarians believe that dinosaur bones are millions of years old, they have to at least be a bit skeptical of any carbon-14 dates that fall into that range.

9.5 It would have told uniformitarians that the meteor which made that crater didn't cause the extinction of the dinosaurs. The meteor would have needed to strike the earth roughly when the dinosaurs went extinct, not 35 million years earlier.

9.6 No. Gymnosperms do not make flowers, so insects that live on nectar from flowers would not live at a time when only gymnosperms existed.

9.7 You would look for pairs that have alternating colors. Varves form in pairs, and each layer of the pair has its own color.

9.8 They would be much larger. When limestone is formed by erosion, the crystals are larger, which is why uniformitarians don't think the White Cliffs of Dover were a result of erosion.

9.9 No. YECs think that the continents sank and oceanic crust rose during the Flood, so the landscape was very different from what it is now.

9.10 Because the sandstone is found over such a large scale, you would need multiple local catastrophes happening all at the same time.

9.11 The second. It is easy for both uniformitarians and YECs to understand fossils from different ecosystems being in different layers. It is hard for a uniformitarian to understand fossil graveyards that contain fossils from different ecosystems all mixed together. However, it's easy to understand that in the catastrophist view.

9.12 The snakes with big heads could eat the cane toads, and that killed them. As a result, snakes with smaller heads are more likely to survive. Natural selection preserves such traits.

9.13 It should have rear legs that are smaller than those of *Rodhocetus*. It lived after *Rodhocetus*, so it should be more whale-like. Whales have no rear legs at all, so smaller rear legs would be more whale-like.

9.14 This is an example of convergent evolution. When similarities cannot be fit into a macroevolutionary sequence, they are interpreted as evolving independently. There are hundreds of examples like this.

9.15 It would reduce the number of new types of animals that would have to evolve in such a short amount of time. Remember, the problem is how diverse everything is in the Cambrian as opposed to the prior era, which is Precambrian. If you can find more diversity in the Precambrian, there is less that macroevolution must do in the time covered by the Cambrian Explosion.

9.16 The wolf, domestic dog, and fox. YECs refer only to microevolution, so the animals must be similar enough to be the result of microevolution. Horses and rats are just too different from wolves, dogs, and foxes to be the result of microevolution.

9.17 Uniformitarians would try to find a process that replenishes it, while YECs would say it supports their view. After all, in the 4.5 billion years that uniformitarians think the earth has been around, an isotope like that would have gone through so many half-lives that it would not be detectable anymore. Thus, there must be something currently making it. However, we would expect such isotopes to exist in an earth that is only thousands of years old.

9.18 No. The YEC view says there should be no magnetic field, and since Pluto rotates slowly, the uniformitarian view says there should not be one. Thus, if we find no magnetic field, it supports both views. If we find a magnetic field, it contradicts both views.

Chapter Review

1. Define the following terms:

a. Calibration
b. Mass extinction
c. Gymnosperms
d. Angiosperms
e. Varve
f. Natural selection
g. Microevolution
h. Macroevolution
i. Convergent evolution
j. Cambrian Explosion
k. Biogeography

2. How do uniformitarians use the potassium/argon dates of rocks in the Hawaiian Islands to support their view? What do YECs say to challenge that?

3. What is odd about the potassium/argon dates of igneous rocks formed from the Mount Saint Helens eruption? What do YECS conclude from this? What do uniformitarians say about it?

4. Lots of people say that carbon-14 dating demonstrates that the earth is billions of years old. What is wrong with that statement?

5. When is carbon-14 dating very accurate and why? How do uniformitarians and YECs address the carbon-14 dating of dinosaur fossils?

6. What is Chicxulub and how does it support the uniformitarian view?

7. How does pollen in the fossil record support the uniformitarian view? How does grass in the fossil record contradict the uniformitarian view?

8. How do the varves in the Green River formation support the uniformitarian view? What do YECs say about them?

9. What kinds of fossils can be found at the tops of many mountains, and how does that support the YEC view?

10. How do layers of rock like the sandstone layer found across most of North America support the YEC view? What about sandstone made from sediment that was transported long distances?

11. How do fossil graveyards support the YEC view? How do uniformitarians address them?

12. How do paraconformities support the YEC view? What about folded sedimentary rocks?

13. Why do evolutionists think the fossil record demonstrates that macroevolution produced whales? What do YECs say about that?

14. Both evolutionists and YECs have an "explosion" that challenges their view. What are these explosions and how do they challenge each view?

15. What problem do animals like kangaroos present to the YEC view?

16. Explain the piece of evidence discussed at the end of this chapter that supports a billions-of-years-old earth. Explain the piece of evidence that supports a thousands-of-years-old earth.

Chapter 10: Water and the Hydrosphere

A Very Important Chemical

We have been spending so much time on the geosphere that you might have forgotten the earth has other "spheres." To understand how amazing this planet is, you need to learn about those spheres as well, so we will spend the next two chapters on the hydrosphere. As you might recall from the Chapter 2, the hydrosphere is defined as all water on the planet, no matter what its phase or location. Ice in the Arctic, water in the ocean, and water vapor in the air are all part of the hydrosphere.

This image shows the earth and the moon as seen from a spacecraft orbiting Mars. Note that the moon's brightness has been enhanced. It is actually quite dim compared to the earth.

In many ways, the hydrosphere is what makes this planet unique in the known universe. In fact, astronomers often call the earth the **blue planet.** You can see why by looking at the image on the right. It was taken by a robotic spacecraft that was in orbit around the planet Mars. On the lower left, you see the earth, and on the upper right, you see the moon. Can you tell where the sun is based on the picture? It is shining on both the earth and the moon from the right side of the image, which is why the right side of the earth and moon are illuminated. Notice that the earth looks like a blue ball with white patches on it. The blue is the result of water in its liquid phase, which is mostly found in the ocean. Since most of the earth's surface (about 71%) is covered with water, it has a distinct blue color when viewed from space. The white comes from clouds, which also contain water.

Water is abundant on the earth because the earth is a haven for life, and based on all the science we currently know, life depends on water. Your body, for example, is about 60% water. You can go without food for 2-3 weeks, but if you cannot drink water, you will die within 3-4 days! Even the most basic organism on the planet depends on water in order to survive. When it comes to living organisms, then, the hydrosphere is probably the most important part of our planet.

A Molecule with Charge

Most people know that the chemical formula for water is H_2O, which means there are two atoms of hydrogen and one atom of oxygen in a water molecule. They are arranged as shown in the illustration on the right. The "rods" that connect the hydrogen atoms (white balls) to the oxygen atom (red ball) represent chemical bonds, and they are made up of electrons that the atoms share between them. However, the oxygen is an "electron hog," and it takes more than its fair share of those electrons. As a result, it has a bit of excess when it comes to negative charges, so the atom has a slight negative charge. Of course, that means the hydrogens get less than their fair share of electrons, so they have a bit of an electron deficit. That gives them a slight positive charge. The molecule, then, has small charges in it.

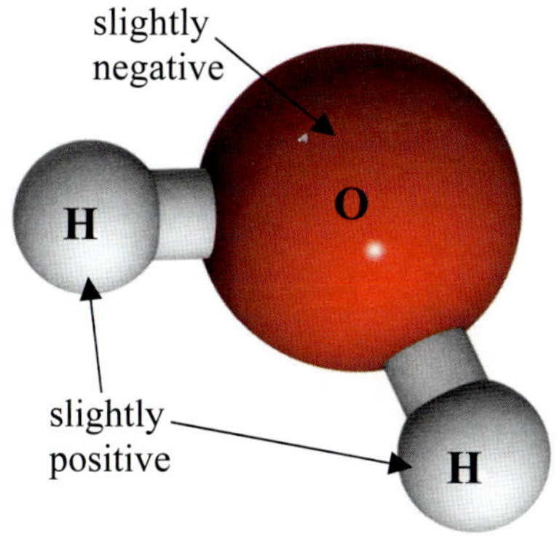

This is what scientists think a water molecule looks like.

We call molecules like water **polar molecules**, because the term "poles" refers to opposites, and positive charges are the opposite of negative charges.

Polar molecule – A molecule with an uneven distribution of electrons, which results in slight electrical charges within the molecule

Because polar molecules have electrical charges on them, they attract one another (and other polar molecules) strongly. After all, if the H's are slightly positive and the O is slightly negative, the molecules can arrange themselves so the H's on one molecule are close to the O of another molecule, as shown in the image below. Since opposite charges attract, the molecules attract one another.

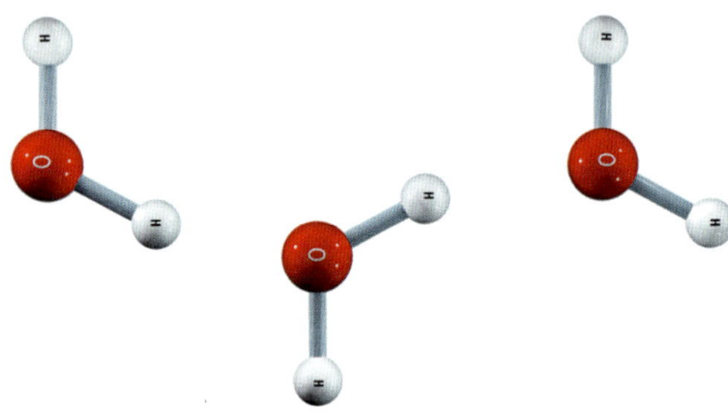

Water molecules can orient themselves so that their slightly negative oxygen atom is close to another molecule's slightly positive hydrogen atom.

Many molecules are not polar. Their electrons are evenly distributed among the atoms, so no atom gets more or less than its fair share of electrons. As a result, there are no slight electrical charges within the molecules themselves. Not surprisingly, they are called **nonpolar molecules**. They are attracted to one another, but the attraction is not nearly as strong as what exists between polar molecules. You can see this for yourself by comparing water (which is made of polar molecules) to oil (which is made of nonpolar molecules).

Experiment 10.1: Floating Paper Clips

Supplies:
- Four standard-size metal paper clips with no plastic covering (You may want a couple more in case things don't go well at first.)
- Two bowls, like soup bowls
- A couple of squares of toilet paper
- A fork
- Water
- Vegetable oil (Any liquid cooking oil will work.)

Instructions:
1. Fill one bowl halfway with water.
2. Drop a paper clip in the water. Observe what happens.
3. Tear a section of toilet paper so that it is larger than the paper clip, but not a lot larger. However, it must be large enough for you to hold it in your hands while the paper clip rests on top of it.
4. With the paper clip resting on top of the paper, lower it onto the water so that the toilet paper and the paper clip both float on the surface. The toilet paper should act like a "raft," keeping the paper clip afloat.
5. Wait about a minute for the paper to get soaked with water. Ideally, it will sink when it becomes saturated.
6. If the paper doesn't sink after a full minute, use the fork to gently push down on a corner of the paper. Don't touch the paper clip. You want to gently "peel" the paper away from the paper clip so that the paper sinks.

7. Observe what happens to the paper clip when the paper sinks. If the paper clip sinks with the paper, try again with another paper clip and another piece of toilet paper. Eventually, you should be able to see the paper sink while the paper clip stays afloat.
8. Use your finger to start making waves on the surface of the water. Start making gentle waves, and then continue to increase the size of the waves until the paper clip sinks.
9. I want you to repeat steps 1-4 with the other bowl, but this time, use vegetable oil instead of water. If you are worried about wasting vegetable oil, don't fill the bowl halfway. Just fill it enough so you can float the toilet paper and paper clip on the surface of the oil.
10. Once the toilet paper and paper clip are floating on the vegetable oil, watch it for a while. Unlike the water, you should not need to use the fork to get the paper to sink. It should sink on its own. What happens to the paper clip when the paper sinks?
11. Clean up your mess. You will need a lot of dish soap to clean the oil out of the bowl. You will probably need to wash your hands with a lot of soap as well, since they probably have oil on them.

When you dropped the paper clip into the water, it sank. You should have expected that. The density of a paper clip is higher than the density of water, and as you already know, if the density of an object is greater than the density of water, it sinks. However, the paper clip floated on the water after the toilet paper on which it was floating sank. Why? Because of the fact that the water molecules are strongly attracted to one another.

In order for the paper clip to sink, it must push the water molecules on the surface away from each other so it can move through them. When you dropped the paper clip in, its motion disrupted the water molecules on the surface, making it easy for the paper clip to push them away from each other. When it was just sitting on the surface of the water after the paper sank, it didn't have any motion, so the water molecules weren't disrupted. As a result, it had to move undisrupted molecules away from each other, and it couldn't. When you disrupted the water molecules with your fingers, however, it once again could. But did the paper clip float on the oil? It shouldn't have. That's because oil molecules (like all nonpolar molecules) are not as strongly attracted to one another. As a result, it was easier for the paper clip to move them away from each other, even when they weren't disrupted.

The attraction between molecules at the surface of the liquid produces what scientists call **surface tension**, which resists the surface of the liquid being broken. The stronger the attraction between the molecules, the stronger the surface tension. Water's surface tension is strong enough to keep a paper clip floating on the surface, even though it should sink. However, since oil molecules are not as strongly attracted to one another, its surface tension is much lower.

The surface tension of water is important for life on earth. An insect like the water strider uses it to walk on the surface of water, even though it should sink. Lots of plant material, etc., ends up floating on water even though it should sink, because of surface tension. This provides an easy source of food for fish. Also, the way tall plants transport water takes advantage of the strong attraction between water molecules. Clearly, the polar nature of water is an important part of God's design!

Comprehension Check

10.1 Most minerals are ionic compounds, which means they are made of positively- and negatively-charged particles called ions. Would you expect minerals to be more attracted to water molecules or oil molecules?

NOTE: The experiment in the next section requires you to wait for 2 hours, so plan accordingly.

An Exceptional Molecule

As you learned in the previous section, water's polarity allows it to have strong surface tension. It also gives rise to another important property that is necessary for life to flourish on earth. You probably know about this property already, but I want you to do the following experiment so you can understand it even better.

Experiment 10.2: Density Change

Supplies:

- The graduated cylinder from the laboratory kit made for this course
- The mass scale from the laboratory kit made for this course
- A medicine dropper from the laboratory kit made for this course
- A freezer with enough space for the graduated cylinder to stand upright inside
- A stick of butter or margarine
- A means by which to melt the butter (a microwave and an appropriate container or a stove and a pan)
- A small glass, like a juice glass
- Paper towels
- A serrated knife, like a steak knife

Instructions:

1. Set the scale on a flat surface and hit the "on/off" button to turn it on.
2. Look at the display. In the upper, right-hand corner, the unit is displayed. If it doesn't show a "g" for grams, hit the "mode" button until it does.
3. Hit the "tare" button, which resets it to zero.
4. Measure the mass of the empty graduated cylinder.
5. Fill the graduated cylinder with water until the surface of the water (or the bottom of the curve on the surface) is even with the 40-mL mark. The easiest way to do this is to add water to the graduated cylinder until it is close to but under the 40-mL mark. Then, add some water to the glass, pull some water into the medicine dropper, and use the dropper to add water drop-by-drop until the surface of the water lines up with the desired mark. When making sure the water is at the right level, remember to look at it so the 40-mL mark is at the level of your eyes.
6. Use a paper towel to dry the outside of the graduated cylinder if it is wet.
7. If the mass scale turned off, turn it back on and make sure it is reading in grams.
8. Hit the "tare" button, which resets it to zero.
9. Measure the mass of the graduated cylinder that now has 40 mL of water in it.
10. Put the graduated cylinder in the freezer and note what time it is. Make sure the graduated cylinder stands in the freezer upright; it cannot be tilted.
11. Calculate the density of the water. Do that by subtracting the mass of the empty graduated cylinder (measured in step 4) from the mass of the graduated cylinder with the 40 mL of water (measured in step 9). That will give you the mass of just the 40 mL of water, which should be close to 40 grams. Divide by 40 mL to get the density in g/mL. Remember, there are sample calculations right after the answers to the "Comprehension Check" questions near the end of the chapter.
12. Use the serrated knife to cut a square of butter roughly equal to a tablespoon's worth from the end of the stick of butter.
13. Remove the paper wrapping (if it exists) from the larger part of the stick of butter and melt it, either in a pan on the stove or in a microwave with an appropriate container.
14. If you put water in the small glass, empty it and dry it with a paper towel.

15. Once the larger part of the stick of butter is completely melted, pour it into the juice glass.
16. Take the paper wrapping (if it exists) off the small square of butter.
17. In the next step, you are going to drop the small square into the melted butter. Do you think it will sink or float?
18. Drop the small square of butter into the melted butter and see if you were right.
19. Empty the melted butter and the small square into the sink, unless your parents want to use it for something. Clean out the glass with a lot of soap and water, and wash your hands with soap and water as well.
20. Check the time again and wait until the graduated cylinder has been sitting in the freezer for two hours. Do other schoolwork or chores so you are using your time efficiently.
21. After it has been in the freezer for two hours, remove the graduated cylinder and wipe the outside with a paper towel.
22. Look at the top of the ice in the graduated cylinder so that it is at the level of your eyes. Read the mark it is at now, which is the volume of the ice. It should be larger than 40 mL.
23. Turn the mass scale back on and make sure it is reading in grams.
24. Hit the "tare" button, which resets it to zero.
25. Measure the mass of the graduated cylinder that now has ice in it.
26. Calculate the mass of ice by subtracting the mass of the empty graduated cylinder (measured in step 4) from the mass you just measured.
27. Calculate the density of ice by dividing that mass by the volume you read in step 22.
28. Clean up the rest of your mess.

How did the volume of the ice compare to the volume of the water? It should have been larger. You have already been told that water expands when it freezes. That's why the ice had a larger volume. As the water froze, it had to expand, taking up more space. However, the mass shouldn't have changed. It might have changed a bit because the graduated cylinder might have collected some frost that didn't get wiped off, or some water might have evaporated before it froze. However, it should have been very close. Since the mass stayed pretty much the same but the volume increased, what happened to the density? It went down. The actual density of water at room temperature is close to 1.0 g/mL, while ice's density is close to 0.9 g/mL.

Why does this happen? Remember that because of water's polarity, a hydrogen atom on one molecule (which has a slight positive charge) gets very close to the oxygen atom on another molecule (which has a slight negative charge). This actually produces a weak bond, which is called a **hydrogen bond**. Now remember, a chemical bond exists between two atoms *within* a molecule. A hydrogen bond exists between two *different molecules*, so this is not a normal chemical bond. However, it is a strong attraction that causes the water molecules to stay very close to one another. The drawing on the right illustrates this point.

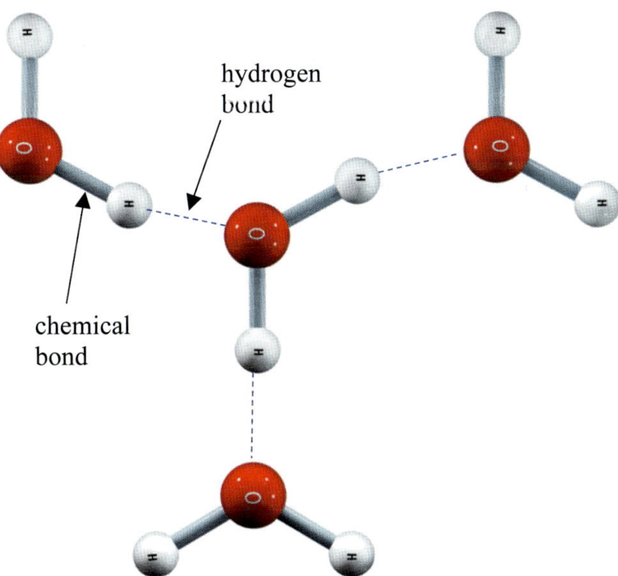

Unlike chemical bonds, hydrogen bonds exist between molecules, pulling them close to one another.

When they are in their liquid phase, molecules can move around, so it is easy to form hydrogen bonds. As a result, the water molecules stay pretty close together. However, to form a solid, molecules must stay in a set pattern, and the pattern required for ice is pretty complicated. The important thing is that the molecules must be a certain distance apart in order to form that pattern properly. That distance is actually *greater* than the distance that exists when the hydrogen bonds are allowed to pull on the molecules with all their strength. Thus, the molecules actually have to move away from one another to form the pattern. This increases the volume, reducing the density.

What does this mean? Since the density of ice is lower than that of water, ice floats on water. You probably already know that, because when you add ice cubes to a drink, they don't sink; they float. Did you also know that this is an exception to the general rule? Usually, a substance shrinks when it freezes. Think about the butter or margarine you used in the experiment. Once you melted the larger part of the stick, you had liquid butter. What happened when you put the square of butter (which was still solid) into the liquid butter? It sank. What does that tell you about the density of solid butter as compared to liquid butter? It's higher. That's because butter, like most substances on the earth, shrinks when it freezes. As a result, its mass is compacted into a smaller volume, so its density increases.

There are a few other chemicals that expand when they freeze, but they are rare. The vast majority of substances on the planet shrink when they freeze. The fact that water is an exception to the general rule is an important part of God's design for nature. Why? What happens to a body of water when it gets really cold? The body of water starts to freeze. However, because water expands when it freezes, the ice that is formed stays on the surface of the body of water. As more freezing occurs, the ice thickens, but it stays on top of the water.

It turns out that when ice reaches a certain thickness, it insulates the water below it from the cold air above it. As a result, no more water freezes. Thus, as long as a body of water is deep enough, it will never completely freeze. This, of course, is good news for the living organisms in the body of water, because that means they won't be frozen solid during the winter! In other words, organisms in lakes and other bodies of water that freeze can live through the winter specifically because water is an exception to the general rule!

Comprehension Check

10.2 A standard stick of butter has a volume of 8 tablespoons, which is 120 mL. Suppose you melted an entire stick of butter and measured its volume. Which of the following volumes could it be: 115 mL, 120 mL, or 125 mL.

A High-Capacity Molecule

Water has another property that makes the earth a haven for life. The best way to explore this property is to start with an experiment.

Experiment 10.3: Heating Things Up

Supplies:

- A rock from the laboratory kit made for this course (It is in a small plastic bag labeled "Mass Activity Rock.")
- The graduated cylinder from the laboratory kit made for this course
- Two small cups or mugs for hot drinks (Styrofoam cups, for example, or coffee mugs)

- A pan for boiling water
- A stove
- Kitchen tongs
- A spoon
- Water
- Ice

Instructions:
1. Use the mass scale like you did in the previous experiment to measure the mass of the "Mass Activity Rock" from your kit in grams.
2. Put the rock in the pan and add water to the pan until the rock is completely covered.
3. Put the pan on the stove and turn on the burner to boil the water. Keep checking the water while you do the next steps.
4. Use the graduated cylinder to measure out 30 mL of cold water from the tap. It doesn't have to be exact. Just get close to 30 mL.
5. Pour the 30 mL of water into one cup.
6. Repeat steps 4 and 5 so that both cups have about the same amount of water in them.
7. Add an ice cube to each cup.
8. Swirl the water in both cups for a full minute, making sure that all the water in the cup comes into contact with the ice cube a lot.
9. Put the graduated cylinder in the sink so that it is standing upright.
10. Once the water in the pan has been boiling vigorously for at least two minutes, turn off the burner.
11. Use the spoon to remove from the cups what remains of the ice cubes. Now your cups have ice-cold water in them but no ice cubes.
12. Use the tongs to remove the rock from the hot water and put it into one of the two cups.
13. Take the pan over to the sink and pour the hot water into the graduated cylinder until the water is above the 50 mL mark.
14. Empty the graduated cylinder until the number of mL in the cylinder is equal to the mass you measured in step 1. If the mass of the rock is 35 g, for example, there needs to be 35 mL of water in the cylinder. Once again, it doesn't have to be exact; it just needs to be close.
15. Pour the hot water in the graduated cylinder into the cup that doesn't have the rock in it.
16. Swirl both cups for 30 seconds.
17. Put both cups on a level surface and stick your index finger into the water that is inside each of them. Do you feel a noticeable difference in the temperature? If so, which is warmer?
18. Clean up your mess.

What happened in your experiment? If things went well, you should have noticed that the cup that did not have the rock in it had warmer water than the cup that did have the rock in it. Remember that water has a density of about 1.0 g/mL. Thus, the very hot water you measured out had the same mass as the rock. After all, if the rock was 35 g, you added 35 mL of water. Since water's density is about 1 g/mL, that means you added 35 g of water. Also, the rock and the water that you boiled were in contact throughout the boiling process. That means the rock and the very hot water were at the same temperature. Your experiment showed that an equal mass of water at the same temperature warmed up the ice-cold water better than the rock.

Why does hot water heat things up better than a hot rock? Well, how do you heat things up? You add energy to them, usually in the form of heat. You boiled the water by putting it on a hot burner. That hot burner produced a lot of heat, and a lot of that heat got added to the water. Since the rock was in contact with the water, some of the heat went into the rock as well. The more heat that went into the rock and water, the hotter they got. When they both got hot enough, the water started boiling.

When you put the hot rock and hot water into the ice-cold water, what happened? Heat left the hot rock and hot water and went into the ice-cold water. Gaining that heat warmed up the ice-cold water, but losing that heat cooled down the hot rock and hot water. Eventually, the hot rock and hot water cooled down to the same temperature that the ice-cold water warmed up to. At that point, **equilibrium** (ee' kwah lih' bree uhm) was reached, and the temperatures stopped changing.

Equilibrium – A state of balance between opposing actions

In this case, you had some parts of the cups' contents increasing in temperature (the cold water), and some parts decreasing in temperature (the hot rock and hot water). Those are opposing actions. When the temperature of both parts became the same, there was a state of balance, so we say equilibrium had been reached.

But why was the temperature at equilibrium lower in the cup with the hot rock? Because the rock contained *less energy*. That means it couldn't give as much heat to the ice-cold water. As a result, the ice-cold water couldn't increase as much in temperature. But wait a minute. The rock and the hot water were at the same temperature. How could they not have the same energy? Remember from Chapter 1: *temperature is not a measure of heat*. It is a measure of how the atoms or molecules in a substance are moving, and different substances have different molecules/atoms. As a result, a given substance needs a specific amount of heat to raise its temperature, and another substance needs a completely different amount of heat. This means that each substance has its own **heat capacity**.

Heat capacity – The amount of heat required to increase the temperature of an object by 1 degree

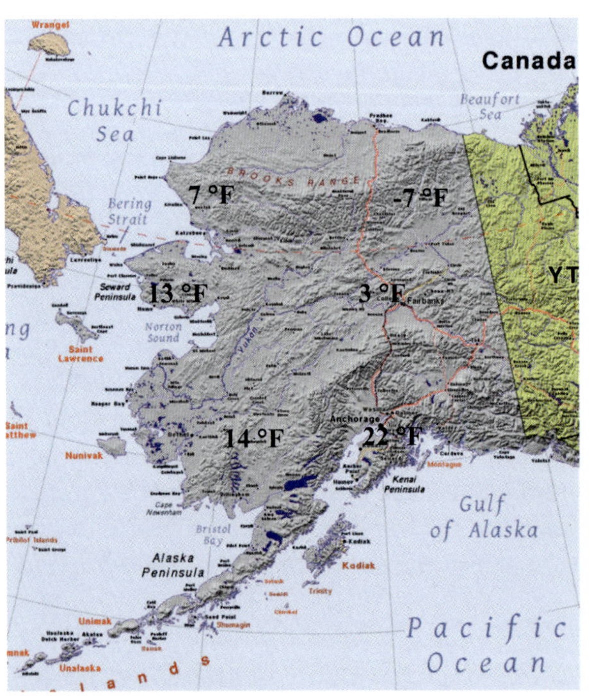

This map of Alaska shows that in the winter, it's warmer near the water.

The rock has a lower heat capacity than water. As a result, to get to the same temperature, the rock required less heat. When it went into the ice-cold water, that meant it had less heat to give, so it couldn't raise the temperature of the ice-cold water as much.

It turns out that water has one of the largest heat capacities of any natural substance. If the rock you used was granite, for example, the heat capacity of water is *seven times larger* than the rock's heat capacity. That means the water had to absorb *seven times as much energy* to heat up! Once again, this is an important part of God's design for nature. Consider, for example, the map of Alaska shown on the left. On that map, you can see the average high temperature in January at several different locations. Notice that the average high temperature in the far north is 7 °F on the coast but -7 °F away from the coast. Those two parts of Alaska are at the same latitude, so they are equally close to the North Pole. Why is it so much warmer on the coast?

Because of water's large heat capacity. The ocean warms up in the summer, and to do that, it must absorb *a lot* of heat. In the winter, it gives up that heat, warming the surroundings. When you are far from the ocean, you don't get that extra heat. Notice that's the same in the middle of the state and in the South – the closer you are to the ocean, the warmer it is in January, on average.

Water, then, is a great moderator of temperature. Since there is so much water on the earth, its average temperature is more constant than it otherwise would be. If most of the earth were covered by land, the earth's summer temperatures would be higher and its winter temperatures would be lower. In addition, remember what happens to gases that are dissolved in liquids. They leave the liquid as its temperature increases. Well, many of the organisms that live in water require oxygen to survive. That oxygen is dissolved in the water. If water's temperature increased significantly in the summer, the amount of oxygen dissolved in it would decrease significantly, making it hard for those organisms to survive. Water's heat capacity is ideal for making the earth a haven for life!

Comprehension Check

10.3 Suppose the illustration on the previous page showed you the average high temperatures of those same places in July. How would the two numbers compare for each latitude?

NOTE: The experiment in the next section requires you to wait for 2 hours, so plan accordingly.

The Opposite of Fresh is…Salt?

Before I discuss the details of the hydrosphere, I want to discuss the two "types" of water found there: saltwater and freshwater. Oceans contain saltwater and are called "marine" environments. Lakes, rivers, and streams mostly contain freshwater and are called "freshwater" environments. If you have ever had the chance to swim in both, you know one big difference: the taste. Saltwater tastes very different from freshwater. There are more important differences, however, which are best explored with an experiment.

Experiment 10.4: Freshwater versus Saltwater

Supplies:

- Water
- Salt
- The mass scale from the laboratory kit made for this course
- The graduated cylinder from the laboratory kit made for this course
- Two small slices of raw potato
- Two small containers that can go in the freezer (Ideally, this would be two rectangles in an ice cube tray. Otherwise, two pill bottles or something of that size.)
- A freezer
- Two small glasses, like juice glasses
- A spoon for stirring
- A knife that will cut through the raw potato

Instructions:

1. Cut the potato so that you have two "discs" that are about one-eighth of an inch thick. Don't measure; just estimate. It there is skin on the edge of the discs, that's fine.

2. Use the paper towel to dry both disks as well as you can. Leave them sitting on the paper towel.
3. Fill each glass about halfway with water.
4. Add a spoonful of salt to one of the glasses and stir a lot. Try to dissolve all the salt in the water. If you can't, that's fine. If you do end up dissolving all the salt, add half a spoonful and try again. Continue to do that until there is at least some salt that cannot be dissolved.
5. Use the mass scale as you did in the previous experiment to measure the mass of the empty graduated cylinder in grams.
6. Using the glass that has only water in it, pour water into the graduated cylinder until the water is between the 40 mL and 50 mL mark.
7. Read the volume. Remember, each little mark is 1 mL.
8. Use the mass scale to measure the mass of the graduated cylinder and the water.
9. Pour about half of the water in the graduated cylinder into one of the small containers. If you can't fit even half of the water in the container, that's fine.
10. Empty the rest of the water in the graduated cylinder into the sink and try to shake out any water that remains inside.
11. Repeat steps 5-10, but this time, use the saltwater and the other small container.
12. Put both small containers in the freezer.
13. Use the mass scale to measure the mass of one potato disc. In your notebook, mark that mass as the mass of the "saltwater" disc, and put it in the glass that has saltwater in it.
14. Use the mass scale to measure the mass of the other potato disc. In your notebook, mark that mass as the mass of the "freshwater" disc, and put it in the glass that has only water in it.
15. Wait for two hours. Do chores or other schoolwork to be as efficient as possible.
16. After two hours, pull the potato slice out of the freshwater and use a paper towel to dry it as much as possible. As you dry it, note whether it is firm or floppy.
17. Use the mass scale to measure its mass. How does it compare to the mass you measured before putting it into the water?
18. Repeat steps 16 and 17 for the potato slice that was in saltwater.
19. Pull the small containers out of the freezer and place them on a flat surface.
20. Push your index finger down on the surface of the frozen freshwater. How solid is it?
21. Do the same thing with the saltwater. Do you notice a difference? You should.
22. Clean up your mess.

Before I discuss the experiment, I want you to calculate the density of the freshwater and saltwater from your data. Do this the same way you calculated density before. Subtract the mass of the empty graduated cylinder from the mass of the graduated cylinder with the liquid in it. That gives you just the mass of the liquid. Then, divide by the volume that you read. That will give you the density in g/mL. Remember, there are sample calculations after the answers to the "Comprehension Check" questions if you need to see how it is done.

How do the densities compare? You should see that saltwater is denser than freshwater. While freshwater has a density of about 1.0 g/mL, saltwater can have a density as high as 1.2 g/mL at room temperature. If you think about it, that should make sense. After all, you added a lot of salt to the water, but its volume didn't increase much. If the mass increases but the volume doesn't change much, the density will be greater.

So the first thing we see is that the water in the oceans is denser than the water in rivers, lakes, and streams. Why does that matter? Well, remember that whether or not something floats depends on its density compared to the density of the liquid it is floating in. What happens, for example, when you jump into a swimming pool? You sink most of the way, right? That's because your density is very close to that of water. Technically, most people float in a pool, but they usually have to concentrate on how they breathe, how they are moving in the water, etc. Even then, most of a person's body is underwater in a pool. The same goes for most lakes or rivers.

But look at the photograph on the right. These three people are clearly floating, and they are not really concentrating on it. Notice how much of their bodies are actually above water. Why? Because they are floating in saltwater that has a density of about 1.1 g/mL. That's about 10% more than the density of freshwater, so people float rather easily in it. That means something which sinks in a lake or stream may very well float on the oceans, since the oceans contain saltwater, which has a higher density than freshwater.

These three people are floating in saltwater.

Moving on to something else you saw in your experiment, you should have noticed that the frozen freshwater was pretty firm. You probably couldn't say the same thing for the saltwater. It was partially frozen, but most likely, your finger broke through the frozen surface, and you probably found that some of the sample was still liquid. Thus, the saltwater did not freeze nearly as easily as the freshwater, despite the fact that they were in the same freezer for the same amount of time.

Why? Well, remember that in order to freeze, the molecules that make up a substance must form a set pattern. What happens when you dissolve something like salt into water? The molecules that make up the salt mix together with the molecules that make up the water. However, the pattern formed by the molecules that make up solid salt is very different from the pattern that must be formed by water molecules to make ice. Thus, the dissolved salt gets in the way of the water molecules, which makes it more difficult for the water to freeze. As a result, saltwater freezes at a much lower temperature than freshwater.

It turns out that this is true for all liquids. If you dissolve something in a liquid, the temperature required for it to freeze goes down. This is called **freezing point depression**, and it is the reason that we spread salt on the roads when we are afraid they might get icy. The salt dissolves in the water, and as a result, it has to get even colder in order for the saltwater to freeze. It is also the reason we add antifreeze to a car's radiator. A car needs water to cool parts of the engine, but in winter, the water might freeze. Because water expands when it freezes, that can really damage the engine! Antifreeze is just a liquid that dissolves really well in water. That makes the temperature at which the water freezes decrease so that it stays liquid all the time, even in the winter.

Even though dissolving salt in water makes it harder for the saltwater to freeze, it is not impossible. In fact, spreading salt on the roads typically only protects them to a temperature of about -20 °C (-4 °F). At temperatures below that, even the saltiest water freezes. In the same way, while freshwater lakes freeze when the temperature stays below 0 °C (32 °F), ocean water doesn't freeze until it gets colder. However, in parts of the world, it does still freeze. The picture below, for example, shows ice that has formed on the ocean. It actually doesn't need to be all that cold for water in the ocean to freeze. Most saltwater in the ocean can freeze at temperatures as "high" as -2 °C (28 °F). However, in reality, it must be colder than that, because waves in the oceans are constantly churning the water, which makes it much harder for the water to freeze. Nevertheless, there are places on the earth where ocean water freezes. Now please understand that icebergs are not formed by ocean water freezing. I will discuss them in the next chapter.

The ice in this photo is formed by ocean water freezing.

Your experiment showed another important difference between freshwater and saltwater, and I will discuss that the next time you do science. Nevertheless, as you can see, saltwater and freshwater are very different from one another. As a result, we often think of the marine environment (which involves saltwater) as the opposite of a freshwater environment.

Comprehension Check

10.4 Suppose you added 1 teaspoon of salt to a gallon of water, and you added 5 cups of salt to another gallon of water. If you tried to freeze both gallons, which would freeze at the higher temperature?

Evening Things Out

What happens when you put a few drops of food coloring in a glass of water? You will see the color start to spread through the glass, as shown in the illustration below. If you wait long enough, the food coloring will be evenly distributed throughout the water, despite the fact that you didn't do anything to stir the water and food coloring together. Why? The molecules that make up both the water and the dye are actually moving around in all directions. As they move, they mix together. Over time, this motion "evens out" the distribution of food coloring and water molecules. In other words, the food color and water don't need any help mixing together; they just need time.

When food coloring is dropped into water, it will spread evenly throughout the water, if it is given enough time.

It turns out that this happens when any chemicals are mixed together and their molecules are free to move in any direction. There is a general tendency in creation to even out concentration. When you first put the drops of food coloring in the water, its molecules are highly concentrated in their drops. Gravity pulls those drops down to the bottom of the glass, and their motion (as well as the motion of the water molecules) causes them to spread throughout the glass, evening out their concentration until it is the same throughout the glass. This process is called **diffusion** (dih fyoo' shun).

Diffusion – The movement of molecules from a region of high concentration to one of low concentration

We would therefore say that the food coloring diffused throughout the water.

While situations like the one pictured on the previous page are great examples of diffusion, it is important to realize that diffusion happens throughout creation. Look, for example, at the picture below. It shows a river (which contains freshwater) emptying its contents into the ocean (which contains saltwater). Why is the river brown? Because it is carrying a lot of sediments. Remember, freshwater is the opposite of saltwater, so there isn't much salt dissolved in it. But that doesn't mean there isn't anything else in the water. Indeed, you do not find pure water in nature. As you will learn later, even raindrops have things dissolved into them! The water you get from your tap isn't pure, even if you use a filter. That's because a filter cannot remove all the dissolved material in water.

The sediments being carried by the river are diffusing into the ocean.

Notice what's happening to the sediments as they reach the ocean. They are spreading out, aren't they? You might think that this is because the river has a current, and that current is continuing to push them out to the ocean. That's true, but the motion given to the sediment by the current is changed by the ocean, so that effect alone doesn't explain what you are seeing in the picture. The sediment is diffusing into the ocean. Over time, it will sink to the bottom, but while that is happening, it will also spread out. In addition, all the chemicals that are currently dissolved in the freshwater will diffuse, moving away from the river (where they have a high concentration) and out into the ocean (where they have a lower concentration).

This is actually very important for life in the ocean. A lot of the chemicals carried by the sediment are used by marine organisms. As the marine organisms take them from the ocean, they need to be replenished. Weathering and erosion puts sediments containing those chemicals into rivers, which then carry them to the ocean. So while you might think at first glance that the river in the picture above is polluting the ocean, it is not. In fact, it is replenishing the ocean with chemicals that are absolutely necessary for some of the organisms found there! Thanks to diffusion, those chemicals will spread throughout the ocean, supplying those organisms' needs.

Why am I talking about diffusion now? Because it relates to one of the things you observed in your experiment. In the experiment, you made saltwater by dissolving salt in water. You actually mixed the salt and water together, but if you had just let the salt sit in the water, it would have eventually dissolved on its own through diffusion. Regardless of how it happens, when you dissolve one chemical into another, you have made what chemists call a **solution**. A solution has two parts: a **solute** and a **solvent**. The solute is what you are dissolving, and the solvent is what you are dissolving the solute into. In the case of the saltwater you made, then, water was the solvent, salt was the solute, and they combined to make a solution of saltwater.

Since the salt was free to move throughout the water, diffusion could happen, so the amount of salt dissolved in the water was pretty much even throughout the glass. However, there are situations in which that cannot happen. In those cases, there is a barrier that allows some material through it but not others. We call that a **semipermeable** (sem' my pur' me uh bul) barrier.

Semipermeable – Allowing passage of some substances but not others

There are all sorts of semipermeable barriers in nature, most of which are constructed by organisms.

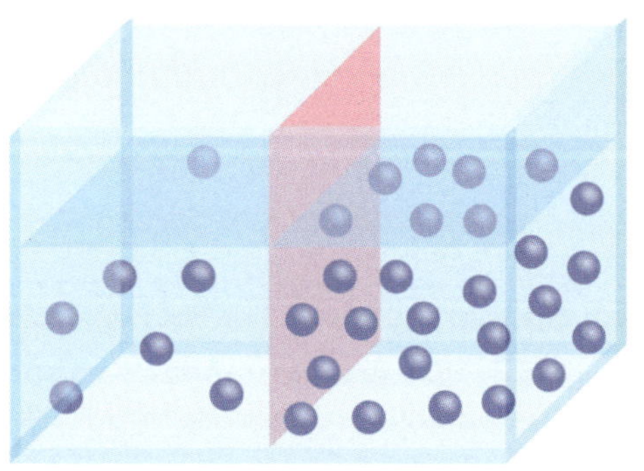

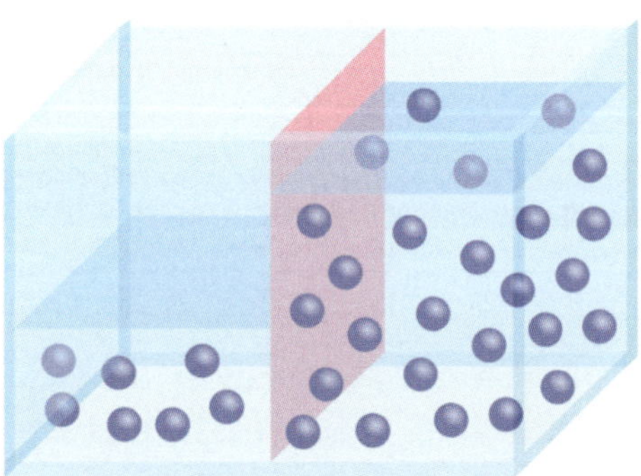

A semipermeable membrane separates a solution of low concentration (top left) from a solution of high concentration (top right). Osmosis will cause the solvent to move into the area of high concentration so that in the end, both solutions have the same concentration (bottom).

Suppose you have a situation like what is shown in the top part of the illustration on the left. In this case, you have two solutions separated by a pink semipermeable barrier. The balls represent the solute molecules, and the blue liquid is the solvent. Notice that on the right side of the barrier, the solute molecules are much more concentrated. The semipermeable barrier will not allow the solute molecules to pass through it, but it will allow the solvent molecules to pass through it. What is going to happen?

Creation tends to even things out so that concentration is the same everywhere. When solute molecules can move, diffusion does that very nicely. In this situation, however, the solute molecules cannot move across the barrier, so they can't do anything to even out the concentration between the solutions. Because of that, the solvent must travel across the barrier to get the job done. But how will the solvent need to travel? It will need to travel *into* the area of high solute concentration. That way, it will dilute the solute on the right side of the barrier. At the same time, since it is leaving the left side of the barrier, the solute molecules will become more concentrated there. In the end, the result is pictured in the bottom part of the illustration on the left. Notice that the concentration of solute molecules is now the same on both sides of the barrier. Thus, the solute concentration got evened out, but not because the solute moved. Instead, it happened because the solvent moved. This process is called **osmosis** (ahz moh' sis).

Osmosis – The movement of solvent molecules across a semipermeable barrier from an area of low solute concentration to one of high solute concentration

This is what you saw in the potato part of your experiment. You put a potato slice into freshwater and one into saltwater. The potato was part of a plant, and so it is made of cells. The cells are surrounded by a semipermeable membrane. Most chemicals (including salt) cannot pass through the membrane, but water can. Each cell, of course, holds a lot of chemicals inside it.

Given that information, what should happen when a slice of potato is put in freshwater? There aren't a lot of chemicals dissolved in freshwater. There are some, but not nearly as much as what is inside a cell. So the concentration of solutes inside the cell is high, and the concentration of solutes outside the cell is low. The solutes can't move to even things out, but the water can. So water travels into the cell to dilute the solutes in the cell and concentrate what little solute is dissolved in the freshwater. That's why the potato slice put in freshwater increased in mass; water traveled into its cells. That increased the mass of each cell, which increased the mass of the potato slice.

Now think about the potato slice in saltwater. You dissolved as much salt as possible into the water, so the concentration of solutes in the water was much greater than the concentration of solutes in the cells. Once again, only the solvent (water) could move, so it had to leave the cells. That way, the solutes inside the cells would become more concentrated, and the solutes in the saltwater would become more dilute. Since water moved out of the cells, they lost mass. That caused the entire potato slice to lose mass. Since plant cells use water to stay firm, the slice also became floppy.

Now remember, whether we are talking about diffusion or osmosis, the result is the same. Creation tends to even out the concentration of solutes. If the solute is free to move, it does so, going from an area of high concentration to an area of low concentration. That's diffusion. However, there are situations in which the solute molecules cannot move but the solvent molecules can. In those situations, the solvent molecules must move to even out the concentration, but they have to move in the opposite way. They have to move out of the area of low solute concentration and into the area of high solute concentration. That's osmosis.

Freshwater organisms (top) and marine organisms (bottom) must deal with osmosis differently.

Like diffusion, osmosis is an important process in creation, especially when it comes to organisms that live in water. Consider, for example, fish that live in freshwater. Their cells have a higher concentration of solutes than the water in which they live. So what does osmosis do? It causes water to move from the lake, river, or stream and into the fish itself. Thus, fish that live in freshwater must constantly remove the water that osmosis is pushing into them. As a result, they urinate quite a bit.

Most fish that live in marine environments have the opposite problem. The saltwater in which they live has a higher concentration of solutes than their cells. As a result, osmosis is constantly pulling water out of their cells and pushing it into the ocean. Thus, they must continually replenish the water they are losing by drinking a lot. They also have mechanisms to get rid of the salt in the water they are drinking. Some marine fish, like sharks, actually increase the concentration of solutes in their blood so that they don't have a problem with osmosis. That way, they don't have to drink as much as other marine fish. However, they must have other mechanisms that help them deal with such high solute concentrations in their blood.

This is why most fish live only in freshwater or saltwater. They can only deal with osmosis going one way, so they live in the water in which osmosis works that way. However, there are organisms that can live in both kinds of water. Atlantic salmon, for example, start their lives in freshwater, but then they eventually swim to the ocean and spend most of their lives there. However, they do come back to the same freshwater stream in which they are born to mate and produce offspring. They have methods for dealing with osmosis in either direction, and they use whatever method matches the water in which they find themselves at any given time.

This is also why people, as well as many animals, cannot drink water from the ocean, at least not in any significant amount. We need water to survive, but drinking ocean water causes the concentration of salt in our blood to rise. If it rises to where the solute concentration is higher than what is found in our cells, osmosis pulls the water from our cells, which can eventually lead to death. If you have ever read the poem "The Rime of the Ancient Mariner" by Samuel Taylor Coleridge, it tells of just that problem. A sailor is on a ship without any freshwater. He is very thirsty, but knows he cannot drink water from the ocean. The poem reads, "Water, water, every where, / And all the boards did shrink; / Water, water, every where, / Nor any drop to drink."

It's a Salty World After All

It should come as no surprise to you that most of the earth's water supply is in the form of saltwater. After all, the oceans are huge, and they contain a lot more water than the rivers, lakes, and streams that contain freshwater. However, you might be shocked at the actual numbers. Freshwater makes up only 2.5% of the earth's hydrosphere! The other 97.5% of the hydrosphere is saltwater! Most of it can be found in the oceans, but there are other sources of saltwater that you will learn about later. Nevertheless, the vast majority of the earth's water is in a form that you and I cannot drink.

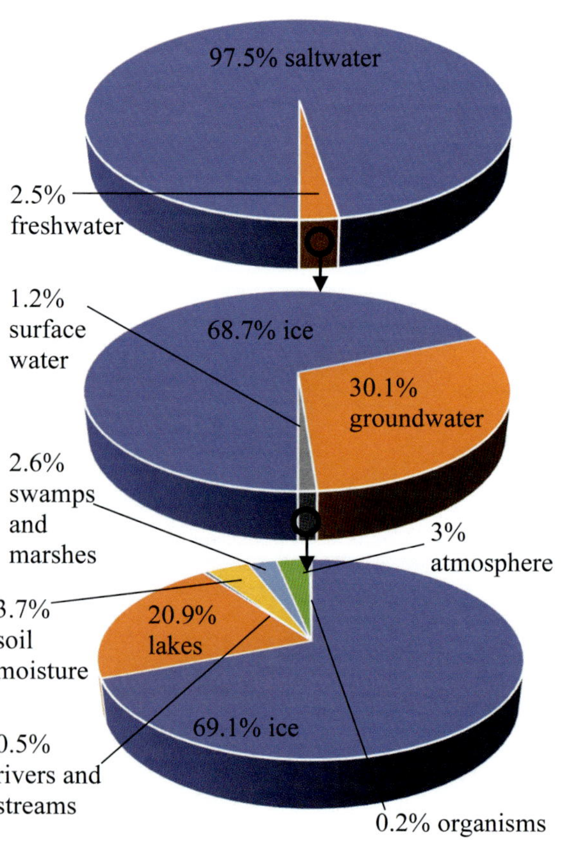

These graphs illustrate the distribution of water on the earth.

Where do you think most of the earth's freshwater is found? It's mostly in lakes, rivers and streams, right? Wrong! In fact, 68.7% of all the earth's freshwater is frozen! Another 30.1% of it is liquid, but it is running underground where you don't see it. Not surprisingly, we call that **groundwater**. Only a mere 1.2% of all the earth's freshwater is found on the surface of the earth! It's called **surface freshwater**.

So 1.2% of the earth's freshwater is found in lakes, rivers, and streams? No! In fact, most of the earth's surface freshwater (69.1%) is frozen in the soil! You find 20.9% of the surface freshwater in lakes and 0.5% of it in rivers and streams. The rest is found as moisture in the soil (3.7%), swamp and marsh water (2.6%), water in the atmosphere (3%) and organisms (0.2%).

This is all illustrated for you in the graph on the previous page. At the top, you see a pie chart showing you that the vast majority of the earth's water is saltwater. The graph below it concentrates on just the freshwater, showing you that most if it is ice, most of the rest is groundwater, and only a tiny portion is the freshwater that you and I come into contact with. Concentrating on just that tiny amount of surface freshwater, the final graph shows that most of it is frozen in the soil. Most of the rest is in the earth's lakes. Soil moisture makes most of the rest, followed by water in the atmosphere, water in swamps and marshes, water in rivers and streams, and finally water in organisms. Throughout the rest of this chapter and the next, I will discuss the details of where you find the earth's water.

Comprehension Check

10.5 Suppose you made a very concentrated solution of saltwater, and you carefully added freshwater to the top of the solution. Would diffusion or osmosis act to even out the salt concentration?

10.6 Suppose you made a very concentrated solution of saltwater and covered it with a thin, semipermeable film. If you then carefully added freshwater on top of the film, would diffusion or osmosis act to even out the salt concentration? Which way would the water travel in the process?

10.7 If a sample of water was randomly taken from somewhere on the earth, would it most likely be freshwater or saltwater?

10.8 If a sample of freshwater were randomly taken from the earth, would it most likely be solid, liquid, or gas?

Why Do Both Exist on Earth?

In the past two sections, you have learned about the differences between saltwater and freshwater. However, you might wonder why the oceans hold the former, while rivers, streams, and most lakes hold the latter. Perform the following experiment to find out.

Experiment 10.5: Purifying Water

Supplies:
- Salt
- Ice
- Water
- A small pot or pan for boiling water
- A pot lid with a handle on top (It doesn't have to fit the pot you are using.)
- A wide bowl that isn't as tall as the pot
- A Ziploc bag
- A stove
- A measuring tablespoon

Instructions:
1. Fill the pot at least halfway with water.
2. Add three tablespoons of salt and stir so that it mostly dissolves. If there is some undissolved salt, don't worry about it.

3. Use the tablespoon to scoop up a small amount of the water, and taste the water. **NOTE: You should never taste anything from an experiment unless you are told to do so by someone who knows a lot about chemistry!** You can taste this because I am telling you to taste it. It's not very pleasant, is it?
4. Put the pot on a burner that has a nice open space next to it where the bowl can go (see the picture below.) Do not turn on the burner yet.
5. Place the bowl next to the pot so it is close to but not touching the pot.
6. Turn on the burner so the water heats up and eventually boils.
7. Put several ice cubes into the Ziploc bag and zip it closed.
8. Once the water starts to boil, use one hand to hold both the handle of the pot lid and the Ziploc bag of ice. The bag should be sitting on top of the pot lid, as shown in the picture on the right.
9. Hold the pot lid so that it is partly over the pot and partly over the bowl.
10. Tilt the pot lid so it is slanted towards the bowl.
11. As time goes on, some of the steam should hit the pot lid, condense back to a liquid, drip down the inside of the pot lid, and then drip into the bowl. As time goes on, then, the bowl should start filling with water.

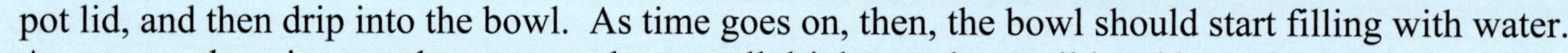

12. As soon as there is enough water to take a small drink, put the pot lid and bag of ice down.
13. Turn off the burner.
14. Lift the bowl to your mouth and drink what is inside the bowl. Once again, you can do this only because I am telling you to drink it!
15. What's the difference between the taste of what is in the pot and what is in the bowl?
16. Clean up your mess.

What was the difference between the water from the pot and the water from the bowl? If things went well, you should have tasted salt in the water from the pot, since it was saltwater. However, you shouldn't have tasted any salt in the water from the bowl. Why? Because the salt didn't evaporate from the pot; only the water did. Thus, when the water vapor hit the cool lid and became liquid again, the water had no salt in it. In other words, the water that evaporated left the salt behind.

The process you followed in the experiment is called **distillation** (dis' tuh lay' shun), and it is a common means by which water is purified.

Distillation – The process by which a solution is boiled and the vapor is then collected and condensed

Remember, all natural sources of water (even rainwater) have chemicals dissolved in them. If you want pure water, you must get rid of those chemicals, and distillation is one way to accomplish that. If you buy distilled water at the store, you are buying water that has been purified in a way similar to how you purified the saltwater in your experiment.

Something similar happens in nature. When water evaporates, it leaves the solids that have been dissolved in it behind. Water in the ocean is evaporating all the time, but the salt doesn't evaporate with it. Thus, it just stays in the ocean. But what about lakes, rivers, and streams? Water evaporates from them, too. Why aren't they salty? In a river or stream, some of the water does evaporate, but the water that remains travels with the current, carrying anything that has been dissolved in it. As a result, the dissolved materials get dumped out at the end of the river.

Where do rivers end? Either at a lake or an ocean. Remember the picture on page 293. It shows a river that ends at an ocean. The river dumps all its contents (sediments, water, dissolved chemicals) into it. The oceans, then, receive a lot of water and other materials from rivers. Of course, those materials include dissolved substances like salt. However, the only chemical that can leave the oceans in any significant amount is water, and it does so mostly by evaporation. As a result, the dissolved substances just stay in the ocean, making it salty.

What about lakes? Many rivers do end in a lake, dumping their contents into it. However, *most* lakes also have water flowing out of them. That means water flows into the lake from rivers (or other sources), but it also leaves the lake in a process that doesn't involve evaporation. Since the dissolved substances travel with the water, they don't build up in the lake. However, there are some lakes that do not have water flowing out of them. Because of that, they are salty as well. One example is the Great Salt Lake in the state of Utah. The picture of people floating in saltwater on page 291 actually comes from there. Another example is the Dead Sea, which is called the "Salt Sea" in many translations of the Bible (see Numbers 34:2 and Joshua 3:16, for example). It is called the "Dead Sea" because it is so salty that fish and other common water-based organisms cannot live in it. So while we think of lakes as sources of freshwater, not all of them are.

There is so much salt dissolved in the water of the Dead Sea that it precipitates out close to the shore.

I want to spend a moment on the chemical nature of the salt found in the ocean and in lakes like the Dead Sea. When you and I use the term "salt," we are usually referring to a specific chemical, which is called **sodium chloride**. However, earth scientists use the term "salt" to refer to any chemical made up of positively-charged particles and negatively-charged particles, which are called **ions**. Sodium chloride, for example is made up of the positively-charged sodium ion and the negatively-charged chloride ion. Where do these ions come from? They come from the minerals in the geosphere. Many minerals are made up of ions, and as water encounters such minerals, they tend to dissolve in the water. Thus, as water flows over minerals, it picks up some of the ions that make up those minerals, taking them with it. Remember, water is polar, so it is attracted to charged particles.

Now it turns out that the most abundant ions found in the ocean are the sodium ion and the chloride ion. In fact, the sodium ion and the chloride ion make up about 86% of the salt found in the ocean. The remaining salts found in the ocean are mostly made of magnesium ions (positive), calcium ions (positive), potassium ions (positive), and sulfate ions (negative). There are a lot of other ions as well, but all the ions I have listed make up more than 99% of the salt in ocean water.

Most of the salt you find in the supermarket is sodium chloride. It is cheap and easy to make, so that's what most people get. However, you can get salt that has been extracted from the oceans. It is still mostly sodium chloride (86%), but it also contains salts made of the other ions mentioned above. Since your body needs all of those ions, some people prefer to use salt from the ocean (sea salt) because they think it provides more of what your body needs. Some companies also produce salt that comes straight from ground-up minerals. Himalayan salt, for example, is a powdered form of the mineral halite that is found in a region of Pakistan. More than 96% of it is sodium chloride, but there are also small amounts of other minerals in it, which usually give it a light pink color.

Comprehension Check

10.9 Suppose you find a lake that has a river flowing into it, but there is no river flowing out of it. You sample the water and find that it is freshwater. What can you conclude from this? (**HINT**: Look at the middle pie chart on page 296).

The Hydrologic Cycle

As you have learned, the oceans are very different from the lakes, rivers, and streams. However, it's important to realize that while they are very different, they are connected because they actually exchange water with one another. In fact, the earth's water is constantly moving around in an incredible process called the **hydrologic** (hi' druh lah' jik) **cycle**, which is also called the **water cycle**.

Hydrologic cycle – The constant motion of water throughout the earth's hydrosphere

It's obvious that water moves around the earth. After all, rivers have currents that move water in a specific direction. However, there is a lot more to water's movement!

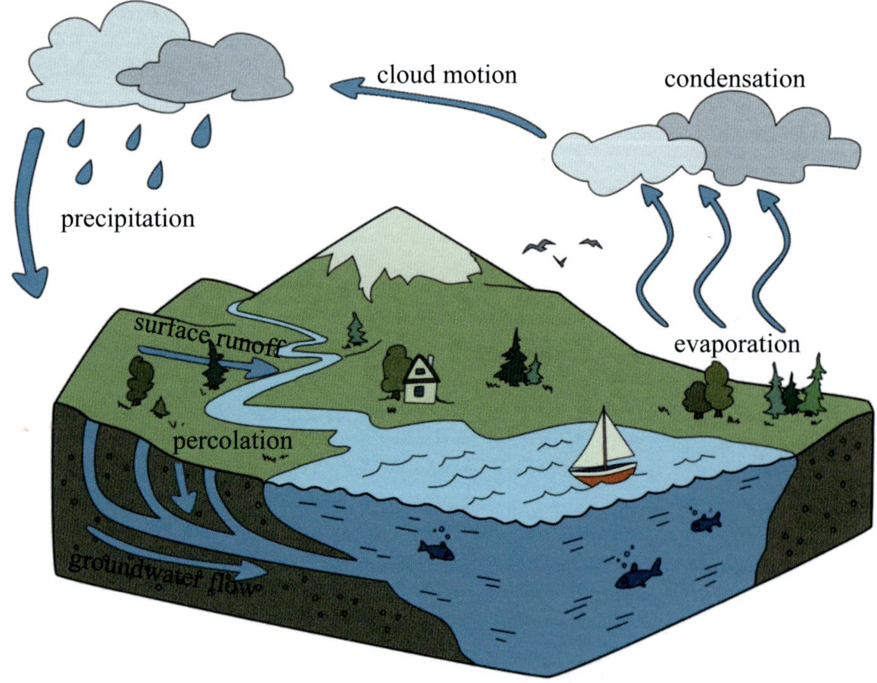

This is a schematic representation of the hydrologic cycle.

The illustration on the left shows you the basics of the hydrologic cycle. Starting on the right side of the illustration, you see evaporation. This is happening to all surface water all around the world. The water in the oceans, rivers, lakes, streams, and even puddles evaporates, sending water vapor (water in its gas phase) into the air. You can't see this happening, because you can't see the water vapor (I will get back to that in a moment). However, you can see its effect. If you leave a glass of water out long enough, what happens to the water level in the glass? It goes down, because the water is evaporating, turning into a gas and moving out of the glass.

Some of that water vapor stays in the air, but most of it ends up condensing to form clouds. You might think that clouds are made of water vapor, but they are not. You will learn a lot more about water vapor in the air and clouds later on in this course, but for now, please understand that clouds are not made of water vapor. They are made of water that has condensed onto tiny particles floating in the sky. If the cloud is high enough, the condensed water freezes. Thus, clouds are made of water in its liquid or solid phase, not its gas phase.

What happens to those clouds? They move! I am sure you have looked up at clouds in the sky and have seen them moving. That motion transports the water away from the place where it evaporated. Eventually, the clouds collect so much water that it falls down, producing **precipitation**.

<u>Precipitation</u> – (meteorology) Water in its liquid or solid phase falling from clouds

You already learned the definition of that word which comes from chemistry. Well, in science, sometimes the same word is used to mean different things in different fields of science. In **meteorology** (the study of weather), precipitation refers to water falling from clouds. The water might be in its liquid phase (rain) or solid phase (snow, sleet, or hail). Since the clouds moved before precipitation occurred, the water ends up in a different place from where it evaporated.

Some of that precipitation will run across the surface of the soil and end up in a lake, river, stream, or the ocean. That's called **surface runoff**, and it usually happens when rain falls close to one of those bodies of water. Most of the liquid precipitation (rain) ends up soaking into the soil. Some of that water is absorbed and used by plants. Interestingly enough, however, some of the water absorbed by the plants doesn't get used by them. Instead, it evaporates from the leaves of the plants. Evaporation really does occur everywhere in creation!

The liquid precipitation that is not absorbed by the plants will either evaporate from the soil or percolate down into the soil, where it eventually joins the groundwater. That water is also in motion, eventually becoming part of a lake, river, or ocean. You don't normally see that motion, because it is happening underground. However, in certain situations where the rocks are arranged in a specific way, the motion of groundwater causes it to erupt from the ground. This is called an **artesian** (ar tee' shun) **well**. A "well" is simply a place where you can get groundwater. If there is no artesian well nearby, you can dig a hole and eventually get to the groundwater, forming the more common version of a well.

This artesian well is a convenient way to access groundwater.

Now remember, a lot of the earth's freshwater supply is in its solid phase as snow or ice. As a solid, it cannot flow, but eventually that snow or ice melts, and the water can flow again. Usually, this happens fairly quickly. In most parts of the world, the snow might stay for a while, but eventually, the weather warms and the snow melts. However, there are places where the water stays solid for a longer period of time. At high elevations, for example, it stays cold enough to keep snow and ice solid for many months, perhaps even years. Even then, however, there comes a time when the weather gets warm enough to melt at least some of the snow and ice that has accumulated. At that point, the liquid water flows down to lower elevations. In mountainous areas, this leads to rivers that have a lot of water flowing through them in the spring, but the flow tapers off as the snow and ice that can melt dwindles away.

You can see this effect by looking at the two photographs below. They both show a tourist attraction called Takakkaw (ta' kuh kaw) Falls in British Columbia, Canada. The picture on the top shows the waterfall in spring, while the one on the bottom shows it in the fall. Do you see the difference? In the spring, the waterfall is wide and has a lot of water flowing in it. In the fall, it is thinner with less water flowing down it. In this case, it is due to the fact that in the spring, there is a lot more snow and ice melting at the top of the mountains. As a result, there is a lot more water flowing down the mountains. By autumn, the amount of snow and ice melting is smaller, so the flow of the waterfall is weaker.

These two pictures show the same waterfall in the spring (top) and fall (bottom).

Now please understand that there are many other factors which affect how much water is flowing in rivers and down waterfalls. The amount of rain in the area also plays a big role. However, when all the other conditions are similar, the amount of snow melting in the mountains will strongly affect how much water is flowing in the rivers of mountainous regions.

As you can see, then, water travels all around the earth. This means water that was once in the ocean might eventually be in a lake that is in the middle of a continent. This is really important, because all life depends on water. If it didn't move around like it does, there would be places on the earth with too much water and other places with too little. The hydrologic cycle ensures that doesn't happen. Also, the hydrologic cycle helps move nutrients around the earth as well. The oceans would be depleted of the chemicals used by organisms that live there if it weren't for the constant flow of dissolved minerals and sediments coming into the oceans from the rivers, groundwater, and surface runoff that feed into them.

Scientists took quite a while to figure all this out. From ancient times, several thinkers speculated about how water finds its way up to the top of the mountains, but it took until the late 1500s for natural philosophers (what scientists were called back then) to start describing the hydrologic cycle properly. However, the Bible talked about it long before scientists did! Ecclesiastes 1:7 says, "All the rivers flow into the sea, Yet the sea is not full. To the place where the rivers flow, There they return again." How does this happen? In discussing God, Job 36:27,28 says, "For He draws up the drops of water, They distill rain from the mist, Which the clouds pour down, They drip upon man abundantly." The first part of the passage describes evaporation, and the second part describes precipitation.

It's not always as obvious as it is in this picture, but rivers flow from high elevations to low ones.

Does it surprise you that the Bible contains scientific information? It shouldn't. Think about the source of the Bible – the very Author of creation. Since the One who made the earth inspired the writers, it is to be expected that they would end up discussing some of the earth's inner workings. Indeed, when I first read the Bible as an atheist, I was surprised by all the accurate science that I found in its pages. In the next few chapters, you will learn three other scientific truths that were revealed by the Bible before they were figured out by science!

The verse I quoted from Ecclesiastes says that all the rivers flow into the sea. That's pretty much true. Lots of rivers flow into lakes, but most of those lakes also have rivers (or groundwater flow) carrying water out of them. Thus, pretty much all rivers flow into the sea or into something that eventually flows into the sea. But what causes that flow? How do rivers "know" where the sea is so that their water can be dumped into it? The answer is surprisingly simple: gravity. Water flows from higher elevations to lower ones. Thus while a river might flow north, south, east, or west, it is really flowing *down*.

How Long Have You Been a Resident?

Because of the hydrologic cycle, water molecules move around the earth. As a result, a specific molecule will not stay in the same place forever. It might start off in the ocean, but it will eventually evaporate from the ocean, become part of a cloud, and then precipitate somewhere else. Now, of course, since most of the earth's surface is covered by the ocean, it is likely that the molecule will precipitate back into the ocean. However, *eventually*, it will precipitate over land and no longer be a part of the ocean.

But if you think about it, the nature of the water source has a lot to do with how long a water molecule will stay in it. Rivers and streams flow, dumping their water into a lake or ocean. That means the molecules of water in rivers and streams will become part of another body of water. How long that takes will depend on how fast the water is flowing. Most lakes have water flowing in them and also out of them. When a water molecule becomes part of a lake, it will probably eventually flow out of the lake as well. However, it will spend some time in the lake before that happens. Most likely, then, a water molecule stays part of a lake longer than it stays part of a river.

Groundwater flows as well, so a water molecule underground will eventually be dumped into another body of water. A water molecule in ice will not move into another body of water until the ice melts. Most water molecules in the ocean leave by evaporation, which happens in all bodies of water. This leads us to the concept of **residence time**.

Residence time – The average amount of time a substance stays in a specific reservoir

In science, the term **reservoir** refers to a body that holds something. So when we talk about the hydrologic cycle, a reservoir is a specific body of water, like a lake or an ocean.

While it isn't really possible to measure the residence time of a water molecule in a specific reservoir, we can make some reasonable assumptions and calculate it. The table below shows you average residence times that have been calculated for specific reservoirs in the hydrologic cycle.

Reservoir	Residence Time	Reservoir	Residence Time
Antarctica	20,000 years	**Shallow groundwater**	100-200 years
Oceans	3,200 years	**Deep Groundwater**	10,000 years
Glaciers	20-100 years	**Lakes**	50-100 years
Seasonal Snow	2-6 months	**Rivers**	2-6 months
Soil Moisture	1-2 months	**Atmosphere**	9 days

If you look at the numbers, they should mostly make sense. The shortest residence time is the atmosphere. That's because once a water molecule evaporates into the atmosphere, it will join a cloud and then be removed from the atmosphere by precipitation. The oceans have a much longer residence time, because evaporation is the only way out, and there are *a lot* of water molecules in the ocean, so it takes a long time for a specific water molecule to make it to the surface and evaporate. If you don't know what glaciers are, you will learn about them in the next chapter.

But what about Antarctica? Why is the residence time so long there? Well, most of the water is ice there, and most of that ice will probably never melt. Thus, those water molecules will probably never leave Antarctica. However, even though most of Antarctica's ice doesn't melt, some of it does, especially during the Antarctic summer. Some summers are warmer than others, and those summers will produce a lot of melting, allowing more water molecules to leave. Remember that the numbers in the table are *averages*, so when we average a lot of molecules that will probably never leave with a few molecules that leave during really hot summers, the result is a very long residence time.

Comprehension Check

10.10 A molecule of water starts out in the ocean and ends up in groundwater flow. What hydrological cycle steps occurred to make that possible?

10.11 Based on the residence times, which flows faster: shallow groundwater or deep groundwater?

Sample Data and Calculations for Experiment 10.2

Mass of the graduated cylinder: 31.2 g
Mass of the graduated cylinder and 40 mL of water: 70.8 g

Mass of the water: 70.8 g – 31.2 g = 39.6 g
Density of the water: 39.6 g ÷ 40 mL = 0.99 g/mL

Volume of the ice: 42 mL
Mass of the graduated cylinder and ice: 70.8 g

Mass of the ice: 70.8 g – 31.2 g = 39.6 g
Density of the ice: 39.6 g ÷ 42 mL = 0.94 g/mL

Sample Data and Calculations for Experiment 10.4

Mass of the graduated cylinder: 31.2 g
Mass of the graduated cylinder and water: 72.4 g
Volume of the water: 42 mL

Mass of the water: 72.4 g – 31.2 g = 41.2 g
Density of the water: 41.2 g ÷ 42 mL = 0.98 g/mL

Mass of the graduated cylinder: 31.3 g
Mass of the graduated cylinder and saltwater: 82.8 g
Volume of the saltwater: 43 mL

Mass of the saltwater: 82.8 g – 31.3 g = 51.5 g
Density of the saltwater: 51.5 g ÷ 43 mL = 1.20 g/mL

Answers to the Comprehension Check Questions

10.1 They should be more attracted to water molecules. The key is the electrical charges. If something has electrical charges, it will be attracted to other things with electrical charges, so ionic compounds are more attracted to polar molecules like water than nonpolar molecules like oil.

10.2 125 mL. Remember, butter is like most substances, it shrinks when it freezes. Thus, solid butter has a smaller volume than the same amount of liquid butter. So the solid version must take up less volume than the liquid version. That would only be the case if the liquid version was 125 mL.

10.3 The temperatures near the coast would be cooler than the temperatures far from the coast. Remember, water *moderates* the surrounding temperature. In the summer, the water must *absorb* a lot of heat to raise its temperature. As a result, it stays cooler than the land. That makes the nearby land cooler. For the two middle temperatures, for example, the one near the coast is 58 °F, while the one away from the coast is 73 °F.

10.4 The gallon with just one teaspoon of salt would freeze at the higher temperature. Adding salt to water lowers the temperature at which it freezes. The more salt you add, the stronger the effect should be. Thus, the more salt the water has, the lower its freezing temperature would be. That means the less salt you have, the higher its freezing temperature will be.

10.5 Diffusion would even out the concentration. Diffusion works as long as there is no barrier preventing the solute from moving.

10.6 Osmosis would act, causing the water to move down from the freshwater and into the saltwater. Since there is a semipermeable film, the solute can't travel. Thus, water must. That's osmosis. To even out concentration, water must move from where there is little solute (freshwater) to where there is a lot of solute (saltwater). That way, the salt becomes more dilute in the saltwater.

10.7 It would most likely be saltwater. A random sample will probably be from the largest source. Since the vast majority of water on the earth is saltwater, a random sample will most likely be saltwater.

10.8 It would most likely be solid. A random sample will probably be from the largest source. Since most freshwater on the earth is frozen, a random sample will most likely be ice, which is solid.

10.9 It must have groundwater running out of it. Since it is not saltwater, there must be a way for water to leave the lake. If there is no river, that means the water must be underground. Remember, as shown in the middle pie chart on page 296, most of the freshwater that is in its liquid phase exists underground.

10.10 Evaporation, condensation into a cloud, precipitation, and percolation. It could only leave by evaporation. The only way it could get to where there is ground is to become part of a cloud that then rained on the ground. Then it would have to percolate into the soil.

10.11 Shallow groundwater flows faster. The shorter the residence time, the faster the water molecules leave. The fastest way for them to leave is through flow, so the flow must be slow to produce a long residence time.

Chapter Review

1. Define the following terms:

a. Polar molecule
b. Equilibrium
c. Heat capacity
d. Diffusion
e. Semipermeable
f. Osmosis
g. Distillation
h. Hydrologic cycle
i. Precipitation
j. Residence time

2. What makes the earth appear to be blue when seen from space?

3. A metal paperclip floats on the surface of liquid A easily, while the same paperclip sinks in liquid B. If the density of the paper clip is significantly greater than the densities of both liquids, which liquid has the higher surface tension? If one is polar and the other is nonpolar, identify the nonpolar one.

4. Your DNA is made of two helices that are held together with hydrogen bonds. Is DNA polar or nonpolar?

5. Bismuth freezes the way water does, but lead freezes like most substances. You are given two densities for bismuth: 10.0 g/mL and 9.8 g/mL. Which is for solid bismuth? You are given two densities for lead: 11.3 g/mL and 10.7 g/mL. Which is for solid lead?

6. Two objects have equal mass and temperature. You give them each the same amount of heat. The first becomes hotter than the second. Which has the higher heat capacity?

7. You compare the temperatures of several locations at the same latitude. Do you expect the summer temperatures to get higher or lower as you approach the ocean? What about in the winter?

8. Suppose you had a layer of saltwater and a layer of freshwater in the same container. Before they mixed together, which would be on top?

9. You have three samples of a clear liquid water that are all in the midst of freezing. You are told that one is freshwater, one is saltwater, and the other is not water at all. Sample A is freezing at 2°C, sample B is freezing at 0 °C, and sample C is freezing at -1 °C. Identify the samples.

10. You measure the concentration of a solute at two different places in a large container that holds a solution. There are no barriers of any kind in the solution. The concentrations you measure are different. If you measured them again a long time later, would you expect the difference to be the same, more, or less than the difference between the first measurements? Which process (diffusion or osmosis) do you expect to have happened?

11. You have two solutions of different concentration separated by a semipermeable membrane. The concentration of solutes in the solution on the right is much higher than the concentration of solutes in the solution on the left. What will happen to the volumes of the solutions over time? What process would cause that to happen?

12. List the following forms of water in order of increasing abundance in the hydrosphere: freshwater ice, freshwater liquid, and saltwater.

13. What phase is most of the earth's surface freshwater found in?

14. If a lake is salty, how can water leave it?

15. A molecule of water is a gas in the atmosphere. What are the hydrologic cycle steps that could happen to put it in the ocean as quickly as possible? Name another set of steps that would take longer but still get the water molecule into the ocean.

16. A water molecule starts off in a cloud, precipitates on land, and ends up in the ocean, but it was never part of a lake, river, stream, or groundwater. How did it get there?

17. List these reservoirs in terms of increasing residence time: a fast-flowing river, a lake, an ocean, and a slow-flowing river.

Chapter 11: More on the Hydrosphere

Introduction

In the previous chapter, you learned about several properties of water, some differences between saltwater and freshwater, and the various places the earth stores its water. I want to continue discussing the hydrosphere, but I will spend most of the rest of my time on the ocean. This should make sense, of course, since most of the earth's water is found in the ocean. Since this is an earth science course, however, I won't be talking about the amazing organisms found there. That would be a fascinating discussion, but it goes beyond the scope of this course. Thus, I will concentrate on what we know about how the ocean operates as a reservoir for saltwater.

Motion in the Ocean, Part 1: Surface Waves

When you think of the ocean, what first comes to mind? For me, it's the waves that you see when you are on the shore. Depending on where you are and what the weather is like, those waves might be large and powerful (as shown in the photo on the near right), or they might be gentle and lapping (as shown in the photo on the far right). While you might find waves on a lake, they are generally not as regular or as large as what you see in the ocean. What is it about the ocean that causes it to have waves like this? Let's start with an experiment

Waves near the ocean shore can be very different, depending on many different factors.

Experiment 11.1: Making Waves

Supplies:

- Water
- A straw
- A rectangular baking pan that is at least 28 cm (11 in) long and 5 cm (2 in) deep (glass is best)
- Food coloring (Blue works best. The gel-based ones will work, but not as well.)
- A flat, level surface
- Someone to help you

Instructions:

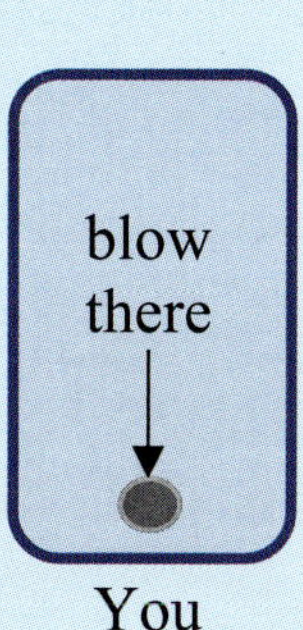

1. Fill the baking pan ¾ full of the coldest tap water.
2. Set it on the flat, level surface and arrange it so that it looks like the diagram on the right.
3. Put the straw in your mouth and lean over the baking pan so that it is pointed straight down at the water at the center of the near side of the pan, as shown in the diagram on the right. Do not put the straw into the water.
4. While watching the surface of the water, blow through the straw. Don't blow as

strongly as you can. You just want a steady stream of air flowing through the straw and hitting the water vertically. What do you see on the surface of the water? You should see waves traveling away from the straw. Since there is more room for them to travel to the end that is far from the straw, most of the waves that you see should be traveling towards the far end of the pan. Do this a couple of times so that you are familiar with the waves that you can make this way.

5. Change the force with which you are blowing to see how that affects the waves that are being made. Try to find a force that will produce a noticeable depression in the water right under the straw but will not make the water splatter.
6. Have your helper hold the food coloring container over the very center of the baking pan near the surface of the water. Tell him or her to drop a single drop of food coloring into the water right before you start blowing.
7. Repeat steps 3 and 4, this time watching the food coloring as it hits the water and begins to spread out. Blow with the force you used in step 5 to make a noticeable depression. You will probably have to inhale and exhale several times to get an idea of what is happening to the food coloring. Notice that the food coloring on the surface acts differently from the food coloring that is below the surface.
8. Stop blowing through the straw and now aim the straw directly at some place where there is still a lot of blue food coloring clumped together.
9. Blow with the same force you were using in step 7. What's the difference?
10. Clean up your mess.

The first part of the experiment shouldn't have been surprising. When you blew into the water, you disrupted it, pushing the water molecules away. This produced waves that traveled away from where the water was being hit. This is the way the waves that we see traveling along the surface of the ocean are made. They are the result of wind hitting the surface, pushing the water molecules on the surface. Because they are formed by wind and travel on the surface of the ocean, they are called either **surface waves** or **wind waves**.

But what happened with the food coloring? Did it travel with the waves? The food coloring on the surface probably did, but the food coloring under the surface did not. It probably spread out a bit, but if it moved one way more than the other, it probably moved towards you rather than with the waves. Why? Well, there are two effects, but the first one is the most important for our discussion.

Even though you see waves traveling from one place to another, the water that makes up those waves *is not* traveling in that direction. Instead, it is traveling in a circle. The water molecules rise as the wave's crest approaches, and they begin to move with the wave. However, gravity starts fighting that rise, pulling them down. As the water molecules fall, they get sucked backwards into the trough, which is the lowest part of the wave. As a result, the water molecules follow a circular path, as shown in the illustration on the left. Because water molecules attract one another, this causes circular motion in the water molecules

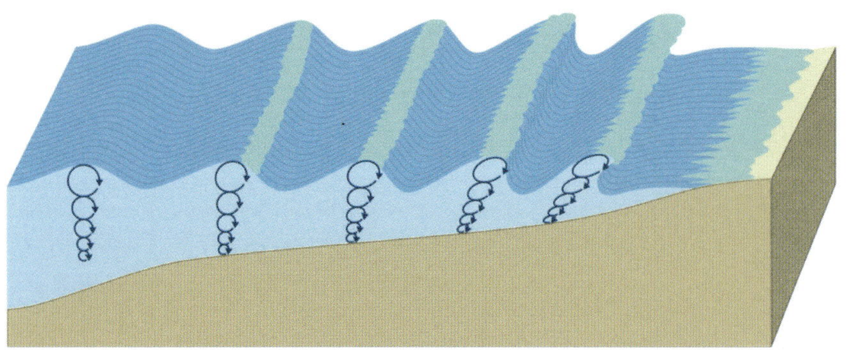

The water in waves doesn't travel with the waves; it travels a circular path within the waves.

below. The effect gets smaller the deeper you go, but in general, there is a "tower" of water molecules stacked on top of each other, all traveling in circles.

The waves you made in your experiment looked like the leftmost wave in the illustration on the previous page. Since the water molecules moved in circular paths, they didn't carry the food coloring in the direction of the wave. Food coloring on the surface moved because some of the wind you made traveled across the surface, pushing it away.

However, there was also a depression right under the straw. The water surrounding the depression tried to fall into it to fill it in. This made a current of water running towards the depression. Since the waves didn't carry the food coloring away from the straw, the current caused by the depression might have gotten strong enough to pull some of the food coloring towards the depression.

If the water in a wave isn't traveling with the wave, what is? It's the *energy* that made the wave to begin with. In your experiment, the wind formed by you blowing through the straw hit the water, transferring energy to it. That energy started spreading out away from the straw, and you saw it happening because it made waves on the water. So water waves don't transport water from one place to another; they transport energy. In the last part of the experiment, you aimed the straw at a clump of food coloring, and it moved in the direction you were blowing, because the wind pushed it that way.

But wait a minute. The waves that crash onto the beach aren't like the ones you made in your experiment. Why? Because your baking pan didn't change depth. Look again at the illustration on the previous page. If the ocean never changed depth, the waves would behave more like they did in your experiment, looking mostly like the wave farthest from the shore in the illustration. However, as ocean waves get closer to shore, they encounter shallower waters. As a result, the water molecules deep under the wave get slowed down when their circular motion drags them against the bottom of the ocean. This deforms their circle. The water molecules that are not near the bottom aren't slowed down, so their circles aren't deformed. Eventually, the circles on the bottom are so deformed that balance can no longer be achieved, and the water molecules on the crest fall. When that happens, we say that the wave breaks against the shore.

So the waves that we see on the ocean's surface are caused by wind, and their behavior near the shore is caused by the change in depth that occurs there. If the change in depth is gentle, the waves crash gently, and you get gentle, lapping waves. If the change in depth is abrupt, the waves tend to be larger and more powerful. However, that's only part of the story. The wind that originally caused the waves gave them their energy, so the stronger the wind that formed them, the bigger and more powerful the waves. Thus, gentle, lapping waves can be larger or smaller, depending on the weather where the waves were formed. In the same way, waves formed by an abrupt change in depth can be even larger and more powerful if they were formed by particularly strong winds.

Comprehension Check

11.1 Two coastlines are drawn on the right. Assuming that the weather conditions are similar in both regions, which you would expect to experience larger, more powerful waves?

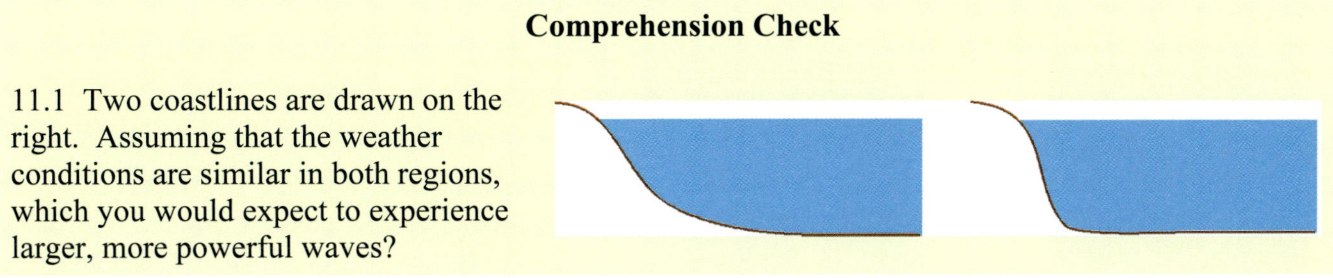

Motion in the Ocean, Part 2: Tides

While surface waves are the most obvious example of how the water in the ocean moves, they aren't the only example. If you live near the ocean, you know that there are times during the day when the ocean covers a large portion of the shore, and there are other times when it does not. Look, for example, at the two pictures below.

Both of them come from a place called Hopewell Cape on the Bay of Fundy, which is on the coast of Canada between the provinces of New Brunswick and Nova Scotia. Notice that in the photo on the left, people are walking around the interestingly-shaped rocks. The arrow in the picture on the left points out where I am in the picture. In the photo on the right, which was taken just six hours later, you see the same rock formations, but the water is so high that a large part of each rock formation is under water. I'm glad I wasn't standing near the rocks then!

What caused this difference? You might think there was some huge amount of rain in the area, but there wasn't. Even if there had been a lot of rain, it wouldn't have resulted in such a dramatic change in the water level. In fact, the difference between the two pictures is a result of the ocean's **tides**.

Tides – The rising and falling of an ocean, usually twice each day, caused by the sun and the moon

When the water level reaches its lowest point, we say that the shoreline is experiencing a **low tide**. The picture on the left, then, shows the Bay of Fundy at low tide. When it reaches its highest level, it is called **high tide**. Thus, the picture on the right was taken at high tide.

Now I have to admit that the pictures above show a rather striking example of the effect of tides. At any ocean shoreline, the difference in water level between high tide and low tide is noticeable. At most beaches, however, it just looks like the water retreats from the beach at low tide and comes in to cover a lot more of the beach at high tide. Because of the way the land around the Bay of Fundy is shaped, the effect is amplified; lots of water gets funneled into or out of the bay. As a result, its difference between high and low tide (as much as 11.7 meters or 38.4 feet) is the largest in the world. There are other places that are similar, however. The port of Avonmouth in England can experience sea level changes as big as 9.6 meters (31.5 feet)

Notice that the definition of tides indicates that this is all caused by the sun and the moon. How can the sun and the moon cause the ocean to rise and fall at the shore? I am sure you know that even though it seems to you and me like the earth is sitting still, it is not. It is moving around the sun, but it is also spinning like a top, as shown in the illustration below. It takes one year (just over 365 days) for the earth to complete its trip around the sun. While we don't notice the motion, we can see its effect: the seasons change throughout the course of the trip. You will learn why later on in this course.

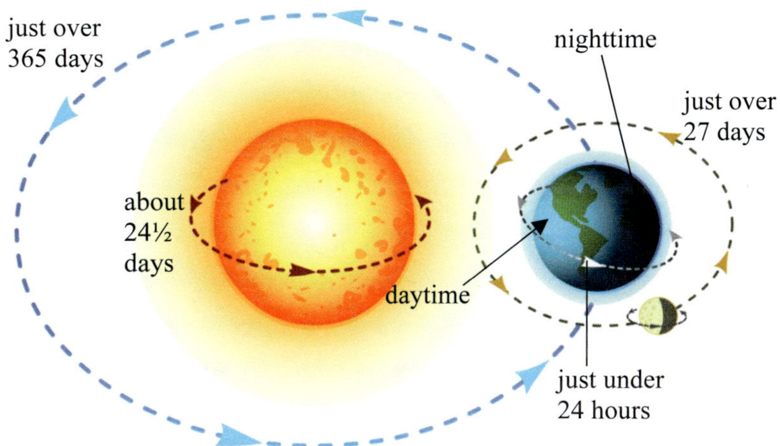

The earth spins on its axis as it orbits the sun, while the moon spins on its axis as it orbits the earth. The sun also spins on its axis.

We can't feel the earth spinning, but we can see its effect: it turns day into night and night into day. Like a top, the earth spins on its **axis**, an imaginary line that runs from the top of the earth (the North Pole) to the bottom (the South Pole). As it spins, the parts of the earth facing the sun change. When you are on the part of the earth that faces the sun well enough so that the sun can shine on it, you are experiencing daytime. When the part of the earth you are on is facing away from the sun so that you cannot get any of the sun's light, you are experiencing nighttime. It takes just under 24 hours for the earth to complete one spin.

The moon orbits the earth, much like the earth orbits the sun. It takes just over 27 days (about a month) for the moon to make one full trip around the earth. It is also spinning on its axis, but it takes the same amount of time to spin on its axis as it does to orbit the earth. Because the time it takes to spin matches the time it takes to orbit the earth, the same part of the moon is always facing the earth. To see the other side of the moon, astronomers had to take pictures of it from robotic spacecraft that traveled behind the moon. The sun also spins on its axis, but that isn't relevant to our discussion here.

Why does the earth orbit the sun? Because the earth and the sun are attracted to one another through the force of **gravity**. That force keeps the earth in its orbit. In the same way, the moon orbits the earth because they are both attracted to one another by the force of gravity. There are two things that affect the strength of the gravitational force: the masses of both objects, and the distance between them. However, distance affects gravity much more strongly than mass. So even though the sun's mass is much greater than the earth's, the moon is much closer to the earth than it is to the sun. As a result, the gravity between the earth and the moon is much stronger than the gravity between the moon and the sun. That's why the moon orbits the earth and not the sun.

The gravitational forces between the sun, moon, and earth are responsible for the tides. Since the moon is so much closer to the earth than the sun, the gravitational force between the earth and the moon has the strongest effect, so let's start with that. Now remember, the earth spins on its axis once every 24 hours, while it takes just over 27 days for the moon to orbit around the earth. As a result, the earth's spinning is really fast compared to the moon's movement around the earth. This means that over the course of a single day, the moon doesn't move very far in its orbit. To make this discussion easier, I will act like it doesn't move at all. Over the course of a given day, then, I will treat the moon as if it is motionless. It's not, but to simplify things, let's just treat it that way.

Now remember, gravity attracts the moon to the earth, and it attracts the earth to the moon. That force applies to *everything* on the earth, including the ocean. Well, because it is a liquid, the water in the ocean can actually move, so the water stretches out towards the moon, since gravity is pulling it that way. However, water molecules are attracted to each other, and they are also attracted to the earth. Thus, the water stretches towards the moon, but the other water molecules and the earth fight that stretch. This deforms the water in the oceans, making it an oval, with the earth at the center of the oval.

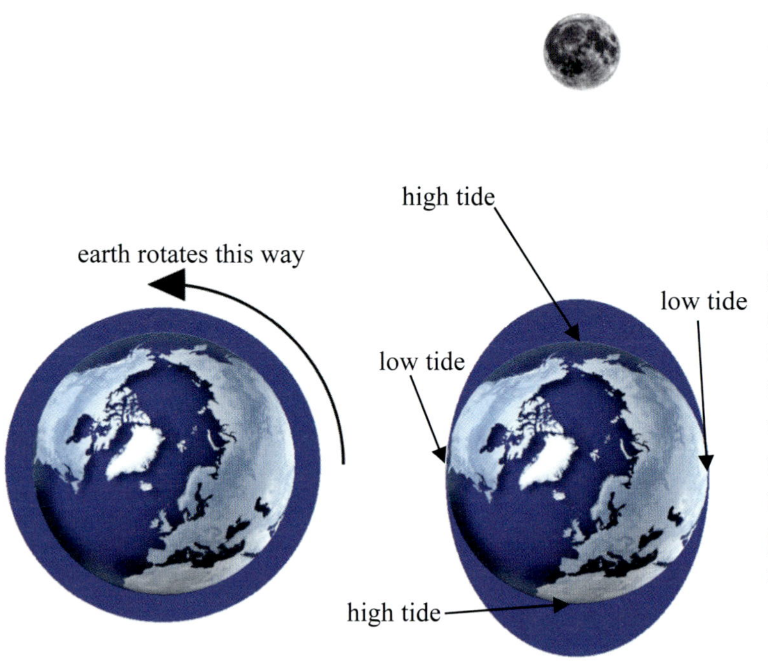

If there were no moon (left), the water in the oceans would be at similar levels everywhere, and the effect of the tides would be very small. Because the moon's gravity attracts the water (right), however, the water forms an oval, producing two high tides and two low tides each day. **NOTE**: The thickness of water is exaggerated for the purpose of illustration.

Think about what this oval means. Where the oval is wide, the water is high. Where it is narrow, the water is low. So any part of the earth that is where the water's oval is wide experiences high tide. Any part of the earth that is where the oval is narrow experiences low tide. So at any given moment, there are two places on the earth experiencing high tide and two places on the earth experiencing low tide. That's shown in the illustration on the left. The moon forms the waters in the ocean into an oval, so there are always two places on earth that are at the widest parts of the oval and two places at the narrowest parts of the oval.

If we think of the moon as not moving at all during a given day, then the wide and flat parts of the oval never move, but the earth spins inside that oval. Thus, the places experiencing high and low tide change as the earth spins. Since the earth makes one complete spin each day, every place on earth passes through both wide parts of the oval and both narrow parts of the oval each day. Thus, in a 24-hour period, every part of the earth experiences two high tides and two low tides. That means the time between high tide and low tide is six hours.

Now remember, we are treating the moon like it stays in one place over the course of the day. That means the oval never changes. However, the moon does, in fact, move a little, which changes the oval a little. As a result, the time between high tide and low tide isn't exactly 6 hours; it's a bit longer. High tide and low tide are actually separated by about 6 hours and 13 minutes.

But what about the sun? The definition says the sun has an effect, too. Well, there is also gravitational attraction between the earth and the sun, which means the water in the ocean also forms an oval that is pointed towards the sun. However, since the sun is farther away, its gravity can't pull on the water as hard as the moon. As a result, the oval produced by the sun is about one-third as wide as the oval produced by the moon. During most of the time, then, the tides caused by the sun are just not noticeable. However, there are times when it is.

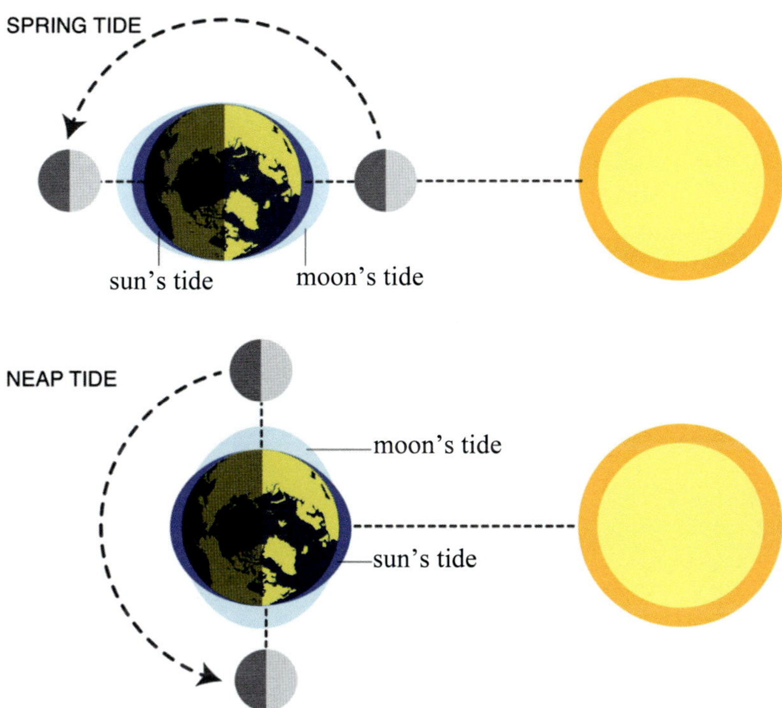

The sun's gravity can either add to the moon's gravity or take away from it, making a spring tide or a neap tide.

When the sun is positioned so that its gravity either pulls with the moon or pulls directly against the moon, its effect can be seen. Consider, for example, the illustration shown on the left. The oval produced by the moon is shown in light blue, while the smaller oval produced by the sun is shown in darker blue. When both ovals are lined up like they are at the top of the illustration, they add. This makes the high tides higher and the low tides lower. In other words, the sun's gravity enhances the effect of the tides. When the sun is off to one side of the earth compared to the moon as shown in the bottom part of the illustration, the sun's oval is opposite the moon's oval. Where the moon makes a high tide, the sun makes a low tide, and vice-versa. This makes the low tides higher and the high tides lower.

When the sun's gravity adds to the moon's gravity to enhance the difference between high and low tide, it is called a **spring tide**.

Spring tide – A tide in which the sun's gravity adds to the moon's gravity, enhancing the extremes

In this definition, "extremes" refer to the high tides and low tides. When the sun's gravity fights against the moon's gravity to reduce the difference between high and low tide, it is called a **neap tide**.

Neap tide – A tide in which the sun's gravity fights the moon's gravity, reducing the extremes

The term "spring" doesn't refer to the season, since it is not a seasonal effect. Instead, "spring" refers to the water "springing forth" in a high tide. The term "neap" is an old English word that means "without power," which refers to the fact that the tides are "weaker."

Now it turns out that the tides are actually more complicated than this. Even though there are two wide parts and two narrow parts to the oval, not all places on earth experience two high tides and two low tides. That's because the water actually has to move into the area during a high tide and out of the area during a low tide. Sometimes, because of the shape of the surrounding land, that can't happen twice each day. So there are some places that experience only one high tide and only one low tide each day. Other places experience two high tides and two low tides, but the sizes of the highs and lows aren't the same. You don't have to worry about that, because most places experience two similar high tides and two similar low tides. I just wanted to make you aware that there is more you could learn about the tides.

When I discussed tsunamis in Chapter 6, I mentioned that they are sometimes mistakenly called **tidal waves**. Now I can tell you what a tidal wave actually is. As low tide turns into high tide, water must move into the area. In most places, this happens slowly, so you don't notice. However, in some places, the motion of the water is resisted in some way. This "piles up" the water, making a wave. Eventually, that wave builds up to the point where it can overcome the resistance, and it moves the way the tides are forcing it to move.

Consider, for example, the picture below. It shows a portion of the River Nith in Scotland, near the ocean. Remember, rivers dump their contents into the ocean, so the current of the river is flowing out towards the ocean. Now think of what happens at the place where the river meets the ocean as low tide turns to high tide. The water from the ocean is trying to move towards land, so it needs to go into the river. But the river's current is moving in the opposite direction. As a result, the current and the tide "fight" one another. This causes the tidal waters to pile up until they become powerful enough to overcome the river's current. At that point, a wave travels up the river. That's a tidal wave. Now it turns out that this is just one kind of tidal wave. There are others, but the key is that tidal waves are caused by the tides. Remember, tsunamis are caused by earthquakes, so tsunamis and tidal waves are two completely different things.

This is a tidal wave running up a river, against the current.

Comprehension Check

11.2 Suppose a day lasted 32 hours rather than 24. Roughly how many hours would exist between low tide and high tide?

11.3 If the moon were significantly less massive, it would exert significantly less gravity. How would that affect low tide and high tide at any given place?

11.4 If the moon were closer to the earth, how would that affect low tide and high tide at any given place?

11.5 Look at the rock formations in the picture of Hopewell Cape on page 312. If you were to look at a drawing of those rock formations from 100 years ago, would they look the same? Why or why not?

Motion in the Ocean, Part 3: Surface Currents

In 1839, a U.S. naval officer named **Matthew Maury** was in a stagecoach accident and suffered a broken leg. This effectively ended his career as a sailor. However, he kept his naval position and began studying weather and its effects on ocean navigation. He eventually became the officer-in-charge of the Navy's Depot of Charts and Instruments, which allowed him to study all the navigation records that were kept there.

Why would Maury spend his time looking through old navigation records? Because of the words found in Psalm 8:3-8:

> When I consider Your heavens, the work of Your fingers, The moon and the stars, which You have set in place; What is man that You think of him, And a son of man that You are concerned about him? Yet You have made him a little lower than God, And You crown him with glory and majesty! You have him rule over the works of Your hands; You have put everything under his feet, All sheep and oxen, And also the animals of the field, The birds of the sky, and the fish of the sea, Whatever passes through the paths of the seas.

Notice what the passage says at the end. It says that there are things that pass through the **paths of the seas**. Because Maury believed that every word in the Bible is true, he decided to use the navigation records to look for the paths that the Bible says are in the ocean.

None of the ship's logs recorded paths in the ocean, but they did record wind speeds, ship speeds, travel times, routes, etc. Based on careful examination of many such reports, he decided that there were winds that produced long-lasting currents which ships could use to their advantage. He produced charts of these winds and currents, and their use reduced travel time for many ships taking long sea voyages. Because of this, he became well-known in naval circles. Using his charts alone, he was even able to tell rescue ships where to look for a vessel that had become lost. The vessel was right where Maury said it would be.

This is a monument to Matthew Maury in Richmond, Virginia. Notice that he is called the "pathfinder of the seas." The book resting by his left foot is the Bible. Although Maury opposed secession, his statue was removed on July 2, 2020 because he joined the Confederacy when the U.S Civil War began.

Because of his remarkable success, the U.S. Navy authorized him to enlist the help of ship captains from around the world. He had them record specific information about winds and speeds during their voyages, and he had them release weighted bottles with messages inside them, noting the time and position where they were released. Each message instructed whoever found the bottle to report the contents of the message along with the day and location where the bottle was found. Maury studied all the information that he could gather this way, and by 1855, he produced what became the century's most important textbook about the ocean, *The Physical Geography of the Sea.* In it, he detailed the currents (paths) that he had found. His discussion was very detailed, but the image below shows you the big picture.

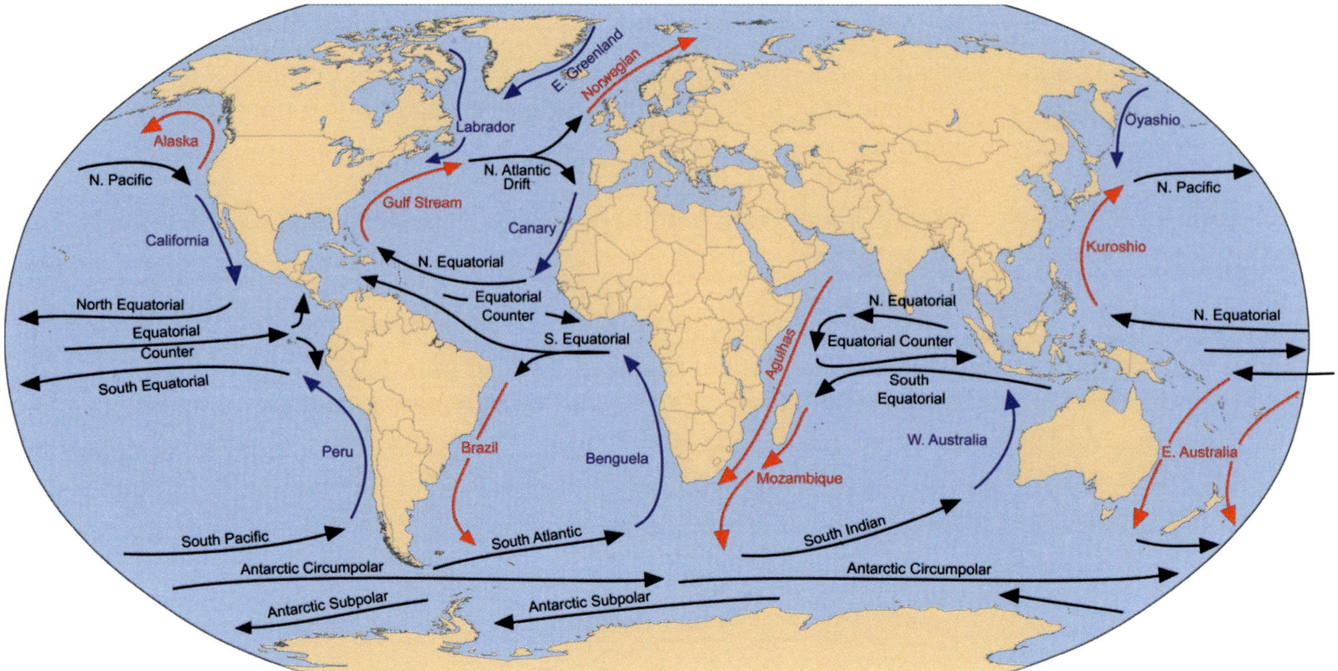

This map of the world shows you the major currents found in the seas. Ships travel significantly faster when they are on these currents. The red currents are usually warmer than the surrounding ocean water, and the blue ones are usually cooler.

Maury's textbook revolutionized the way ships traveled on the ocean. By following the currents, they could reduce their travel time, even though they had to go out of their way to stay on the currents. So like the hydrologic cycle, while science has finally figured out the paths of the seas, the Bible was the first to document that they exist.

As he became more famous among scientists, Maury's views about the Bible became known. He would talk about how there are scientific truths in the Bible, such as the paths of the sea. Many scientists didn't like that and tried to discourage him from connecting the Bible to science. In a speech given at the founding of The University of the South, he gave those scientists a stern rebuke: "I have been blamed by men of science, both in this country and in England, for quoting the Bible in confirmation of the doctrines of physical geography. The Bible, they say, was not written for scientific purposes, and is therefore of no authority in matters of science. I beg pardon! The Bible is authority for everything it touches." (Diana Fontaine Corbin, *A Life of Matthew Fontaine Maury*, Samson, Lowe, *et. al.*, 1888, p. 192: Note that "physical geography" meant "study of the ocean" back then.)

One thing you might notice about the illustration above is that many of the currents run roughly in ovals. Why? To understand that, you need to do an experiment.

Experiment 11.2: Bending the Path

Supplies:

- A pie pan
- The cardboard tube from the center of a roll of toilet paper
- A marble or small ball that easily fits in the tube
- Scissors

- Tape
- Someone to help you

Instructions:
1. Use the scissors to cut the tube in half lengthwise.
2. Tape one half of the tube securely to the edge of the pie pan so that it will not move and is tilted into the pan (see the picture on the right).
3. Set the pie pan on a level surface and have your helper hold the ball at the top of the tube half.
4. Have your helper release the ball so it rolls down the tube half and into the pan. Not surprising, right?
5. Hold onto the pan with both hands, one on each side with the tube half in between them.

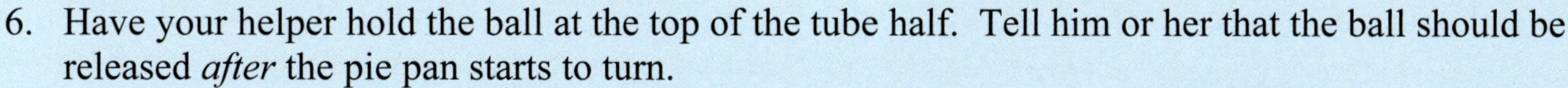

6. Have your helper hold the ball at the top of the tube half. Tell him or her that the ball should be released *after* the pie pan starts to turn.
7. Rotate the pie pan clockwise slowly and watch how the ball rolls. Stop rotating the pan when the ball hits the edge. How does the ball's path compare to its path in step 4?
8. Repeat steps 5-7 but increase the speed at which you rotate the pan. How does that affect the way the ball rolls?
9. Repeat steps 5-7, but rotate the pie pan *counterclockwise* slowly. How does that affect the way the ball rolls?
10. Repeat steps 5-7, rotating the pie pan *counterclockwise* quickly. How does that affect the way the ball rolls?
11. Clean up your mess.

What happened in the experiment? The ball rolled straight across the pan when it wasn't turning, as you probably expected. However, what happened when you rotated the pan? The ball followed a curved path, didn't it? The faster the rotation, the tighter the curve. Also, the direction of the rotation affected the direction of the curve, as shown in the drawing on the right.

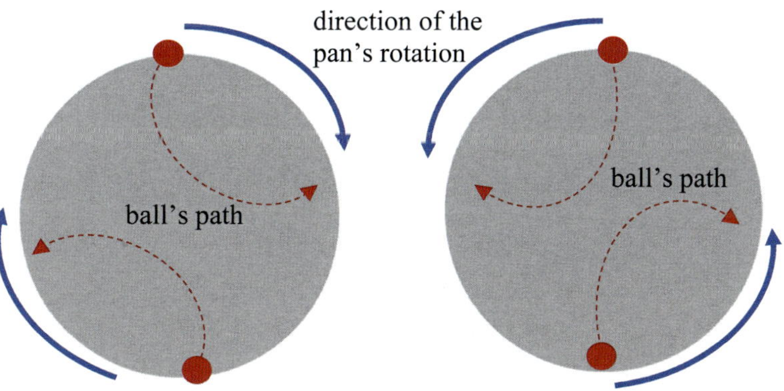

In the experiment, the direction of the pan's rotation (blue arrow) caused the ball's path to curve (dashed red arrow).

Why did that happen? Think about the pie pan rotating in a circle. If it takes 5 seconds to make one full rotation, each point on the pie pan must make a complete trip around the circle in 5 seconds. However, the points near the edge of the pie pan must travel much farther to make the trip than the points near the center. Thus, the farther it is from the center of the pie pan, the faster a point on the pan must travel in order to make the complete trip in 5 seconds.

The ball was the farthest from the center of the pie pan, so it was traveling the fastest. As it rolled towards the center, it started rolling over parts of the pie pan that weren't traveling as fast. As a result, it moved ahead of those parts of the pie pan. In other words, it "outran" the parts of the pan over which it was rolling. From the ball's point of view, it was just rolling straight. However, because the parts of the pan it rolled over were moving more slowly than it was, its path in the pan curved in the direction it was rotating. If this is a bit confusing, don't worry. The next time you do science, we will pick up this discussion, and it will become clearer.

Comprehension Check

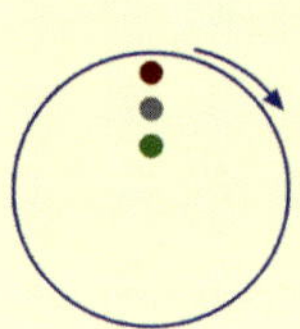

11.6 Suppose you have a circle with dots inside, as shown on the left. If the circle starts rotating around its center, which dot moves the slowest? Which moves the fastest?

More on Surface Currents

In the previous experiment, you saw that the rotation of the pie pan caused the path of the marble to bend. The rotation of the earth causes the same thing to happen when things change their latitude. Look, for example, at the illustration below. It shows the earth with its lines of longitude and its lines of latitude. The dashed line is the Prime Meridian. Concentrate on the two lines of latitude that I have highlighted in red. Now remember, the earth makes a complete rotation in 24 hours. Thus, every point on those two red circles must return to its initial position every 24 hours. Look, for example, at where the Prime Meridian intersects with the lower red circle (somewhere in northwestern Africa). That point in Africa must travel around the red circle, and it must make its complete trip in 24 hours. Now look at where the Prime Meridian intersects with the top red circle (somewhere in the ocean near the North Pole). Once again, that point on the earth must make a complete trip around the red circle in 24 hours.

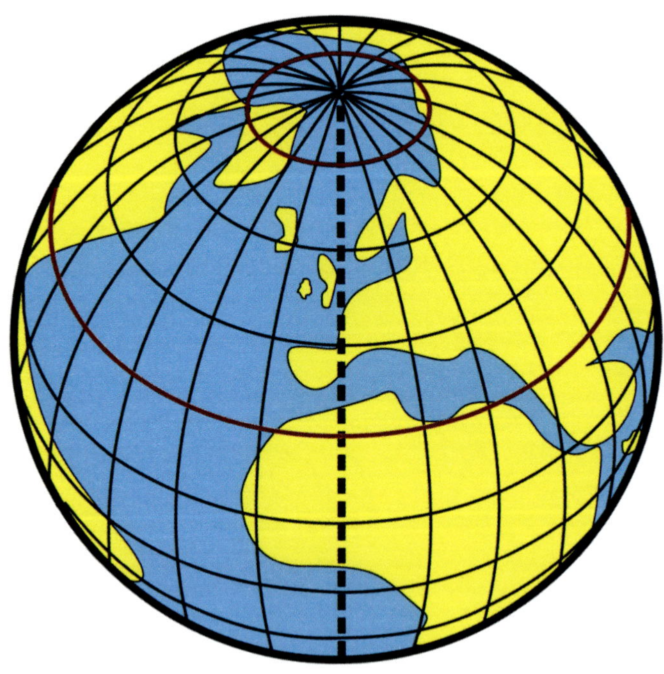

Because the circles swept out by the lines of latitude get smaller as you move away from the equator, points on the earth farther from the equator move more slowly than points on the earth nearer to the equator.

But look at the sizes of the circles! The lower red circle is much larger than the upper red circle. What does that tell you? Each point on that lower red circle must travel farther in 24 hours than each point on the upper red circle. So the points on the earth that fall on the lower red circle must travel faster than the points on the earth that fall on the upper red circle. The largest circle is swept out by the equator, so in general, we can say that all points on the equator move fastest during the earth's rotation. As we move away from the equator, the circles get smaller, so we can also say that the farther you are from the equator, the slower those points move as the earth rotates.

So what? Well, think about the experiment you did. When the pie pan wasn't rotating, the marble fell straight down the tube and rolled straight across the pan. However, when you rotated the pie pan, the marble had to sweep out a larger circle than any of the points in the pan. That means the marble was moving with the pie pan at a faster speed than any point in the pie pan. As the marble rolled down the tube half and into the pan, it was still moving in its circle, but it was moving faster than the points over which it was rolling. That means it "outran" the parts of the pan over which it was rolling, so its path bent towards the direction in which it was rotating.

The same thing happens when something changes latitude on the earth. Think about firing a long-range cannon due north from Louisiana (the purple dot on the globe to the right). If the shell from the cannon traveled due north, it would follow the gray line of longitude and hit somewhere in Wisconsin (the black dot). However, that's not what happens. The cannon moves with the earth as it spins, so the cannon is traveling in the direction pointed out by the red arrow (east), along with all other points on the globe. However, because it is close to the equator, it is moving east quickly. As the shell starts moving north, it flies over land that is traveling east, but more slowly. Thus, it "outruns" the land. This means it travels farther to the east than the land over which it is traveling. As a result, it "outruns" the land and ends up landing east of Wisconsin, following the path of the dashed purple arrow.

Because of the rotation of the earth, the path of something moving north or south through the air bends relative to the ground over which it is traveling.

But suppose you were shooting that same long-range cannon south from the state of Washington (green dot in the illustration). If the shell fired from the cannon traveled directly south, it would hit southern California (gold dot). However, the cannon is traveling to the east with the spinning of the earth, but since it is far from the equator, it is traveling in that direction slowly. As the shell moves south, it travels over land that is moving to the east faster. As a result, the land "outruns" the shell, traveling farther to the east than the shell. As a result, the shell ends up following the dashed green arrow, landing in the Pacific Ocean.

In the Southern hemisphere, the curves are opposite. A shell fired south from near the equator is traveling to the east faster than the land over which it is traveling. Thus, it "outruns" the land, ending up farther to the east, following the dashed brown arrow in the illustration. If the shell is fired north from a point far from the equator, it is moving to the east more slowly than the land over which it is traveling, so the land ends up traveling farther to the east than the shell. Thus, the shell follows the path of the dashed orange arrow.

If the earth were not spinning, this would not happen. However, because the earth spins, any object that is not firmly attached to the earth and changing latitude over a long distance will follow a curved path. This is called the **Coriolis** (kor' ee oh' lus) **effect.**

Coriolis effect – The deflection of the path of any object that changes latitude over a long distance and is not firmly attached to the earth

It is named after the French mathematician Gaspard Gustave de Coriolis, who studied spinning systems like waterwheels in the early 1800s. He never applied what he had learned to a study of the earth, but other scientists did.

The Coriolis effect is very real, but it only affects things that travel a long distance. After all, it happens because different points on the earth travel at different speeds as the earth rotates. In order for the Coriolis effect to be noticeable, then, an object must travel far enough for that difference in speed to become noticeable. Have you ever heard that the water in a basin will swirl counterclockwise in the Northern Hemisphere and clockwise in the Southern Hemisphere? That idea is based on the Coriolis effect, but it isn't true. A water basin is not large enough for the Coriolis effect to become noticeable, because there isn't a significant difference in the speed at which the different parts of a water basin rotate with the earth. As a result, the Coriolis effect cannot influence the way water drains from a basin. Generally, the shape of the basin and the direction in which water is injected will determine whether the water swirls clockwise or counterclockwise around a drain.

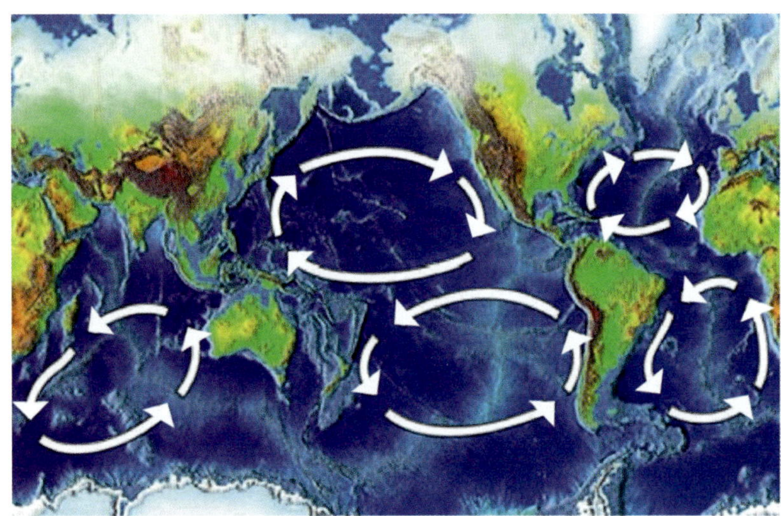

The oval currents shown above are called gyres, and the direction in which they move is determined by the Coriolis effect.

Now, of course, the winds that shape the ocean currents cover a long distance, so the Coriolis effect shapes them significantly. I showed you a large map of the major ocean currents on page 318, but instead of having you look at it again, I want you to look at the image on the left. It shows you the broad strokes of just a few of those currents. As you can see, the currents form ovals. The curves of those ovals are the result of the Coriolis effect. Look at the oval on the top left. It is made of currents found between the western coast of the United States and the eastern coast of Asia. Notice that the directions of the north/south currents in the ovals follow the Coriolis effect. The current that is near the western United States curves to the west, as you expect for something traveling from a northern latitude towards the equator. In the same way, the current closest to Asia curves to the east, as you expect for something traveling away from the equator to a northern latitude in the Northern Hemisphere.

Notice that the currents in the Southern Hemisphere (the three lower ovals) travel in the opposite direction. The currents moving from southern latitudes towards the equator curve to the west, while the currents traveling away from the equator towards southern latitudes curve to the east. Once again, the curves are what you expect for the Coriolis effect. These currents in the ocean are called **gyres** (jye' urz).

<u>Gyre</u> – A giant ocean current that forms an oval

Now please understand that gyres (and all other surface currents in the ocean) are formed by wind. Periodic blowing of wind like you did in Experiment 1.1 forms waves. A more constant blowing of wind forms currents. So the Coriolis effect shapes the way the winds blow, which in turn, shapes the currents. You will learn more about wind when you study the atmosphere.

The gyres affect anything that is floating in the ocean. When ships use them properly, they can be used to speed up ocean voyages, as Matthew Maury's work showed. However, they can also "trap" things that float in the ocean and cannot move of their own accord. Consider, for example, a piece of plastic floating in the ocean. What would happen if it enters a gyre? It will start spinning around with the current. Since it has no way of moving out of the current, it simply keeps spinning around with the current.

Because of this, garbage that has been thrown into the ocean tends to collect inside gyres. Since the western United States and Asia both pollute the ocean a significant amount, the gyre between them has collected an enormous amount of garbage, mostly in the form of plastic. It is often referred to as the **Great Pacific Garbage Patch**. It is difficult to know how much garbage is found there, because as the plastic floats in the ocean it gets broken down into tiny bits that are hard to see with the naked eye. However, some studies estimate that it might contain as much as 80,000 tons of garbage! That's a small amount of weight compared to the weight of the water found in that part of the ocean, but it is still distressingly high! A group called The Ocean Voyages Institute has been trying to net and collect the garbage so it can be removed from the patch, but the process is difficult due to the small size of the pieces. Nevertheless, as of 2020, they managed to remove over 170 tons.

Surface Currents Causing Vertical Currents

So far, I have been discussing currents that occur on the surface of the ocean. Most of the time, these currents don't affect the water below the surface. However, under certain conditions, they can. Consider, for example, a wind that blows south along the west coast of a continent in the Northern Hemisphere. Because of the Coriolis effect, the wind will curve to the west. As shown in the illustration on the right, that will pull the water away from the coast. But water stays level, so if water from the surface is being pulled away, water from below the surface will rise up to take its place. This is called **upwelling**.

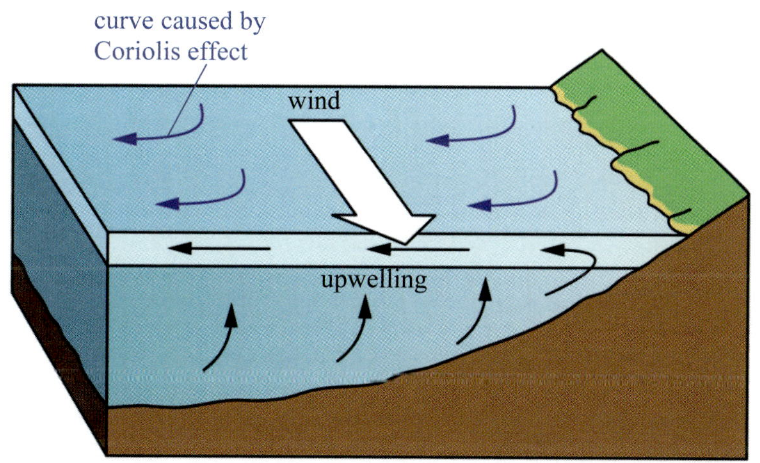

Surface currents that pull water away from the coastline will cause water from below the surface to rise up.

Now if you think about it, upwelling happens near a coast. After all, in the middle of the ocean, there is water all around. If surface water is pushed to the west, surface water from the east can replace it. However, at the coast, there is an edge to the water. When you pull water from that edge, there is no surface water to replace it. As a result, the water must come from below the surface.

It turns out that upwelling is an incredibly important process, because the water near the seafloor is rich in chemicals that living organisms need. After all, dead things may float in the ocean for a while, but they eventually sink and decompose on the seafloor. The decomposed matter has lots of chemicals that are useful to various organisms, but most ocean organisms live at fairly shallow depths, because there is a lot of sunlight there. Upwelling can take the chemical-rich water near the seafloor and move it to shallower depths, which allows the organisms there to use those chemicals.

Of course, the opposite can occur as well. Consider a north wind blowing on that same coastline. The Coriolis effect will cause the wind to curve east, pushing water into the coastline. In the open ocean, that would just cause a current traveling east. However, when the water hits the coast, it will pile up there, especially if the coast is steep. The weight of that piled-up water will force the water underneath it to move down. Not surprisingly, this is called **downwelling**, which allows the surface water (which isn't rich in chemicals) to go down towards the seafloor, where it can pick up chemicals.

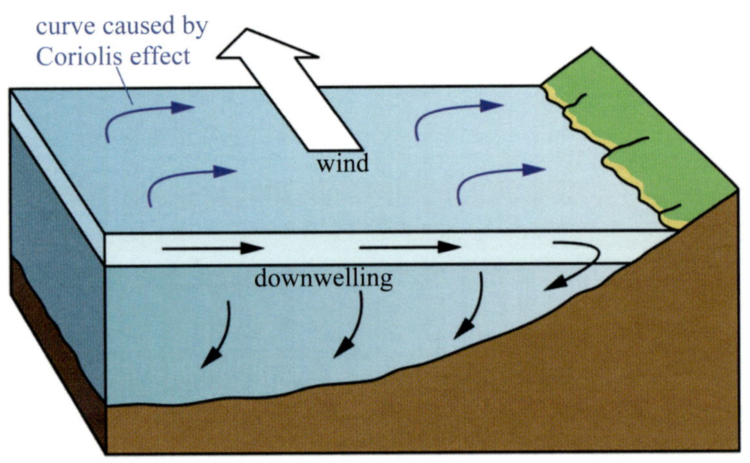

Surface currents that push water towards the coastline will cause water from the surface to sink.

Upwelling and downwelling, then, are part of the ocean's "recycling system." Without them, many vital chemicals would end up at the bottom of the ocean and would be unavailable to the organisms that live in shallower waters. However, upwelling brings those chemicals up into the shallower waters, and downwelling brings new water down to the seafloor so that it can become enriched in chemicals. The ocean has another important recycling system, which I will discuss the next time you do science.

Comprehension Check

11.7 Suppose you are in the Southern Hemisphere and have a long-range cannon. You want to hit a target that is 200 miles directly south of your location. In order to hit that target, in which direction would you aim your cannon (south, southeast, or southwest).

11.8 An ocean surface current flows north but curves east. Which hemisphere is it in?

11.9 On the east coast of a continent in the Southern Hemisphere, upwelling is occurring. Is the wind blowing to the north or south?

Thermohaline Currents

Upwelling and downwelling are important vertical currents, but they happen near the shore. There are vertical currents that happen away from the shore, but they are not produced by winds. Instead, they are produced by changes in the seawater's density. How can seawater change its density? One is by changing its salt content. You already saw that saltwater is denser than freshwater. Well, the more salt that is dissolved, the denser the saltwater becomes.

How does the amount of salt dissolved in seawater change? In cold regions, seawater can freeze. However, in order for water to freeze, it must remove the dissolved salt so that the salt doesn't get in the way of the pattern that the water molecules must form. Thus, the salt is pushed out of the freezing water. That's what makes the temperature at which saltwater freezes lower than the temperature at which freshwater freezes. As the salt is pushed out, some of it goes into the surrounding seawater. The rest gets pushed into small pockets of water inside the ice. Those pockets of water get so rich in salt that they never freeze, so frozen sea ice is made of freshwater ice surrounding tiny pockets of very concentrated saltwater. If you put a cube of frozen seawater in your mouth, then, it would taste salty because of those pockets of saltwater inside the cube.

But remember, some of the salt gets pushed into the seawater surrounding the ice. That makes the seawater saltier. The amount of salt dissolved in water is often referred to as the water's **salinity** (suh lynn' ih tee).

Salinity – A measure of how much salt is dissolved in water

So in places where the oceans are cold enough to freeze some of the seawater, the salinity of the seawater increases. That makes it denser. In addition, the very fact that it is cold makes it denser still. Perform the following experiment to see what I mean.

Experiment 11.3: Temperature and Density

Supplies:

- Two paper or Styrofoam cups
- A glass baking dish or bowl that is wide enough for the two cups to sit side-by-side in the dish (The dish shouldn't be wide enough to fit three of the cups side-by-side. See the picture below.)
- Two different colors of food coloring
- A pen or other object that can make a hole in the cups
- Water
- Ice
- Someone to help you

Instructions:

1. Poke a hole in each cup close to the bottom. Each hole should be about the same size and small enough to cover with your finger but large enough for water to flow out of it in a stream when each cup is filled with water.
2. Adjust the tap so that roughly room-temperature water is flowing, and add water to the glass baking dish until the water level is noticeably higher than the holes you put in the cups.
3. Set the baking pan on a level surface near the sink you are using to get the water.
4. Add ice to one cup until it is about ½ full.
5. Hold one cup in one hand and the other cup in the other hand, covering each hole with your middle finger. Keep them both over the sink.
6. Have your helper add several drops of food coloring to the cup with ice in it.
7. Have your helper add several drops of the other color of food coloring to the empty cup.
8. Have your helper run water from the tap until it is as cold as it can get.
9. Keeping the hole covered with your finger so that very little (if any) water comes out, fill the cup of ice with the cold water. As you fill it, the food coloring should spread out.
10. Keeping that cup over the sink in case some water comes out of the hole, have your helper run water from the tap until it is as hot as it can get.
11. Keeping the hole covered (and the other cup still over the sink), fill the cup that didn't have ice in it with the hot water, and the food coloring should spread out.
12. Carefully set the cups in the baking dish so that they are each near one end and very close to the side. Each of your middle fingers should be touching the side of the baking pan so each hole points to the side it is near (see the picture on the right).

13. At the same time, remove your fingers from the holes so that colored water flows out of each cup. Because of the way you positioned them, water should be flowing out of each cup and hitting the nearest side of the dish.
14. Hold each cup down, because as water leaves the cups, they will start to float.
15. Watch the colored water through the end of the baking pan that the cups are near. You should see a definite pattern to where each color of water moves.
16. Where is the colored water that came from the cup with ice cubes compared to the colored water that came from the cup of hot water?
17. Clean up your mess.

What happened as the colored water started moving through the baking dish? You should have noticed that the ice-cold water flowed under the hot water. Why? Because cold water is denser than hot water. Water is different from most substances when it freezes, but before (and after) it freezes, it behaves like any other substance – it expands when it warms up and contracts when it cools down. Thus, cold water is more compact than hot water. Since density is a measure of the mass divided by volume, the smaller the volume, the higher the density.

Now think about seawater in parts of the ocean that are cold enough for the water to freeze. The water surrounding the frozen sea ice is saltier, which makes it denser. Thus, it begins to sink. However, it is also cold. That makes it denser still. So the cold, salty water that is near sea ice will sink. As that water sinks, however, other water must come in to replace it. Thus, water is pulled towards the sea ice, and then it sinks. This sets up a current that pulls surface water towards the poles (where it is very cold) and pushes water down. Because the current is produced by changes in the amount of salt in the water and changes in the water's temperature, it is called a **thermohaline** (thur' moh hay' line) **current**.

Thermohaline current – An ocean current produced by changes in the salinity and temperature of seawater

The "thermo" part of the word refers to temperature, and "haline" refers to salt, because the mineral name of sodium chloride is "halite."

The basic pattern of thermohaline currents in the ocean is shown in the illustration below. The red stripes represent warmer waters, and the blue stripes represent colder waters. Notice that the warmer waters run above the colder waters. That should make sense based on the fact that cold water is denser than warm water. Notice also where those warm waters travel. They travel towards the poles. When they get to the poles, they cool down, and their salinity increases as they encounter water that is freezing. The lower temperature and increased salinity increase their density, and they sink. Once they sink, they become part of the cold-water current that is flowing deep in the ocean.

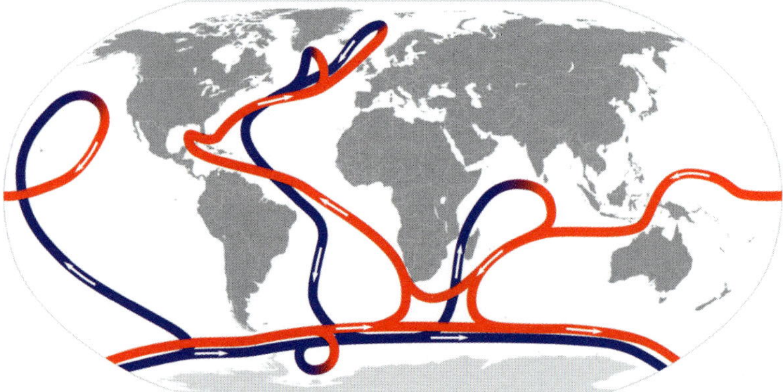

Thermohaline currents, driven by the temperature and density of seawater, mix seawater throughout the ocean.

As you can see, the thermohaline current moves seawater throughout the ocean. As a result, it is often called the **global conveyor**. Like downwelling and upwelling, the global conveyer is part of the ocean's "recycling system," mixing surface water with deep water so that chemicals get more evenly distributed. As I mentioned previously, many important chemicals needed by organisms are less concentrated in the shallow waters, so as those waters sink in the thermohaline current, they gain those chemicals.

While deep waters tend to have more useful chemicals in them, there is one very useful chemical that is more concentrated in shallow waters – oxygen. Most living things in the ocean need oxygen. It is made by certain organisms (like algae), but those organisms tend to live in shallower waters where there is a lot of sunlight. Oxygen from the air can also dissolve into the ocean, but that will only happen at the surface. Thus, shallow waters are rich in oxygen, while deeper waters are not. The global conveyor helps bring oxygen down into the deep waters, where the organisms living there can use it.

More than any other current in the ocean, the global conveyor demonstrates that there really is just one ocean. All the water in the ocean is connected, and it all mixes together. As a result, even though we often talk about the Pacific Ocean, Atlantic Ocean, Arctic Ocean, etc., those are artificial distinctions based mostly on geography. Yes, the Atlantic Ocean separates North and South America from Europe and Africa, while the Pacific Ocean separates North and South America from Asia and Australia. However, they (and all the other "oceans" of the world) are connected, and their waters mix. So while we often talk about the earth's ocean**s**, there is really just one big ocean that we subdivide based on geographic location.

Comprehension Check

11.10 Suppose the poles got warmer and less sea ice froze there. Would that affect the thermohaline current? If so, how?

Freshwater in the Hydrosphere, Part 1: Glaciers

I spent most of this chapter discussing the ocean, since it makes up the vast majority of the earth's hydrosphere. However, we cannot exist without freshwater, so it is also important to cover the freshwater in the hydrosphere. As you learned in the previous chapter, most (68.5%) of the earth's freshwater is frozen. It makes sense, then, to start with that.

About 90% of the earth's frozen freshwater is found in Antarctica. When precipitation occurs there, it comes in the form of snow. Most of that snow doesn't melt, so it just starts to accumulate. However, snow has weight, so as more snow accumulates, the lower layers of snow get compacted. As the snow gets compacted more and more, it becomes ice. But remember, snow is still accumulating above the ice, so the weight continues to increase.

Suppose you have a big ball of cookie dough or bread dough. If you push down on it, what happens? The dough gets compacted together, but it also spreads out. The same thing happens to the ice that forms underneath all that snow. Once it is compacted, it starts to spread out. In other words, the ice actually *flows*, moving so that it covers all the land in the area. At that point, you have an **ice sheet**.

Ice sheet – A permanent layer of ice covering a very large area of land

That's what you find in Antarctica, and about 90% of the ice in the hydrosphere is found in the Antarctic ice sheet. Most of the rest of the ice in the hydrosphere is found in the ice sheet on Greenland.

Computer generated images of the Antarctic ice sheet (top) and the continent over which it has formed (bottom).

So even though we don't see the land that makes up Antarctica, it is there, under the flowing ice sheet. In fact NASA has been flying missions over Antarctica and using radar to map the land that is found underneath the ice sheet. The result is shown in the two images on the left. They aren't pictures. They are computer-generated images based on the results of NASA's analysis. The top image shows what we can see – a continent covered in an ice sheet. The bottom image shows what the radar found: the land underneath the ice sheet. This, of course, is the major difference between the Arctic and Antarctica. Antarctica is a continent; the Arctic is not. One common way scientists distinguish the two is to say that Antarctica is a continent surrounded by an ocean, while the Arctic is an ocean surrounded by continents.

Now remember, the ice sheet in the image is spreading out. As it does, its edges reach the ocean. Since ice is less dense than water, it starts floating in the ocean, but it is still connected to the ice that is on land. As a result, it doesn't float away. This is called an **ice shelf**.

Ice shelf – A floating sheet of ice that is permanently attached to land

Antarctica, then, is an ice sheet that has spread out enough to form ice shelves. This means the ice that we see in Antarctica covers a lot more area than the continent itself.

An ice sheet is a specific example of a more general structure called a **glacier**, which is just any mass of slowly-moving ice.

Glacier – A slowly-moving mass of ice caused by the accumulation of snow

While ice sheets are (as their name implies) fairly flat, many glaciers are more like "rivers" of ice that start out at a high elevation. As you are probably aware, when a mountain is high enough, it can reach elevations that stay cold enough to keep snow all year long. As a result, the snow accumulates like it does on an ice sheet. However, as the compacted ice begins to move, it moves down the mountain. This happens very slowly, but nevertheless, it does happen. As a result, the "river" of ice flows (very slowly) down the mountain until it eventually reaches a body of water, forming an ice shelf. The picture below shows a glacier in Norway. Notice how it looks like there was a river flowing down the mountain, and it suddenly froze. That never happened. It is a glacier, so it has been frozen the entire time. Pressure from newly-formed snow at the top of the mountain has been slowly pushing the ice down until it hits water, at which time, it becomes part of an ice shelf.

The ice shelf will eventually travel far enough into the water that it will start to become less and less stable. Eventually, parts of the ice shelf will break away in a process known as **calving** (kav' ing). Look at the lower picture on the right. It comes from a different glacier (one in Argentina). In the picture, you are only seeing a small portion of the glacier. In fact, you are just seeing part of the ice shelf at the end of the glacier. The piece of ice that is falling is in the process of calving.

The top photo shows a glacier in Norway, while the bottom shows the ice shelf from a glacier in Argentina, which is in the process of calving.

The mass of ice that you see calving in the photo ended up floating on its own, since frozen water floats on liquid water. At that point, it is called an **iceberg**. Now think about what that means. Even though there are a lot of icebergs floating in the cold parts of the ocean, they are not made from frozen seawater. They are made of frozen freshwater and are the result of an ice shelf calving. That's much different from the sea ice that is pictured on page 292, which is the result of ocean water freezing.

Now remember, most of the world's freshwater supply is frozen, and most of that is in glaciers. Thus, it shouldn't surprise you to learn that there are *a lot* of glaciers in the world. Estimates indicate that there are at least 200,000 glaciers on the earth; probably more. So while you and I might not have much experience with glaciers, the world is packed with them!

With all of this ice in the hydrosphere, you might wonder what would happen if a lot of it started to melt. Well, that depends on where the ice is. If the ice is on land, the liquid that comes from melting will flow more quickly. Eventually, it will flow into the ocean or into another body of water that empties into the ocean. As a result, if a lot of the ice that is sitting on land melts, the ocean levels will rise, because they will have a lot more water in them.

However, imagine filling a glass with ice cubes and then adding enough water so that the ice cubes are right at the very top of the glass. If those ice cubes melt, what happens to the water level in the glass? Nothing! What happens if you fill a glass partway with water and then add ice? The water level rises. That's because about 90% of the ice goes underwater, and it must push the liquid water out of the way to do that. The liquid has nowhere to go but up, so the ice cube makes the water level rise. In fact, it makes it rise to exactly the same level it will have when the ice cube melts.

So in the end, any ice that is floating in the ocean will have no effect on the ocean's level when it melts. However, most of the hydrosphere's ice is in glaciers (including ice sheets), which sit on top of land. Thus, if a lot of the ice in the hydrosphere melts, the level of the ocean will rise. In fact, that's happening right now. Over time, scientists have kept detailed records of the ocean's water level, and it has been rising. I will discuss this more when I discuss the earth's atmosphere.

Freshwater in the Hydrosphere, Part 2: Groundwater and Soil Moisture

While most of the hydrosphere's freshwater is in the form of ice, there is also a lot of freshwater found in the ground. How does it get there? Mostly through precipitation and percolation. Remember, when rain falls on the ground, it might run across the surface of the land and drain into a lake, river, stream, or ocean. If that's the case, it is called surface runoff. However, it might also soak into the soil, traveling downward. At that point, it is part of the soil moisture. Not surprisingly, plants can absorb that soil moisture with their roots and either use it in photosynthesis or allow it to evaporate from their leaves. However, the water might also continue to travel deeper into the soil.

Of course, as you learned in Chapter 2, soil isn't a solid mass. Instead, it is a collection of particles with pores in between. A fertile soil has pores that are sized so that they hold on to water. However, they don't *fill* with water. That would be bad for the roots of plants. Plants make food through photosynthesis, which also makes oxygen. However, when the plants actually use that food for energy, they consume oxygen. As a result, all parts of a plant (including the roots) need to be exposed to oxygen. So the pores in soil need to have air in them, but they must also have water in them. When this is the case, we say that the soil is **unsaturated** with water. In other words, because there is air in the pores, there is room for more water.

The roots from most plants are found in unsaturated soil. That way, they can absorb the water, but they also are exposed to oxygen in the air. But wait a minute. The water is percolating down into the soil because gravity is pulling it down. Why does it "stop" and stay in the unsaturated soil? Because the water is attracted to chemicals in the soil, and that attraction can overcome gravity, keeping the water "stuck" to the soil. Think about wetting a towel and then hanging it out to dry. If the towel is soaking wet, the water will drip down out of the towel and onto the ground. Eventually, however, the towel will stop dripping, but it will still be wet. That's because the water is being held to the towel by its attraction to molecules in the towel. In the same way, water is held in the pores of unsaturated soil by its attraction to chemicals in the soil.

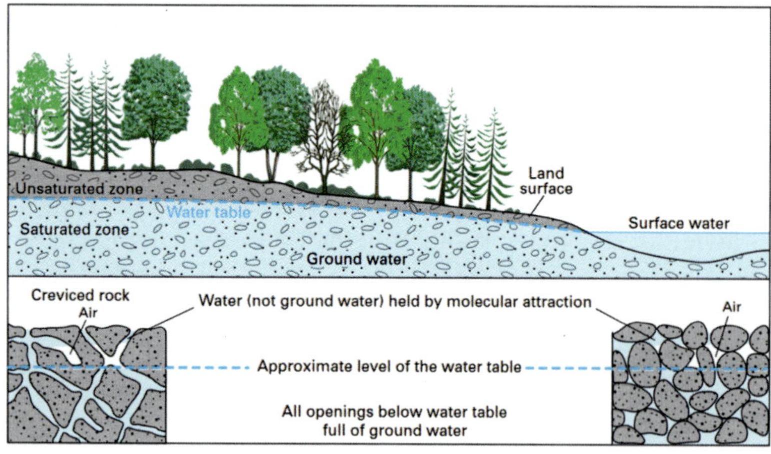

Soil moisture is found above the water table, while groundwater is found below it.

If the soil is wet enough, some of the water will continue to percolate deeper into the soil, just like water dripping from a soaked towel. If it percolates deep enough, it will eventually encounter soil in which all the pores are completely filled with water. This is called **saturated** soil, because there is no room for any more water. The separation between the unsaturated and saturated soil is called the **water table**.

Water table – The level below which the soil is saturated with water

The water table also applies to rocks, since they have crevices and pores in them which can hold water. When soil moisture reaches depths below the water table, it is part of the hydrosphere's groundwater. If it stays above the water table, it is part of the hydrosphere's soil moisture, even when it is in rocks.

Now if you think about it, you should realize that the depth of the water table changes, depending on how much rain there is in the area. If there is a lot of rain, there is a lot of water percolating down into the soil, and the soil will become saturated at shallower depths. That means the water table will be shallower. Conversely, if there hasn't been a lot of rain for a long time, the groundwater will have flowed away (remember, groundwater flows to where it can be dumped into a river, lake, stream, or ocean) and not been replaced. That means soil which was once saturated will now be unsaturated, which means the water table will be deeper. In most regions, the water table is shallower in the spring because spring brings a lot of rain and, at the same time, there is a lot of snow and ice melting. The water table then gets deeper in the summer, when there is less rain and no more snow and ice to melt.

Even when we are considering soil moisture, we cannot get away from the fact that there is a lot of ice in the hydrosphere. In fact, some parts of the planet are so cold that the soil moisture remains frozen all year round. We call that **permafrost**, like "permanent frost." Of course, it's not really permanent. If there are several years of warm weather in a row, some or all of the frozen soil moisture will melt. Nevertheless, it is more permanent than most frost, since it tends to stay frozen through the summer.

How much permafrost do you think exists on earth? You might be surprised to find out that about one-fourth of the entire Northern Hemisphere's land is covered in permafrost. It is mostly found in the far north latitudes, like the northern parts of Siberia, Canada, Greenland, and Alaska. The picture of permafrost on the right, for example, was taken in Alaska. That's not snow clinging to the side of the hill. It is soil moisture that has been frozen and was exposed by weathering and erosion. There is less permafrost in the Southern Hemisphere, being found mostly in the mountains of South America and New Zealand. There is a lot of permafrost in the soil of Antarctica as well. Now don't confuse the ice sheet of Antarctica with its permafrost. The ice sheet sits on top of the ground in Antarctica. The permafrost is in the frozen soil of Antarctica, which is under the ice sheet.

Soil moisture that stays frozen for at least two years is called permafrost.

Comprehension Check

11.11 You have two samples of ice. You are told that one is from an iceberg, and one is from frozen seawater. How could you determine which is which?

11.12 You have a sample of soil, and you are told that it was taken from in between a bunch of tree roots. Was the soil probably taken from above the water table or below? Would the soil have been saturated or unsaturated?

11.13 The water table in a particular region is measured to be at a depth of 3.1 meters. In a few months, it is measured to be at a depth of 2.5 meters. Most likely, what was the rainfall like in that region between the times the two measurements were taken?

Freshwater in the Hydrosphere, Part 3: Water in the Atmosphere

One part of the hydrosphere that can escape your notice is the water found in the air around you. Some of it is in the form of gas, which comes from evaporation. As you learned when you studied the water cycle, surface water is evaporating into the air all the time. This means there is water vapor (water in its gas phase) in the air. Now, of course, this varies depending on where you are on the earth. In the state where I live (Indiana), we have a lot of water vapor in the air. In states that are part of a desert (like Arizona), there isn't much water vapor in the air. We usually refer to the amount of water vapor in the air as **humidity** (hyoo mid' uh tee).

Humidity – A quantity that describes the amount of water vapor in the air

You will study humidity in more detail when you learn about weather. For now, just remember that even though water vapor is in the atmosphere, it is part of the hydrosphere.

The other water you find in the atmosphere is in the form of clouds. While clouds are very familiar, they are often misunderstood. Some students think that clouds are made of water vapor. They are not. Perform the following experiment to see what they are actually made of.

Experiment 11.4: Making a Cloud

Supplies:
- A small, clear plastic bottle, like the ½-liter bottles water comes in, and its lid (The flimsier the plastic bottle, the better.)
- Scissors (You only need these if the bottle has a label on it.)
- Matches
- Water

Instructions:
1. If there is a label on the bottle remove it by cutting it with scissors and tearing it off the bottle. Don't worry if some of the label stays on the bottle. You just want the bottle to be mostly clear.
2. Let the water run until it gets warm. It doesn't have to be really hot, but it needs to be warmer than room temperature.
3. Add water to the bottle until there is about 3 centimeters (about 1 inch) of water in it.
4. Put the lid on the bottle and close it so it is airtight.

5. Grab the bottle with one hand and hold it out in front of you at eye level.
6. Squeeze the bottle as hard as you can and hold it for a few seconds. If the lid is airtight, you won't be able to completely collapse the bottle. If the lid is not airtight, go back to step 4 and try again.
7. Look at what is inside the bottle. It's probably not very interesting.
8. After you have squeezed the bottle has hard as you can, relax your grip so that the bottle expands again. Keep watching the inside of the bottle, and don't drop it!
9. Did you notice any difference inside the bottle while you were squeezing and after you relaxed your grip? Probably not. If you did notice a difference, it was pretty small.
10. Remove the bottle's lid.
11. The bottle is probably still a bit crushed. If so, use your mouth to blow air into it so that it is less crushed.
12. Light a match and allow it to burn so that you have a strong flame.
13. Drop the match into the bottle so that it goes out when it hits the water at the bottom of the bottle.
14. Put the lid back on the bottle so it is airtight.
15. Repeat steps 5-9. Now you should see a big difference in the contents of the bottle after you stop squeezing. Feel free to do this several times.
16. Remove the lid and dump out the water and match.
17. Once again, if the bottle is crushed, use your mouth to blow air into it to partially uncrush it.
18. Hold the bottle in one hand and put your thumb (or a finger) in the opening. Stick it so far into the bottle that you form an airtight seal between the base of your thumb and the bottle.
19. Once again, squeeze the bottle as hard as you can, concentrating on the temperature your thumb feels.
20. Relax your grip, once again concentrating on the temperature your thumb feels.
21. Feel free to repeat steps 19 and 20 several times.
22. Clean up your mess.

What did you see in the experiment? Let's start with the end of the experiment. When you squeezed the bottle, your thumb should have felt a bit warmer. When you relaxed your grip, your thumb should have felt cooler. Remember, temperature is a measure of the energy of molecules in motion. When you squeezed the bottle, you were giving the molecules of air inside the bottle more energy, so the temperature increased. When you relaxed your grip, they moved to fill the extra volume, and they had to use up energy to do that. Thus, the amount of energy in the molecules decreased, which reduced the temperature.

The same thing happened when you squeezed the bottle with water in it. The air inside the bottle had water vapor in it, since water was evaporating from the liquid at the bottom. When you squeezed the bottle, the air got hotter. This caused more water to evaporate, since the warmer it is, the faster water evaporates. When you relaxed your grip, the temperature reduced. What happens when you cool water vapor? It condenses, forming a liquid.

When you first did this, you might not have seen any condensation. That's because condensation works best when there is a surface onto which it can happen. You might have seen some condensation on the walls of the bottle, since they are a surface onto which water can condense, but that is about it. What happened in the second part of the experiment? You should have seen a "cloud" form in the bottle. That's because there were all sorts of surfaces onto which the water vapor could condense. What were those surfaces? They were the smoke particles formed by the match when it was extinguished by the water.

What happened in the bottle was a good model for what happens in the atmosphere in order to form a cloud. Evaporation produces water vapor, and that water vapor rises in the air. As it rises, it cools. You will learn why later on. For right now, just understand that for the part of the atmosphere in which clouds form, the higher you go, the colder it is. Thus, as water vapor rises, it cools enough to condense.

For the condensation to occur, however, there must be surfaces onto which the water can form droplets. Fortunately, there is a lot of fine dust floating in the air, and just like the smoke in your bottle, the dust particles provide a surface onto which the water vapor can condense. These particles are called **cloud condensation nuclei**. As long as there is a lot of water vapor in the air as well as cloud condensation nuclei high in the air where it is cool, water droplets will form. As that continues to happen, a cloud is formed. If the clouds form high enough, it will be cold enough for the water droplets to freeze. At that point, there will be ice on the cloud condensation nuclei.

The colors you see are caused by sunlight hitting ice particles in the clouds. When the angles are just right, the ice produces a "flat rainbow."

So clouds are *not* composed of water in its gas phase. They are composed of water in its liquid or solid phase. Sometimes, this can produce a dramatic effect. The picture on the left, for example has not been edited. If you were to view the same kind of clouds in the same sunlight, you would see the same thing. The clouds in the picture are composed of ice. The sun (out of the photo on the left) is shining its light on them, and the light hits the ice crystals in just the right way to form a "flat rainbow." If the cloud were made of liquid, you would not see the colors. You see them only because the clouds are made of ice!

But wait a minute. Water is denser than air. It should fall, right? Ice is less dense than water, but it is still more dense than air. Why don't clouds fall from the sky? Well, first, in order to fall, the water or ice in the clouds must push the air molecules away. The water droplets and ice particles in clouds are so small that they don't do that very well. They do it a bit, but not much. If that were the end of the story, they would fall – very slowly.

But remember that there is water vapor rising as well. There is also air rising. That forms a gentle current pushing upward. As long as the water droplets or ice crystals are small enough, that gentle upward current stops their slow downward fall. However, as more water accumulates in the cloud, the droplets get bigger, and the gentle current can't stop their fall anymore. At that point, the clouds produce precipitation.

Comprehension Check

11.14 You have two containers holding water vapor. You blow a gentle stream of cold air through one of them. In the other, you insert a flat piece of glass at the same cold temperature as the air you blew into the other container. Which of those two situations will produce more condensation?

Answers to the Comprehension Check Questions

11.1 The one on the right. The change in depth is more abrupt there, which should produce larger, more powerful waves.

11.2 8 hours. There are two low tides and two high tides per day, for a total of four tides. They must be spread out over the day, so if a day is 32 hours, there must 8 hours in between them.

11.3 High tide would be lower and low tide would be higher. With less gravity, the oval wouldn't be as stretched out, so the wide part would not be as wide, and the narrow part would be less narrow. That means there would be less difference between low and high tide.

11.4 High tide would be higher and low tide would be lower. Remember, the closer the object is, the stronger the gravitational force. Thus, the oval would be even more stretched out, making the wide part wider and the narrow part narrower.

11.5 They would not look the same because of weathering and erosion. Remember, the water moves in and out twice per day. That will weather the rocks and carry the sediments produced by that weathering away. Indeed, a scientific paper, "Stacks and notches at Hopewell Rocks, New Brunswick, Canada" (*Earth Surfaces, Processes, and Landforms*, **23**: 975–988, 1998) analyzed old records of Hopewell Cape and showed that the rock formations are different now from what they were 100 years ago.

11.6 Red moves fastest, and green moves slowest. Remember, they all have to make a trip around the circle in the same amount of time. Red's circle is largest, so it has the longest trip. Thus, it travels the fastest. Green's circle is smallest, so it travels the slowest.

11.7 You would aim southwest. In the Southern Hemisphere, going south means moving away from the equator. So your cannon is traveling east faster than the ground over which the shell will be moving. Thus, the shell will curve to the east. Look at the drawing on page 321. The dashed brown curve is moving south in the Southern Hemisphere. It curves east. If you want to hit something directly south, you will have to aim to the west to counteract the curve that will go to the east.

11.8 It is in the Northern Hemisphere. To curve east, the thing that is traveling must start moving with the earth faster than the part of the earth into which it is moving. If you move north in the Southern Hemisphere, you are moving towards the equator, which means you are moving into parts of the world that are traveling faster with the earth's spin. In the Northern Hemisphere, moving north causes you to go into parts of the world that are traveling slower. Look at the drawing on page 321. Only the dashed purple curve goes north and east, and it is in the Northern Hemisphere.

11.9 It is blowing to the south. For upwelling to occur, water must be pulled from the coast. Since this is an east coast, water must be pulled to the east. In the Southern Hemisphere, currents traveling south are the ones that curve east (the dashed brown curve in the illustration on page 321).

11.10 Yes. It would reduce the current. Remember, the salinity of the water near the poles increases as sea ice forms, pushing salt into the water. Less sea ice means less salt being pushed into the water, which means lower salinity. That means less density, which would reduce how much the water sank. Also, the water wouldn't get as cold, which would have the same effect.

11.11 See if there is salt in it. The one with salt is the sea ice. Salt has to be removed from water for it to freeze, but remember that it can be either pushed out of the ice or pushed into tiny packets of unfrozen water inside the ice. That means there will be salty water inside sea ice. Since icebergs come from glaciers, there is almost no salt in them.

11.12 It was probably taken from above the water table, where the soil is unsaturated. Remember, roots need oxygen in the air, so there needs to be air where the roots are. If there is air, the soil is not saturated, which means it is above the water table.

11.13 There must have been a lot of rain. A lot of rain will cause the shallower soil to become saturated, so the water table will rise to shallower depths.

11.14. The one that has the glass will have more condensation. The first one has air at a temperature that would allow condensation to happen, but condensation needs a surface to happen effectively. The piece of glass provides a surface, so lots of condensation happens. This is why a cold glass develops water droplets on its outside surface. The temperature is not only low, but there is a surface upon which the water can condense.

Chapter Review

1. Define the following terms:

a. Tides
b. Spring tide
c. Neap tide
d. Coriolis effect
e. Gyre
f. Salinity
g. Thermohaline current
h. Ice sheet
i. Ice shelf
j. Glacier
k. Water table
l. Humidity

2. If you were able to watch a single water molecule in the ocean, how would you see it moving when it is part of a surface wave?

3. Why are surface waves near the shore very different from surface waves that are far from the shore?

4. What is the main cause of the ocean's surface waves?

5. What two bodies influence the earth's tides? Which is more important?

6. You watch the ocean at a beach for 24 hours. How many high tides should you see? How many low tides?

7. What is the main difference between a tidal wave and a tsunami?

8. Where did Matthew Maury find inspiration for his discovery of the ocean's surface currents?

9. What effect causes surface currents in the ocean to form ovals?

10. Suppose you measure the speed at which the land is moving as it rotates with the earth. The first place you measure the speed is near the North Pole. The second place is near the equator. How do those speeds compare?

11. What is the Great Pacific Garbage Patch, and what causes it?

12. Where does upwelling and downwelling occur? Why is it important?

13. You have two samples of water. The salinity of the first is higher than the salinity of the second. If they are in layers, which is on top?

14. For thermohaline currents, in what part of the earth does ocean water sink? What makes it sink?

15. Which part of the thermohaline current is warmer: the part that runs in shallow water or that part that runs in deep water?

16. What is the difference between an ice sheet and an ice shelf?

17. Where is the most important ice sheet in the hydrosphere?

18. Think about the terms "glacier" and "ice sheet." Which is the more general term?

19. What is calving? What does it produce?

20. Suppose all the earth's icebergs suddenly melted, but none of the glaciers did. Would the ocean level rise, fall, or remain the same?

21. What happens to the depth of the water table when an area experiences a long period without rain?

22. Where is groundwater in relation to the water table?

23. When soil moisture stays frozen for a long time, what is it called?

24. What are the possible phases for water that makes up a cloud?

25. Besides water, what else must a cloud have?

Chapter 12: The Atmosphere

Introduction

So far, you have learned a lot about the geosphere and the hydrosphere. The last "sphere" I want to discuss is the atmosphere. As you learned in Chapter 2, the atmosphere is the collection of gases that surround a planet. On earth, we call this collection of gases **air**, and it is absolutely essential for life. Despite the fact that we are surrounded by air and are constantly breathing it in and out, it took scientists a long time to truly understand air and the many effects it has on us and our surroundings. In this chapter, you will learn about the composition of air, its effects on the surroundings, and how it changes with altitude.

Air Has Weight and Takes Up Space

Move your arm up and down. It moves easily, doesn't it? Suppose you were underwater and made the same motion. Would your arm move as easily? Of course not. The water would resist your arm's motion. Because it is so easy to move through air, you might think that, unlike water, air doesn't have any weight. Also, since you can't see it, you might think that it doesn't take up any space. Let's dispel both of those notions with a couple of experiments.

Experiment 12.1: Air Has Weight and Takes Up Space

Supplies:

- A stick that is at least 30 centimeters (1 foot) long and would be very easy for you to break in two with your hands (It can be a dead stick from a tree or a thin piece of wood, but it must be much longer than it is wide.)
- Two full sheets from a newspaper, or two of any sheet of paper that is about 60 centimeters x 60 centimeters (23 inches x 23 inches). You can also tape pieces of paper together to make two sheets that size
- A table where the newspaper can be completely spread out
- A balloon
- A plastic bottle (or jug) that has a small enough opening for the balloon to fit over it (see picture on the next page)
- The nail from the laboratory kit made for this course
- A facial tissue or a couple of squares of toilet paper
- Something to cover your eyes, like safety glasses or goggles (for the first part of the experiment)

Instructions:

1. Put the stick on the table so that slightly more than half of it rests on the table, and the rest hangs out over the edge.
2. Ball up one of the newspaper sheets and rest it on the end of the stick that is on the table.
3. Put on your eye protection.
4. Hit the part of the stick that is hanging over the edge of the table hard with your hand. What happens?
5. Put the stick back on the table the same way as before.
6. Put the sheet of newspaper over the stick so the stick is under the center of the sheet and the newspaper comes up to the edge of the table (see the picture on the right).

7. Repeat step 4. What's different this time? If the stick didn't break, try again, and hit it even harder.
8. Lower the balloon into the bottle and then pull the opening of the balloon over the opening of the bottle so that the balloon hangs upside down in the bottle (see the photo below).

9. Put your mouth over the openings of the bottle and balloon and try to blow up the balloon.
10. Remove the balloon from the top of the bottle.
11. Use the nail to poke a hole near the very bottom of the bottle. It needs to be on the side, very close to the bottom.
12. If the bottle didn't change shape at all, skip the next two steps.
13. Put your mouth over the opening of the bottle and cover the hole you just made with a finger.
14. Blow into the bottle to inflate it back to its original shape.
15. Repeat step 8 so the balloon is hanging upside down in the bottle once again.
16. Set the bottle on the table.
17. Wad up the tissue and lay it right next to the hole at the bottom of the bottle. The wad will relax; that's okay.
18. Keeping the bottle where it is, once again put your mouth over the openings of the bottle and balloon and try to blow up the balloon.
19. What happens?
20. Clean up your mess.

What happened in the first part of the experiment? When the paper was wadded up, the stick should have easily flipped off the table, throwing the wad of newspaper into the air. However, when the newspaper was spread out, the stick should have been harder to flip off the table. Most likely, it actually broke! Why? It couldn't have been the weight of the newspaper, since the weight of the wadded-up newspaper was roughly the same as the weight of the spread-out newspaper.

The difference was the amount of air that had to move in order for the stick to move. After all, the stick and paper were surrounded by air. For them to move, the air around them had to be pushed out of the way. When the paper was wadded up, only a small amount of air had to move. When the newspaper was spread out, *a lot* more air had to move. The stick wasn't strong enough to move all the air over the spread-out newspaper, so the stick broke.

Why wasn't the stick strong enough? Well, think about putting a tiny rock on the stick instead of the wadded-up newspaper. It would be easy for the stick to move that tiny rock, so the results would have been the same as what you saw with the wadded-up newspaper. However, what would have happened if you used a heavy brick instead of a tiny rock? The stick would have broken, just like it did with the spread-out newspaper. Why? Because of the brick's *weight*. The stick wouldn't have been strong enough to move the brick, since the brick would have a lot of weight. The stick broke with the spread-out newspaper for the same reason. It wasn't strong enough to move the air above the newspaper, since all that air had a lot of weight.

So this tells us that air has weight. For most of the history of science, people didn't understand that. After all, you don't feel air's weight like you feel the weight of a brick. Thus, for most of science history, scientists thought it had no weight. However, by the 1600s, enough evidence had accumulated that scientists began to understand that air does, indeed, have weight. Interestingly enough, however, the Bible has always taught that air has weight. Job 28:25 says, "When He imparted weight to the wind and meted out the waters by measure." In other words, God gave the wind (air) weight. Since weight requires mass, we can also say that air has mass. This is yet another example of a scientific truth that the Bible taught long before science figured it out. Why don't you feel the weight that God imparted to air? You will learn that the next time you do science.

What happened in the second part of the experiment? You shouldn't have been able to blow up the balloon much before you poked a hole in the bottle. Why? Because in order for the balloon to blow up, it needed to take up more space in the bottle. However, even though you couldn't see it, there was air in the bottle, and it was taking up space inside the bottle. Since there was nowhere for the air to go, there was not much space in the bottle, so the balloon couldn't expand much.

When you poked a hole in the bottle, the air had somewhere to go. You should have seen that by watching the tissue. As you blew up the balloon, the tissue moved away from the bottle. That's because the balloon forced air out of the bottle so that it could take up the space the air was taking up. The wind created by the air moving out of the bottle pushed the tissue away from the bottle.

The blue haze around the earth represents its atmosphere.

Air surrounds the planet, forming the earth's atmosphere. The image on the right attempts to illustrate this with the blue haze around the earth. Why doesn't this blue haze just dissipate out into space? Specifically because air has mass. The earth's gravity attracts anything that has mass, including air. That holds the air close to the earth. If air didn't have mass, the earth wouldn't be surrounded by an atmosphere!

Comprehension Check

12.1 Suppose the earth's gravity were a lot stronger. Would you expect the amount of air in the atmosphere to be the same as it is now? If not, would there be more or less air in the atmosphere?

Air Pressure

Suppose you need to slice some vegetables in order to make dinner. You look at the knives that are clean, and you find a dull butter knife and a sharp steak knife. Which do you choose? The sharp one, right? Why? Because you know the sharp knife will cut through the vegetables better than the dull one. But what is it about a sharp knife that makes it cut better than a dull one? Mostly, it's about the **pressure** that you can exert with each knife. While we use that word in a variety of different ways in everyday speech, scientists have a very specific definition for it:

Pressure – The force applied to an object divided by the area over which it is applied

As you already learned in Chapter 1, area is the length of something times its width. For a knife, then, the area of the knife edge is the length of the knife times the width of the edge. The sharper the knife, the thinner the edge, so the smaller the area.

When you are cutting something, you are using your muscles to push down on the knife. Your strength determines the force you can use; the stronger you are, the more force you can push with. However, the pressure you exert depends on the force *divided by* the area of the knife. Since sharper knife edges have smaller areas, that means the pressure you can use with a sharper knife is calculated by taking the force your muscles can exert and dividing it by a small number. What happens when you divide by a small number? The result is big. That means the sharper the knife, the larger the pressure you can exert with it. The larger the pressure, the easier it is to cut something up.

Now think about the unit you would need for pressure. Remember that in the English system, the unit for weight is the pound. In the metric system, it is the Newton. Well, weight is simply a measure of the force with which gravity pulls on an object, so pound and Newton are also the units for weight. Since area is length times width, its units are square inches (in^2) in the U.S. system and square meters (m^2) in the metric system. To get pressure, you divide force by area, so in the U.S. system, the unit for pressure is **pounds per square inch (pounds/in^2)**, or **psi** for short. In the metric system, it is **Newtons per square meter (Newtons/m^2)**, which is called the **Pascal** in honor of a famous scientist named Blaise Pascal, who taught us a lot about pressure.

Remember that air has weight which pushes on objects, exerting a force. If you divide by the area over which the weight is pushing, you get the **air pressure**, which is also called the **atmospheric pressure**. On average, the air around you is exerting a pressure of 14.7 psi. That means there is 14.7 pounds of air pushing down on every square inch of your body! Does that sound like a lot? It *is*. To get an idea of just how strong this pressure is, perform the following experiment.

Experiment 12.2: Air Pressure

Supplies:

- An empty thin aluminum can with a small opening, like the kind soda comes in (More than one is even better.)
- A stove
- A pan (Make sure it's an old pan, because this experiment can damage its finish.)
- Kitchen tongs
- A bowl
- Water
- Ice

Instructions:

1. Run the water from the tap until it is as hot as it can get.

2. Add a small amount of hot water to the can, so that there is about a centimeter (under half an inch) sitting at the bottom of the can.
3. Put the pan on a burner and turn it on high.
4. Put the can in the center of the pan. You want to get the water in the can to boil.
5. While you are waiting for the water in the can to boil, add several ice cubes to the bowl.
6. Run the water from the tap until it is cold, and add that to the bowl so it is mostly full of ice water.
7. Set the bowl of ice water on the counter near the stove but not so close that it will get warm.
8. Wait for the water to be boiling vigorously inside the can. The best way to see that is to look for steam rising up and out of the opening of the can. When you get a noticeable amount of steam coming out continuously, you know the water is boiling vigorously.
9. Use the tongs to grab the can and lift it out of the pan and over the bowl of ice water.
10. Quickly turn the tongs so that the can turns upside down, and lower it into the bowl of ice water so as much of the can as possible is under water. What happens? If nothing dramatic happens, you should probably try again, waiting for more steam to come out of the can's opening.
11. **The pan will be *very* hot**, so be careful while you clean up your mess. You might want to just turn off the burner, move the pan to another part of the stove, and wait before you do the cleanup.

If things went well in your experiment, you should have seen the can get crushed the moment it came into contact with the ice water. Why? Think about what is going on with air in the experiment. When you have an empty, open can, there is air outside the can pushing in on it, but there is also air inside the can pushing out. The pressure pushing from inside the can is the same as the pressure pushing from outside. Since the two pressures are equal and pointed opposite of each other, they cancel out, and the can feels no net pressure, as shown in the near side of the illustration on the right.

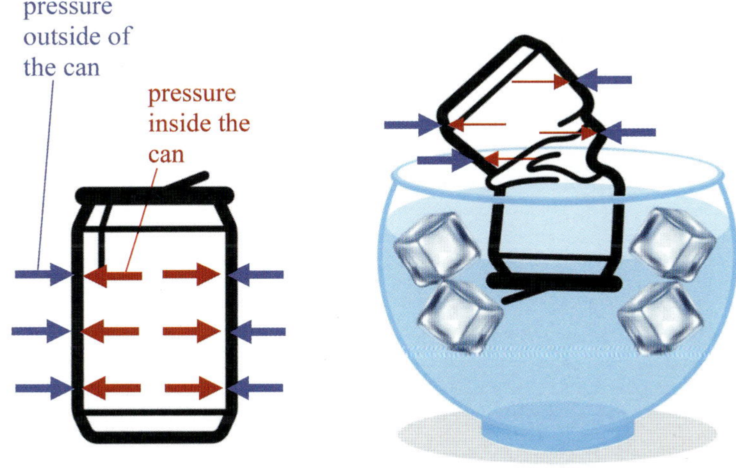

In your experiment, when the water vapor in the can condensed, the pressure inside the can got very small, and it could no longer cancel out the pressure that air was exerting on the outside of the can.

When you boiled the water inside the can, the pressure situation remained the same, but the water vapor that was being produced by boiling replaced the air in the can. It still had the same pressure as the air, but it was all hot gaseous water, not air. Once again, the can felt no net pressure, because the water vapor was pushing out on the can with the same pressure as the air pushing in on the can. However, when you inverted the can and put it in the ice water, the water vapor condensed back into a liquid. That meant there was very little water vapor, so there was almost no pressure inside the can. Since there was very little pressure pushing out but a lot of pressure pushing in, the can was crushed.

The left-hand side of the illustration also explains why you don't feel the weight of air or its pressure. There are 14.7 pounds of air pushing in on every square inch of your body. However, your body has an opening (your mouth) that connects it to the air outside your body. As a result, there is air inside your body, and it is pushing out with the same pressure as the air outside your body is pushing in. Those two pressures cancel out, and like the soda can, you don't feel any net pressure. That's what fooled scientists for so long. You don't feel weight or pressure from air, but that's not because air has no weight and exerts no pressure. It's because there is equal pressure pushing out from inside your body, which cancels the pressure air is exerting from outside your body. If the air inside your body were suddenly removed, you would be crushed, just like the can.

Since air pressure is produced by the weight of air pushing against things, it actually changes depending on where you are on the earth. Consider, for example, the beautiful landscape pictured below. It comes from the Tara River Gorge in Montenegro. Think about the amount of air pushing down on the surface of the river. It's essentially all the air directly above it, right? If you divide the weight of all that air by the area of the river's surface, you get the pressure that the river is being exposed to.

The air pressure at the peak is lower than the air pressure on the river.

Now think about the amount of air pushing down on the big peak pointed out in the picture. How would it compare to the weight of air pushing down on the river? Well, since the peak is higher, there is less air above it. Thus, there is less weight pushing down on the peak. If you divide that weight by the area of the peak, you will find that the pressure is *lower* on the peak than it is on the river, because the weight of the air is lower. As a result, the higher you are above the surface of the earth, the lower the air pressure.

Comprehension Check

12.2 Suppose you are at the top of a very high mountain. You drink a bottle of water, but like a good hiker, you don't just leave the empty bottle there. You put the lid back on it and take it down to the bottom of the mountain, where you can dispose of it properly. Assuming the lid is airtight, will the bottle look different when you get to the bottom of the mountain? If so, what will it look like?

The Composition of Air

One of the reasons air is so important for life is because it contains oxygen. Most animals need oxygen in order to live, and even plants need oxygen to live. Yes, they produce oxygen when they make food for themselves using photosynthesis, but they then use oxygen in order to burn that food for energy. They produce more food than they ever burn, however, so on balance, they make more oxygen than they use. Nevertheless, without oxygen plants and most animals would die.

While oxygen is the main thing that we need from the air, we can't have too much of it. Like most chemicals, whether or not oxygen is good or bad for us depends on its concentration. If there is too much oxygen in the air and we breathe it for a long time, our lungs get damaged, our eyes get damaged, and our brains stop functioning at their peak efficiency. Thus, while there must be oxygen in the air, there cannot be *too much* oxygen in the air.

Well, if there can't be too much oxygen in the air, there must be one or more other gases that dilute the air's oxygen content. For the air you are breathing, that gas is nitrogen. It is a very unreactive gas, so it doesn't really affect our body in any way (once again, as long as there isn't way too much of it). Thus, it is the ideal gas to dilute the oxygen content of the atmosphere. However, there are many other gases in the air as well, and some of them are also absolutely necessary for life on this planet.

One gas whose concentration varies a lot from place to place is water vapor. As you learned in the previous chapter, the amount of water vapor in the air determines the humidity, and it can vary significantly from place to place. So, for right now, we are going to ignore it. The illustration on the right, then, shows the composition of **dry air**, which has no water vapor in it. As you can see, dry air has a lot more nitrogen in it than oxygen. This is important, because not only is breathing too much oxygen for an extended time bad for our bodies, the amount of oxygen in the air determines how easy it is to start a fire. The more oxygen, the easier fires can start. Calculations indicate that for every 1% increase in oxygen content, there is a 70% increase in natural forest fires. Thus, having a limited amount of oxygen in the air is not only important for our health, it is important for our safety!

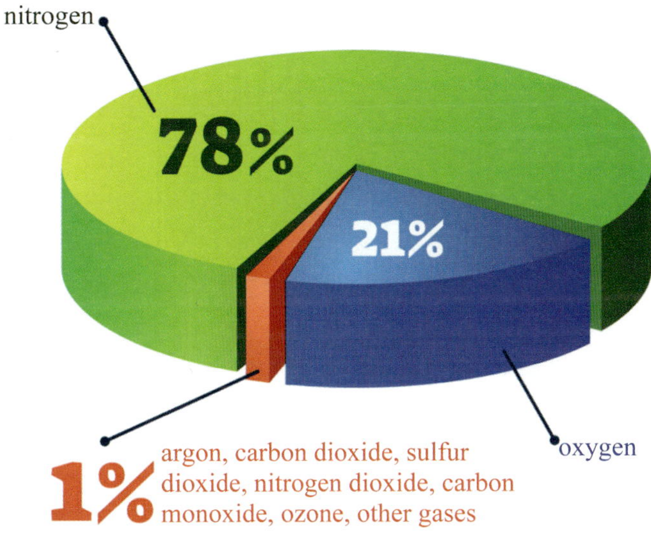

This pie chart shows the composition of dry air.

Notice that 99% of dry air is made of nitrogen and oxygen. However, as the illustration indicates, there are *a lot* of other gases that make up the final 1%. Surprisingly, most of that 1% is made of argon, another unreactive gas. The other gases listed in the illustration exist in such low concentrations that a different unit is often used to measure their concentration. That unit is called **parts per million (ppm)**.

Parts per million (ppm) – The amount of the chemical of interest found per million units of a mixture

So, for example, the concentration of carbon dioxide in the air you breathe was about 415 ppm in 2020. That means if I gathered a million liters of dry air in the year 2020, 415 of those liters would have been carbon dioxide gas. Obviously 415 liters out of a million isn't much, but you will soon see that for certain chemicals, such concentrations can have an effect.

Now it turns out that "percent" actually just means "parts per hundred." If I gathered 100 liters of dry air, then, 78 of those liters would be nitrogen, 21 of those liters would be oxygen, and the last liter would be all the other gases. Not surprisingly, then, there is a conversion relationship between percent and ppm:

$$1 \text{ ppm} = 0.0001 \text{ percent}$$

We can use this conversion relationship, along with the factor-label method, to convert between percent and ppm.

Example 12.1

The concentration of argon in the atmosphere is 9,300 ppm. What is the percent of argon in the atmosphere?

Remember that the factor-label method allows us to convert between any two units that have a relationship. Well, the relationship between ppm and percent is:

1 ppm = 0.0001 percent

To use that relationship, we set up the conversion like we are multiplying fractions. The original measurement we have goes over 1:

$$\frac{9{,}300\ \text{ppm}}{1}$$

Now we multiply by a fraction made from the conversion relationship. However, we need to get rid of the "ppm," so the "1 ppm" needs to go on the bottom of the fraction. That way, the ppm's will cancel:

$$\frac{9{,}300\ \cancel{\text{ppm}}}{1} \cdot \frac{0.0001\ \text{percent}}{1\ \cancel{\text{ppm}}} = 0.93\ \text{percent}$$

So a concentration of 9,300 ppm is the same as 0.93%. Notice that this is very close to 1%, which shows you that all the other gases that make up the rest of the atmosphere total only 0.07%. That's why their concentrations are often measured in parts per million!

Divisions of the Atmosphere

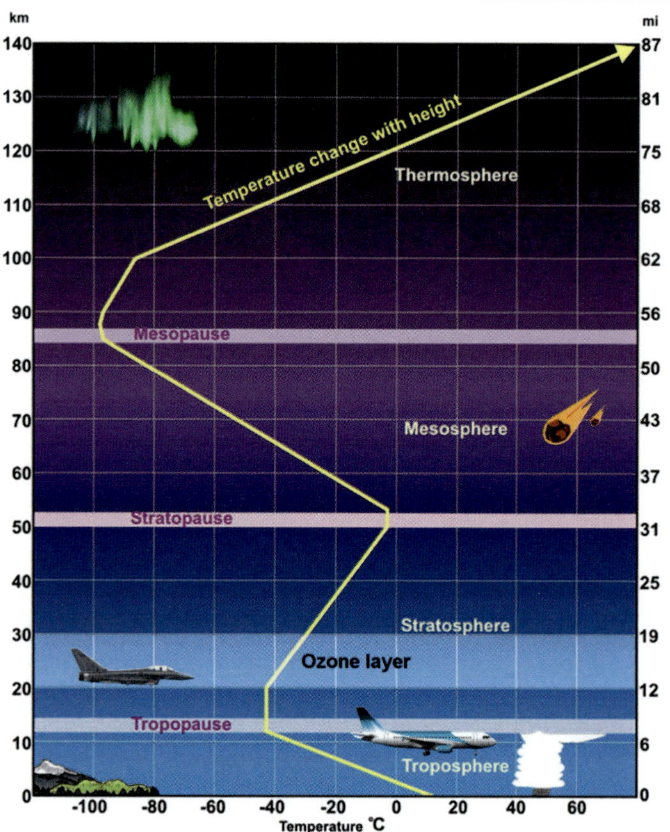

This graph shows how temperature changes with altitude.

The atmosphere can be divided into several different sections, based on the way temperature changes the higher you go. Typically, your height above sea level is called your **altitude**, and initially, the temperature decreases as you increase your altitude. This can be seen in the graph on the left. The vertical axis shows the altitude in both kilometers (left) and miles (right). The horizontal axis is the temperature, and the yellow line gives you the average temperature at any altitude. At an altitude of 0 km (in other words, at sea level), for example, the average temperature is 10 °C, which is the same as 50 °F. As altitude increases, the temperature decreases, until you get to an altitude of just above 10 km. At that point, the temperature is just under -40 °C. After that, the temperature no longer decreases with increasing altitude. This marks the division between the **troposphere** (trohp' uh sfear) and the **stratosphere** (strad' uh sfear). That division is called the **tropopause** (trohp' uh pawz).

For you and me, the troposphere is the most important part of the atmosphere. As shown in the graph, even mountains don't rise above it, so the air that we breathe is all part of the troposphere. It's also the part of the atmosphere that contains almost all the weather we experience.

Troposphere – The lowest region of the atmosphere, which contains almost all of the weather we experience

Most commercial airplanes fly at the top of the troposphere or even in the tropopause, but only specialized aircraft, like fighter jets and supersonic aircraft, fly above the tropopause.

As you rise in altitude above the tropopause, you enter the stratosphere. For a while, the temperature stays pretty constant, at just below -40 °C. Then it starts to increase. At that point, you are still in the stratosphere, but you have entered a region called the **ozone layer**.

Stratosphere – The layer of the atmosphere above the troposphere and tropopause

Ozone layer – The layer of the atmosphere that contains the highest concentration of ozone

Ozone is a very important chemical, and I will discuss it and the ozone layer in more detail later. As you rise above the ozone layer, it continues to get warmer, until you reach the **stratopause** (strad' uh pawz), which separates the stratosphere from the next layer of the atmosphere, the **mesosphere** (me' zuh sfear).

Mesosphere – The layer of the atmosphere above the stratopause and stratosphere

In the mesosphere, the temperature once again decreases with increasing altitude. If you see "falling stars" in the night sky, those are actually rocks from space that are falling into the earth's atmosphere. Their interaction with the mesosphere produces a lot of friction, which makes a lot of heat. Some of that heat gets converted into light, which makes the meteor look like a star that is falling in the sky.

From the troposphere through the mesosphere, the composition of the air remains the same: roughly 78% nitrogen, 21% oxygen, and 1% a variety of other gases. These regions are often collectively called the **homosphere**, since the term "homo" means "same." Thus, the homosphere is the part of the atmosphere in which the mixture of gases remains the same.

Something does change, however. The *amount* of air that exists continually decreases as you increase in altitude. Some refer to this as the air getting "thinner" at higher altitudes. So even though the air throughout the troposphere is 21% oxygen, the total amount of air decreases as you increase altitude, which means the total amount of oxygen to which you are exposed decreases. As a result, the higher you go in the troposphere, the harder it is to get the oxygen you need to survive. If you aren't used to being at high altitudes, your body might not be used to the lower amount of oxygen it is getting, and

At high elevations, you can get altitude sickness, which can be treated by breathing extra oxygen from a canister.

you can get altitude sickness, which can cause headaches, nausea, muscle aches, and dizziness. It can typically be treated by inhaling pure oxygen for brief periods to supplement what little oxygen you are getting from the thin air.

As you rise in altitude, you reach the **mesopause** (meez' uh pawz), which separates the mesosphere from the **thermosphere** (therm' uh sfear).

Thermosphere – The layer of the atmosphere above the mesopause and mesosphere

This is where the composition of air begins to change, because the molecules and atoms in the air start being hit by high-energy light coming from the sun. This changes the molecules and atoms into charged particles, which are called ions.

This is a picture of the Aurora Borealis, as seen in Canada.

The higher you go, the more molecules and atoms get turned into ions. As a result, this region is also called the **ionosphere**. The ions found here can interact with high-energy particles coming from the sun to produce brilliant light shows, called **auroras**. While ions can be found throughout the thermosphere, the high-energy charged particles from the sun are deflected by the earth's magnetic field, and they tend to be funneled to the poles. Thus, auroras are best seen when you are at northern latitudes (where they are called the Aurora Borealis) or southern latitudes (where they are called the Aurora Australis).

The temperatures found in the thermosphere are a bit strange. A thermometer would read very cold temperatures, because a thermometer determines the temperature through the collisions between it and the surrounding particles. Since there are only a few molecules or ions in the thermosphere, the thermometer wouldn't experience many collisions, so it would read a cold temperature. However, the molecules and ions that are in the thermosphere have a lot of energy in their motion. Remember that this is how temperature is actually defined. Thus, despite the fact that a thermometer would read lower and lower temperatures the higher it went in altitude, the definition of temperature tells us that the temperature actually increases with higher altitudes in the thermosphere. In fact, this odd effect is where the name "thermosphere" comes from.

Comprehension Check

12.3 Suppose you were to determine the percentage of nitrogen in air that has a lot of humidity. Would be it higher, lower, or equal to 78%?

12.4 If a gas makes up 0.00012% of the air, what is its concentration in ppm?

12.5 You are looking at temperatures being measured by a craft that is rising in altitude. If the temperature is steadily increasing but then levels off and starts decreasing, what region of the atmosphere did the craft enter when the temperature leveled off?

Why Does the Temperature Vary Like That?

The divisions of the atmosphere are based on the way temperature changes with altitude, but what causes those changes? Let's start with a quick experiment.

Experiment 12.3: Blowing Away the Temperature

Supplies:
- Yourself

NOTE: You will be doing a lot of hard exhaling in this experiment. That can make some people dizzy. If you start to get dizzy, take a break before continuing with the experiment.

Instructions:
1. Open your mouth wide and put your flat palm in front of it, with your fingers touching the tip of your nose and your palm very close to but not touching your lips.
2. With your mouth open wide, exhale hard so that your breath falls on your palm. Does your breath feel warm, cold, or at about room temperature? Repeat this a couple of times so you have a good idea of what it feels like.
3. Purse your lips like you are going to blow out a candle. If your lips touch your palm, tilt your hand so that your palm is close to, but not touching, your pursed lips.
4. Exhale hard through your pursed lips, like you are blowing out a bunch of candles. Does the temperature feel different than before? If you can't tell, alternate between exhaling with your mouth wide open and exhaling with pursed lips. Eventually, you should notice that there is a difference in the temperature that your palm feels.
5. With your lips pursed again, hold your flat palm about 15 cm (6 inches) from your face. Your palm should be at the same height as it was before, just farther from your mouth.
6. Once again, blow hard through your pursed lips.
7. Keeping your lips pursed, bring your palm back close to your face like it was in step 4 and blow again.
8. Repeat steps 5-7 until you can determine the difference in temperature produced by your palm being farther from your face.

What happened in the experiment? With your mouth wide open, your breath should have felt warm. That should make sense. After all, the room you are in is probably at room temperature, which is about 25 °C (77 °F). However, your body temperature is about 37 °C (99 °F). So your breath is hotter than the air that is hitting your palm. As a result, it makes your palm feel warm.

But what happened when you pursed your lips and blew? The air coming out of your mouth was still at 37 °C, but when it hit your hand, it should have felt cooler than when your breath was coming out of your open mouth. Why? It's tempting to say that it's cooler because wind cools things down, but that's not really true. Wind *can* cool things down, but it can also warm things up, depending on the temperature of the air blowing in compared to the temperature of the air surrounding you. When the wind brings in warmer air, it warms you up. The wind only cools you down when it brings in cooler air. Since the "wind" coming out of your mouth is the same as your body's temperature, it should warm your palm up, like what you felt when you exhaled with your mouth open.

So why did your breath feel cooler through pursed lips? The last part of the experiment answers that question. When you held your hand farther from your lips, you should have felt less "wind" hitting your palm, but the wind should have felt even cooler. All the temperature changes you felt in your experiment weren't caused by the fact that you were making wind. They were caused by what the air had to do when it left your mouth.

When your mouth was wide open, your breath started expanding in all directions. However, since your mouth was wide open, it didn't expand much before it hit your palm. When you pursed your lips and exhaled, your breath came out in a thin stream. Once again, it started expanding in all directions, and since it started out in such a thin stream, its expansion was more significant than when your mouth was wide open. When you held your palm farther away from your mouth, it had even more time to expand, so the expansion was even more significant. The more significant the expansion, the cooler your breath was.

Why did expansion cool your breath down? Remember how temperature is defined. It's a measure of the energy in the random motion of the molecules or atoms that make up a substance. Well, it takes energy for something to expand, and that energy must come from somewhere. In the case of the air leaving your mouth, it came from the random motion of the molecules. Energy was taken away from that random motion and converted into energy that allowed the air to expand. That means the molecules had less random energy, so their temperature went down.

This is Mount Cook in New Zealand. The green plants in the foreground tell you it's summer, but the mountain is so high that it is still cold enough for snow near the top.

This is the reason the air gets colder the higher you go in the troposphere. In order to travel to higher altitudes, the air must expand. The energy required for that expansion comes from the random motion of the molecules and atoms, so that reduces their energy of random motion, which reduces their temperature. The higher the air goes, the more it must expand, so more energy is removed from the random motion of the molecules and atoms. As a result, the temperature goes down even more. This, of course, is why water vapor condenses into clouds high in the troposphere and why there is snow on the tops of very high mountains, even during the summer.

Now don't get confused by the fact that air cools when it expands. You have probably been taught that when air heats up, it expands. That's very true. After all, when you heat up air, you are giving it more energy. Some of that energy can go into the random motion of the molecules, which causes an increase in temperature. At the same time, the rest of it can go into expansion, allowing the air to expand. So when heat is *added*, temperature can rise, and air can expand. However, when air expands *without energy being added*, the air cools down.

But wait a minute. The air still has to expand to get into the stratosphere. Why does the temperature level off in the stratopause and then start rising in the stratosphere? That's because of the ozone layer. Ozone is an interesting chemical. The oxygen you breathe is made of molecules that each have two oxygen atoms linked together (O_2). Ozone is made of three oxygen atoms linked together (O_3), and that extra oxygen atom makes it behave very differently from O_2. At certain concentrations, ozone can damage your lungs. However, since the ozone layer is in the stratosphere, you don't really have to worry about that, because no one is breathing the air at that altitude.

The other interesting thing about ozone is that it can absorb high energy light rays coming from the sun, which are called **ultraviolet** (ul' truh vye' uh let) **rays**. These rays have so much energy that they can kill living tissue. You have experienced this damaging power if you have ever gotten a sunburn. When you stay out in the sun too long, you are getting exposed to too many ultraviolet rays. That kills a lot of your skin cells, which creates a burn. If it weren't for the ozone in the ozone layer, *a lot more* ultraviolet rays would reach the surface of the earth, killing *a lot more* living tissue. In fact, with no ozone layer, most living things on the surface of the earth would die. However, because ozone absorbs ultraviolet rays, it keeps them from hitting the surface of the earth, protecting the life that is here.

Once again, remember what temperature is. It is a measure of the energy in the random motion of molecules or atoms. Well, when ozone absorbs ultraviolet rays, it absorbs their energy as well. That energy is converted into random motion, which increases the temperature of the ozone. So the stratosphere warms as you increase in altitude because the ozone layer is absorbing energy from the ultraviolet rays of the sun.

Now think about that for a moment. Ozone is toxic to us at certain concentrations, but those concentrations are necessary in order to protect us from the ultraviolet rays produced by the sun. Thankfully, that concentration of ozone exists in a part of the atmosphere where no one is breathing! Thus, its toxic effects on the human body aren't an issue. I don't think this is a coincidence. God designed the earth to be a haven for life. Life depends on the sun, but the sun's ultraviolet rays can kill living things. Thus, there needs to be an ozone "shield" to protect us from those rays, and God put it in a part of the atmosphere where it could do its job and we wouldn't suffer from its toxic effects.

Eventually, the heating effects of the ozone layer taper off, and the temperature levels off and starts to decrease again as what little air is left continues to expand as it rises. However, as the altitude increases, the air that is there starts interacting with all the high-energy light coming from the sun. That gives the few molecules there a lot of energy, which increases their temperature. Once again, a thermometer would read a continually decreasing temperature in the thermosphere because there are so few molecules, atoms, and ions there. However, the energy of those molecules, atoms, and ions increases with increasing altitude, so based on the definition of temperature, their temperature does as well.

There is actually one more layer of the atmosphere above the thermosphere. It's called the **exosphere**, and the air is so thin there that it is very similar to outer space. In fact, most of the satellites that we launch into orbit around the earth are in the exosphere. Once again, what few atoms or ions are there are moving with a lot of energy, so the temperature is very high, even though a thermometer would say it is very cold.

Comprehension Check

12.6 You are looking at a range of mountains. Some of the peaks are closer to you than others, so it is hard to judge which peaks are the tallest. However, some of the peaks have snow on them, while others don't. What does that tell you about the peaks' relative heights?

What About the 1%?

As you've already learned, dry air in the homosphere is 78% nitrogen and 21% oxygen. Most of the remaining 1% is argon. Nitrogen gas doesn't react with other chemicals easily, so it doesn't affect us much. Argon gas is even less reactive than nitrogen gas, so once again, that doesn't affect us much. However, most of the other gases that make up the rest of that 1% can be very reactive, and they can affect us in many different ways.

To understand these gases, you first have to realize that we burn *a lot* of different things. Burning releases a lot of energy (think about the heat from a fire), which can be converted into many useful things. Burning gasoline in an engine releases energy that can then be used to make an automobile go. Burning coal, natural gas, or any other fuel in an electrical power plant releases energy which can then be converted into electricity. Burning wood or natural gas can produce heat that warms our homes or cooks our food. Scientifically, the process of burning is called **combustion**.

Combustion - A chemical reaction that occurs when oxygen combines with other substances to produce heat and usually light

In short, our lives are made significantly better and more enjoyable because we can burn things. However, the process of combustion can have some serious unintended consequences.

Consider, for example, the picture below. It shows the city of Shanghai in China. You can see the sun in the background of the picture, but all the buildings are obscured by a haze. That haze is called **smog**, and it used to be a pretty common sight in many major cities around the world. Nowadays, it is less common, but as the picture shows, it can still be seen. What causes smog? Mostly, it is one of the unintended consequences of combustion.

This picture shows Shanghai, China obscured by smog.

From the definition, you can see that combustion involves oxygen combining with other substances. This is why fires need oxygen in order to burn. In order for combustion to occur, oxygen must be present, and it must chemically react with the substance being burned. Coal, for example, is mostly made of carbon atoms (C). When it burns, carbon reacts with oxygen (O_2), producing **carbon dioxide** (CO_2):

$$C + O_2 \rightarrow CO_2$$

The problem is that coal isn't pure carbon. There are some contaminants in it, such as sulfur. The sulfur burns with the coal, producing **sulfur dioxide**.

$$S + O_2 \rightarrow SO_2$$

That's one of the gases that makes the smog you see in the picture. At the right concentration, sulfur dioxide can be toxic to breathe. It can also further react with oxygen to make sulfur trioxide (SO_3), which is also toxic at the right concentration. These two gases are often grouped together and called **sulfur oxides**.

Those aren't the only gases that make up the smog in the picture. Whenever you burn something, the nitrogen in the air heats up. While nitrogen isn't normally reactive, at high temperatures, it can react with oxygen in the air to produce **nitrogen monoxide** (NO) and **nitrogen dioxide** (NO_2). These gases are often grouped together and called **nitrogen oxides**. Like sulfur oxides, nitrogen oxides can also be toxic at the right concentrations.

Under certain circumstances, combustion can also produce ozone (O_3). Now remember, ozone is absolutely necessary to protect us from the ultraviolet rays of the sun, but it is also toxic to us at certain concentrations. Thus, it is good to have ozone in the ozone layer (where no one is breathing), but it is not good to have it near the surface of the earth, where people breathe.

Finally, when oxygen is in short supply, combustion can also produce **carbon monoxide** (CO). This is usually called **incomplete combustion**, because there wasn't enough oxygen available to burn the fuel completely. While carbon dioxide is only toxic to us at high concentrations, carbon monoxide is toxic at much lower concentration. It takes the place of oxygen in our blood, so even though we are breathing, our tissues don't get enough oxygen. As a result, you can suffocate by breathing air with too much carbon monoxide.

The chemicals I have discussed so far are important pollutants in the air around us. Since we burn a lot of fuel, we make a lot of these pollutants. If that were the end of the story, they would build up in the air to the point where they would cause a lot of health problems. In fact, that's really what happened. As people started burning more and more fuel, more and more pollutants were produced. They started building up in the air. By the 1950s, scientists in the U.S. recognized that this buildup could produce health problems, so in 1955, the U.S. Congress passed the Air Pollution Control Act with the goal of figuring out how big the problem was and how to solve it. Over the subsequent years, many laws were passed regulating how fuels were burned, what kinds of fuels could be burned, and what had to happen to the products of that combustion.

What has been the result of all these laws? See for yourself. The graphs below show the concentrations of the pollutants I discussed from 1980 to 2019:

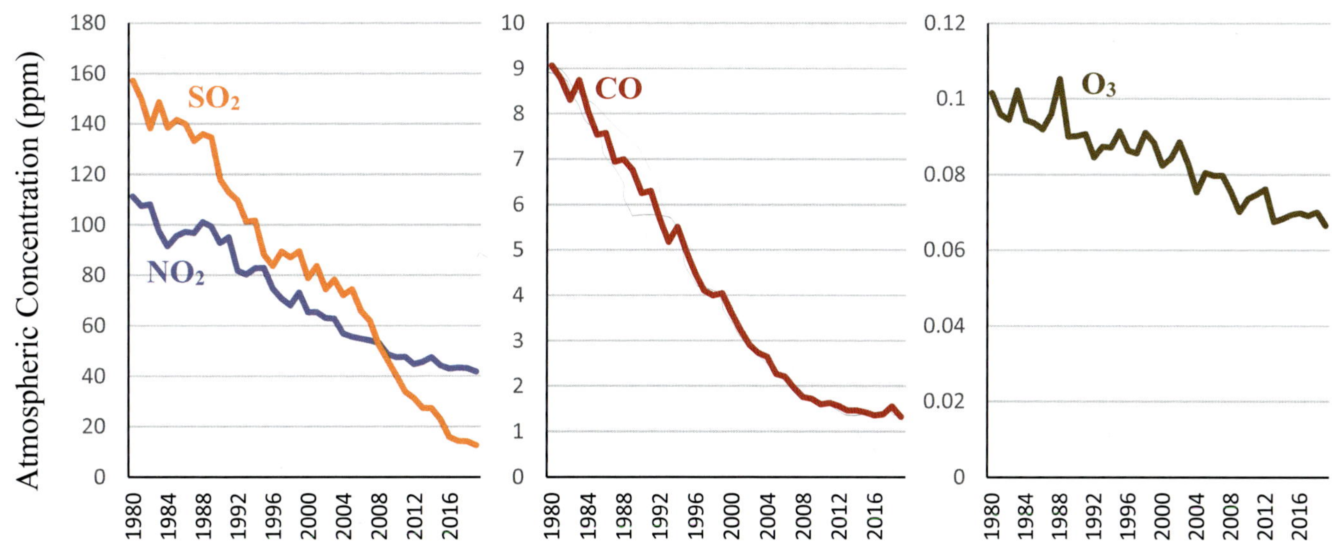

As you can see, the concentration of each pollutant (measured in parts per million) has decreased significantly since the 1980s. Does that surprise you? Depending on the sources from which you get your news, you might have thought that the air has been getting more and more polluted as time has gone on. That's simply not true. As I said previously, smog-obscured cities used to be a common sight, but because of legislation that has been enacted over the years, it is not nearly as common today. That's because the air we are breathing now is significantly cleaner than it has been in 40 years!

Is it possible for the concentration of air pollutants to become so reduced that they are essentially zero? No. That's because there are natural sources for each of the pollutants shown in the graphs on the previous page. Volcanoes produce sulfur oxides, nitrogen oxides, and carbon monoxide. Some microscopic organisms produce sulfur oxides and nitrogen oxides. Lightning also produces nitrogen oxides and ozone. Forest fires produce carbon monoxide. The nitrogen oxides that are made naturally can form ozone when they interact with light coming from the sun. Thus, there is no way to eliminate air pollutants. However, we can reduce the human-made sources so that they become as low as possible.

Before I move on, I want to point out a couple of things. First, go back and look at the graphs on the previous page. There is a reason I showed you three different graphs. Look at the vertical axis of each graph. In the one that contains SO_2 and NO_2, the numbers range from 0 ppm to 160 ppm, but the one that contains CO has numbers from 0 ppm to 10 ppm, and the one that contains ozone has numbers ranging from 0 ppm to 0.12 ppm. SO_2 and NO_2 have concentrations in the air that are similar enough to show on one graph, but the CO concentrations are much lower, and the ozone concentrations are even lower. Had I tried to show them all on the same graph, the CO and O_3 data would be all crammed in at the bottom, and you wouldn't have been able to see how they changed over the course of 40 years.

This is important because each chemical has its own concentration at which it becomes toxic. NO_2, for example, can produce harmful effects on the body if its concentration is about 150 ppm, while SO_2 can produce harmful effects at about 100 ppm. The body is much more sensitive to carbon monoxide, however. It can produce harmful effects at a concentration of 10 ppm. Ozone is even more toxic, potentially producing harmful effects at 0.1 ppm. When assessing the level of pollution, then, each pollutant must be addressed individually, because the human body has different sensitivities to each one.

Notice that the current concentrations are much lower than the potentially toxic concentrations listed above. That, of course, is good. However, notice that for SO_2, the concentration didn't fall to below its potentially toxic level until about 1994. Thus, some people were suffering ill effects by just breathing the air from 1980 (and probably earlier) until 1994. In the same way, ozone didn't stay below its potentially toxic concentration until about 1987. This demonstrates how important the legislation was back in those days.

Notice also that with the possible exception of CO, all the pollutants shown on the previous page are still decreasing in their concentrations. That, of course, is also good. At the same time, however, it is important to note that all the legislation that has been produced to lower these pollutant levels comes at a cost. In order to produce less SO_2, for example, coal-burning plants must install devices called "scrubbers" that absorb the gas before it is released into the atmosphere. Those scrubbers cost money to maintain and install, so that increases the cost of whatever is being produced by the coal-burning plant.

When deciding whether or not to pass new legislation, lawmakers need to decide whether the potential benefit outweighs the cost. When the pollutant concentrations were above or near their potentially toxic levels, the benefit was large, so it was worth the cost. The lower the pollutant levels get, the less benefit we get from lowering them even more. Also, in general, the lower the pollutant levels are, the more costly it becomes to lower them even more. So while it sounds like a good idea to completely eliminate air pollution, that's not really a responsible goal. At some point, the cost is simply not worth the tiny benefit (if there is any at all).

More About Carbon Dioxide

One of the gases that makes up less than 1% of the air we breathe is carbon dioxide. You know that we exhale carbon dioxide. That's because our bodies do combustion as well. They burn the food we eat to release the energy we need to survive. The combustion done by our bodies is heavily controlled, however, so there aren't little fires going on in our bodies. Nevertheless, oxygen does combine with the products of digestion to release energy. You also know that through the process of photosynthesis, plants absorb carbon dioxide from the air and use it to make the food that they will later burn for energy.

But carbon dioxide has another important function that makes life possible on the earth. It helps to keep the earth warm enough to support life. The sun gives a lot of energy to the earth, which is illustrated by the yellow arrows in the drawing below. A small amount of that energy bounces off the atmosphere, but most of it passes through and hits the earth. The earth absorbs most of it and reflects the rest. The absorbed energy warms the planet. However, the earth also releases a lot of the energy it absorbs, which is represented by the red arrows in the drawing. If that were the end of the story, earth would end up releasing too much energy into space, making earth very cold. It would also make the temperature difference between night and day more severe. To keep that from happening, the earth's atmosphere has certain gases that trap a lot of the energy the earth releases. These gases are called **greenhouse gases**, and carbon dioxide is one of them. When the greenhouse gases absorb that energy, it warms the atmosphere, which in turn, warms the planet. This is often called the **greenhouse effect**, and it is one of the many design features the earth has to make it a haven for life.

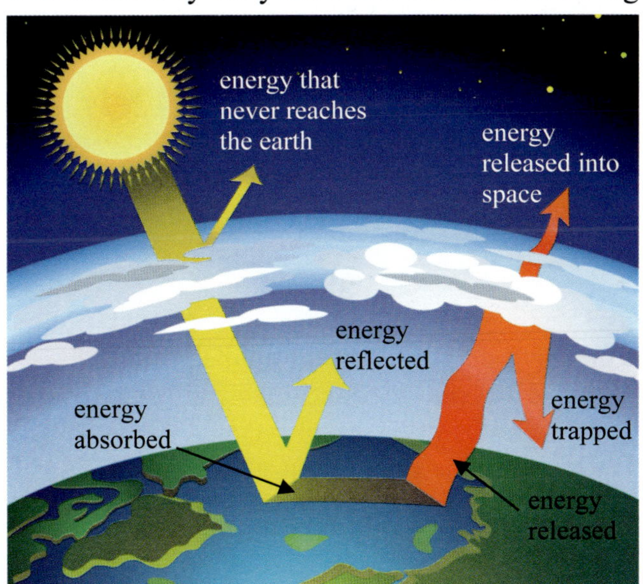

The greenhouse effect makes the earth's temperature warmer and more even so that it can support life.

Now please understand that there are a lot of greenhouse gases. Remember ozone? It's also a greenhouse gas. The energy it absorbs is responsible for the fact that the temperature in the stratosphere increases with increasing altitude. Water vapor is also an important greenhouse gas. So is natural gas (CH_4), whose chemical name is **methane**. All of these greenhouse gases work together to produce the greenhouse effect, but the one that is changing the most right now is carbon dioxide. Its concentration in the atmosphere has been steadily increasing from about 290 ppm in the year 1900 to 415 ppm in 2020. That's a significant increase!

What's causing the increase? The fact that we are burning so much fuel. Nearly every fuel we burn has carbon atoms in it, and when the fuel burns, carbon dioxide is produced. Gasoline, for example, is a mixture of carbon-containing compounds, but a good example of what you find in gasoline is called "octane" (ok' tayn – C_8H_{18}). When an automobile burns octane in its engine, the result is:

$$2C_8H_{18} + 25O_2 \rightarrow 16CO_2 + 18H_2O$$

Actually, that's only when complete combustion happens. Incomplete combustion also occurs in an automobile, so carbon monoxide is also produced. However, most of what is done in an engine is complete combustion, so there is a lot more carbon dioxide produced than carbon monoxide.

Well, if the carbon dioxide concentrations in the atmosphere are increasing, that means the greenhouse effect is increasing, right? Not necessarily. The earth has lots of design features that regulate systems like the greenhouse effect, and we haven't come close to understanding all of them. I want to discuss this in more detail, but there is still one more thing you have to know about carbon dioxide, which you will learn the next time you do science.

Comprehension Check

12.7 A good friend of mine grew up in Los Angeles, California in the 1970s. He says that when he would ride his bicycle for a long time, his lungs would feel like they were burning. He recently went back to his hometown and rode a bicycle for several miles, but his lungs never felt uncomfortable. Why?

12.8 Fireworks have gunpowder in them to produce their explosions, but they have other things in them that produce pretty colors when they burn. Fireworks that produce blue colors, for example, have copper in them. What do you think would be the name of the chemical produced when that copper burns during the explosion?

12.9 You are working in a plant, and an analysis of the air inside the plant is done. There is a pollutant found with a concentration of 1 ppm, and the plant is immediately closed down until the pollutant concentration can be lowered. If the pollutant is one of those discussed in this chapter, which one is it?

12.10 The atmosphere of Venus is 96% carbon dioxide. Compare the greenhouse effect on Venus to the greenhouse effect on the earth.

Carbon Dioxide, Acids, and Bases

In order to understand another role that carbon dioxide plays in the atmosphere, you need to learn a bit about **acids** and **bases**. When you hear the term "acid," you probably think about a caustic chemical that "eats through" things. While strong acids are like that when they exist at high concentrations, acids are found throughout creation, and many of them are not caustic and cannot really eat through anything. For example, if you enjoy citrus fruits (oranges, lemons, limes, etc.) or their juices, part of what gives them their flavor is citric acid. Acids tend to taste sour to us, so the more concentrated the acid, the more sour the taste. Thus, lemons have a higher concentration of citric acid than limes and oranges. Most soda pops contain acid as well and once again, the acid contributes to their taste.

Bases also exist throughout creation, and while you do ingest them from time-to-time, you do so less frequently than you do acids. Bases taste bitter to us. Thus, they are often used in spices, but they are not very concentrated in most of the things we consume. Acids and bases react well together, each negating the other's properties. Just like acids, a base's chemical nature and concentration affect what it can do. Strong bases that are concentrated can be just as caustic as acids, and they can also "eat through" things. However, most bases in creation are not strong and are rarely at high concentrations.

When something contains a strong acid at high concentration, we say that it is very acidic, and you need to be careful with it. When something contains a strong base at a high concentration, we say it is very **alkaline** (al' kuh lynn), and you need to be careful around it. However, many things that contain acids or bases are neither strongly acidic nor strongly alkaline, so most of them are not dangerous. To help gauge how acidic or alkaline a substance is, scientists have developed the **pH scale**. On that scale, a "7" means the substance is neither acidic nor alkaline. The lower you go on the scale, the more acidic the substance is, and the higher you go, the more alkaline. The illustration below gives you a visual of the scale and where several common substances fall on it.

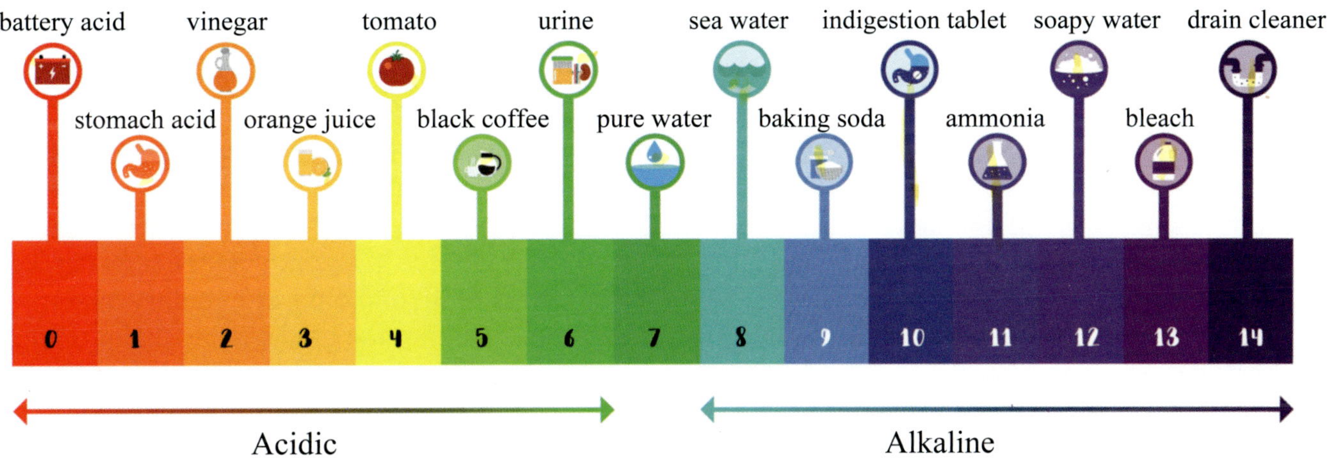

As you can see, battery acid is very acidic, as is the acid in your stomach. Thus, they can be dangerous. While vinegar is very sour, it is generally not dangerous. Orange juice, tomatoes, black coffee, and urine are all acidic, but they are also not dangerous. In the same way, sea water, baking soda, indigestion tablets, and soapy water are all alkaline, but not really dangerous. The alkalinity of ammonia doesn't make it dangerous, but its other chemical properties do. Bleach and drain cleaner, however, are very alkaline and can be dangerous.

Notice that sea water has a pH of around 8. That means it is slightly alkaline, but not at all dangerous. The organisms in the ocean are designed to live in that slightly alkaline environment, and that can be affected by many things, including the level of carbon dioxide in the atmosphere. How? Perform the following experiment to find out.

Experiment 12.4: Carbon Dioxide and pH

Supplies:

- The graduated cylinder from the laboratory kit made for this course
- The pH paper from the laboratory kit made for this course
- The mass scale from the laboratory kit made for this course
- A plastic bottle, like the kind water comes in
- A bendy straw

- Play-Doh or other soft modeling clay
- Two small glasses, like juice glasses
- Toilet paper
- A small plate
- A spoon for stirring
- Vinegar
- Baking soda
- Ammonia (Sold in supermarkets with the household cleaners. You need only a drop or two.)
- Water

Instructions:
1. Add vinegar to the graduated cylinder until it reaches the 50 mL mark or just a bit above it.
2. Pour the vinegar into the plastic bottle.
3. Repeat steps 1 and 2 so there are about 100 mL of vinegar in the bottle.
4. Rinse out the graduated cylinder a couple of times to remove any vinegar left behind.
5. Wrap some Play-Doh around the straw on the short end below the bend. The wad of Play-Doh needs to be larger than the opening in the bottle so it can seal the bottle (see the picture below).
6. Add just a tiny amount (a drop or two) of ammonia to the juice glass and immediately cap the ammonia bottle. **Caution: it smells foul. Avoid inhaling it.**
7. Use the graduated cylinder like you did before, only this time, add 100 mL of water to the tiny amount of ammonia in the glass. Stir the contents of the glass with the spoon.
8. Dip a single strip of pH paper into the contents of the glass. If it turns green, you are ready to proceed. If it turns blue, dump about half of the contents out of the glass, add 50 mL of water, stir, and try again. If it doesn't turn color, add another tiny amount of ammonia, stir, and try again.
9. Once you have a solution that turns the pH paper green, pour it into the graduated cylinder until it reaches the 20 mL mark or just over.
10. Pour that 20 mL into the juice glass that is still empty.
11. Turn on your mass scale, make sure it is reading grams, and hit the tare button.
12. Put a single square of toilet paper on the scale.
13. Add baking soda to the square until the scale reads about 5 grams.
14. Wrap the baking soda into the square of toilet paper so that it makes a "pill" that you can put into the bottle. Don't do that yet. I am just telling you the goal.
15. Read the next step completely and look at the picture so you know what you are going to do before you actually do it.

16. Drop the baking soda/toilet paper "pill" into the bottle, put the Play-Doh over the opening, and hold the bottle and glass so that the long end of the straw is sticking in the 20 mL of solution, as shown in the picture on the left. Bubbles will form in the bottle, and most of the gas will travel through the straw and make bubbles in the 20 mL solution.
17. As the bubbles in the juice glass slow down, gently move the bottle back and forth to mix the baking soda and vinegar. That should make more bubbles. Do that until you think there are no more bubbles to be made.
18. Put everything down and get a new strip of pH paper.

19. Put the strip of paper in the glass and compare the color to the green that you got before. If there is no change, empty the bottle, put in new vinegar, and repeat steps 11-18.
20. Clean up your mess.

What happened in the experiment? If you look at the color codes on the inside flap of the pH paper container, you can see that you started off with a solution that had a pH of somewhere between 7 and 9, depending on the shade of green you saw. Let's say 8. That's the pH of sea water. The reaction between baking soda and vinegar produced carbon dioxide, which bubbled through the solution. After all those bubbles, the color of the paper indicated the pH was between 5 and 6.

Why did the pH change? When carbon dioxide reacts with water, it forms an acid known as **carbonic acid**. That's the same acid in soda pop, and it is made the same way. The "fizz" of soda is carbon dioxide, and when that carbon dioxide reacts with the water in the soda, carbonic acid is formed. The carbonic acid that formed in your experiment reacted with the ammonia, which is a base. The acid negated the properties of the base, and eventually built up enough in the water to reduce the pH to make the solution slightly acidic.

Carbon dioxide can do that to any water it touches, including sea water. The air above the surface of the ocean has carbon dioxide in it, and the higher the concentration of carbon dioxide, the more it can react with the sea water, and the more it can reduce the pH. This is called **ocean acidification**.

Ocean acidification – The reduction of the ocean's pH due to increasing carbon dioxide in the air

While we honestly don't know anything for certain, it is thought that if the pH of seawater reduces enough, it can start to harm the organisms that live in the ocean, which are designed to live in a slightly alkaline environment.

Comprehension Check

12.11 Suppose you open a fresh bottle of soda and measure its pH. Then, you set it out on the counter without the cap on and forget about it. A couple of days later, you see it again and measure its pH. Would you expect the pH to have changed? If so, how?

The Effect of the Atmosphere's Increasing Carbon Dioxide on Global Temperature

Now that you know carbon dioxide is not only a greenhouse gas but can also cause ocean acidification, I can discuss the possible consequences of the fact that its concentration in the atmosphere is increasing. The first one is something you have undoubtably heard about before, and that's **global warming**. Since we know carbon dioxide is a greenhouse gas, it makes sense that the more carbon dioxide in the atmosphere, the warmer the earth will get. Thus, the fear is that the entire earth (the globe) will warm due to the increasing carbon dioxide.

So what? In some places, like Alaska, a bit more warmth will be welcome! That's true, but there are other places, like Arizona, which are already really hot. With global warming, they will get even hotter. The consequences for the ocean are even worse. Remember that the global conveyor works because of all the sea ice near the poles. If the globe gets warmer, there will be less sea ice, and the global conveyor will not work as well. That could cause a serious disruption to the necessary recycling of the ocean's nutrients. In addition, if the earth gets warm enough to melt a lot of the glaciers, ocean levels will rise, which could cause flooding in communities that are near the shore.

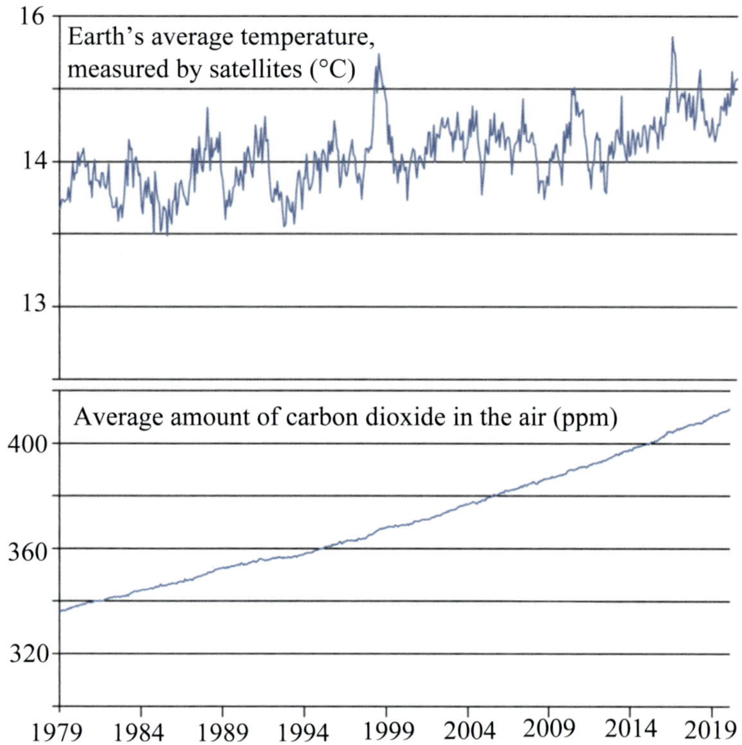

The carbon dioxide levels by year come from the National Oceanic and Atmospheric Administration, and the global temperature comes from University of Alabama, Huntsville.

Of course, the first thing we have to investigate is whether or not the globe is warming because of the increased amount of carbon dioxide in the air. The answer to that question is, "We don't know." It is very hard to measure the average temperature of the earth. You might think we could just put a bunch of thermometers around the earth and average their temperatures. That's a good idea, but it's difficult, because the earth is really big, and it's hard to get an even coverage of thermometers all over the earth. The best way to measure the average global temperature is with satellites that orbit the earth and can measure the temperature everywhere. The problem is, we have only had satellites doing that since 1979. Nevertheless, the average temperature of the earth as measured by satellites is shown in the graph on the left, along with the measured amount of carbon dioxide in the air.

As you can see from the bottom graph, the amount of carbon dioxide in the air has been rising steadily since 1979. It has been rising for longer than that, but since the satellite data start in 1979, it only makes sense to start the carbon dioxide data at that time, too. During that same time period, the average temperature of the earth has risen just slightly, from about 13.7 °C (56.7 °F) to about 14.5 °C (58.1 °F). Does that mean the increase in carbon dioxide is warming the planet? Not really. The average temperature of the earth fluctuates naturally, and the small changes we have seen since 1979 could easily be just part of the natural fluctuation.

How can we determine whether or not the temperature changes we are seeing are being caused by the increasing carbon dioxide levels? Well, one thing we can try to do is determine the average temperature of the earth long before people started putting a lot of carbon dioxide in the atmosphere. After all, if the higher temperatures we are seeing are due to the extra carbon dioxide, then the earth should have been cooler before all that carbon dioxide was produced.

How do we do that? After all, satellites that measure the average temperature of the earth didn't exist before 1979, and even thermometer measurements around the globe weren't tabulated until the early 1900s. How can we determine global temperature before those measurements were made? Well, there are other things we know about the past that are affected by temperature. For example, we can determine the age of any tree ring by counting backwards from the outside of a tree, since trees usually form just one ring each year. It turns out that the characteristics of a tree ring, like its width and density, are affected by the temperature. So is the growth of coral, the amount of pollen found in a region, etc. These things are referred to as **climate proxies**, because they are ways of determining the climate of a region before records of the climate were tabulated.

Climate proxy – A preserved physical characteristic used to determine past climate conditions

While climate proxies are by no means infallible, when several climate proxies are studied, and they all lead to the same conclusion, it is at least reasonable to believe what they say.

Now, of course, climate proxies tend to be more accurate when you restrict your study to a particular region of the planet. After all, tree rings are affected by the local weather, not the global weather. Thus, climate proxies can tell us what a particular portion of the globe experienced in the past, so to determine what was happening globally, you need to look at lots of different regions to see if you can find a pattern.

Not surprisingly, this has been done, and a definite pattern has emerged. The graph below shows a reconstruction of what is thought to have happened to the temperature of the Northern Hemisphere over the course of the past 2,000 years. It is based on using lots of different climate proxies in many different places across the Northern Hemisphere. The scientific paper from which the data come, "A New Reconstruction Of Temperature Variability In The Extra-Tropical Northern Hemisphere During The Last Two Millennia," is by Fredrik Charpentier Ljungqvist and was published in 2010 in the Swedish Society of Anthropology and Geography's national journal.

This is how climate proxies indicate the average temperature of the Northern Hemisphere changed from Christ's birth to the year 2,000.

When using proxies, it is hard to get an actual temperature; it is easier to measure a change, so that's what is graphed on the vertical axis. Any time the line is below zero, it is colder than normal, and any time it is above zero, it is warmer than normal. Notice that the warmest time in the past 2,000 years was in the year 950! The warm years around that time are referred to as the **Medieval Warm Period** or the **Medieval Climate Optimum**. Eventually, however, the Northern Hemisphere cooled, and the years from about 1500 to 1700 are referred to as the **Little Ice Age**, since the Northern Hemisphere was unusually cold during that period. Notice that the Northern Hemisphere has been steadily warming since the year 1700.

While I have shown you only the data for the Northern Hemisphere, scientific papers have been published showing that this general trend (a Medieval Warm Period, followed by a Little Ice Age, followed by warming) also existed in South America, Antarctica, and Australia/New Zealand. Thus, this seems to be a global trend. If that's the case, then the warming that we see right now is probably unrelated to the carbon dioxide buildup in the atmosphere. It looks like the earth is heading towards another warm period, like the Medieval Warm Period. If nothing else, we can say that if these climate proxies are reliable, the global temperatures we are experiencing right now are not unusual. It was just as warm in the years 50 and 160 as it is today, and it was warmer during the peak of the Medieval Warm Period.

One way we can check to see if these conclusions are reliable is to look at how ocean levels have changed over the same time period. After all, if the globe warms, more of the glaciers will melt, and the ocean will rise with the new addition of water. Also, the water in the ocean will expand, which will once again cause the ocean to rise. If the globe cools, the opposite will happen. By looking at how sediments accumulate in areas near the coast, scientists can estimate the sea level in the past, and they see a similar thing. The sea level rose during medieval times, then sank, and then began to rise again. The last rising of the sea levels started near the end of the Little Ice Age, indicating that the climate proxies are giving us a reasonable picture of what happened to global temperatures over the past 2,000 years.

But what if these proxies are wrong? We know carbon dioxide is a greenhouse gas, so increasing carbon dioxide must warm the planet. Isn't it possible that increasing carbon dioxide is adding to the natural warming that is occurring? Absolutely. However, scientists are having a very difficult time figuring out how much (if any) it is adding. Climate scientists have defined a very important term called the **climate sensitivity**:

Climate sensitivity – The change in temperature that results from a doubling of the concentration of carbon dioxide in the atmosphere

The higher the climate sensitivity, the more worried we should be about carbon dioxide's effect on the temperature of the globe.

Remember, the carbon dioxide levels have risen from 290 ppm in the year 1900 to 415 ppm in 2020. Thus, we are about 43% of the way to doubling the amount of carbon dioxide. If we knew the climate sensitivity, we could at least estimate how much of the warming that we have seen so far is due to carbon dioxide levels. Unfortunately, we don't know that number. Early climate studies said it could be as high as 4.5 °C, which would indicate that a lot of the warming we have seen so far is the result of carbon dioxide buildup in the atmosphere. However, later studies have lowered that value to somewhere between 0.2 °C and 3.5 °C. Obviously, if the value is in the lower end of that range, then global warming caused by carbon dioxide is not a serious issue at all, and the warming we have seen so far is mostly a result of natural variation. If it is on the high end of that range, global warming is more serious, and the warming we have seen so far is at least partially due to carbon dioxide buildup in the atmosphere.

The Effect of the Atmosphere's Increasing Carbon Dioxide on Ocean pH

As your previous experiment showed, carbon dioxide can also lower the pH of any solution to which it is exposed, including sea water. We know that this is already happening, at least in certain parts of the ocean. For example, the University of Hawaii has been measuring the amount of carbon dioxide dissolved in the sea water as well as the pH in a section of the Pacific Ocean near Hawaii. As you can see from the graphs on the next page, they found that since 1989, the concentration of carbon dioxide in the ocean has bounced up and down, but there is a general upward trend, as shown in the top graph. Correspondingly, the pH of that part of the ocean has bounced up and down, but there is a general trend downward. Notice how perfectly the two graphs match. Each time dissolved carbon dioxide concentration goes up, pH goes down. The red dots and line in the top graph show the amount of carbon dioxide in the atmosphere at a nearby laboratory on Mauna Loa.

Now, of course, you do have to look at the range, as shown by the numbers on the vertical axis of the pH graph. The line representing the trend starts near a pH of 8.11, and it ends at a pH of 8.08, so the overall trend is about a 0.03 drop in pH. Nevertheless, it is something that can be measured. Also, please note that like the Richter scale, the pH scale is logarithmic. A drop of 1 pH point means the water is *ten times* more acidic. A drop of 2 points means the water is 100 times more acidic. Thus, the change in acidity is more significant than the measurement makes it seem.

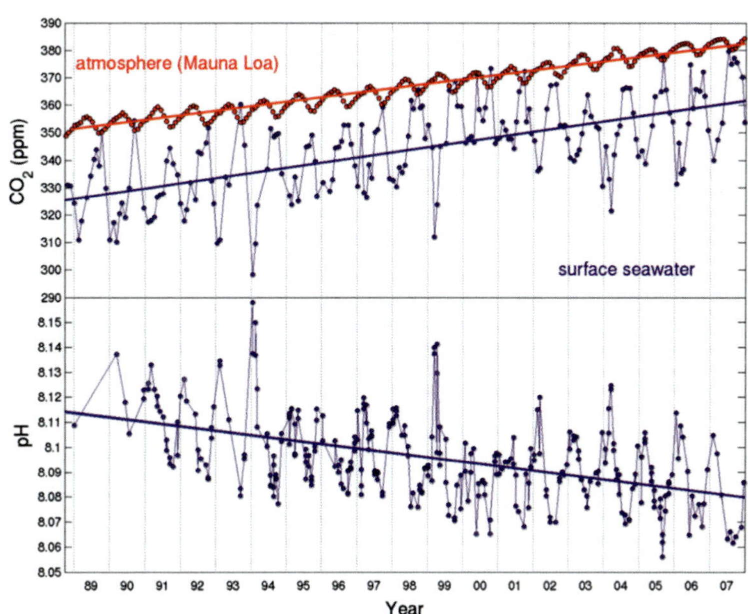

These graphs show how the pH of the ocean (bottom) is affected by the increasing amount of carbon dioxide dissolved in the ocean (top blue), which is caused by the increasing carbon dioxide in the atmosphere (top red).

One important thing to point out is that pH is also affected by other issues, so different parts of the ocean will show different values of pH and probably a different trend. Nevertheless, the graph does show that the increasing carbon dioxide concentration in the atmosphere is having a measurable effect on lowering the pH in one part of the ocean. Since the entire atmosphere is increasing in carbon dioxide, it is reasonable to think that the entire ocean is lowering in pH.

Can a change in pH negatively affect the organisms that live in the ocean? We don't know. Experiments in the lab indicate that some organisms are harmed when the pH of their aquariums drop by even just 1 pH point. However, other studies indicate that these effects are short-term, and that the organisms adapt to a lower pH. Thus, it is just not clear whether or not the changes we are seeing in the pH of the ocean are going to be a problem.

But wait a minute. We know that the pH of the ocean is lowering, even though the effect is small. We also know that even if the climate sensitivity is only 0.2 °C, the extra carbon dioxide is causing *some* warming. Why don't we reduce the amount of material we burn to stop the buildup, even if it is doing only a small amount of damage? While that idea sounds good, there is a big problem: The processes that produce carbon dioxide end up saving people's lives!

Consider, for example, your home. If you live far from the equator, the reason winter isn't deadly for you is that your home is heated. Some people heat their homes by burning natural gas in a furnace. That, of course, produces carbon dioxide. Some people heat their homes with electricity, and a lot of electricity is made by burning fuel, which also produces carbon dioxide. Either way, heating most homes ends up putting carbon dioxide into the atmosphere.

Of course, there are other ways to heat your home, but they tend to be more expensive, and when it gets expensive for people to heat their homes, they tend to use less heat. Unfortunately, that can lead to serious problems. For example, do you remember the nuclear power plant in Japan that was destroyed by a tsunami in 2011? I told you about it in Chapter 6. Because of the loss of that power plant, and because of other things that happened afterwards, the price of electricity in Japan increased. A recent scientific analysis showed that in the three years after the accident, 1,280 people died because the price of electricity made heating their homes too expensive. As a result, they didn't heat their homes as much as they needed to, and they died.

It goes well beyond heating homes. Most of the technology that we use to make our lives safer and healthier is built using electricity and other forms of power that increase the amount of carbon dioxide in the air. There are other ways of producing power that do not involve making more carbon dioxide, but right now, they are more expensive and less reliable. If we use more expensive, less reliable forms of energy, we won't be able to produce as much of that technology. That means it will be used less, which will cost lives in the end.

Now, of course, when we can figure out a way to make electricity and other forms of power without burning fuels and without it costing more money, we obviously should do that. However, right now, we don't know how. As a result, if people are forced to start using more expensive energy in order to reduce the amount of carbon dioxide in the atmosphere, it will result in more people dying. So we have to be very careful when it comes to proposing ways to "fix" the amount of carbon dioxide in the atmosphere. Yes, it would be nice to reduce the amount of carbon dioxide in the air, but we probably shouldn't do so if it is going to cost a lot of people their lives!

This, of course, brings me back to the costs and benefits that I discussed in relation to air pollution. Every time a decision is being made about what the people of a country should or shouldn't do, we need to consider both the costs and the benefits. Right now, we don't know the benefits related to reducing the carbon dioxide in the atmosphere. However, we do know the cost. Any decisions that will make energy more expensive and less reliable will cost people their lives. Thus, until we know more about the benefits, or until we can produce reliable energy at the same cost without burning things, we just can't make a good scientific decision about reducing the carbon dioxide we are producing. Hopefully, more scientific research will bring us closer to one of those two goals.

Comprehension Check

12.12 Greenland is currently covered in ice, and the temperature rarely goes above 10 °C (50 °F). However, there are remains of forests in the southern parts of Greenland, under the ice. When might those forests have grown?

12.13 Suppose scientists determine that the equilibrium climate sensitivity is 2 °C. Does that mean the temperature of the earth will be 2 °C warmer when the amount of carbon dioxide in the atmosphere doubles?

12.14 Fill in the blank: Some scientists who study the ocean say that ocean acidification is a bad term to use, because the ocean is not really becoming more acidic. They say that a better way to express what is happening right now is that the ocean is becoming less __________.

Answers to the Comprehension Check Questions

12.1 It would not be the same. There would be more air in the atmosphere. Since the earth's gravity holds the air in place, more gravity could hold more air. In the same way, less gravity holds less air. This is why the moon doesn't have much of an atmosphere. Its gravity is too weak to hold onto one.

12.2 It will be crushed. Remember your experiment. When the air pressure inside the can was less than the pressure outside, the can got crushed. If you closed the bottle at the top of the mountain, the pressure inside would be equal to the pressure at the top of the mountain, which is lower than the pressure at the bottom. Thus, when you get to the bottom, if air couldn't leak into the bottle, the pressure inside the bottle will be lower than the pressure outside, so the bottle will be crushed. I do this when I go hiking in the mountains with friends, and most of them are surprised by the amount the bottle had been crushed.

12.3 It would be lower than 78%. Remember, humidity refers to the amount of water vapor. So this is air that also has water vapor in it. Well, the total amount of nitrogen in the air doesn't change just because water vapor is there, so the total amount of nitrogen is the same. However, the total amount of in the air is higher, since it also has water vapor in it. The percentage of nitrogen is the amount of nitrogen divided by the total amount of gas in the air times 100. If the total amount goes up, then you are dividing by a larger number, which means the percentage goes down.

12.4 1.2 ppm. The relationship between ppm and percent is:

$$1 \text{ ppm} = 0.0001 \text{ percent}$$

To use that relationship, we set up the conversion like we are multiplying fractions. The original measurement we have goes over 1:

$$\frac{0.00012 \text{ percent}}{1}$$

Now we multiply by a fraction made from the conversion relationship. However, we need to get rid of the "percent," so the "0.0001 percent" needs to go on the bottom of the fraction. That way, the percents will cancel:

$$\frac{0.00012 \cancel{\text{percent}}}{1} \cdot \frac{1 \text{ ppm}}{0.0001 \cancel{\text{percent}}} = 1.2 \text{ ppm}$$

12.5 It entered the mesosphere. If the temperature was rising with altitude, it couldn't have been in the troposphere, since the temperature decreases with altitude in that region. However, it increases with increasing altitude in the stratosphere, then levels off, and then decreases in the mesosphere.

12.6 The peaks with snow on them are higher than the ones without snow on them. Closer peaks may look taller, simply because they are closer to you. However, the peaks with snow on them must be the tallest ones, since the presence of snow indicates it is colder on those peaks. Colder peaks must be taller.

12.7 The pollution in Los Angeles is much less concentrated now compared to when he was young. Pollutants can damage the lungs, and when my friend was young, they were concentrated enough in the air around Los Angeles to do exactly that. Nowadays, they are not.

12.8 Copper oxide. Remember that burning means chemically combining a substance with oxygen. If the copper burns, then, it will make a molecule that has copper and oxygen in it. If you said something like "copper oxygen," you were on the right track. When you take chemistry in high school, you will learn how to name chemicals properly.

12.9 It's ozone. That concentration is not enough for the other pollutants we discussed to be toxic, but it is 10 times the level at which ozone is potentially toxic.

12.10 The greenhouse effect is greater on Venus. Indeed, Venus is the warmest planet in the solar system, despite the fact that it is not the closest one to the sun. The reason is its greenhouse effect.

12.11 The pH will be higher. Since carbon dioxide makes carbonic acid in soda, as the soda goes "flat," there will be less carbon dioxide, so there will be less carbonic acid. Thus, the solution will be less acidic, which means a higher pH. You can try this yourself, but the pH change will probably be too small to see with the pH paper you have.

12.12 They might have grown during the Medieval Warm Period. It must have been warmer when the forests grew, and climate proxies indicate it was warmer during that time period. This would also correspond to when the Vikings settled in Greenland, which explains why they could flourish there.

12.13 No, it does not. It means that the carbon dioxide levels will have caused a 2 °C rise in temperature. However, there are clearly other things that affect the earth's temperature, since it varied a lot before people were making much carbon dioxide. Those things could cause the overall temperature change to be lower or higher.

12.14 alkaline. Remember, something isn't acidic until it has a pH under 7. Since it is currently above 8, it is alkaline, but becoming less so.

Chapter Review

1. Define the following terms:

a. Pressure
b. Parts per million
c. Troposphere
d. Stratosphere
e. Ozone layer
f. Mesosphere
g. Thermosphere
h. Combustion
i. Ocean acidification
j. Climate proxy
k. Climate sensitivity

2. What holds the atmosphere in place around the earth?

3. You have a water bottle that has no more water in it. You put the top on, which is airtight. Is it possible to crush the bottle so that there is absolutely no space left inside the bottle? Why or why not?

4. What two units for pressure did you learn in this chapter?

5. How does the air pressure change with altitude?

6. List the two main components of dry air. Which is more concentrated?

7. NO_2 can start having toxic effects on the body when it reaches a concentration of 150 ppm. What percent of the air would it make up at that concentration?

8. If a gas makes up 0.5% of the air, what is its concentration in ppm?

9. List the main sections of the atmosphere and the regions that separate them, starting with the lowest altitude and then continuing with higher and higher altitudes.

10. In which regions of the atmosphere does temperature decrease with increasing altitude? In which regions does it increase with increasing altitude?

11. Why is the ozone layer important to life, and why is its position in the atmosphere also important?

12. In which region of the atmosphere do you find the ozone layer?

13. When air expands without the addition of energy, what happens to its temperature?

14. Have the concentrations of pollutants in the air increased or decreased over the past 40 years?

15. What are the four main pollutants we discussed in this chapter? Is it possible to completely eliminate them from the atmosphere?

16. What is the greenhouse effect, and why is it a good thing?

17. What are the potential consequences of the increase in carbon dioxide concentration in the atmosphere?

18. What is the pH scale? What value on the scale represents something that is neither acidic nor alkaline? What numbers represent something acidic? What numbers represent something alkaline?

19. You have two substances. One has a pH of 13, while the other has a pH of 8. Based solely on that information, which would be safer to put on your skin?

20. Generally, has the average temperature of the earth been increasing, decreasing, or staying the same over the past 40 years?

21. What do climate proxies indicate was the warmest period in the past 2,000 years? You don't need to know the dates, just the word that refers to that time period.

22. Generally, has the pH of the ocean increased, decreased, or stayed the same over the past 30 years?

Chapter 13: Weather, Part 1

Introduction

"What's the **weather** going to be like today?" You have probably asked that question several times. If you are going outside, you need to know what kind of clothes to wear. If you are planning to do something outdoors, like hiking or camping, you might change your mind if the weather is severe. In the end, the weather affects our lives, so it is important to learn about it.

One of the first things you need to get straight in your mind is the difference between weather and **climate**. Weather is related to climate, but the two are not the same. The best definition of climate is:

Climate – The general weather conditions prevailing in an area over a long period of time

Think, for example, about the desert. In general, you don't expect it to rain very much in a desert. You also expect the humidity to be low. In most deserts, you expect it to be sunny and hot during the day. So if you are planning to take a trip to the desert, you pack light clothes, sunscreen, and extra water.

However, that doesn't mean it is always sunny and never rains in the desert. Once, when I was riding horses in the Arizona desert with my daughter, we got caught in a sudden downpour. It started raining so hard that we had to stop our horses, turn our backs to the wind, and simply wait it out. I still recall shivering with cold and thinking that the rain was going to leave bruises on my back because it was hitting me so hard. It didn't, but it felt like it might.

Professor Andrew John Herbertson of Oxford university said it best. In his book *Outlines of Physiography*, he wrote, "Climate is what on an average we may expect, weather is what we actually get." The climate in the desert is dry and hot during the day. That's what my daughter and I expected when we went riding. What we got was a painful, cold downpour! While we will touch on climate from time to time in the next two chapters, our focus is going to be on weather.

You Are My Sunshine

The earth gets nearly all the energy it uses from the sun. As a result, the sun is one of the most important factors when it comes to the weather we experience. As I already discussed, the earth absorbs light that comes from the sun and then releases a lot of it. Greenhouse gases trap a lot of the energy that is being released, which not only allows the planet to be warm enough to support life, but it also evens out the temperature differences between night and day. It is important to realize, however, that when I say "light" that comes from the sun, I am not just talking about the light that you see. The sun sends energy to the earth in different forms, and collectively, we call it **radiation**.

Radiation – Energy that is emitted from a source

When I have used that term so far, I have been talking about radiation that comes from radioactive isotopes. That is one form of radiation. After all, when a radioactive isotope emits an alpha particle, that particle has energy, and it is being emitted from the isotope. However, that is just one form of radiation. In fact, the light that illuminates your home is also radiation, because it is energy that is being emitted from the light bulb. The light that comes from the sun is also radiation.

Since there are different forms of radiation, it's important to make distinctions among them. Most of the energy that comes from the sun is **electromagnetic** (ih lek' troh mag net' ik) **radiation**, which comes in the form of transverse waves that are called **electromagnetic waves**. Now remember from Chapter 6 that a wave can be characterized by its wavelength, which is the distance between the crests. When the crests are far apart, the wavelength is large, and when they are close together, the wavelength is small. Well, there is a wide range of wavelengths found in electromagnetic radiation, and the sum total of all those wavelengths is often referred to as the **electromagnetic spectrum** (spek' trum), which is illustrated below.

The Electromagnetic Spectrum

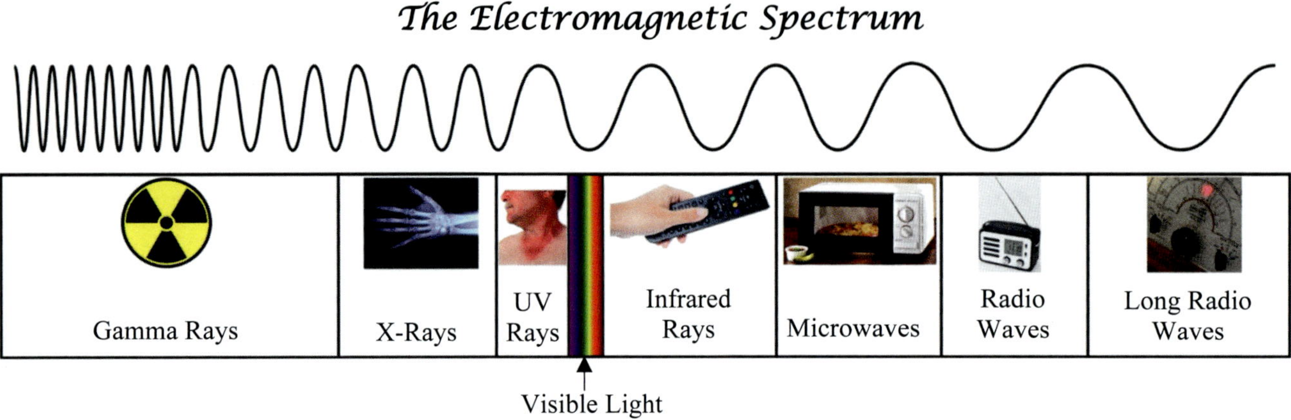

The top part of the illustration shows a transverse wave whose wavelength gets longer and longer. This represents all the different wavelengths that electromagnetic waves can have. The shorter the wavelength, the more energy the electromagnetic wave has. On the left side of the illustration, then, you find the electromagnetic waves with the shortest wavelength and therefore the highest energy. They are called **gamma rays**, and they are usually produced by radioactive isotopes and other processes related to the atomic nucleus. These waves have enough energy to do serious damage to living tissue. Nevertheless, they can be useful. Tumors, for example, are composed of living tissue, but they are unhealthy. Some tumors can be killed by exposing them to gamma rays.

Electromagnetic waves with longer wavelengths are referred to as **X-rays**. They have enough energy to pass through many types of tissues, damaging them as they do so. However, they don't have enough energy to pass through bone tissue. Thus, if you shine X-rays on a person and put film or an X-ray detector on the other side, an image can be formed, showing you the bones that the X-rays couldn't pass through. Even though this does kill some tissue, as long as you don't expose a patient to too many X-rays, the dead tissue will be replaced, and a doctor can diagnose problems with a patient's bones without having to do surgery to actually look at the bones.

You already learned about ultraviolet (UV) rays. They have enough energy to kill living tissue, but not as much energy as X-rays. Nevertheless, exposure to too many ultraviolet rays will produce a sunburn. The ozone layer blocks most of the UV rays coming from the sun, so as long as you are careful, you don't have to worry about getting a sunburn. Also, you can put sunscreen on your body, which absorbs the UV rays before they hit your skin. That way, you won't get a sunburn even if you spend a lot of time out in the sun. UV rays can also be useful, because they can be used to kill bacteria and other disease-causing microorganisms.

The light that we can actually see is called **visible light**, and it has longer wavelengths and lower energy than UV rays. As you may already know, the different colors that we see actually correspond to different wavelengths of visible light. The shortest wavelengths make up violet light, and as the wavelengths get longer, the color becomes indigo, then blue, then green, then yellow, then orange, and then red. Those are the only wavelengths that our eyes can detect when it comes to electromagnetic radiation. Obviously, the energy of the waves is low enough in visible light that it doesn't cause significant harm to living tissue.

As the wavelengths get longer, the electromagnetic waves have even less energy. **Infrared rays** have longer wavelengths than visible light. Most remote controls send infrared light to whatever device they are controlling. You can't see the light with your eyes, but with the right kind of detector, the device can sense it and respond to the commands that the remote is sending.

This image was taken with an infrared camera, showing the infrared rays being emitted by things that are warmer than the surroundings.

Also, when something is warmer than its surroundings, it releases a lot of its heat in the form of infrared rays. You, for example, are usually warmer than your surroundings, so you emit a lot of infrared rays. With the right kind of detector, you can see those rays, so infrared cameras can take images of warm objects, even when there is no visible light. The image on the right, for example, was taken at night. The black colors indicate things that are cooler than the surroundings. Blue represents things a bit warmer than the surroundings, green is warmer, yellow is warmer, and red is warmest. Notice that the two people in the image show up in red, since they are the warmest things in the area.

In my discussion about global warming, I told you that the earth absorbs energy from the sun, but then it releases a lot of that energy. It turns out that most of the energy the earth releases is in the form of infrared radiation. Greenhouse gases can absorb that infrared radiation, which gives them energy. That makes them move faster, raising their temperature, which in turn, raises the temperature of the atmosphere.

Microwaves have longer wavelengths than infrared rays. That means they have even lower energy, and it turns out that this amount of energy can be absorbed by water molecules. When microwaves are shined on food that contains water molecules, the molecules absorb the energy of the electromagnetic waves, which causes them to rotate. As they rotate, they rub up against the surrounding molecules, which results in a lot of friction. That friction produces heat, which can cook the food. Cell phones also use microwaves to communicate with the phone network. As a result, cell phones heat up anything nearby that contains water. This includes your body tissues. However, the intensity of the waves is much, much lower than what is used in a microwave oven, so the heating caused by a cell phone is not enough to warm your tissues significantly above your normal body temperature.

Radios and televisions use **radio waves** to communicate. They have the longest wavelengths and therefore the lowest energy. That turns out to be useful for long-range communication. As you can see, then, we have managed to find uses for all wavelengths of electromagnetic waves. That's particularly interesting, because the man who actually demonstrated the existence of electromagnetic waves we cannot see thought that they were "…of no use whatsoever." (Anton Z. Capri, *Quips, Quotes, and Quanta: An Anecdotal History of Physics*. World Scientific 2007, p 93.)

It's Like Night and Day

The sum total of the radiation that the earth receives from the sun is referred to as **insolation** (in' suh lay' shun), which is short for **incoming solar radiation**.

Insolation – The amount of solar radiation received by the earth

However, if you think about it, that can only happen during the day. After all, the only part of the earth that can get insolation is the part that faces the sun. If the sun's rays can't hit a place on the earth, that place cannot receive any insolation. Since insolation is what gives the earth most of its energy, we can also say any given part of the planet receives energy from the sun during the day, but it gets no energy from the sun at night.

This, of course, explains why it is usually cooler at night than it is during the day. During the day, you are being hit by a lot of insolation, so everything around you gets warm. At night, you aren't getting any insolation. As a result, everything around you tends to cool off. Consider, for example, the difference between day and night on the moon. During the day, the surface of the moon has a temperature over 100 °C. In other words, it's hot enough to boil water! At night, however, the temperature can get as low as -173 °C! On the moon, then, the difference in day and night temperature is nearly 300 °C!

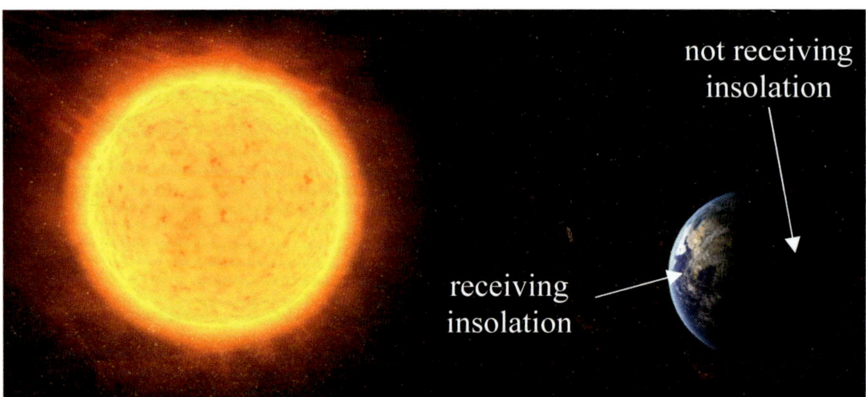

The earth gets insolation only on the regions that face the sun.

Of course, the difference in daytime and nighttime temperature is not nearly as dramatic on the earth as it is on the moon. Why? There are two main reasons. You already learned about one of them – the greenhouse gases in the earth's atmosphere. Remember that the earth emits a lot of the energy it absorbs from insolation, and that happens more at night than during the day. Rather than losing all that energy, a lot of it is trapped by the greenhouse gases. As a result, the earth's temperature doesn't decrease at night as much as it would without the greenhouse gases.

The other main reason is related to what actually turns night into day: the rotation of the earth. As the earth spins on its axis, some regions move out of the sunlight, while others move into it. So the regions that were warming from the insolation stop receiving it and start to cool off, while the parts that were cooling start to receive insolation and start warming up.

If you think about it, the longer any region on earth is receiving insolation, the warmer it will get. In the same way, the longer a region is not receiving insolation, the cooler it will get. Well, the earth completes a rotation every 24 hours, which means that on average, most parts of the earth spend 12 hours receiving insolation and 12 hours not receiving it. Compare that to the moon, which takes just over 27 days to make one rotation. That means on average, every region on the moon receives insolation for 14 days and doesn't receive it for the next 14 days. That gives each region on the moon a lot longer to heat up during the day, which produces a very high temperature. Similarly, each region on the moon has 14 days to cool off, which results in a very low temperature.

Even though we say that the earth makes one rotation every 24 hours, that's not really true. It actually makes a rotation every 23 hours, 56 minutes, and 4.09 seconds, which is called its **sidereal** (sigh' dir ee uhl) **day**. However, that's not the length of a day according to what we see, because while the earth is rotating, it is also orbiting the sun. As a result, the earth has to make slightly more than one rotation to get the same part of the earth pointing towards the sun in exactly the same way. That means the **solar day**, which is what we actually perceive as a day, is a bit longer. In fact, it is very close to 24 hours. Thus, a day really is very close to 24 hours long, but it is slightly longer than the time it takes for the earth to make one full rotation.

Believe it or not, however, the speed at which the earth spins does change by a small amount. There are many reasons for this. For example, the moon's gravitational force causes changes in shape not only to the ocean (which results in tides) but also to the shape of the earth as well. In addition, large tectonic movements can change the way the earth's matter is distributed, which changes its speed of rotation. As a result, the length of the day varies. During 2019, for example, the shortest solar day was 0.00095 seconds less than 24 hours, and the longest solar day was 0.00168 seconds more than 24 hours. The average solar day lasted 24 hours and 0.00039 seconds.

The amount of time a part of the earth is exposed to insolation isn't the only thing that affects how warm the earth gets. We will discuss the other ways insolation affects the earth's temperature the next time you do science.

Comprehension Check

13.1 Which color of visible light has the most energy? Which has the least?

13.2 Certain snakes, called pit vipers, have pits in their head that can sense infrared light. Suppose a pit viper is hunting at night. Will it be more likely to find a mouse (which is a warm-blooded) or a lizard (which is cold-blooded)?

13.3 The planet Mercury has the highest difference between daytime and nighttime temperatures (427 °C during the day and -173 °C at night). It has no atmosphere. How does the speed of its rotation compare to that of the moon?

It's Not Even

When we think about the insolation a particular region of the earth gets during the day, it should be clear that it is variable. At dawn, it's not very bright, because the earth has just rotated you towards the sun enough to start seeing its light. You first see the sun as being low in the sky, and you say the sun is rising. At this point in the day, you aren't pointed directly at the sun, so the sun's light is not shining directly on you. As a result, you aren't getting as much energy as you could be getting. However, as the day wears on, the earth starts rotating you so that you are facing the sun more directly. That means you get a more direct view of the sun, and the sun appears to move higher in the sky. Around noon, the sun reaches its highest point in the sky, and at that point, you are getting as much light as you can from the sun on that day. After that, the earth's rotation starts moving you out of direct view of the sun, and you see that as the sun moving lower in the sky. Eventually, you see the sun very low on the other side of the sky, and you say the sun is setting. The image below attempts to illustrate that, by showing how the sun travels across the sky over the course of a day.

The sun's position in the sky changes throughout the day, which changes the amount of insolation being received.

In the end, then, the amount of insolation at the very beginning of the day is low, but as the day goes on, the amount of insolation increases. However, after the sun passes its highest point in the sky, the amount of insolation decreases. Nevertheless, you are still getting insolation throughout the day, so the sun is warming your part of the earth throughout the day. As your surroundings warm up, however, they also start to release some of that energy. The hotter they get, the more energy they release, so eventually, the insolation cannot warm your surroundings faster than they are being cooled by the release of energy, so your region cools off. As long as the weather stays the same throughout the day, it gets warmer throughout the day until around 3-5 o'clock, at which point, it starts to cool off.

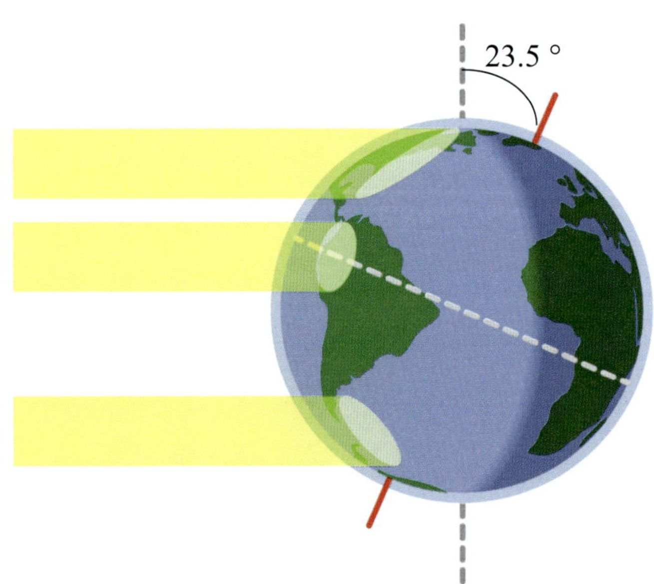

The earth's axial tilt means that the amount of insolation each hemisphere receives is different.

The insolation you get not only varies during the day, but it also varies with your position on the earth. This happens because the earth is tilted relative to the sun. Look, for example, at the drawing on the left. It shows three beams of light coming from the sun and hitting different parts of the earth. Notice first that if the earth were not tilted relative to the sun, the North Pole and South Pole would be where the gray dashed lines are in the drawing. However, that's not the way it is. The earth is tilted relative to the sun. Since the imaginary line that runs from the North Pole to the South Pole is called the earth's **axis**, we say that the earth has an **axial tilt**. The angle of the tilt is about 23.5 degrees.

Now think about what that tilt does to the electromagnetic radiation coming from the sun. At the equator, the light rays hit the earth nearly straight on. As a result, a beam of light will spread out in a circle, but that circle will be about as wide as the beam of light. What happens to the part of the earth tilted away from the sun? In the illustration above, that's the Northern Hemisphere. The light hits at an angle, which cause the beam of light to spread out into an oval. The oval is much larger than the circle at the equator, so the light is spread over a larger area. If the same amount of insolation is spread over a larger area, that means each part of the earth in that region gets less insolation than each part of the earth on the equator.

But what about the part of the earth that is tilted towards the sun? In the illustration on the previous page, that would be the Southern Hemisphere. Well, the light hits that part of the earth at an angle, but because that part of the earth is tilted towards the sun, the oval the light spreads into is smaller than the oval in the part that is tilted away from the sun. It's still larger than the circle at the equator, but not as large as the oval in the part of the earth that is tilted away.

In the end, then, the equator gets the most insolation. That's why it is warmer at the equator (on average) than anywhere else on the earth. The equator is getting the most direct sunlight, so it is getting the most insolation and therefore warms up the most. As you move away from the equator, the sunlight arrives at an angle. This causes the sunlight to spread out, reducing the insolation received at any given place. The farther you are from the equator, the more effect this has. Thus, as you move farther from the equator, on average, the earth gets colder. However, there is a difference between the hemispheres. The hemisphere that is tilted away from the sun gets less insolation than the hemisphere that is tilted towards the sun. Thus, on average, the cooling that you experience as you travel away from the equator is strongest in the hemisphere that is pointed away from the sun.

The fact that insolation varies depending on your distance from the equator is one of the most important factors in understanding the earth's weather. However, before I can give you the complete story, you need to learn something about how the earth orbits the sun. The best way to do that is with an experiment.

Experiment 13.1: It's Not Really a Circle

Supplies:
- Two standard sheets of white copy paper (without any lines)
- A 15-cm (6-inch) length of string that is stronger than thread
- A ruler
- A pencil

Instructions:
1. Use the ruler to find the very center of the paper and make a mark there.
2. Make two marks, each of which is 5 cm (2 inches) away from the center mark.
3. Use your fingers on the hand that you don't write with to hold one end of the string to one of the marks you just made and the other end to the other mark.
4. Use the point of the pencil to stretch the string so that it is tight. Your setup should look like the picture on the right.
5. Move the pencil so that the string stays tight the entire time and the pencil draws a curve. The curve should stretch over and a bit beyond both of the fingers you are using to hold the string.
6. Flip the paper over so that the curve is now under the marks you made.

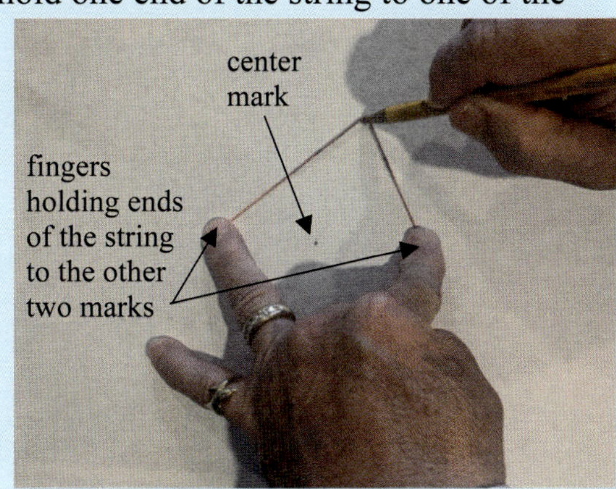

7. Repeat steps 3-5 so that you make the same curve above and slightly beyond your fingers.
8. Remove the string and use the pencil to connect the two curves you made smoothly. You should see that you have made an oval.
9. Repeat the entire experiment with a new piece of paper, but this time, put the two marks each 1-cm (½ inch) from the center mark.
10. Compare the two shapes that you drew. They are both ovals, but one should look a lot more oval than the other.
11. Clean up your mess.

In your experiment, you drew two examples of an **ellipse**, which is defined by two points, each of which is called a **focus**. An ellipse is the collection of all points where the sum of the distances from each focus to the ellipse is the same. In your experiment, you held each end of the string at a focus of the ellipse. Since the string was 15 cm long, the ellipses you drew contained all the points where the sum of the distances from the point to each focus was 15 cm. What did you see when you compared the two ellipses you drew? The one where each focus was only 1 cm from the center mark should have been noticeably less oval than the one where each focus was 5 cm from the center mark.

If you don't really understand what an ellipse is, don't worry. You will study it more in math. Right now, you can think about an ellipse as an oval that is defined by two points, each of which is called a focus. As your experiment showed, the closer each focus is to the other, the more circular the oval is. When we discuss ellipses, then, we often talk about their **eccentricity** (ek' sen tris' uh tee), which is how oval they are. The higher the eccentricity of an ellipse, the more oval and less circular it is.

Why am I telling you all this? Because the earth and all the other planets actually orbit the sun in an ellipse. The sun sits at one of the two focus points. Now the eccentricity of the earth's orbit is very low, so it is almost a circle. Nevertheless, it is an ellipse, which has an important consequence when it comes to the insolation that the earth receives. You will learn more about that the next time you do science.

Comprehension Check

13.4 The eccentricity of Mercury's orbit around the sun is greater than the eccentricity of the earth's orbit around the sun. Which orbit is less circular?

We Take This Trip Every Year

On average, the earth is 149,597,870 kilometers (92,955,807 miles) from the sun. However, since its orbit is an ellipse with the sun at a focus, it is sometimes closer to and sometimes farther from the sun. When the earth is closest to the sun, it is "only" 147,098,074 kilometers (91,402,506 miles) away, and we say that it is at **perihelion** (pehr' uh hee' lee un).

Perihelion – The point in a body's orbit when it is closest to the sun

When the earth is farthest from the sun, it is 152,095,295 kilometers (94,507,635 miles). At that point, we say that the earth is at **aphelion** (uh fee' lee un).

Aphelion – The point in a body's orbit when it is farthest from the sun

The easiest way to remember this is to think about the "a" that begins "aphelion." It is also the first letter in the word "away." So the earth is **a**way from the sun when it is at **a**phelion.

While these distances are really large, it is important to put them in perspective. The earth is only 1.7% farther from the sun at aphelion than it is at perihelion. This is important, since the distance between the earth and the sun affects the insolation the earth receives. If its aphelion distance were much larger than its perihelion distance, it could get dangerously cold at aphelion and dangerously hot at perihelion. The planet Mercury, for example, is 52% farther from the sun at aphelion, at which point it can get 150 °C colder than when it is at perihelion! So while the earth's orbit is technically an ellipse, it is so close to a circle that the earth's temperature doesn't vary much between aphelion and perihelion.

The illustration below puts all this information together. Please note that the earth's orbit is not nearly as elliptical as is shown below. It has been exaggerated for the purpose of the discussion. Also, the relative sizes of the sun and the earth are not correct.

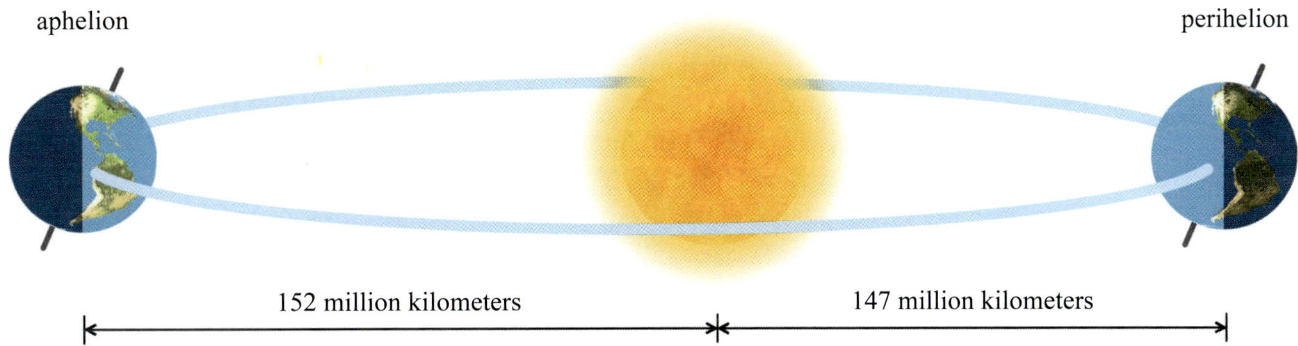

In this illustration of the earth's orbit around the sun, the ellipse has been exaggerated so that you can see it easily.

Notice that the sun isn't exactly at the center of the ellipse. Remember, it is at a focus, and neither focus is at the center. Notice also that the earth's axial tilt means that the Northern Hemisphere is pointed towards the sun when the earth is at aphelion, while the Southern Hemisphere is pointed towards the sun when the earth is at perihelion.

Now think about where you live. Most likely, it is warmer in the summer and cooler in the winter. When do you think summer happens: when the earth is at aphelion or perihelion? You might be surprised to find out that it depends! If you live in the Northern Hemisphere, you experience summer at aphelion, but if you live in the Southern Hemisphere, you experience summer at perihelion. Why? Because the effect of the earth's axial tilt on insolation is much greater than the effect of aphelion and perihelion.

Remember, the equator gets the most insolation, because it is always pointed more or less directly at the sun. As a result, the temperature stays very stable throughout the year. When the earth is at perihelion, the Southern Hemisphere is pointed towards the sun. As the illustration on page 374 shows, the solar radiation doesn't spread out as much when it hits the Southern Hemisphere, so each part of the Southern Hemisphere gets more insolation. Alternatively, the Northern Hemisphere is pointed away from the sun, so each point there gets less insolation. This means that on average, the

Southern Hemisphere is warmer than the Northern Hemisphere. In other words, the Southern Hemisphere experiences summer while the Northern Hemisphere experiences winter. At aphelion, the Northern Hemisphere is pointed towards the sun, so it experiences summer while the Southern Hemisphere experiences winter.

Even though the earth does experience a change in its distance from the sun, that doesn't affect insolation nearly as much as the earth's axial tilt. As a result, the seasons that you experience when you live away from the equator are caused by the axial tilt, not the elliptical nature of the earth's orbit. Thus, when it is summer in the Northern Hemisphere, it is winter in the Southern Hemisphere. This is important to remember:

The seasons are caused by the earth's axial tilt, and they are opposite for each hemisphere.

Now remember, the equator is always pointed towards the sun, so the effect of aphelion and perihelion is a bit easier to see there. While local weather affects the daily temperature more than the earth's distance from the sun, the equator is, on average, slightly warmer when the earth is at perihelion. For example, the city of Quito, Ecuador is pretty much right on the equator. At aphelion, the average temperature is 19 °C, and it "rises" to 20 °C at perihelion. That small change in temperature illustrates how little the equator's insolation changes throughout the course of the year.

Now if you think about it, when the Southern Hemisphere is tilted towards the sun, the earth is at perihelion. When the Northern Hemisphere is tilted towards the sun, the earth is at aphelion. Even though the difference in insolation between aphelion and perihelion is small, you might expect that the Southern Hemisphere's summers are warmer than the Northern Hemisphere's summers. While that might make sense, it's actually the opposite. The Northern Hemisphere summers are warmer. Why? The percentage of the Southern Hemisphere covered by the ocean is much greater. Remember what water does. It moderates temperature. So even though the earth gets more insolation when the Southern Hemisphere experiences summer, the Southern Hemisphere's summer is cooler than the Northern Hemisphere's summer. In the same way, even though the Southern Hemisphere experiences winter when the earth is getting the least amount of insolation, its winter is warmer than the Northern Hemisphere winter!

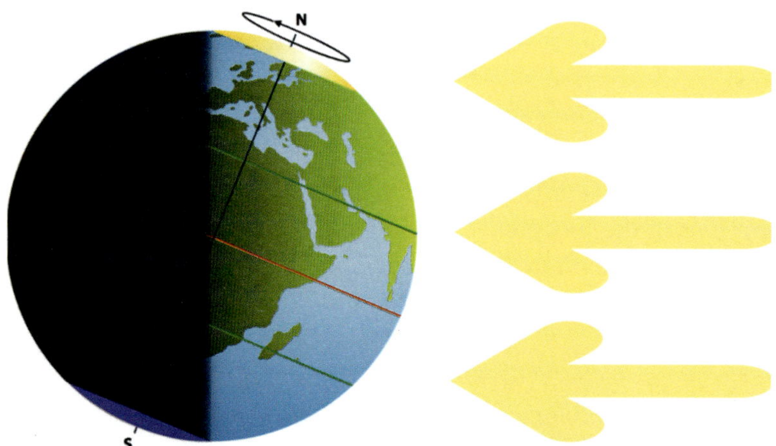

The earth's axial tilt causes 24 hours of daylight at a pole that is pointed directly towards the sun and 24 hours of night at the pole that is pointed directly away from the sun.

But temperature isn't the only change that comes with the seasons, is it? You probably know that the days tend to be longer in the summer and shorter in the winter. Once again, that's because of the earth's axial tilt. The easiest way to think about it is to imagine what happens when the North Pole is pointed most directly at the sun, as shown in the illustration on the left. The arrows indicate where the solar radiation is coming from. Now think about the earth as it spins on its axis. Over the course of one full rotation, the area near the North Pole (highlighted in yellow) will never be dark. In the same way, the area near the South Pole (highlighted in violet) will never get into the light. In other words, when the North Pole is pointed most directly at the sun, it stays light for 24 hours a day, and the South Pole stays dark for 24 hours a day.

When the earth moves in its orbit, however, the North Pole will move away from being tilted towards the sun, and the South Pole will become tilted towards the sun. After six months, the North pole will be pointed most directly away from the sun and will experience 24 hours of darkness per day. At the same time, the South Pole will be pointed most directly at the sun and will have 24 hours of daylight. What happens in between those times? Well, as the North Pole moves away from being pointed most directly at the sun, the amount of daylight it sees decreases. As the South Pole becomes more directly pointed at the sun, the amount of daylight it sees increases. This happens over the entire globe, except right on the equator. It's just easiest to see when it comes to the poles.

What does this look like from your perspective sitting on the earth? When the part of the earth you are on points as directly as possible at the sun, it appears to be higher in the sky. When the part of the earth you are on is tilted away from the sun, it looks lower in the sky. Over the course of the day, you observe this as the sun rising in the east and setting in the west, as shown in the picture on page 374. Over the course of the year, you see it as the overall height of the sun in the sky changing. When your hemisphere is facing the sun, it goes higher in the sky over the course of the day. When your hemisphere is facing away from the sun, it doesn't go as high.

Consider, for example, the image on the right. It represents a series of pictures taken over the course of a year at noon from the same position in front of Bell Laboratories in New Jersey. Notice how the sun's position in the sky at noon changes dramatically as the days move on. When the sun is high in the sky at noon, it is summer. When the sun is low in the sky at noon, it is winter. Between summer and winter, the sun gets progressively lower in the sky. That results in less insolation, producing cooler, shorter days. From winter to summer, the sun starts rising higher in the sky over the course of a day. This results in more insolation, producing longer, warmer days. As long as you are away from the equator, you can produce an image similar to this one, where the maximum height of the sun in the sky changes from low in the winter to high in the summer. The details of the pattern made by the sun will vary based on latitude, but it will always sweep out a similar shape, which is called an **analemma** (an' uh lem' uh).

This is a composite of pictures taken at noon each day during different times of the year. When the sun is high in the sky at noon, the area is experiencing summer. When it is low in the sky at noon, the area is experiencing winter.

In the end, then, the earth's axial tilt affects not only the temperatures we experience at latitudes away from the equator, but it also affects the length of the day at those same latitudes. Because tracking the seasons is important, certain positions in earth's orbit are used as "benchmarks" in determining where the earth is in its seasons. First, we have the **solstices** (sol' stusez). These are the days in which a hemisphere is pointed most directly toward or away from the sun, resulting in the sun reaching its maximum or minimum height in the sky.

Solstice – The day at which the sun reaches its maximum or minimum height in the sky

Based on the definition, you should realize that there are two solstices: one where the sun is at its maximum height and the other where it is at its minimum height. In the Northern Hemisphere, it is at its maximum height somewhere between June 20th and June 22nd, depending on the year. Of course, at that time, the sun is at its minimum height in the Southern Hemisphere. In the Northern Hemisphere, the sun is at its minimum height between December 21st and 22nd, at which time it is at its maximum height in the Southern Hemisphere.

Now, when you live in a specific hemisphere, you often refer to the solstices according to the season. So in the Northern Hemisphere, the solstice in June is often called the **summer solstice**, and the solstice in December is often called the **winter solstice**. The problem, of course, is that in June, the Southern Hemisphere is experiencing winter, so they would call the solstice in June the winter solstice. Thus, it is best to refer to the solstices as the **June solstice** and the **December solstice**.

What's important about the solstices? Well, when the sun reaches its maximum height in the sky, the daylight lasts the longest. Thus, the Northern Hemisphere experiences the longest days during the June solstice. When the sun is at its minimum height in the sky, the days are the shortest, so the Northern Hemisphere experiences its shortest days during the December solstice. Of course, it is opposite for the Southern Hemisphere, which experiences its longest days during the December solstice and its shortest days during the June solstice.

Since the days go from being very long during one solstice to being very short during the other solstice, there must be a time when the days are exactly 12 hours long. In other words, there must be a point at which there is exactly as much daylight as there is night. Those are called the **equinoxes** (ek' wuh nahksez).

Equinox – The two times each year when the length of the day is equal to the length of the night

Not surprisingly, these happen roughly halfway in between the solstices: September 20th–23rd and March 20th–21st. In the Northern Hemisphere, the **September equinox** is often referred to as the **autumnal equinox**, because it ushers in autumn, and the **March equinox** is often called the **vernal** (or **spring) equinox**, because it ushers in spring. This is all summarized in the illustration on the left.

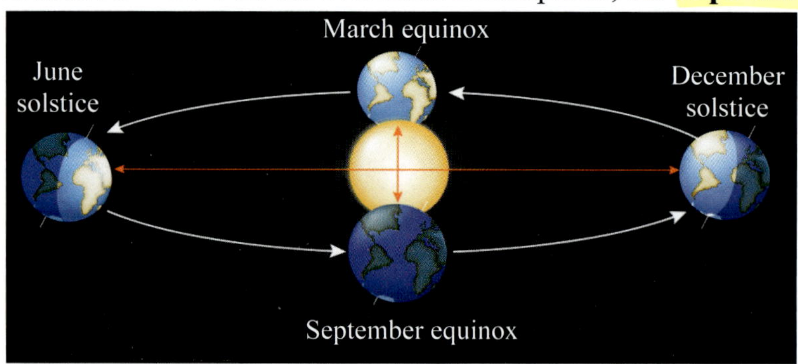

This illustration shows the position of the earth in its orbit during the equinoxes and solstices. Once again, the ellipse is exaggerated, and the relative sizes of the earth and sun are incorrect.

If you are in the Northern Hemisphere, then, you experience the longest days during the June solstice, and that is considered the first day of summer. After that, the days start to get shorter, and at the September equinox, the day is 12 hours long. That's the first day of fall. After that, the days continue to get shorter until the December solstice, which has the shortest day of the year. That starts winter. After that, the days lengthen until they get back to being

12 hours long at the March equinox, which begins spring. After that, the days continue to get longer until you once again reach the June solstice. The opposite progression happens in the Southern Hemisphere.

Before I end this discussion, I want to make it clear that the amount of insolation is not the only factor that affects the earth's temperature. The **albedo** (al bee' doh) of an area is also very important.

Albedo – A measure of how much light is reflected from a surface instead of being absorbed

When sunlight hits the earth, it can be reflected, absorbed, or both. Have you ever noticed that the day is brighter when the ground is covered in snow? That's because the snow has a very high albedo, reflecting as much as 95% of the light that hits it. The reflected light is not absorbed by the earth, so it doesn't do much to heat it up. Compare that to the asphalt that makes up a road. It's dark, because it absorbs most of the light that hits it. As a result, its albedo is very low, and it gets hot when exposed to sunlight.

Human activity tends to change the albedo of an area, changing the way the surroundings are heated by the sun. Consider, for example, the city of Phoenix. It is in the middle of the desert and gets a lot of sun. The pavement and buildings have a low albedo, so they absorb a lot of energy during the day. As a result, they are very hot by the time the sun sets. This increases the nighttime temperatures significantly. There are nights when the city is 6 to 8 °C (10 to 15 °F) warmer than the surrounding countryside. This is often called the **urban heat island effect**, because cities create "islands" of heat that would not otherwise be there. At least part of the reason the average temperature of the earth has increased over the past century is because more and more larger cities have been built, producing more and more urban heat islands.

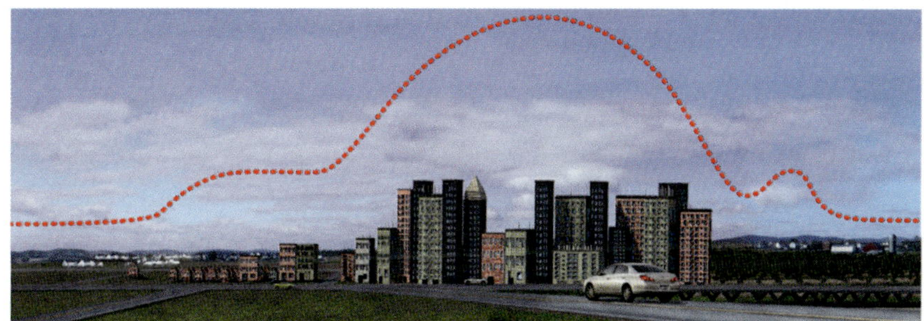

In this illustration, the red curve represents the temperature. It is higher in the city and lower in the rural areas because of the urban heat island effect.

Comprehension Check

13.5 Mars is 20% farther from the sun at aphelion than at perihelion. Is its orbit more or less oval than the earth's orbit?

13.6 The height of the sun in the sky over the course of a year is measured in two cities. The maximum height is lower in the first city. Which city is closer to the equator?

13.7 If you are in the Southern Hemisphere, when would you see the sun reach its maximum height in the sky: at noon on the June solstice, noon on the September equinox, noon on the December solstice, or noon on the March equinox?

13.8. You are in the Northern Hemisphere. The days have been getting shorter, but then on a specific day, the day is just as long as the night. Have you reached the June solstice, the September equinox, the December solstice, or the March equinox?

Wind and Temperature

The differences in temperature caused by differences in insolation are responsible for another very important aspect of weather. To find out what that is, let's start with an experiment.

Experiment 13.2: Making Wind

Supplies:

- The tea bags from the laboratory kit made for this course
- Matches or a lighter
- A plate made out of something that will not catch fire
- Scissors

What you should do:

1. Cut the tea bag as shown in the picture below.
2. Unfold the teabag as shown in the right picture below.
3. Empty the tea out of the bag.
4. Put the plate on a surface that is far from anything else that can burn. A stove is a perfect place to put it.
5. Form the tea bag into a tube that stands up on its own, and stand it up on the plate. Your experiment should now look like the bottom picture.
6. Light the match or lighter.
7. Touch the flame from the match or lighter to the top of the tea bag so that the material catches on fire.
8. Watch the tea bag burn all the way to the bottom of the plate.
9. If the tube falls over (which happens sometimes), you will need to redo the experiment. The effect is best seen when the tube burns down without falling over.
10. You should see something interesting happen when the bag is very nearly burned to the bottom.
11. If you don't see something interesting, try again with another tea bag. The brand of tea is important, and the bags in the kit have been chosen with that in mind. You can try other brands, if you like.
12. Clean up your mess.

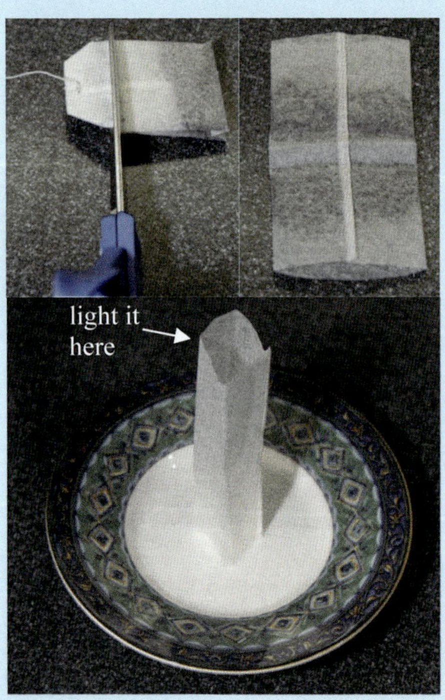

What did you see in the experiment? If it went well, you should have seen the last remains of the tea bag launch into the air and float away once the bag was nearly burned out. As I stated in the experiment, sometimes the effect is not dramatic if you don't have the right kind of tea bag, but generally you'll see a small part of the bag lift off the plate and into the air. What causes this effect?

Remember, air is composed of many molecules of gas. When you lit the tea bag, the air near the flame heated up, causing it to expand. That reduced its density. So, what does that mean? It means that hot air floats on top of cold air, just like hot water floats on top of cold water. So, the air that was heated up by the flame floated up and away from the tea bag. Think about what that means in terms of air pressure. A lot of molecules rose from where they normally were. That left fewer air molecules, and the fewer the air molecules, the lower the air pressure. As a result, the place where the air was before it rose away became an area of lower air pressure. What happened as a result? Air rushed in from the surrounding areas, filling the low-pressure area with more molecules.

Now, think about what that means, then. Air was rising from where it was being heated, but because of the loss in pressure, air was rushing in to fill the area where the air had just left. In other words, air was moving up, and as a result, air from other places was moving in to fill the void. This created a constant motion of air, which we call wind. So, the high temperature of the flame started making air move in a constant direction (up and away from the plate), which created wind.

Even though the wind started blowing as soon as the tea bag caught on fire, it was not a very strong wind. At first, the tea bag was heavy enough to be unaffected by that weak wind. However, the more it burned, the lighter the tea bag became. Near the end of the burning process, it was mostly ashes, which are really light. In fact, they were light enough to be lifted off the ground by the wind that was formed by the flame.

So, wind forms because of a pressure difference. Sometimes, that pressure difference is caused by pushing molecules around. For example, when you force air out of your lungs, muscles like your diaphragm compress your lungs, increasing the pressure inside. The air pressure outside your body is lower, so the air inside your lungs moves from the area of high pressure (your lungs) to the area of low pressure (outside your body). However, pressure differences can also be caused by differences in temperature, as you saw in your experiment. On the earth, that's how most pressure differences that produce wind occur.

If you have ever been near the ocean on a warm, sunny day, you might have noticed a pleasant breeze that blows from the ocean into the shore. That wind is caused by a temperature difference. The sun heats up land very quickly, but as you've already learned, it takes much longer for the sun to heat up water. As a result, on a warm, sunny day, the shore is warmer than the ocean. This means that warm air is rising from the shore, and that leaves an area of low pressure on the shore. The air above the ocean is at a higher pressure because it hasn't been heated as much, so air moves from the ocean to the shore. This produces a **sea breeze** that blows from the ocean to the shore. Since that wind brings in cooler air, it tends to cool you down, which is why sea breezes are so pleasant on warm, sunny days.

But that's not the end of the story. As the air from the ocean replaces the air on land, a low-pressure area is created over the ocean, so air from above rushes in to fill that low-pressure area. But that moving air leaves a low-pressure area above the ocean, so air from the shore rushes to fill that low-pressure area. This produces a "loop" of wind, with cool air blowing from the ocean to the shore near the surface of the earth, and warm air blowing from the shore to the ocean high above the surface of the earth.

During a warm, sunny day, wind near the surface of the earth blows from the ocean to the shore because the shore is warmer than the ocean.

At night, wind near the surface of the earth blows from the ocean to the shore because the ocean is warmer than the shore.

That's what happens during the day, but what about at night? The land cools off pretty quickly, but once again, water doesn't change temperature very easily. As a result, the ocean ends up being warmer than the land. This, of course, causes the situation to reverse. Now warm air rises from the ocean and is replaced by cooler air blowing in from the shore. High above the surface of the earth, the warm air that was rising from the ocean rushes to the land to fill the low-pressure area high above the shore. This makes a **land breeze** blowing cool air out to the ocean near the surface of the earth and blowing warm air into the shore high above the surface of the earth.

Notice the pattern that occurs in both a sea breeze and a land breeze. The wind forms "loops," which are typically called "cells." These loops result in wind blowing one direction near the surface of the earth and the opposite direction high above the surface of the earth. This also happens on a global scale, but we will discuss that the next time you do science.

Comprehension Check

13.9 Imagine a city that has been exposed to the sun all day. You are standing in a grassy field outside the city and feel a breeze. If the breeze is caused by the urban heat island effect (as shown on page 381), would you feel the breeze blowing towards the city or away from it? If someone were measuring the wind high above your position, would it be blowing towards the city or away from it?

Global Wind Patterns

Because there are varying amounts of insolation across the globe, there are temperature differences across the globe. They lead to pressure differences, which end up producing a pattern of global winds. Consider, for example, the equator. On average, it is the warmest region on the earth, because it gets the sun's direct light throughout the year. Overall, then, what happens to the air at the equator? It rises, because it is very warm. Now think about the poles. It is cold at the poles, and since cold air is denser than warm air, the air at the poles tends to sink.

Based on what you learned about land breezes and sea breezes, then, you might think that there is a global pattern where wind blows along the surface of the earth from the poles to the equator. However, that doesn't happen, because there are other factors at play. Humidity is one of those factors. Remember, humidity is a measure of how much water vapor is in the air. At the equator, the air is very humid, because the warm temperatures cause a lot of evaporation, which puts a lot of water vapor in the air.

This is important, because humid air is less dense than dry air. That might sound strange to you, since you think of water as being much heavier than air. However, you are thinking about *liquid* water, and that's not what humidity measures. Humidity measures the amount of water *vapor* (water in its gas phase) in the air, and the more water vapor air has in it, the less dense it becomes. At the same temperature, then, humid air is less dense than dry air.

So at the equator, there is a lot of warm, humid air, and it is rising. But what happens as it rises? It cools. What happens when humid air cools? The water condenses, making clouds. Those clouds produce a lot of rain. So as the air is rising, it produces a lot of rain. That's why there are rainforests at the equator. The weather is ideal for plant growth all year long, and there is a lot of rain being produced as the warm, humid air rises and cools off.

Now remember, as the warm, humid air rises, the cooler air near the surface rushes in to take its place. That causes low pressure where the cooler air was, so the air above rushes in to take its place. That causes less pressure where the air above was, so the rising air rushes to take its place. That makes a "loop" of air, much like what happens with a sea breeze or a land breeze, only on a much larger scale.

How big is the scale? First, the warm, humid air rises essentially to the tropopause, and then it travels away from the equator. Of course, it is producing a lot of rain, so eventually, that air dries out. It also cools down, since it is now away from the equator and higher in the atmosphere. By the time it is about a latitude of 30 degrees north or south of the equator, it has dried out so much that it is now denser than the air it is above. As a result, it starts to sink. This loop, which is called a **Hadley cell** after the scientist who figured this out, covers the entire troposphere and covers about one-third of each hemisphere!

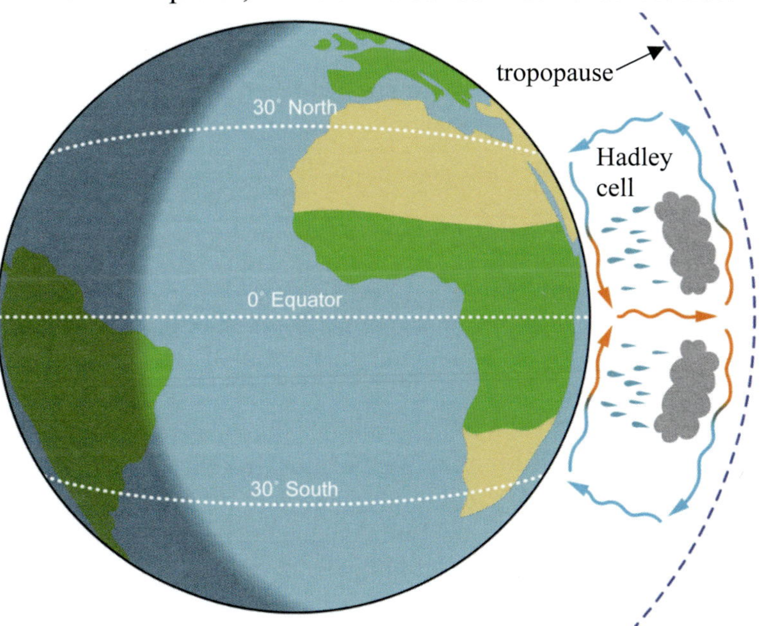

The global wind pattern known as the Hadley cell is responsible for the earth's rainforests and most of its deserts. Note that the height of the tropopause is exaggerated in the illustration.

But think about what kind of climate the sinking air would produce. It is very dry air, because it has lost most of its humidity. There will be some evaporation of water in the area, but because the air is so dry, that evaporation will not produce many clouds, so it will not produce much rain. What do we call a region of the earth that has very little rain? A desert! So where the cooler, dry air sinks, it produces deserts.

If you think about it, the Hadley cell explains why there are rainforests at the equator and why most of the earth's deserts are located around 30 degrees north or south of the equator. There are other factors that can produce lush forests away from the equator and deserts at latitudes other than about 30 degrees north or south, but the Hadley cell is the earth's predominant "rainforest maker" and "desert maker."

Of course, the Hadley cell covers "only" about one-third of a hemisphere, so there must be other cells that form global winds as well. Let's consider what happens at the poles. The air at the poles is cold, but as you travel away from the poles, the air gets warmer. Thus, the air at the poles is denser than the air farther from the poles. That means the air farther from the poles will rise, and the air at the poles will rush across the surface of the earth to replace it. That will set up another cell, with winds moving along the surface of the earth away from the poles and winds high in the air moving towards the poles. Not surprisingly, this is called a **polar cell**. The polar cell also covers about one-third of a hemisphere, so the polar cells go from the poles to a latitude of 60 degrees north or south.

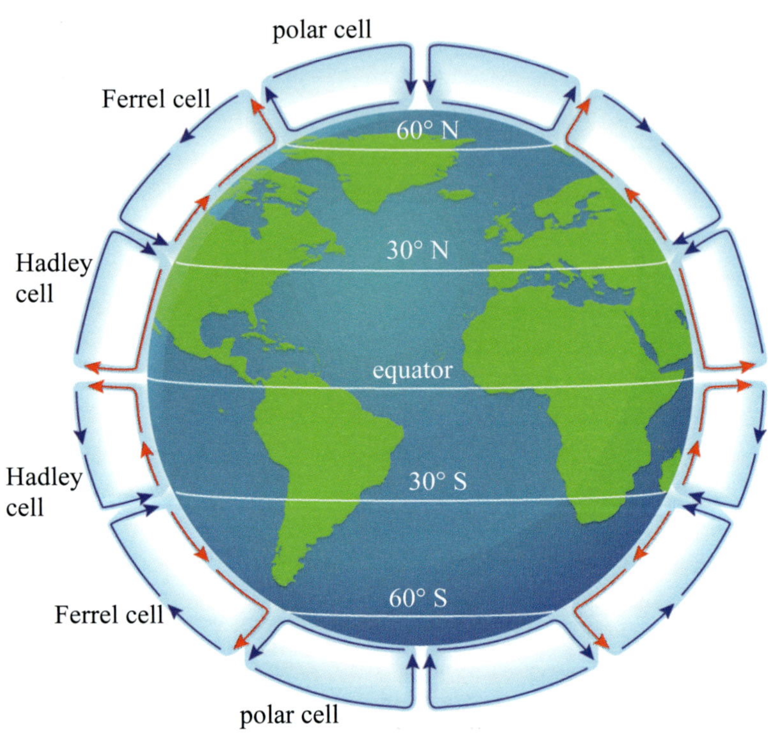

The three overall wind cells for each hemisphere of the earth are shown in the illustration above. The red arrows represent air that is warmer than the air represented by the blue arrows.

What happens in the other third of each Hemisphere? A third cell develops, which is called the **Ferrel cell**, after the scientist who first identified it. This cell ends up following the airflow at its edges. Thus, the cool, dry air at high altitudes and 30 degrees latitude falls, just as the air in the Hadley cell does at the same latitude. The air near the surface of the earth at 60 degrees latitude rises, just as the air in the polar cell does. We can visualize all of these cells with the help of the illustration on the left, where warmer air is shown in red, and cooler air is shown in blue. In the polar regions, some of the air is warmer and some is colder, but overall, it's all cold, so the entire polar cell is blue. In the Hadley cell, the warm air is rising at the equator, and even though the wind traveling along the surface is cooler than the air at the equator, it is still warm overall. As the air rises, it cools off, so all the air at the top of the cells is cold.

Believe it or not, we aren't quite finished yet, because global wind patterns are still more complicated than this. Remember the Coriolis effect from Chapter 11? It affects anything that is moving a long distance over the surface of the earth. The surface of the earth rotates with the earth to the east, but the speed at which the land is traveling depends on the latitude. The farther it is from the equator, the slower it travels. If anything travels in the air and changes latitude, the land over which it travels changes speed, and depending on the direction, either the land will outrun what is traveling over it or vice versa, causing the path of what is traveling to bend.

Let's start by thinking about the wind that is traveling along the surface of the earth in the Hadley cell. In which direction does it blow? Look at the two Hadley cell loops in the Northern Hemisphere. The wind near the surface of the earth is blowing south. Since it starts north of the equator and moves south, it starts on land that is moving more slowly and starts passing over land that is moving more quickly. That means the land will "outrun" the wind, moving more to the east than the wind does. As a result, the wind will bend towards the west. In the Southern Hemisphere, the same thing happens. The wind is blowing to the north, but it starts above slower-moving land and travels over faster-moving land. Thus its path bends to the west as well.

In the Ferrel cells, the winds near the surface of the earth are traveling the opposite direction, so they get bent to the east instead of the west. However, in the polar cells, the winds near the surface of the earth are traveling in the same direction as they are in the Hadley cells, so they get bent back to the west. In the end, then, the winds that blow along the surface of the earth (the winds we feel) end up blowing as illustrated in the drawing on the right. Notice that between the equator and the latitudes of 30° N and 30° S, the winds blow towards the equator but are bent west. In the Northern Hemisphere, they are called the **northeast trade winds**, and in the Southern Hemisphere, they are called the **southeast trade winds**. They are called "trade winds" because they allowed ships to travel across the ocean to trade. They are referred to as "east" winds because they blow from the east.

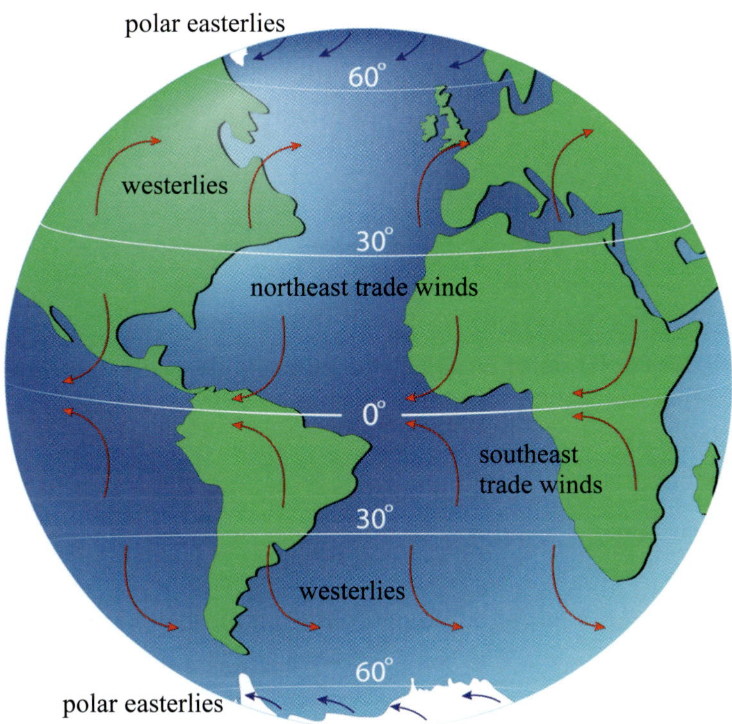

This illustration shows the global wind patterns found near the surface of the earth.

Now think about what happens right at the equator. Remember, the trade winds come from the Hadley cell, and in that cell, air rises at the equator. In addition, the trade winds are blowing towards the equator but to the west. As a result, there aren't many global winds at the equator. If a sailing ship were traveling using the trade winds and ended up at the equator, there wouldn't be much wind to help it sail along. As a result, it could get stuck there, so the waters near the equator were often referred to as the **doldrums**.

Between the latitudes of 30° and 60° (both north and south), the winds blow away from the equator, and they bend to the east. Since that means they blow from the west, they are called the **westerlies**. From 60° (north and south) to the poles, the winds blow towards the equator again but bend west. That means they blow from the east and are called the **polar easterlies**.

There are two important things you must understand about the global wind patterns illustrated in the drawing above. First, they apply only to winds that blow along the surface of the earth. Remember, all these global winds are part of a cell, and the cell extends to the tropopause. That means there are different wind patterns near the top of the troposphere. People don't feel those winds, however, so they aren't as important to this discussion. Second, these global wind patterns aren't the end of the story. There are lots of processes that cause winds to develop in specific regions of the world at specific times, producing **local winds**. The sea and land breezes you learned about earlier, for example, are local winds. Those local winds add to the global winds in any given area, so while the trade winds in the Northern Hemisphere blow generally southwest, there will be places between the equator and latitudes of 30° where the wind blows in a completely different direction for a while, because local winds have added to the global winds, producing new winds blowing differently than the trade winds.

Air Masses

While air moves with the winds, there are times when the winds aren't blowing strongly, and the air tends to stay in one place for a while. When that happens, the air can pick up the temperature, pressure, and humidity of the area in which it is sitting, becoming an **air mass**.

Air mass – A large body of air with uniform temperature, pressure, and humidity at a specific altitude

If air sits at the equator long enough, for example, it will get warm, humid, and have a low pressure. The temperature, pressure, and humidity will all decrease with increasing altitude, but at any specific altitude, the temperature, pressure, and humidity will be roughly the same. When that happens, the air tends to stay together, so when the wind blows, all the air in the air mass will move in the same direction.

Not surprisingly, the characteristics of an air mass depend on the region in which it forms. An air mass might form over the ocean. If that's the case, it's called a **maritime** air mass, and it is high in humidity because of all the evaporation that takes place over the ocean. If it forms on land, it is called a **continental** air mass, and it is typically lower in humidity, because there isn't as much evaporation happening over land. Of course, where it forms in relation to the equator will determine its temperature. **Equatorial** air masses form near the equator and are hot. If an air mass forms close to the equator but not really close, it is a **tropical** air mass. It is warm, but not as hot as an equatorial air mass. If it forms very near the poles, it is very cold and is called an **arctic** air mass. If it forms far from the equator but not very near the poles, it is a **polar** air mass and is cold, but not as cold as an arctic air mass.

To identify a specific air mass, then, you use either "continental" or "maritime" to indicate whether it formed over land or the ocean, and then you add the designation that indicates where it formed relative to the equator. The illustration below shows you sources for different air masses on the earth. Notice that small letters are used to indicate the air mass is continental or maritime, and then it is followed by a capital letter to indicate its relationship to the equator. Thus, the pink shape between Africa and South America is a maritime equatorial (mE) air mass, because it forms over the ocean (maritime) and near the equator (equatorial).

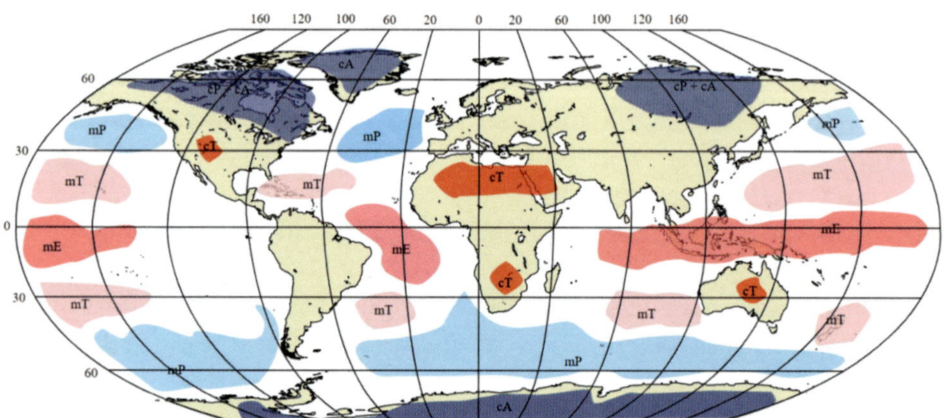

This illustration shows the major sources of air masses and what kinds are formed at those sources. Maritime air masses start with an "m," while continental ones start with a "c." The letters E, T, P, and A indicate position relative to the equator.

You might notice that there are no continental equatorial (cE) or maritime arctic (mA) air masses on the map. That's because the map is showing you where air masses commonly form, and it is rare for a continental equatorial air mass to form. Look at the equator. There isn't a lot of land right on the equator, so most equatorial air masses are maritime. Also, Antarctica is a continent, so any air mass formed there is a continental arctic air mass. Air masses can form in the Arctic ocean, but there is a lot of sea ice there, so the air masses that form there tend to be dry, like those that form over Antarctica. As a result, they are considered continental air masses.

Comprehension Check

13.10 You are watching data from a weather balloon that starts at the surface of the earth in the Northern Hemisphere and rises to the tropopause. Initially, the balloon says that the winds near the earth are blowing mostly towards the north, but the winds near the tropopause are blowing mostly to the south. Is the balloon in the Hadley cell, Ferrel cell, or polar cell?

13.11 A wind-driven ship is caught in the doldrums and has no oars or other mechanical means of propulsion. Is there any chance it can get out of the doldrums?

13.12 The westerlies in your area are blowing south and east. Which hemisphere are you in?

13.13 You are comparing four air masses: cT, cA, mE, and mP. Which two are the most humid? Order all four of them according to temperature, starting with the coldest and ending with the warmest.

Why Is the Sky Blue?

I have spent a lot of time discussing how the sun's insolation affects the air. However, there is something else that the sun does to the air, and the best way to understand it is to start with an experiment.

Experiment 13.3: Orange and Blue

Supplies:

- A tall glass (It can be plastic, but it must be clear so that you can see through it.)
- An LED flashlight (If you don't have an LED flashlight, use a flashlight app on a smartphone. Ideally, the face of the flashlight should be smaller than the bottom of the glass.)
- Black construction paper, scissors, and tape if the flashlight face is bigger than the bottom of the glass
- Milk
- Water
- A medicine dropper from the laboratory kit made for this course
- A spoon for stirring

Instructions:

1. If the face of your flashlight is larger than the bottom of the glass, cut a hole in the black construction paper that is a bit smaller than the bottom of the glass, and cover the face of the flashlight with it so the light that escapes comes out in a smaller circle than the bottom of the glass. Use the tape to secure the construction paper to the flashlight.
2. Add 3 drops of milk to the glass.
3. Add water until the glass is nearly full but not so full that spilling becomes easy.
4. Use the spoon to stir the water and milk so that you have a slightly cloudy solution in the glass.
5. Make the room dim (it doesn't have to be dark) and turn on the flashlight.

6. Hold the flashlight against the bottom of the glass so that the light shines straight up through the glass.
7. Keeping hold of the flashlight, lower the glass and flashlight so you can look down from the top of the glass and see the light shining through the cloudy water.
8. You should be able to see the bulb (or bulbs in the case of some LED flashlights) as a bright light at the bottom. If you just see a wash of light and can't make out the individual bulb or bulbs, dump half the mixture out, add water until the glass is almost full again and stir. Repeat steps 6-8.
9. Once you can see the bulb or bulbs, what color do you see? If it is white, add a small amount of milk to the mixture, stir, and check again. Repeat steps 6 and 7. You want to see a distinct bulb or bulbs, but the color should not be white. It should look orangish or yellowish.
10. Once you can see a distinct bulb or bulbs but not white light, hold the glass out at arms' length and look at it from the side. What color do you see? It might be mostly white, but there should be distinct hue of a color other than white.
11. Pull the flashlight away and look at the glass, then put the flashlight back and look again. How does the color change with the flashlight?
12. Clean up your mess.

What explains the different colors you saw? When light left the flashlight, it had to travel into the cloudy milk and water solution. As it traveled up through the solution, it hit droplets of milk. Sometimes, it would hit a droplet of milk and be deflected, traveling a new direction. When that happens, we say that the light is being **scattered** off the milk droplets. When you looked down through the top of the glass, most of the light that you saw coming from the bulb was not scattered. It traveled straight up through the liquid and hit your eyes. The unscattered light looked orange or yellow.

White light from the sun is passing through the prism, and as a result, it is being separated out by its color, which depends on the wavelength.

Why did it have that color? Remember that the color of light is determined by the wavelength. Red light has long wavelengths, violet light has short wavelengths, and the other colors have wavelengths in between. In order of decreasing wavelength, it's red, orange, yellow, green, blue, indigo, and violet. Remember also that the red wavelengths have the lowest energy, and the blue wavelengths have the highest energy. The others are in between. The higher the energy, the easier they are scattered. So, when you looked down the glass, you were looking at lower-energy light that wasn't scattered, so it had longer wavelengths. Thus, it was a mixture of red, orange, and yellow – the lower-energy wavelengths.

What did you see when you looked at the glass from the side? You should have seen a blue hue. There might have been a lot of white, but it should have been bluer than what normally comes from the flashlight. Why? Remember, short wavelengths are good at being scattered. Well, when you look at the glass from the side, you can only see the light that was scattered. Any unscattered light would travel up through the cloudy water and leave the glass. When you were looking at the side of the glass, then, the light you saw was light that had been scattered so that it would leave through the side. Since higher energies are better at scattering, there was more blue, indigo, and violet light – the higher energy wavelengths.

So the unscattered light was a mixture of red, orange, and yellow. What did that remind you of? A sunset (or sunrise), right? That's why a sunset (or sunrise) is made of reds, oranges, and yellows. The sun's light must travel through the atmosphere, and it can be scattered by things in the atmosphere. When you look at the sun, the light you see is coming straight from the sun. Thus, you are seeing mostly unscattered light, which is made up of lower-energy wavelengths. Now please understand that this happens throughout the day. However, the lower the sun is in the sky, the farther its light must travel through the atmosphere to reach you. Thus, the lower the sun is, the more scattering its light experiences. As a result, the effect is emphasized at sunrise and sunset.

What do you see if you look at a part of the sky that doesn't have the sun in it? If there aren't any clouds, you see blue. Why? Because the shorter wavelengths are good at scattering. They end up bouncing around all over the atmosphere. In order to hit your eyes, they have to be scattered several times, so any light that hits your eyes from any part of the sky away from the sun will have been scattered several times. That means it will be the higher-energy wavelengths - blue, indigo, and violet. As a result, the sky appears blue.

This is very similar to the reason the ocean is blue. When light hits the ocean, it won't get back to your eyes unless it scatters off water molecules. The higher-energy wavelengths tend to scatter more, so that's what you mostly see when you look at the ocean. This is only true when there isn't a lot of other stuff in the water. If there are sediments or colored organisms in the water, other wavelengths will bounce off them and hit your eyes, so you will see colors other than blue. That's why many lakes aren't blue. They have a lot of other things floating in the water that tend to scatter other colors as well. However, lakes that are deep enough and have clear water are blue.

Look at the picture on the right. It was taken on the moon. You can see shadows from the astronaut taking the picture as well as other structures behind him. Thus, the sun is shining on that part of the moon. However, notice that the sky is black. That's because there is no atmosphere on the moon. That means there is nothing to scatter the light. As a result, you can only see light that is coming directly from the sun, and that light is white, no matter where the sun is in the sky. If the moon had an atmosphere like the earth's, there would be scattering, and you would see blue light coming from the sky in the picture. Without an atmosphere, however, you see no light coming from the sky. That makes the sky black. You and I associate a black sky with night, but it is not night in the picture, because the sun is shining brightly, illuminating the astronaut, the flag, and the ground. It is also casting shadows. Thus, it is daytime on the moon, but the sky is black, because the moon has no atmosphere.

The sky in the picture is black even though the sun is shining because there is no atmosphere, so light cannot be scattered. The only light comes directly from the sun, so the rest of the moon's sky is black.

But wait a minute. In a black sky, you expect to see lots of stars. Why don't you see any stars in the picture? Well, there is actually one dim star, in the upper right portion of the sky. However, it's dim, and it's the only one I can see. Why isn't the sky filled with stars? It is, but the stars are obscured by the bright light coming from the sun. A camera captures light in order to make an image. The more light there is, the brighter the image becomes. In order to keep the the astronaut and flag from being washed out in very bright light, the camera captured light from the scene for a very short amount of time. Thus, only the brightest things show up in the image. The dim star you might be able to see in the upper right of the sky is the brightest of all the stars, and you can barely see it, because its light is very weak compared to the light reflecting off the astronaut, flag, and ground.

Comprehension Check

13.14 Light from an LED flashlight or a smartphone has a lot more blue in it than light that comes from an old-style flashlight. How would the experiment change if you used an old-style flashlight?

Clouds

The beautiful blue of the sky is often obscured by clouds, and clouds have a profound effect on the weather. When you studied the hydrosphere, you learned about how clouds form and that they are not composed of water vapor. Instead, they are composed of water in its liquid or solid state. Depending on how they form, their appearance and their effect on the weather can be very different, so I want to do a brief survey of the different kinds of clouds you tend to see in the sky. That way, it will be easier to understand the weather fronts that you will learn about in the next chapter.

When you and I think of clouds, the picture that usually comes to mind is a white, fluffy cloud that resembles a cotton ball. Those are called **cumulus** (kyoom' yoo lus) clouds, from the Latin word *cumulo*, which means to pile up. They form in much the same way the cloud formed in Experiment 11.4. As air rises, it spreads out in all directions. This causes it to cool off, because it takes energy for the air to spread out, and that energy comes from the motion of its molecules. Depending on the humidity of the air, it can reach a temperature where the water vapor must become liquid. If there are cloud condensation nuclei in the air, the water vapor will condense on the cloud condensation nuclei, forming a cumulus cloud. The clouds pictured on the left are examples of cumulus clouds.

The fluffy clouds pictured here are cumulus clouds.

The height of a cumulus cloud depends on how much air is rising. When there are strong, upwards winds, air can travel up very high, causing cumulus clouds to grow taller and taller. Very tall cumulus clouds are called **cumulonimbus clouds**, which are usually dark at the bottom. Most people call them "thunder clouds," because they will eventually produce a thunderstorm. Amazingly, cumulonimbus clouds can become so tall that they stretch all the way to the tropopause. If they reach that high, an interesting effect occurs. Remember that the stratosphere is warmer than the troposphere, because of the ozone layer. This means the air that is forming a tall cumulonimbus cloud is denser than the air in the stratosphere. Thus, it can't rise into the stratosphere. This causes the top of the cloud to flatten and spread out, forming an anvil shape, such as the one shown in the picture on the right. Notice that the bottom of the anvil-shaped cloud is dark, but the higher you go, the whiter it becomes. The higher portions of the cloud (including the top) are composed of ice, while the lower portions are still made of water in its liquid phase.

Cumulonimbus clouds that are tall enough form an anvil shape.

Now please understand that not all cumulonimbus clouds become anvil-shaped. Look at the smaller cloud that is just to the left of the anvil-shaped cloud. It is also a cumulonimbus cloud, but it has not grown tall enough to flatten out against the tropopause. Anvil-shaped cumulonimbus clouds are associated with the most violent of thunderstorms. You will learn more about that in the next chapter.

Based on their wisps, these cirrus clouds indicate that the winds at their altitude are blowing to the right.

Since we are discussing the upper reaches of the troposphere, we might as well cover **cirrus** (seer' us) clouds next. The name for these clouds comes from the Latin word *cirrus*, which means a curling lock of hair. As you can see from the middle picture, these clouds are wispy. Because they form high in the troposphere, they are composed entirely of ice particles that have been deposited on cloud condensation nuclei. They get their wispy appearance from high-altitude winds. Their wisps tend to point in the direction that those winds are blowing.

Stratus clouds like the ones you see in this picture form in layers.

Moving from the high reaches of the troposphere to low altitudes, we find the next type of cloud: **stratus** clouds. You should recognize what the name means. It means the clouds form layers, just like sedimentary rocks form strata. They form when a layer of cold air moves underneath a layer of warm air. The cold air cools the warm air above, and that makes some of the water vapor in the warm air condense, forming a cloud. Stratus clouds are different from cumulus clouds because they are formed over a large area, making a "blanket" of clouds in the air. This means that often, you can't really make out the layers that are formed. They just cover the entire sky. If you look up at the sky and see a blanket of clouds that don't seem to have much of a shape to them, they are probably stratus clouds.

There are also interestingly-shaped clouds called **lenticular** (len tik' you lur) clouds, which comes from the Latin word *lenticularis*, which refers to the shape of a lentil. They tend to form in oval "stacks," as shown in the picture below. They form because winds hit obstructions like mountains and end up being deflected up and back. However, more winds continue to blow toward the mountain, colliding with the winds that are being deflected back. These winds interfere with one another and a wave of air is produced. When the crest of that wave reaches a height where water vapor can condense out, a cloud is formed. The clouds formed as these wave crests stack together, forming the interesting pattern shown in the picture on the left. Because of the wave nature of the winds, the air can be pulled down into a trough, which can cause the water in the cloud to become vapor again, making the cloud disappear. Thus, some lenticular clouds can appear and then disappear rather quickly.

Lenticular clouds are formed by wind waves deflected off obstacles like mountains.

Each type of cloud tends to form at a specific altitude, but there are many exceptions. As a result, we can modify the names of a cloud type to indicate if they form at an unusual altitude. The drawing on the left summarizes how we do that. Cumulus and stratus clouds tend to form at low altitudes. If they form at mid altitudes, they are higher than normal, so we add an "alto" to the beginning of the name. If they form at high altitudes, they are made of ice crystals, like cirrus clouds. As a result, we add "cirro" to the beginning of the name.

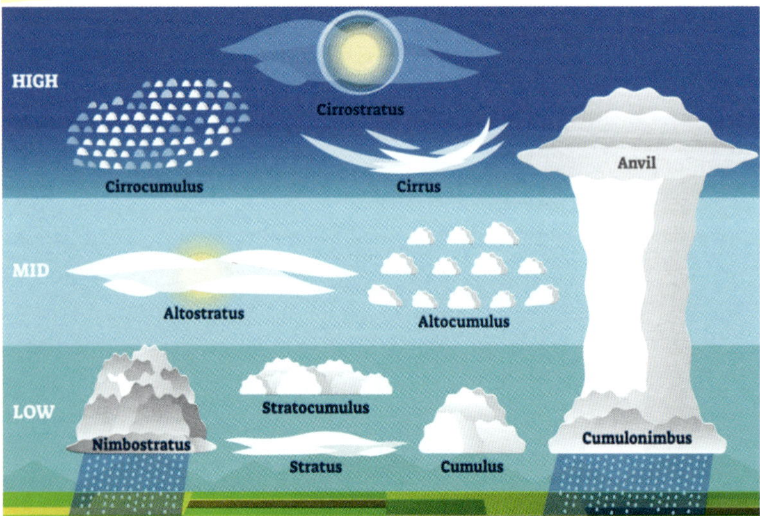

Cloud names can be modified to indicate an unusual altitude or a mixture of characteristics.

In addition, clouds can have a mixture of characteristics. If so, we once again modify the name. When clouds form in layers but appear fluffy, they are called **stratocumulus** clouds. Also, when clouds are dark, that usually means they are about to produce precipitation, so we add "nimbo" to their name. Thus, a dark stratus cloud is a **nimbostratus** cloud, while a cumulus cloud with a dark bottom is called a **cumulonimbus** cloud.

The ice crystals in cirrostratus clouds can produce a halo around the sun.

You might wonder why the sun has a ring around it in the drawing on the previous page. That's because it is something we often see when the sun is behind cirrostratus clouds. When the ice crystals form in a very specific way, they can act like prisms (like the one shown on page 390), separating the sun's light into its different wavelengths. In cirrostratus clouds, the ice crystals can form that way, and if the sun is behind them, a colored ring forms around the sun, as shown in the picture on the right. Notice how the cloud in the picture isn't wispy like a cirrus cloud. Instead, it is widespread like a blanket, which is typical of a stratus cloud. Nevertheless, the cloud is near the top of the troposphere and made of ice crystals, so it is a cirrostratus cloud.

In addition to the natural cloud types that we have been discussing, there are also artificial clouds that are produced by human activity. These are typically called **homogenitus** (ho' moh jen' uh tus) clouds. The prefix "homo" refers to the scientific name for people (*Homo sapiens*), and the "genitus" refers to "genesis," which means "origin." So a homogenitus cloud is a cloud that originated from human activity. Certain industrial facilities can produce a lot of steam. Often, that steam evaporates and simply becomes water vapor in the air. However, it can also form clouds that remain in the air for quite a long time.

The picture below shows a form of homogenitus cloud with which you are probably familiar. It is often called a **contrail**, and it forms from jet planes that are flying through the sky. Jet planes burn fuel, and one of the byproducts is water vapor. However, there are other byproducts as well, and some of them can serve as cloud condensation nuclei. As the water vapor in a jet plane's exhaust leaves the heat of the engine, it becomes a solid due to the low temperatures in the upper troposphere, and the ice crystals form on the cloud condensation nuclei that are also a part of the exhaust. This produces a long, thin cirrus-like cloud that follows the path of the jet plane through the sky, as shown in the picture on the right. As you can see, it is easy to distinguish from the natural cirrus clouds, some of which are cirrostratus clouds, which are producing the colored ring that is also in the picture. Depending on the conditions where they are formed, contrails may be visible for only a few minutes, or they may persist for a long time. If they do, they tend to spread out, eventually coming to resemble natural cirrus clouds.

The long, straight line in this picture is a contrail from the jet that is in the upper right of the picture. You can also see a halo being formed by natural cirrostratus clouds.

As you are well aware, clouds can produce precipitation, but they can also cool the earth. Think about what happens on a hot, sunny day when a cloud gets in front of the sun. It keeps the sun's light from hitting you, which means you aren't receiving as much insolation. As a result, you feel cooler. The same thing happens to the earth as a whole. During the day, clouds block insolation from hitting the earth, which makes the earth cooler than it otherwise would be. Remember, however, that the earth releases a lot of the heat it gets from the sun. Some of that heat is absorbed by greenhouse gases, but some of it is also blocked from leaving by clouds. So clouds both cool the earth and warm the earth, depending on exactly where they are and the conditions at the time. Low, thick clouds tend to do more cooling than warming, because they block a lot of insolation. High, thin clouds tend to produce more warming, because they block more of the heat escaping the earth than the insolation coming to the earth.

As the earth warms, more water will evaporate, but the effect that will have on clouds is not well understood. You might think that more water vapor in the air would mean more cloud formation, but remember that the air gets warmer as well, which means it is harder to form low, thick clouds. That might mean that more high, thin clouds will form, which will lead to even more warming. However, the increased water vapor might be more important than the increased warmth, so it's also possible that more low, thick clouds will form, producing a cooling effect.

Right now, scientists don't know which (if either) will happen. Some scientists predict that as the earth warms, more high, thin clouds will form, making the warming even worse. Others predict that as the earth warms, more low, thick clouds will form, making the warming less severe. In the previous chapter, you learned that one of the most important questions related to global warming is the climate sensitivity, and scientists don't know what it is. Clouds are one of the big reasons for this. Since scientists can't agree on what will happen to the kinds of clouds that form as the earth warms, they don't know how sensitive climate is to carbon dioxide. If warming produces clouds that enhance warming, that means climate sensitivity is higher. If warming produces clouds that produce cooling, climate sensitivity is lower. Until scientists can definitively state what effect warming will have on the types of clouds that form, we will not be able to determine climate sensitivity, which means we will not know the real consequences related to the excess carbon dioxide that we are putting in the air.

Comprehension Check

13.15 You are looking at some clouds, and a meteorologist says you are seeing both cumulus clouds and altostratus clouds. Are all the clouds at the same altitude? If not, which is higher?

13.16 A cumulus cloud is composed entirely of ice crystals. What should it be called?

13.17 Compare how cirrostratus clouds affect the earth's temperature as compared to stratus clouds.

Answers to the Comprehension Check Questions

13.1 Violet has the most energy, while red has the least. Remember that the shorter the wavelength, the higher the energy. In the electromagnetic spectrum illustration, the violet light is shown to have shorter wavelengths than the red light.

13.2 It will be more likely to find the mouse. Warm-blooded animals are warmer than their surroundings, so they emit infrared light. Cold-blooded animals have a body temperature close to that of the surroundings, so they will not emit much infrared light. If the snake can sense infrared light, it will sense warm-blooded things.

13.3 Its rotation is slower than the moon's. Since neither the moon nor Mercury has an atmosphere, the other major effect on temperature is rotation. The slower the rotation, the more the temperature change. Mercury's solar day is 176 earth days long, as compared to just over 27 earth days for the moon.

13.4 Mercury's orbit is less circular. The higher the eccentricity, the more oval and less circular the ellipse is.

13.5 It is more oval. The more oval, the larger the difference between aphelion and perihelion.

13.6 The second is closer to the equator. The equator is pointed directly at the sun, and that means the sun's maximum height is the greatest at the equator. The closer you are to the equator, then, the higher the sun's maximum height in the sky.

13.7 At noon on the December solstice. Remember that the solstice is when the days are their longest or shortest. They will be longest when the hemisphere is pointed directly at the sun, which is in summer. So this is the summer solstice for the Southern Hemisphere, which happens in December.

13.8. The September equinox. If the days are getting shorter, you are between summer and winter. They reach the same length on an equinox. Thus, this is the equinox between summer and winter in the Northern Hemisphere – the September equinox.

13.9 You would feel the breeze blowing towards the city, but high above, it would blow away from the city. The urban heat island effect makes cities warmer. Warmer air rises, so the air in the city is rising. That means air from the surroundings will rush in to replace it, so the wind you feel is blowing towards the city. Winds form loops, however, so high above, it is traveling the opposite way.

13.10 It is in the Ferrel cell. In the Northern Hemisphere, the Hadley and polar cells have winds blowing south near the surface and north near the tropopause.

13.11 Yes. There are no global winds in the doldrums, but there is always the possibility of local winds.

13.12 You are in the Southern Hemisphere. In the Northern Hemisphere, they blow north and east.

13.13 The mE and mP are most humid, and the temperature order is cA, mP, cT, mE. Continental air masses start with "c" and tend to be dry, while maritime air masses start with "m" and tend to be humid. Arctic air masses end with "A" and are the coldest, followed by polar air masses (P), then tropical (T), and then equatorial (E).

13.14 You would not see nearly as much blue from the side. You would still have seen the bulb get orange or yellow from the top, but the side view would not be nearly as blue. The problem with the experiment is you are trying to simulate the atmosphere with cloudy water, and you are using a light source that is really large for to the size of the simulated atmosphere. As a result, a lot of white light escapes from the sides. The white light tends to overwhelm the blues with a regular flashlight, which is why I asked you to use an LED flashlight or an app from a smartphone.

13.15 They are not at the same altitude. The altostratus clouds are higher. Remember, cumulus and stratus clouds are formed at low altitudes. If a stratus cloud forms higher than usual, it is an altostratus cloud. Thus, the "alto" tells you they are higher than normal stratus clouds, which means they are also higher than cumulus clouds.

13.16 It is a cirrocumulus cloud. Cirrus clouds are composed of ice crystals. Thus, this is a cumulus cloud that is as high as the cirrus clouds.

13.17 Cirrostratus clouds warm the earth, while stratus clouds cool the earth. Remember, low, thick clouds tend to cool the earth, while high, thin clouds warm the earth. Stratus clouds are low and thick, while cirrostratus clouds are high and thin.

Chapter Review

1. Define the following terms:

a. Climate	d. Perihelion	g. Equinox
b. Radiation	e. Aphelion	h. Albedo
c. Insolation	f. Solstice	i. Air mass

2. What is the difference between climate and weather?

3. What do we call gamma rays, X-rays, ultraviolet rays, visible light, infrared waves, microwaves, and radio waves? If we order them in terms of wavelength, what is that called?

4. What is the relationship between wavelength and energy for electromagnetic waves?

5. Which has more energy: blue light or red light?

6. Why is it warmest at the equator, at least on average?

7. What causes the seasons that you experience if you don't live on the equator?

8. When the earth is farthest from the sun, which hemisphere is experiencing summer?

9. Which has more effect on the insolation received in the earth's different hemispheres: the change in distance between the earth and the sun or the axial tilt of the earth?

10. You experience a 24-hour period where you can always see the sun in the sky. What season is it? Roughly where are you on the earth?

11. For any region on the earth that is not on the equator, compare the maximum height the sun reaches in summer to the maximum height it reaches in winter.

12. You have been experiencing summer in the Northern Hemisphere, but the days begin getting shorter, while the nights begin getting longer. Which month is the next equinox? Which month is the next solstice?

13. You are in the Southern Hemisphere and it is the March equinox. Will the days start becoming longer or shorter?

14. You have two objects the same size and the same shape. One is black, while the other is white. Which has the higher albedo? Which would be warmer if they both sat in the bright sun for a while?

15. It has been a warm, sunny day at the beach, but now it is night. As you sit on the beach, do you feel the breeze blowing towards the shore or towards the ocean?

16. Using the terms "away from the equator" and "towards the equator," indicate the way the winds near the surface of the earth blow in the Hadley cell, the Ferrel cell, and the polar cell.

17. Which of the three cells in the previous question is responsible for most of the earth's deserts? At what latitudes are those deserts found?

18. What causes the winds near the surface of the earth to curve?

19. If you are on a wind-driven ship and are experiencing the doldrums, what part of the earth are you on?

20. You are comparing a maritime equatorial air mass to a continental tropical air mass. Which is less humid? Which is warmer?

21. Which light is more likely to scatter off of things in the air: red light or blue light?

22. Without an atmosphere, what color would the sky be?

23. Use the terms "cirrus," "cumulus," "stratus," "nimbostratus" and "cumulonimbus" to answer the following questions:

a. In which type of cloud is all the water in its solid phase?
b. Which clouds form layers low in the sky but do not indicate rain will come soon?
c. Which clouds are the tallest?
d. Which clouds form layers in the sky and are very dark?
e. Which clouds form low in the sky, are fluffy, and resemble big cotton balls?

Chapter 14: Weather, Part 2

Introduction

In the previous chapter, you learned a lot about insolation and how it affects climate and winds. Along the way, you learned about air masses, which take on the weather characteristics of the region where they form. Once that happens, the air mass tends to move as a unit, and it will eventually encounter another air mass with different characteristics. This is the formation of a **weather front**.

Weather front – A boundary that separates air masses with different characteristics

When a weather front forms, the area where it forms experiences a change in weather as the two air masses begin to interact along the boundary. To understand how weather changes, then, you have to understand weather fronts.

Weather Fronts

There are four basic kinds of weather fronts, and they each end up producing different changes in the weather. I will start with a **cold front**, where cold air moves into an area occupied by a warmer air mass. Think about what would happen there. The cold air is denser, right? So as the cold air moves in, it will push itself under the warm air. That means the warm air will rise along the cold front. But what would happen as the warm air rises? Any water vapor in the air would begin to condense and form clouds. Most of the time, cumulus clouds are formed, but stratus clouds and stratocumulus clouds can form as well, especially in the warm air that is just beyond the front.

Cold fronts tend to move pretty quickly, so the warm air that they encounter tends to rise quickly. If there is a lot of humidity in the warm air, thick clouds will form at the boundary between it and the cold air that is moving in. In other words, the clouds will be nimbus, usually cumulonimbus. Remember, those are the really tall clouds that are generally referred to as storm clouds. Thus, cold fronts are the ones that can bring in stormy weather. Of course, if the warm air is not very humid, clouds will form, but they won't be thick, so it might not even rain. Most of the clouds and the rainy weather form along the front where the warm air is being pushed upward, so once the front passes, the weather tends to clear. Since cold fronts move rather quickly, that means the unpleasant weather passes pretty quickly. If you have ever experienced a quick, intense thunderstorm, you might have noticed that the temperature cooled afterwards. That's because the thunderstorm was produced by a cold front.

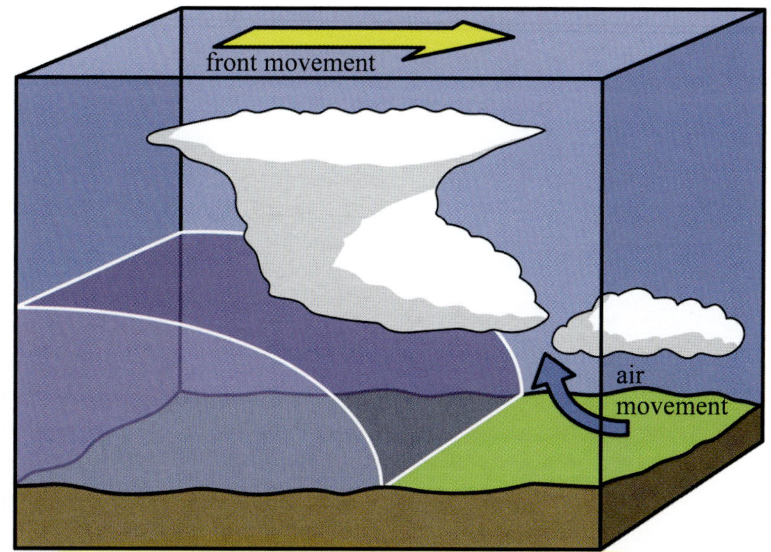

A cold front moves cold air into a region, lifting up the warm air it replaces.

One other important thing to consider is air pressure, which meteorologists often call **atmospheric pressure**. Since warm air tends to rise, warm air masses are associated with lower atmospheric pressures. As a cold front moves in, the warm air is pushed up faster, which means the pressure will get even lower. However, cold air is denser, so cold air masses themselves are associated with higher air pressures. Thus, if you track the pressure while a cold front moves in, it will be low in the warm air mass, and as the front comes into the area, the air pressure will get even lower. However, once the cold air mass has taken over, the air pressure will go up to a value that was higher than what it was in the warm air mass.

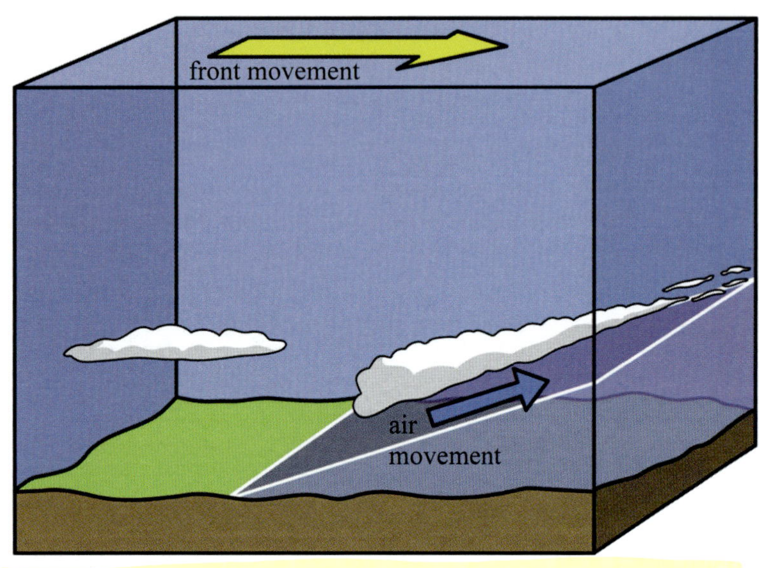

A warm front usually brings in less severe weather than a cold front.

Not surprisingly, the opposite of a cold front is a **warm front**. In this situation, warm air moves into a region where there is cold air. The cold air is still denser, so the warm air still must rise, but warm fronts don't move as quickly as cold fronts. Because of this, the clouds tend to spread out instead of piling up. Generally, cirrus clouds form high in the sky, followed by cirrostratus, altostratus, and stratus clouds. If the warm air has enough humidity, nimbostratus clouds can form and bring rain. Generally, the weather associated with a warm front is less severe than what can develop in a cold front. In addition, a warm front's weather lasts longer, because it develops more slowly.

As you might expect, the atmospheric pressure changes associated with a warm front are rather different from those of a cold front. Since the warm air rises above the cooler air, there is usually no dramatic shift in the air pressure, because the air mass near the surface of the earth is still cold air. However, once the warm air mass takes over, the pressure falls, since warm air masses usually have lower pressure.

A stationary front can cause the weather to remain the same for days.

Cold air takes over an area in a cold front, and warm air takes over an area in a warm front. In each case, the air mass that is taking over is stronger than the air mass that is being overtaken, so the air mass after which the front is named ends up "winning." But that doesn't always happen. It's possible for a warm air mass and a cold air mass to encounter one another without either being strong enough to move the other. As a result, the air masses don't move. Not surprisingly, this is called a **stationary front**.

In a stationary front, the winds typically blow up and down along the front rather than across it. Usually they blow up along the front on one side and down on the other. If neither air mass is very humid, few clouds will form along a stationary front. However, if the warmer air mass is humid, stratus clouds will usually form. If the warm air is humid enough, nimbostratus clouds will form, producing precipitation. Because stationary fronts don't move much, the weather and atmospheric pressure can stay the same for days until the winds change enough to allow one of the air masses to "win." At that point, the stationary front changes to a cold front or a warm front, depending on which air mass overtakes the other.

Now remember, cold fronts move faster than warm fronts. So imagine a situation in which a warm front is moving into an area, but there is a cold front moving in the same direction. Since the cold front moves faster, it can catch up to the warm front and start mixing with it. At that point, you have an **occluded front**. In a front like that, all the different kinds of clouds can be present, because the warm front has made various cirrus and stratus clouds, while the cold front coming in from behind can form cumulus clouds. The atmospheric pressure usually starts by falling, since a warm front is moving in. However, as the cold front overtakes it, the atmospheric pressure levels out.

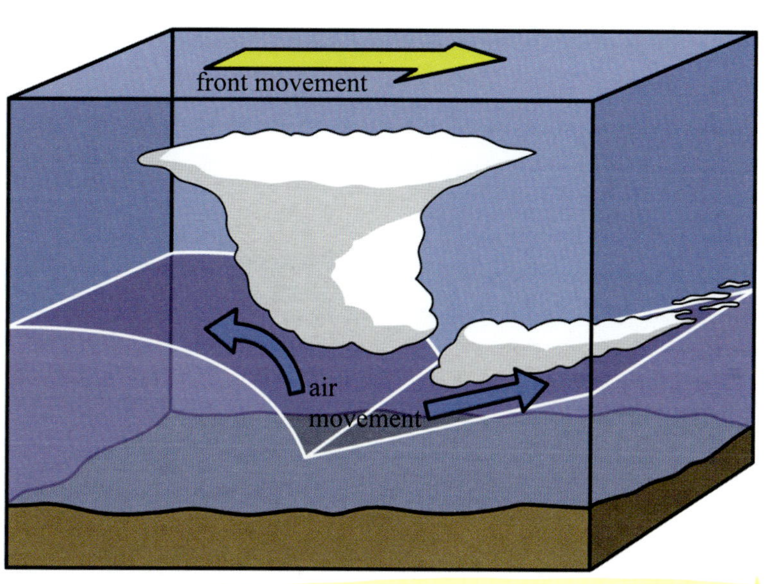

An occluded front is caused by a cold front overtaking a warm front.

Weather Maps

Now, of course, these fronts move around, bringing in weather changes wherever they go. We track them using **weather maps**, but in order to understand those maps, you need to know the symbols that represent the different kinds of fronts. They are shown in the illustration below. A blue line with triangles represents a cold front, and the triangles indicate the direction in which the cold front is moving. The red line with circles represents a warm front. Like the triangles, the circles represent the direction in which the warm front is moving. The line with both circles and triangles represents a stationary front. There is no motion to represent, since the front is stationary. However, the circles are on the side where the warm air mass is, while the triangles indicate the side of the cold air mass. The purple line with both circles and triangles represents an occluded front, and both the circles and triangles indicate the direction in which the front is moving.

These are the symbols used to abbreviate weather fronts on weather maps

Depending on the weather map being used, there can be a lot more information than just the weather fronts and how they are moving. However, one of the other very common things you see on a weather map is the atmospheric pressure, since it is another important characteristic of an air mass. The illustration below shows a weather map that focuses on the continental United States, southern Canada, and northern Mexico. Notice the H's and the L's. Those refer to high atmospheric pressure and low atmospheric pressure. Next, notice the thin, white curves. Those are called **isobars**, and they trace out where the atmospheric pressure is the same. If you travel along one of those curves, the atmospheric pressure will not change. For example, right under the lower "H" on the map, you should be able to see a white number, 1100. That's a measurement of atmospheric pressure. If you travel around the deformed oval that is labeled 1100, you will experience that atmospheric pressure no matter where you are. Notice that the next deformed oval is labeled with 1080. That indicates a slightly lower pressure.

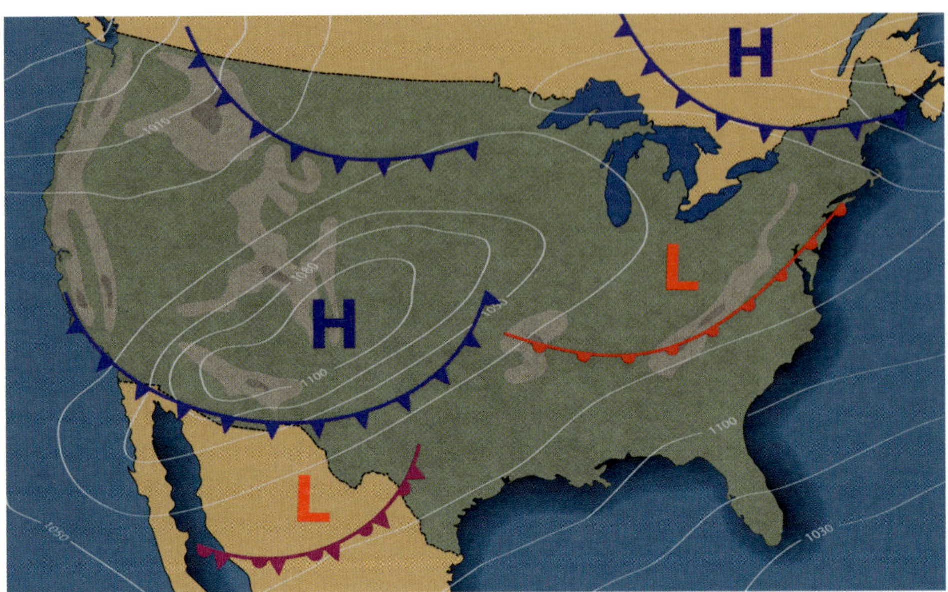

This weather map shows hypothetical atmospheric pressures and weather fronts in the continental United States, southern Canada, and northern Mexico.

Now look at the weather fronts shown on the map, starting with the cold front at the upper right. The curve shows you where the front is, and the triangles tell you it is moving to the southwest. Notice what is behind the cold front. High atmospheric pressure. That's what you expect, because a cold front is bringing in a cold air mass, which has high pressure. That cold front will continue to move until the air masses mix, so we can predict that pretty soon the Great Lakes are in for cumulus clouds that might eventually grow to be cumulonimbus clouds. Depending on the humidity the cold front encounters, thunderstorms might also develop.

Now look at the warm front below the cold front I have just been discussing. As the circles tell you, it is moving southeast. What's behind it? Low atmospheric pressure, since a warm front brings in a warmer air mass, which means a lower pressure. Along that warm front, we can expect cirrus and stratus clouds. Depending on the humidity, those stratus clouds might become nimbostratus clouds and might bring some precipitation. However, whatever rain might develop along that front will probably not be as severe as the rain that could form along the cold front above it.

Finally, look at the occluded front that is over Mexico and parts of Texas. The circles and the triangles tell you it is moving to the southeast. There is low pressure behind it. Often, the pressure behind an occluded front is high because of the cold air mass that is overtaking the warm front. In this case, however, there is a cold front moving in the same direction behind it, and it is the main thing determining atmospheric pressure in that region.

While following the weather fronts, temperatures, and atmospheric pressures is an important part of tracking the weather, there is more to it than that. After all, the clouds formed at weather fronts can produce precipitation, and that's important to track as well. Thus, meteorologists have other tools to map out the weather. For example, they have **radar**, which stands for "**ra**dio **d**etection **a**nd **r**anging." A radar system emits electromagnetic waves whose wavelengths are either in the microwave region of the electromagnetic spectrum or the radio wave region. Most weather radar uses electromagnetic waves that have a wavelength between 2.5 and 15 centimeters. As the waves travel through the air, they bounce off things and return to where they were emitted. Analyzing the properties of the electromagnetic waves that come back allows meteorologists to find out if there is precipitation forming, what kind of precipitation is forming, and how the clouds producing it are moving. The data can be compiled in a map, such as the one shown below:

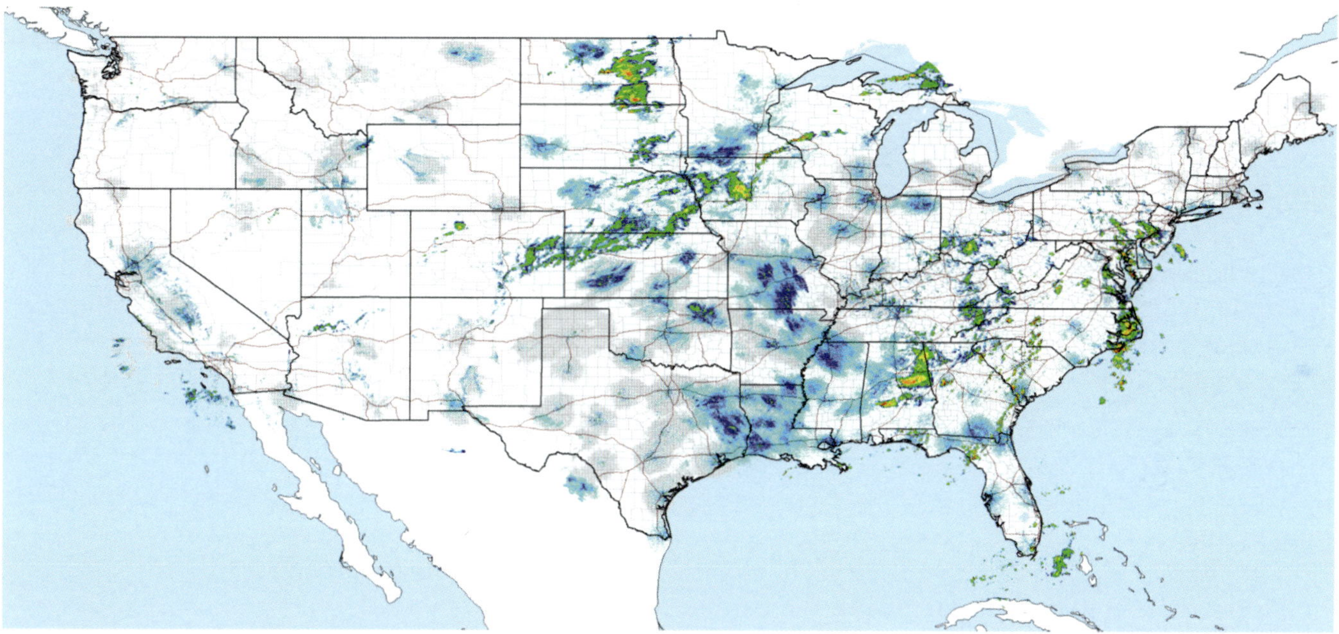

The colors represent areas that are experiencing precipitation, and each color represents the precipitation's intensity: blues indicate the precipitation is light, greens represent light to moderate, yellows represent moderate, and reds heavy. Weather radar can also determine if big chunks of ice, called **hail**, are falling. Areas getting a little bit of hail are usually represented with darker red, while heavy hail is usually signified with purple. Since radar data can be collected quickly, images like these can be put together in an animation that allows you to see how the precipitation moved over a certain amount of time. Most internet weather sites have animations like that so you can see how the precipitation is moving, what it looks like in your area, and where they predict it will go.

Comprehension Check

14.1 As the day progresses, you see cirrus clouds forming, followed by altocirrus clouds, followed by stratus clouds. What's happening to the temperature? Is it increasing, decreasing, or staying the same?

14.2 The atmospheric pressure in your area is low, but over the course of the day, it rises significantly. What's happening to the temperature? Is it increasing, decreasing, or staying the same?

14.3 In the weather map on page 404, there is an occluded front over northern Mexico and southern Texas. However, another occluded front could potentially form. Based on the other fronts shown, where could that happen: central United States, northern United States, or southern Canada?

14.4 Based on the weather radar map given above, where is it raining the heaviest: Indiana, Mississippi, Alabama, or Texas?

Precipitation

As the radar map on the previous page shows, clouds can produce precipitation. How does that actually happen? Remember, clouds are made of water in its liquid or solid phase, along with cloud condensation nuclei. Even though liquid and solid water are denser than air, the clouds can float around because the water droplets are small, and rising air pushes them upwards, slowing or stopping their downward fall. However, as the water droplets or ice crystals get bigger, that gentle push isn't enough, and precipitation occurs.

But what makes the water droplets or ice crystals get bigger? For water droplets, this is best demonstrated with an experiment.

Experiment 14.1: Skating Drops

Supplies:

- A pan for boiling water. The flatter the bottom of the pan, the better, because the bottom of the pan needs to be evenly heated. Please note that this pan will be heated without anything in it for a while, which can be bad for some pans. Check with your parents to make sure you have a pan that is okay to use in that way. An oiled pan will not work for this experiment.
- A cup
- Water
- A stove

Instructions:

1. Turn one of the stove burners on high and place the empty pan on that burner.
2. Allow the pan to heat up for at least three minutes.
3. While the pan is heating, put some water in the cup. You don't need much.
4. Once the pan is very hot, pour just a little water (about ½ teaspoon) into the pan. **Be careful! Don't touch the pan, because it is really hot!**
5. What happens? If all the water immediately boils away, or if you don't hear a sizzle when the water hits the pan, you need to let the pan sit longer before trying the experiment again.
6. Eventually, you should see that a small drop of water remains in the pan, despite the fact that the pan is very hot. Watch the water to see how it behaves.
7. Tilt the pan so that the water drop moves up the side of the pan. What happens then?
8. Put the pan back on the burner. If there is any water remaining in the pan, watch what happens. If not, add a bit more from the cup.
9. Once you have another drop of water that doesn't go away, add another small amount of water to the pan. You want to get several water droplets "skating" around the pan. What happens?
10. Play with this for a while to observe how the water drops behave. Continue to add small amounts of water as needed.
11. Turn off the burner, allow everything to cool down (that will take a while), and then put everything away. Let your family members know to avoid the pan until it cools down.

What happened in the experiment? As mentioned in the instructions, you should have seen drops of water "skating" around the pan. Why? When the water hit the hot pan, it got really hot where it was touching the pan. In fact, it got so hot that it started to boil, making water vapor. But water vapor doesn't conduct heat very well, so once a layer of water vapor formed underneath, the rest of the water didn't heat up nearly as fast. As a result, it stayed liquid for a while. It eventually got hot enough to boil, but it didn't happen right away. That means you had a drop of water that was sitting on top of a layer of water vapor.

Since the drop of water wasn't touching the pan, it could move around really easily. That's why the water drops "skated" around the pan. But what happened when two water drops "skated" into each other and collided? They joined together, producing a bigger drop. Why? After all, when most things collide, they tend to break apart. Why did the colliding water drops join to become a bigger water drop? Because water molecules are strongly attracted to one another. As a result, they want to stay together. This is referred to as **cohesion** (koh he' shun).

Cohesion – A measure of how strongly molecules of a substance are attracted to one another

Water has a high cohesion, so when water molecules interact, they generally try to stay together. If water had a lower cohesion, the drops in your experiment might have broken apart rather than joining together.

While water drops don't "skate" in clouds, they are easily moved around by winds. Remember, one thing that keeps a cloud in the air is the push from air that is rising. This upward breeze is called an **updraft**. As the updraft pushes on a tiny water drop, it will move upward in the cloud. However, it will start to collide with other water drops, so it won't go straight up. It will bounce around, but with a general upward motion. In those collisions, the drops will join together to make a bigger drop, just like the "skating" drops in your experiment did. This process is called **collision coalescence**, because "coalescence" means "joining together." As more collisions occur, the water drop gets bigger. Eventually, it gets big enough that the updraft cannot keep pushing it up. As a result, it begins to fall. However, it will still be colliding with other drops as it falls, so it will get even bigger. Eventually, it will fall out of the cloud as rain.

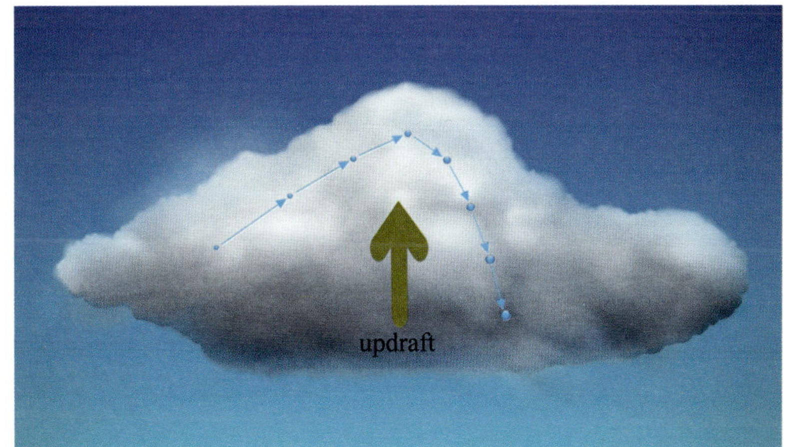

In the collision-coalescence process, a water drop grows by colliding with other water drops as the updraft pushes it upward in the cloud. Eventually, its weight forces it downward.

While the collision coalescence process is important in producing precipitation, it only occurs when the water is in its liquid phase. However, when a cloud (or a portion of the cloud) is high in altitude, the temperature is cold enough for water to freeze. As a result, the cloud doesn't contain water drops. It contains ice crystals. Some ice-crystal clouds, like cirrus clouds, don't produce precipitation, but imagine a tall cumulus cloud. The lower portions of the cloud are at temperatures that are warm enough to keep the water in its liquid phase. However, the higher portions are much farther up in the troposphere, where the temperature is low enough for water to freeze. At that point, another process is responsible for precipitation forming.

To understand this process, the first thing you need to know is that water doesn't always freeze when its temperature reaches 0 °C (32 °F) or below. That's because in order to freeze, the molecules must have something to freeze onto. Remember cloud condensation nuclei? They are necessary for clouds that contain liquid water to form, because water vapor needs something to condense onto in order to become liquid. In the same way, liquid water needs something to freeze onto in order to become solid. That something is called a **nucleation center**, and without it, water will not freeze. Very pure water, then, doesn't freeze when its temperature reaches below 0 °C (32 °F), because it has no nucleation centers. When you have water below 0 °C (32 °F) that is still liquid, we call it **supercooled water**.

In a tall cloud, there are some water drops that do not have nucleation centers, so they become supercooled. Of course, there are also crystals of ice where water found nucleation centers upon which to freeze. Now remember that at high altitudes, the pressure is low. At low pressures, water can evaporate from the supercooled water drops. The vapor produced can find its way to an ice crystal and deposit onto the crystal, increasing its size. As time goes on, then, the supercooled water droplets get smaller, which helps them to stay high in the cloud. However, the ice crystals get bigger, and they fall.

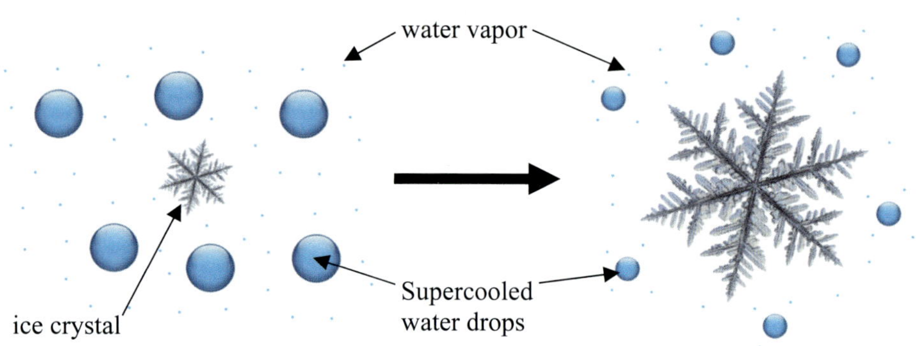

In tall clouds, small ice crystals can be surrounded by supercooled water drops. As water evaporates from the drops, it can deposit on the crystal, making the water drops smaller and the ice crystal larger.

This process is known as the **Bergeron** (bur jer' ahn) **process**, and it occurs whenever a portion of a cloud is at altitudes where the temperature is below 0 °C (32 °F). While it produces large ice crystals, those ice crystals can melt if they fall to a part of the cloud where the temperature is higher. If that happens, they can grow even more through collision coalescence and leave the cloud as rain. However, if the cloud is cold enough, they can stay ice crystals even after they leave the cloud, and then they fall as snow.

Comprehension Check

14.5 In a tall cumulonimbus cloud, which of the two processes discussed in this section makes the water drops heavier near the bottom of the cloud? Which makes them lighter near the top?

More on Humidity

I want to discuss precipitation in more detail, but to do that properly, I need to make sure you understand humidity in more detail. Remember, humidity is a reference to the amount of water vapor in the air. If air is really humid, it will form a lot of clouds as it rises and cools. If it is not very humid, it will not form nearly as many clouds. There are other humidity-related considerations when it comes to weather, so it is important to get a detailed understanding of what it is.

Have you ever heard someone say, "It's not the heat, it's the humidity"? Living in Indiana, I hear that multiple times every summer. Why? Because Indiana summers are pretty humid, and a lot of humidity on a hot day makes it more uncomfortable. To understand why, perform the following experiment.

Experiment 14.2: Cool and Warm Fingers

Supplies:
- A plastic bottle, like the kind water or soda comes in (the smaller the better)
- Two small glasses, like juice glasses
- Someone to help you
- Rubbing alcohol (If you have denatured alcohol, that's even better.)
- Water

Instructions:
1. Pour some alcohol into the plastic bottle so that there is about ½ inch of liquid at the bottom.
2. Swirl the liquid in the bottle quite a bit. As you swirl, tilt the bottle so that the alcohol touches the sides of the bottle as far up as you can without spilling anything. Ideally, you want the entire inside of the bottle to be wet with alcohol.
3. Set the bottle on a counter or table so it is standing upright.
4. Pour some alcohol into one of the juice glasses. You just need enough so that when you tilt the glass, you can stick your index finger in the alcohol and wet it to the second knuckle.
5. Run water from the tap until it is close to room temperature.
6. Add about as much room-temperature water to the other glass as you have alcohol in the first glass.
7. Have your helper tilt both glasses so that each liquid collects in a pool that is deep enough to stick your index finger into up to the second knuckle.
8. Stick your left index finger into the water and your right index finger into the alcohol.
9. Pull both fingers out of the liquid at the same time.
10. Compare the temperatures you feel on each index finger. How do they feel compared to each other? How do they feel compared to the fingers that aren't wet?
11. Repeat steps 8 and 9.
12. Wave your hands in the air, once again, concentrating on the temperatures of your index fingers and your other fingers. What do you feel now?
13. Put the bottle close to your right hand so that it is still standing upright.
14. Put your right index finger into the alcohol again, pull it out, and wave it around a bit so that you feel the same temperature change you felt before.
15. Push your right index finger into the bottle's opening so that it is now inside the bottle. Did the temperature of your index finger change? If so, how? Repeat the process if you didn't feel a change. It should be noticeable.
16. Clean up your mess.

What happened in the experiment? If things went well, both index fingers should have felt a bit cooler than your dry fingers, especially when you were waving them around in the air. Why? Because liquid was evaporating from each of them. Your left index finger had water evaporating from it, and your right index finger had alcohol evaporating from it. What does a liquid need in order to evaporate? It needs energy. As the liquids evaporated, then, they took energy away from the skin on your fingers. That means your skin suddenly had less energy, which made it feel cooler.

Which finger felt cooler? The one with alcohol on it. Why? Because alcohol evaporates more quickly than water. That means it took energy away from your finger's skin faster, so that finger felt cooler. This, in fact, is why you sweat when you are hot. Your skin has sweat glands that produce a mixture of water and a few other chemicals. When you get hot, the sweat glands release that mixture onto your skin. When the water in your sweat evaporates, it takes energy from your skin, and as a result, your skin feels cooler. Sweating, then, is your body's attempt to cool your skin when it gets too warm.

But wait a minute. Sometimes when you sweat, the water doesn't evaporate. It just runs along your skin, dripping around. It also soaks into your clothes. Why? Well, think about the second part of the experiment. When you put your alcohol-soaked index finger into the bottle that had alcohol in it, what did you feel? You should have felt your index finger get warmer. It was cool before you put it in the bottle, but it stopped being cool and might have even felt a bit warm when you put it in the bottle. That happened because the alcohol in your finger could no longer evaporate.

When it is very humid, your sweat doesn't evaporate well, so it doesn't cool your skin much.

By swirling the alcohol around and then letting it sit, you allowed it to evaporate, which filled the air inside the bottle with alcohol vapor. In fact, there was so much alcohol vapor in the bottle that the air could not accept any more alcohol vapor. As a result, the alcohol on your finger stopped evaporating, which stopped the cooling effect that evaporation has. When you sweat and it is very humid out, the air around you is like the air in the bottle. There is already so much water vapor in the air that your sweat cannot evaporate well, so it collects on your skin, making drops and soaking your clothes. You also don't get the cooling effect that evaporation provides, so it makes you feel even warmer. That's why people in Indiana say, "It's not the heat, it's the humidity." If it weren't for the high humidity, the hot weather would not be as miserable!

This leads us to three important terms related to weather. When air holds so much water vapor that it cannot accept any more, we say that the air is **saturated** with water vapor.

Saturated – Completely full of a substance so that no more can be accepted

In your experiment, the air inside the bottle was saturated with alcohol vapor. Because of that, no alcohol could evaporate from your finger. In weather, we worry about whether or not air is saturated with water vapor. When it is, we say the **relative humidity** is 100%.

Relative humidity – The amount of water vapor in the air as a percentage of its saturation level

This is different from the **absolute humidity**, which is a measure of the actual amount of water vapor that you find in a sample of air.

Absolute humidity – The mass of water vapor found in a given volume of air

To illustrate the difference between absolute and relative humidity, suppose I collect a liter of air that is 20 °C, and I find out that there are 0.01 grams of water in it. That's its absolute humidity. However, suppose I also know that at 20 °C, air is saturated when it has 0.02 grams of water in it. That means the air I measured has 50% of the water that it *could* hold. That means its relative humidity is 50%.

Now if you think about it, relative humidity is what's important to you and me (especially on warm days), because it affects our comfort level. The higher the relative humidity, the harder it is for our sweat to evaporate, so the less cooling effect we get when we sweat. That's why the weather report usually gives the humidity in percent. It is giving you the relative humidity, because that will give you some idea of how well your sweat will cool you down.

Of course, to understand relative humidity, you have to know something about when air becomes saturated with water. It turns out that the warmer the air is, the more water vapor it can hold before it becomes saturated. At 0 °C (32 °F), for example, a liter of air becomes saturated when it holds 0.005 grams of water. At 30 °C (86 °F), however, a liter of air becomes saturated when it has 6 times that amount, or 0.03 grams of water. This will become important as you learn about the details of precipitation, which we will discuss the next time you do science.

Comprehension Check

14.6 One day the temperature is 30 °C (86 °F), and the relative humidity is 75%. You leave a bowl of water outside for your dog. The next day, the temperature is the same, but the relative humidity is 30%. You refill your dog's bowl. Assuming the dog drinks the same amount of water each day, on which day is it more likely for the bowl to become empty?

More on Precipitation

Now that you know what produces precipitation and how humidity is affected by temperature, you can learn some more details about precipitation. There are different kinds of precipitation, and they each form under different conditions. The most familiar one, of course, is liquid water falling to the ground in the form of **rain**. While it is tempting to think that the rain falls from the clouds in its liquid state, that's not necessarily true. It's possible that when it leaves a cloud, the water is in its solid state (ice crystals), but as it falls, it encounters warmer temperatures that melt it. So while the collision coalescence process can form rain, it can also be formed by the Bergeron process, if the ice crystals that are formed end up melting either in the lower portions of the cloud or in the air as they fall.

Even though you are probably familiar with rain, it might surprise you to learn a couple of things about it. First, rain is not composed of pure water. You might think it is, since it falls from the sky, but remember that for a cloud to form, there must be cloud condensation nuclei in the air. That means there are other substances besides just water in a cloud. In the same way, there are many substances in the air, including some pollutants. As rain falls through the air, it mixes with those substances.

For example, have you heard of **acid rain**? It was a real problem in the 1980s and 1990s, because of the concentration of pollutants in the air. As rain fell through the air, the pollutants would react with the water, forming various acids. While acids can be good (like the citric acid in citrus fruits), too much acid can be bad. Acid rain added extra acid to the land and water upon which it fell, which could cause problems for the life found there. As you already learned, however, pollutant concentrations have been steadily falling over the past few decades, so acid rain has become less of a problem over time. However, even with no pollutants, rain has a pH of about 5.5, because carbon dioxide reacts with raindrops to make carbonic acid. Thus, all rain is technically acidic, since all rain has a pH of less than 7. However what scientists call "acid rain" has more acid than it should, making the pH even lower.

Sometimes the substances in the air interact with the drops of water to change its color. In 1957 and again in 2001, the rain that fell on part of India's western coast was actually red! While an event like that is rare, it has been reported throughout history. In ancient times, it was called a "blood rain," because the rain looked like blood. While we don't know the details regarding the ancient events, the red rains in 1957 and 2001 were due to a specific form of algae that produces red spores. When the conditions are right, those spores can become heavily concentrated in air, and if rain falls through that air, the spores mix with the water, turning it red.

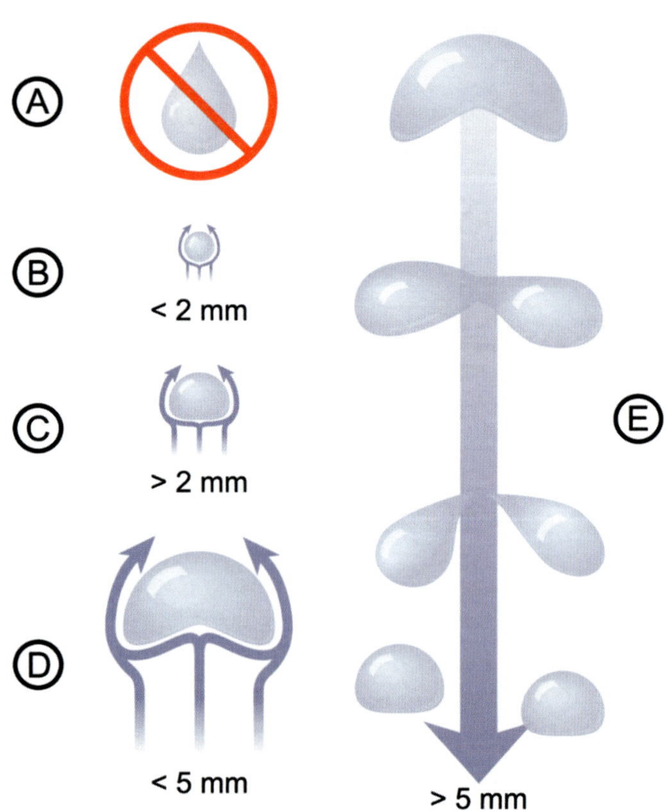

Raindrops are not tear-shaped. Their shape is complex, and it depends on the size of the rain drop.

Another surprising thing you might not know about rain is that it isn't the shape you see drawn in most illustrations. When an artist wants to illustrate raindrops, they are usually drawn to look like tears, as shown in part "A" of the illustration on the left. As indicated, however, that's not what a drop of rain looks like. When raindrops are very small (less than 2 millimeters in diameter), they are spherical (part "B" in the illustration). That's because they fall through the air so that it passes evenly over all sides. However, if the drop is bigger (more than 2 millimeters), the air resists the drop's motion more. As a result, the bottom of the drop flattens out, and the sphere gets deformed (part "C"). If the drop is even bigger, the air actually pushes hard enough against the bottom of the drop to make a depression in it, and the drop becomes shaped like a bean (part "D")! If the drop is even bigger, that bean shape changes as the drop falls. The depression gets larger, eventually forming two lobes with a small line of water connecting them. As the drop continues to fall, that line of water can break, forming two smaller drops (part "E").

The amount of rain that an area receives is measured with a rain gauge, which is really just an open cylinder that collects rain as it falls. The depth of the water in the cylinder is measured, and that measurement (in inches or centimeters, depending on the country) tells you how much rain fell. The rainiest state in the U.S., for example, is Hawaii. It usually gets 460 inches of rain in a year. By contrast, Death Valley in California receives an average of just a bit over 2 inches of rain in a year. Death Valley is a part of the Mojave (moh' ha vee) Desert.

Now remember that a lot of rain actually starts out frozen. The Bergeron process causes ice crystals to grow in a cloud where the temperature is below 0 °C (32 °F), but as those ice crystals grow, they get heavy enough to fall. As they fall, they might encounter warmer air, since the air in the troposphere is usually warmer the closer it is to the ground. If the falling ice crystals reach warmer parts of the cloud, they melt into rain. They may still be ice crystals when they leave the cloud, but as they fall even closer to the ground, they might end up in air that is warm enough to melt them, so they end up forming rain. However, if the air is cold for the entire trip down, the ice crystals fall onto the ground as **snow**.

While it is common for people to use the term **snowflake** to refer to the individual ice crystals that make up snow, that's not the best term to use. "Snowflake" can mean many things. When you see white stuff falling gently to the ground during the winter, you call those white things snowflakes, right? Well, those white things are actually made up of many, many **snow crystals** stuck together. The snow crystal is what forms in the cloud. As those crystals fall, they collide and stick together, forming the white snowflakes that you see falling.

Snow crystals have a hexagon geometry.

As I am sure you are aware, if you look at individual snow crystals, you see some gorgeous shapes. The pictures on the right show snow crystals that have been magnified to about 20 times their normal size. Notice that while they look similar, they are definitely not the same. Can you tell me what's similar about them? They kind of look like stars, but count the "arms" that extend from the center of each snow crystal. How many are there? There are six. Notice the diamond shapes around the center of the crystal on the right. How many of them are there? Six. Snow crystals are generally formed so that there are six repeating structures emerging from the center. A six-sided shape is called a **hexagon**, so we say that snow crystals have a hexagonal geometry.

Why do they have a hexagonal geometry? It actually reflects the way the individual water molecules are arranged in the crystal. When water freezes, the molecules arrange themselves into hexagons. The molecules are too small to see, but remember that a snow crystal grows because water vapor deposits onto an existing crystal. The molecules in the existing crystal are arranged in hexagons, and the molecules in the water vapor must arrange themselves in hexagons in order to freeze. In other words, hexagons are being added to other hexagons. As a result, the resulting shape is based on a hexagon.

Look at the pictures again. While these snow crystals look very similar, they are clearly different, aren't they? The one on the left doesn't have the diamond shapes near the center like the one on the right. Also, the "arms" on each crystal are branched, but the branching is different. For the crystal on the left, each "arm" ends in three branches. For the other one, there are five branches. This is because while snow crystals are based on a hexagonal shape, the way that shape developed depends on the specific conditions surrounding the snow crystal. Since there are many, many different possibilities for the conditions around a snow crystal, there are many, many different variations on the hexagon.

This fact has led to the phrase, "No two snowflakes are alike." While that is a very poetic thing to say, it's not really true. Identical snow crystals have been found. However, because there are so many variations on the basic hexagonal geometry, they are rare. Indeed, an American farmer named Wilson Bentley was the first person to take pictures of snowflakes under the microscope, and he started doing that in 1885. He cataloged 5,000 different snowflakes and found no identical ones. Others did the same, but it took until 1988 for a scientist, Nancy Knight, to find two identical snow crystals. Thus, while it is not *impossible* to find two identical snow crystals, it is *very improbable*.

You already know that ice is less dense than water in its liquid phase. Thus, you would expect snow to be less dense than rain. However, go back and look at the snow crystals on the previous page. Notice that there is a lot of empty space in each snow crystal. That means a collection of snow crystals is even less dense than ice. This has a very practical consequence. If snow crystals remain in their solid phase from the time they form to the time they hit the ground, we call it a **dry snow**, and it can take up as much as 50 times the volume of an equivalent amount of rain. So let's suppose you have an intense snowstorm and the dry snow piles up 10 inches above the ground. That same amount of water would produce only 0.2 inches of rain!

Not all snow is dry. Remember, the snow crystals leave the cloud as ice, but as they fall, they encounter warmer air. That warm air can start to melt the crystals. However, suppose the partially-melted crystals encounter colder air when they get closer to the ground. Remember, a warm front has warm air piled on top of cold air, as does an occluded front. So a snow crystal can partially melt as it passes through the warm air of one of those fronts, but once it sinks into the cold air mass below, it will start to freeze again. This forms **wet snow**. Wet snow is stickier than dry snow, and it is denser as well. Wet snow can take up as "little" as 10 times the volume of an equivalent amount of rain. Thus, 10 inches of wet snow is equivalent to 1 inch of rain. Skiers prefer dry snow, while wet snow is best for making snowballs and snowmen.

Dry snow (top) is powdery and ideal for skiing, while wet snow (bottom) is sticky and best for making snow sculptures and snowballs.

Think about that for a moment. There is as much as 50 times more water in rain than there is in dry snow, and there is as "little" as 10 times as much water in rain than there is in wet snow. Either way, it takes *a lot* of snow to produce the same amount of precipitation as rain. In my state of Indiana, it is not unusual to get 2 inches of rain in a day. That amount of water would make 20-100 inches of snow, depending on how wet or dry the snow is. Fortunately, we *never* get that much snow in a day. In fact, the heaviest snowfall recorded in Indiana was just over 10 inches in one day. Why? Remember that to make any kind of precipitation, the air must be humid. The more humid the air, the thicker the clouds, and the more precipitation that is produced. But remember that warmer air has more absolute humidity than colder air. Snow generally produces less precipitation because snow requires cold air, and cold air is, on average, less humid than warm air. Cooler, less-humid air produces less precipitation.

Now, of course, the humidity of the air can be increased if there is a lot of water nearby. For example, snow is much heavier near large lakes, because there is water evaporating from the lake. That means the air is more humid, which means it will produce more snow. The snow produced in that way is often called **lake-effect snow**, because a lot of the water vapor that produced it came from the unfrozen parts of the lake.

While rain and snow are the two "extremes" when it comes to precipitation, there are some types of precipitation in between. Suppose rain is falling through warm air and then encounters a cold air mass below. If the cold air mass is cold enough, it will start to freeze. It takes time for water to freeze completely, so imagine a situation where the cold air isn't very far above the ground. The rain will start to freeze, but it won't have much time to do so before it hits the ground. As a result, it will be mostly water when it hits. Well, if the cold air right above the ground can freeze water, the ground will be cold enough to freeze it, so the water will finish freezing when it hits the ground. We call that **freezing rain** (or glaze), since it technically hits the ground as rain but then freezes. The plant pictured on the right, for example, was hit with freezing rain. When the water hit the branches or berries, it froze to them quickly, surrounding them in ice.

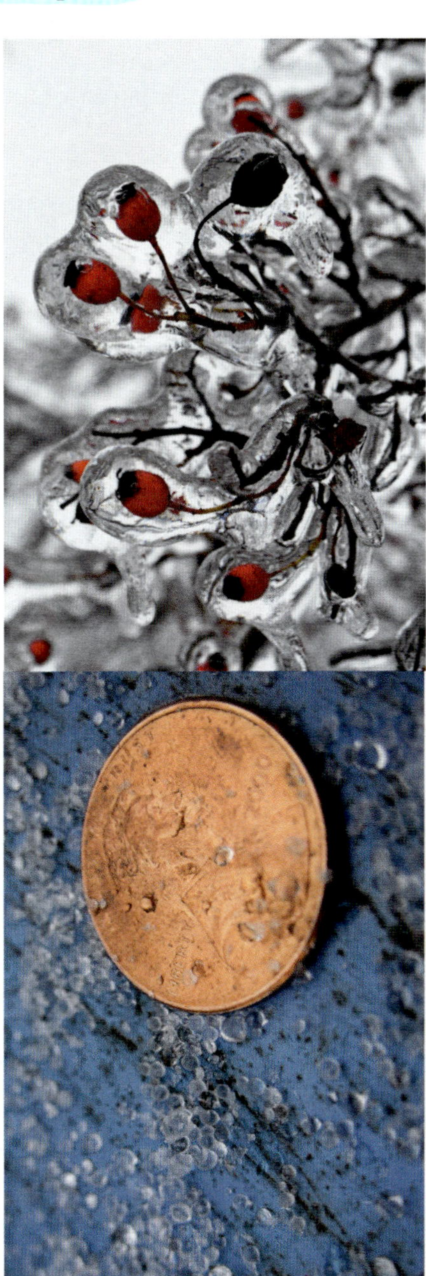

Freezing rain (top) and sleet (bottom) are the result of rain falling through air that is cold enough to freeze water.

Freezing rain can cause real problems for plants, especially trees. As the rain freezes on a tree, for example, its weight presses down on the branches. That causes them to bend, and if they bend too much, they can break. That not only damages the tree, but the broken branches can cause damage to anything they fall on. This can be especially bad for power lines. Because of that, freezing rainstorms are often accompanied by people losing their electricity for a while. Freezing rain is also very dangerous for people who are driving, because the water drops flatten out as they freeze, forming a very slick surface.

Now, of course, if the rain spends enough time falling through cold air, it can freeze completely. That doesn't make snow, however. Snow is composed of snow crystals packed together. If a drop of rain freezes, it forms a small chunk of ice, which is called **sleet**. The picture on the right shows sleet on the ground and uses a penny to give you an idea of the size of the ice chunks that are formed. Typically, sleet is about 0.2 inches (0.05 centimeters) in diameter. Sleet can also be dangerous for drivers, since the ice chunks can form slick surfaces as well.

While each of the kinds of precipitation I have been discussing are formed in a different way, they can also be mixed together. After all, since it takes time for rain to freeze as it falls through cold air, some of what hits the ground might be sleet, and some of it might be freezing rain. If snow is falling into warm air, it takes time to melt, so some of it might hit the ground as snow, and some might hit the ground as rain. We usually refer to such mixtures as a **wintry mix**, and it can be any combination of rain, snow, sleet, or freezing rain.

There is another kind of precipitation that can form, but it is generally associated with thunderstorms, so I will discuss it after I discuss how thunderstorms form.

Comprehension Check

14.7 Precipitation falls out of a cloud as rain. What kinds of precipitation could it be by the time it hits the ground?

14.8 There are parts of Antarctica that receive very little snow each year. On average, do you think those parts of Antarctica are warmer, colder, or the same temperature as the parts that receive a lot of snow every year?

14.9 You see the ground covered in white, but you don't know whether it is snow or sleet. If you have a microscope, how could you tell which it is?

Thunderstorms

There are many ways that weather can become dangerous, but thunderstorms are probably the most common severe weather, at least in most parts of the world. As you already learned, cumulonimbus clouds are responsible for thunderstorms and are therefore often called thunderclouds. They are associated with cold fronts, because as a cold front moves in, the warm air it is overtaking rises, making an updraft. Remember, warm air can become very humid, so when warm, humid air rises in an updraft, a cloud forms. If the updraft is strong, a tall cumulonimbus cloud begins to form, as shown in the far left of the illustration. This is called the **towering cumulus stage** of a thunderstorm.

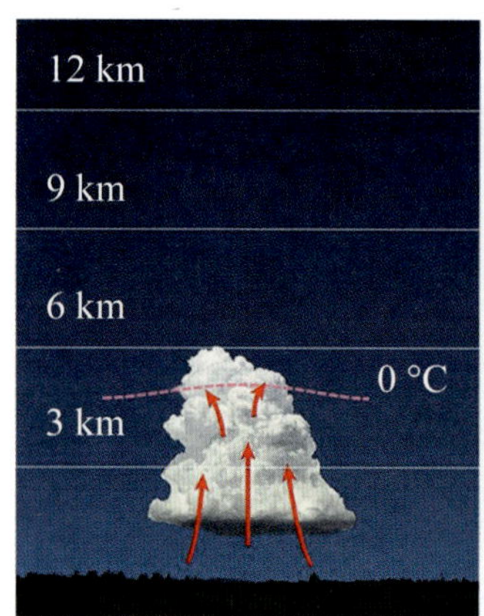

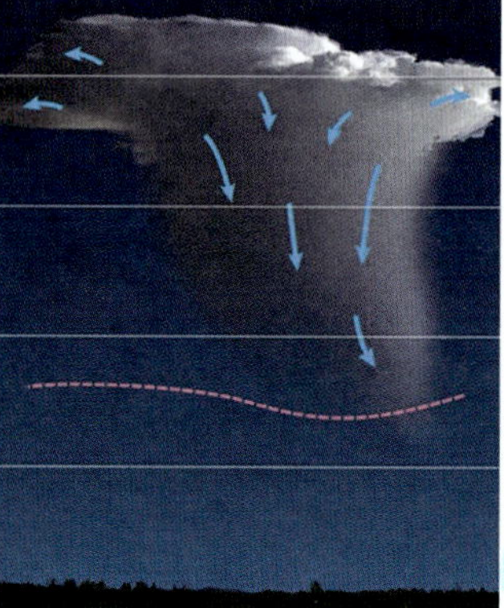

A thunderstorm forms in three stages, which are based on the updrafts (red arrows) and downdrafts (blue arrows) that exist in the cloud. The white lines indicate altitude. Water freezes above the dashed pink curve, and ice melts below it.

As the updrafts continue to blow, the collision coalescence process makes the water drops larger, and the ice crystals in the higher parts of the cloud grow by the Bergeron process. Because they get heavier, the ice crystals begin to fall, creating a downdraft in the parts of the cloud that have been around the longest. As the ice crystals fall into the warmer parts of the cloud, they become liquid and leave the cloud as rain. That marks the **mature stage** of the thunderstorm. Eventually, all the warm air is used up, so there is no more updraft. As a result, the cloud can't replenish the water it is losing, and the thunderstorm has reached its **dissipation stage**, which means it is coming to an end.

What I just described is called a **storm cell**, and it exists for about 30 minutes. A thunderstorm must have at least one cell in it, but if there is only one, it is a relatively weak, short-lived storm. Most thunderstorms are composed of many cells.

Storm cell – The smallest unit of a storm-producing system

A storm cell not only produces rain, but also thunder and lightning. To understand those aspects of a storm cell, let's start with a simple experiment, which works best when the humidity is low.

Experiment 14.3: Knuckle Sparks

Supplies:

- A balloon
- Clean, dry hair
- A room that can be made very dim but not completely dark (A small closet or bathroom with no windows, for example, or a room with windows that have shades which allow only a small amount of light through.)

Instructions:

1. If it isn't already inflated, inflate the balloon to its full size and tie it off so it stays inflated.
2. Go into the room and make it very dim. The goal is for you to be barely able to make out the shape of the balloon as you hold it out at arm's length. I used a windowless room and mostly closed the door, leaving a small crack that allowed a little light into the room. Then, I turned my back to the door.
3. Rub the balloon in your hair vigorously so that it becomes charged.
4. Hold the balloon out in front of you at arm's length so that the part you rubbed in your hair is pointed down.
5. Hold out your other hand so it is far below the balloon. Make a fist and point your knuckles at the balloon.
6. Slowly raise your fist so that your knuckles get closer and closer to the balloon. You should eventually hear a crackle and, if the conditions are right, you will see a spark form between the balloon and one of your knuckles.
7. If you don't see or hear anything, try it a few more times. The more you rub the balloon in your hair, the stronger the crackle and spark should be.
8. Leave the room and put away the balloon.

The result of the experiment probably didn't surprise you. As I am sure you are aware, when you rub a balloon in your hair, it strips electrons away from the molecules that make up your hair. Those electrons give the balloon a negative charge. As you brought your fist near the balloon, the negative charges on the balloon repelled the electrons in the molecules of your skin, so your skin became positively charged near the balloon. Initially, nothing else happened, because even though positive and negative charges attract one another, there was air in between those charges, and air doesn't conduct electricity very well.

However, the strength of the attraction between positive and negative charges grows the closer they get, so as you raised your hand more, the attraction became stronger. Also, your skin near the balloon became even more positive, which increased the attraction even more. Eventually, the attraction became so strong that it overcame the air's inability to conduct electricity, and the charges raced towards each other. Those fast-moving charges collided with the gases in the air, releasing a lot of energy. Some of that energy became light, and the rest heated up the air so much that the air started moving away, forming a sound wave. That sound wave produced the crackle you heard when it reached your ears.

This is actually a great model of how thunder and lightning form. During the mature stage of a thunderstorm, there are ice crystals and water drops moving around. I will call them all "particles." Some particles are being pushed up the cloud by the updrafts, and others are falling down the cloud, making a downdraft. When the downward-moving particles collide with the upward-moving particles, they strip electrons from the upward-moving particles. That makes the downward-moving particles negative, while the upward-moving molecules become positive. Over time, this makes the top of the cloud positive and the bottom of the cloud negative. The longer this goes on, the more positive the top of the cloud becomes, and the more negative the bottom of the cloud becomes.

These positives and negatives are attracted to one another, but once again, air doesn't conduct electricity very well, so at first, they can't move towards each other. Eventually, however, the charges build up to the point where something happens. If the charges can move to each other inside the cloud, a spark is formed within the cloud. If the negative charges in one part of a cloud and the positive charges in another part of the cloud (or another cloud entirely) build up enough, a spark forms between the clouds.

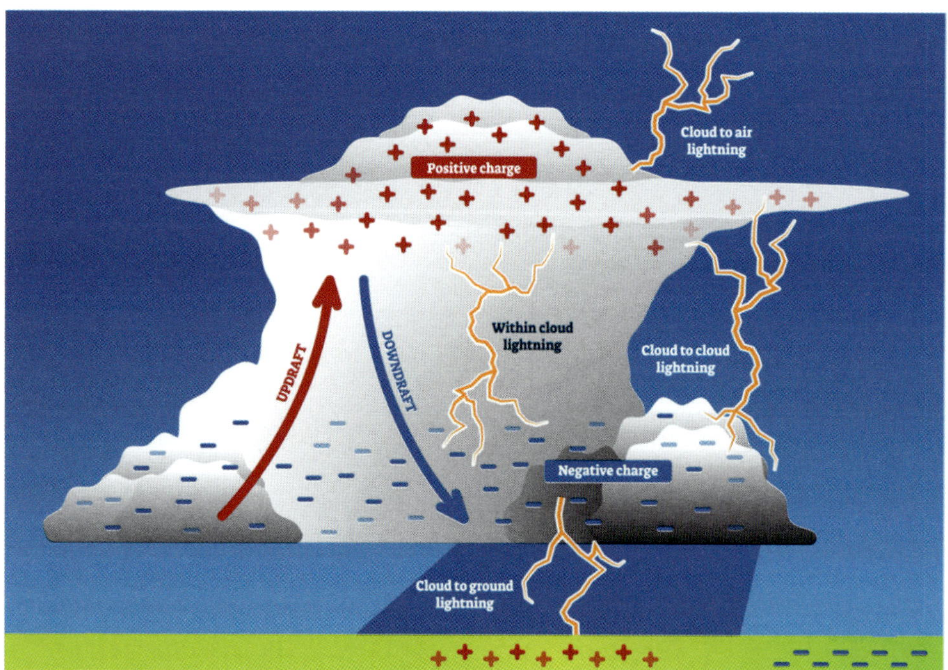

Buildup of charges produce sparks within a cloud, between clouds, between the clouds and the ground, or between the clouds and the air. Those sparks are lightning.

The negative charges at the bottom of the cloud can also cause the land near the cloud to become positively-charged, like the balloon did to your skin. Just like what happened in the experiment, those charges can build up to the point where a spark forms between the cloud and the ground. At the top of the cloud, the positive charges can attract negative charges in the molecules of the air, making the air near the cloud negatively charged. If those charges build up enough, a spark forms between the cloud and the air.

A lightning strike between a cloud and the ground happens in stages. As the negative charges start to pile up at the bottom of the cloud, they begin to push their way through the air in order to reach the positive charges below. Since they are colliding with air particles as they go, they tend to follow a jagged path down, and that jagged path often has many branches to it. This is called the **stepped leader**, because it is the leading part of the lighting strike, and the electrons advance in steps of about 50 meters (160 feet) at a time. This is actually the weaker part of the lightning strike. It is often dim and therefore hard to see.

As the stepped leader gets close to the ground, the positive charges that have formed rise up to meet the bottom of the stepped leader. When those positive charges touch the stepped leader, a conductive path is formed, and that allows the charges to flow freely. This is called the **return stroke**, and it's the powerful part of a lightning strike. Look at the pictures below, which were pulled from a slow-motion video. The first one was taken when the stepped leader got close to the ground. The second one shows the return stroke. Notice how thin and dim the path of the stepped leader is compared to the return stroke. That's because the real power of a lightning strike is in the return stroke. When you see a lightning bolt hit the ground, you are seeing the return stroke, because that's what makes most of the light.

A stepped leader (left) and return stoke (right) of the same lightning strike. Notice how much more powerful the return stroke is.

But wait a minute. What about the thunder? The thunder is the same as the crackle you heard in the experiment. It's a sound wave caused by the fact that the air around a lightning bolt is heated and starts moving outward, forming a sound wave. Since there is a lot of energy produced very quickly in a lightning bolt, the crests of the sound wave are packed with *lots of gas molecules*, and that makes the thunder really loud.

Comprehension Check

14.10 As you might already know, light travels much faster than sound. If you are close to a lightning strike, you will hear the thunder and see the lightning at the same time. However, if you are far from the lightning strike, you will experience one before the other. Which will you experience first?

Hail in Thunderstorms

Thunderstorms can be dangerous, but other things can add to them to make them even more dangerous. Remember that there are strong updrafts in a cumulonimbus cloud. When those updrafts are strong enough, water droplets that are falling near the bottom of the cloud can be blown upward to the point where the cloud is cold enough to freeze them again. The water droplet doesn't form a snow crystal. Remember, snow crystals are produced when water vapor deposits on nucleation centers, then more water vapor deposits on the newly-formed crystal, growing it slowly. When a water droplet is forced up to the cold parts of a cloud, it simply freezes, like water in a freezer. As a result, you don't have a nice snow crystal. You have a ball of ice.

The ball of ice grows when it collides with water droplets that freeze onto it. Remember, there are supercooled water droplets in the cold part of the cloud. They don't normally freeze because they have nothing to freeze onto. However, a ball of ice is a perfect place for them to freeze, so as the ball of ice rises, it grows. It will eventually fall again, but it might be caught in another updraft, which will repeat the entire process, making the ball of ice grow even more. Eventually, however, even the strongest updraft won't blow it up anymore, and it will fall to the ground. When that happens, it is called **hail**, and the balls of ice are called **hailstones**.

The size of hailstones produced in a thundercloud depend on the strength of the updrafts in the cloud. Some hailstones aren't any larger than a pea. However, some can get much larger than that. For example, the picture on the left shows a hailstone that fell in Vivian, South Dakota on July 23, 2010. At the time, it set the record for the largest hailstone that had been found and analyzed. As you can see from the measuring tape, it is about 8 inches wide at its widest point. It weighed almost 2 pounds! Even significantly smaller hailstones can produce a lot of damage and can injure people. The National Oceanic and Atmospheric Administration (NOAA) estimates that more than 20 people in the United States are injured each year by hailstones.

This record-setting hailstone weighed nearly 2 pounds!

Tornadoes

Tornadoes are another form of severe weather that can be associated with a thunderstorm. While we don't completely understand how and when they form, we can at least describe the broad outlines, which are illustrated on the next page. Often, the winds near the ground are blowing in a different direction from the winds above the ground. When air in between begins moving in a third direction because of the motion of weather front, the air can begin rotating as it travels. This is called **wind shear**.

Wind shear – A variation in wind velocity that tends to produce a rotating motion in the air

Wind shear produces a rotating "tube" of air, which can be oriented in nearly any direction. However, the rotating wind tubes that form tornadoes start out parallel to the ground, as shown in the illustration on the next page.

Now imagine a cumulonimbus cloud above that rotating tube of air. The cloud is being formed by an updraft, and if the updraft is strong, it can push the rotating tube of air up into the cloud. That means its orientation changes. As time goes on, it can change so much that the tube runs up and down the cloud. Of course, it is still rotating, so the cloud now has a rotating column of air in it. This starts the entire cloud rotating. When a cumulonimbus cloud starts rotating like that, it is called a **supercell**, and the rotating column of air is called a **mesocyclone** (me' zoh sye' klohn).

Mesocyclone – A rotating air mass at the center of a supercell

Mesocyclones are usually about 3 to 10 kilometers (2 to 6 miles) across. When they are detected, a tornado warning is usually issued, because a tornado could be formed as a result.

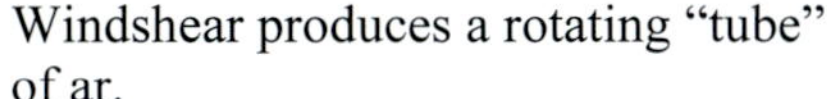

Windshear produces a rotating "tube" of ar.

Updrafts push the tube up the cloud so that it becomes a column.

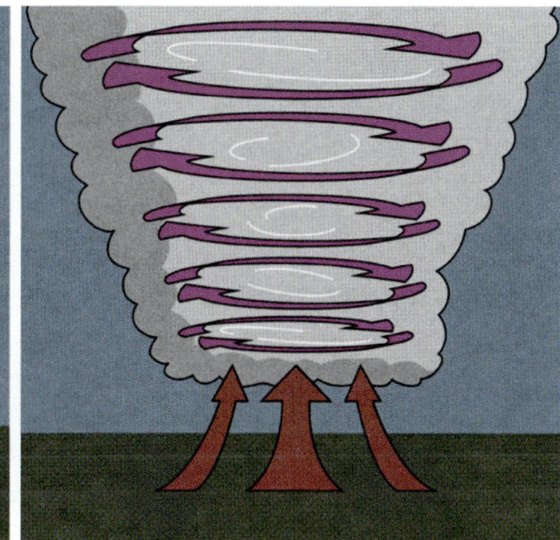

The rotating column becomes a mesocyclone, making a supercell.

Notice that I said a tornado *could* be formed from a supercell. It doesn't always happen. A supercell might remain a large, rotating air mass that produces a thunderstorm. However, sometimes, a supercell will develop a funnel-shaped projection going down to the ground. That's called a **funnel cloud**. If that funnel cloud touches the ground, a tornado is formed.

To illustrate this, the picture on the right is a series of eight shots taken of the same supercell. You can see on the left that there is a funnel-shaped projection going down. That's the funnel cloud. Not all funnel clouds reach the ground, but as you can see in the next frame, this one did. You can see that it is touching the ground because there is dark cloud rising from the ground. That's caused by dirt and debris being pulled up into the rotating column of air. You can see that over time, the tornado becomes darker and thicker, because it is picking up more and more debris.

This is a series of photos taken while a tornado developed.

Why do some supercells produce funnel clouds, while others don't? We aren't really sure. Why do some funnel clouds touch the ground and become tornadoes, while others don't? Once again, we aren't sure. Meteorologists are trying to answer these questions so they can better predict when and where tornadoes form. While many of the details about tornado formation are still unknown, one thing is for certain – tornadoes can be both destructive and deadly. The United States experiences about 1,000 tornadoes per year on average. They result in about 70 deaths and 1,500 injuries each year, and cause about a billion dollars of property damage annually. This is why meteorologists would love to learn enough about tornadoes to be able to accurately predict when and where they will form.

Why are tornadoes so destructive? First, the winds that form the rotating motion can be powerful enough to knock down what gets in their path, including buildings and trees. Anything that is unable to withstand the power of the winds gets lifted up and carried by the tornado. For example, a strong tornado can lift a car off the ground and carry it for 800 meters (half a mile) before the car falls back to the ground! The destructiveness of a tornado can be measured by the speed of the winds that form the rotating column of air. They can range from as "low" as 105 kilometers per hour (65 miles per hour) to more than 320 kilometers per hour (200 miles per hour)! As a result, there is a scale called the **Enhanced Fujita (EF) Scale** that categorizes tornadoes from EF0 (low wind speed, low damage) to EF5 (very high wind speed, incredible damage). That scale is shown in the illustration on the left. Fortunately, 70-75% of all tornadoes in the United States are EF0 or EF1 tornadoes. All tornadoes are dangerous, but at least the majority of them are the weaker ones. About 15-19% are EF2, 6% are EF3, 1% are EF4, and very, very few are EF5.

Rating	Wind Speed	Damage
EFO	65-85mph	minor roof, branches
EF1	86-110	broken windows
EF2	111-135	roofs off, large trees
EF3	136-165	homes damaged
EF4	166-200	homes leveled
EF5	200+	incredible damage

TORNADO RATING
Enhanced Fujita Scale

The Enhanced Fujita (EF) scale categorizes tornadoes based on their destructive ability.

Because of their potential for destruction, weather forecasters try to warn people when it is possible that a tornado might form in their area. If the correct conditions of wind shear and updraft are present in an area, a **tornado watch** is issued. This simply means that the conditions make it possible for a tornado to form and you should be on alert. If a funnel cloud or tornado is spotted by people or on radar, then a **tornado warning** is issued, which means a tornado is imminent. I live in Indiana, which experiences frequent tornadoes. In the city where I live, there is a warning siren that you can hear from anywhere in town. When it goes off, that means a tornado warning has been issued for the city.

The best thing to do in a tornado is get underground. That way, you avoid the winds and the things being thrown around in the wind. Thus, if you can get to a basement or an underground shelter, that is the best place to be in a tornado. When my wife and I hear the tornado siren, we gather our cats and go to the basement, waiting for the end of the tornado warning before we go back up into the main house. If you are in a building that doesn't have a basement, try to get to a small room in the interior of the house, away from windows, doors, etc. If you are outside, find the lowest possible location and lie as flat as possible. The idea is you need to avoid all the flying objects that will be produced by the tornado's winds.

Now remember, most of the earth is covered by water. As a result, tornadoes can form over water as well as over land. They can even form over land and then move to the water. Once it is on water, it is called a **tornadic waterspout**. "Tornadic" refers to the fact that it is really a tornado, and "waterspout" refers to the fact that the tornado looks like a spout of water rising from the surface. However, there is another kind of waterspout, it is called a **fair-weather waterspout**, which is not associated with thunderstorms and is weaker than a tornadic waterspout. In fact, in some ways, it is the opposite of a tornadic waterspout.

Remember, water takes a long time to heat up and cool off. Thus, when the air above a large body of water cools, the water below it stays warmer. When a cold front moves over a body of water, then, a large temperature difference between the water and the cumulus clouds forming above can occur. This enhances the updraft of air. If that updraft is exposed to wind shear from a breeze blowing along the body of water, it will start to rotate. Think about how this is different from a tornado. A tornadic waterspout starts in a cloud as a tornado and then goes down towards the surface of the water. A fair-weather waterspout, on the other hand, rises from the surface of the water and reaches the clouds above. These kinds of waterspouts are generally weak and do not cause much damage.

Another weather phenomenon that involves a rotating column of air is the **dust devil**. Rotation is pretty much all it has in common with a tornado, however. It doesn't form in a thundercloud. Instead, it tends to form on warm, sunny days. When the sun heats up a small part of the ground faster than the rest of the ground surrounding it, the warm air rises rapidly. This pulls the cold air in from around it, and when the right kind of breeze blows, the air coming in can start rotating. Like a fair-weather waterspout, then, a dust devil doesn't come down from a cloud. It rises up from the ground.

It is called a dust devil because the rotating column of air picks up dust and dirt, as shown in the picture on the right. If you examine the details of the picture, you can see why the dust devil formed. First, there is hardly a cloud in the sky, so it is sunny. Also, notice that there is a lot of bare dirt, but you can also see some green. Most likely, there was a bare patch of dirt that got some pretty direct sunlight. Around that bare patch of dirt there was ground that had plants on it. The plants hold water, so they don't heat up nearly as fast as the bare dirt. Thus, the bare dirt provided the hot air, and the plant-covered surroundings provided the cooler air.

While a dust devil looks a bit like a tornado, it is very different.

Dust devils are generally small and not very dangerous. My wife and I were sitting in a car at a stop sign in Arizona when a dust devil crossed the road, passing right in front of us. It felt like we were being hit by gusts of strong wind. The stop sign vibrated back and forth, and dust obscured our vision for a moment, but then it was over. While it was exciting, we never felt like we were in any real danger.

Comprehension Check

14.11 Suppose you could measure the air pressure at the center of a tornado. How would it compare to the air pressure right outside the tornado?

14.12 A supercell is rotating rapidly but doesn't produce a funnel cloud. Will it form a tornado?

14.13 A friend tells you about being caught in a "weak tornado." What could you ask him or her to determine whether it was a weak tornado or a dust devil?

Hurricanes

No study of destructive weather would be complete without a discussion of hurricanes. They are one of the most destructive weather phenomena on the planet, simply because of their size. While tornadoes tend to be hundreds of meters (hundreds of yards) wide, hurricanes tend to be many kilometers (many miles) across. Also, while tornadoes can cause destruction for several minutes, hurricanes can cause destruction for several hours.

Hurricanes start out innocently enough. Warm ocean waters near the equator heat up the air above them, which is also humid because warm water evaporates quickly, producing a lot of water vapor. As the warm air rises, it cools, and the water condenses to form a cloud. That sounds familiar, right? It's essentially the same kind of process that produces thunderclouds over land. The difference, however, is that a cold front isn't moving in, so the warm waters stay warm and continue to feed the clouds. Rather than experiencing a dissipation stage as the rain continues, the thunderstorms produced in this way can last for a long time. When many thunderclouds are formed over a large area and persist for more than a day, meteorologists say that there is a **tropical disturbance**, which is illustrated in the drawing on the left. Notice that it shows five separate cumulonimbus clouds. Each is its own storm cell, producing rain. Together, they form the tropical disturbance.

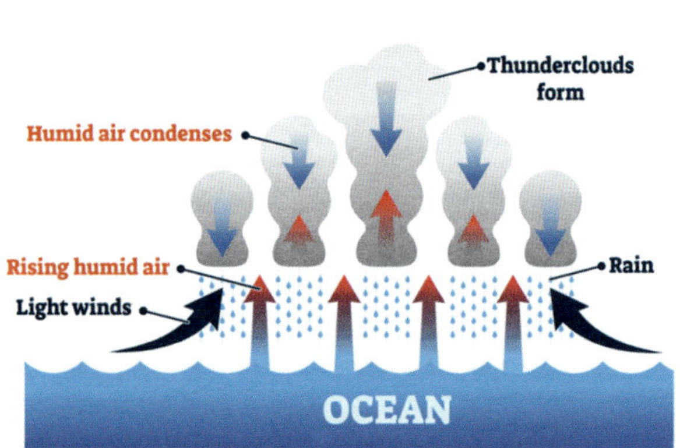

A tropical disturbance is a large, long-lasting group of storm cells formed over warm ocean water.

Because of winds and weather fronts, most tropical disturbances eventually dissipate, so they are nothing more than large, long-lasting thunderstorms over the ocean. However, if the warm waters feed the thunderclouds with enough humid air, the thunderclouds will continue to grow to the point where they form large anvils with flat tops that spread out along the tropopause, away from their centers. Now remember, this covers a large area, so the tops of the anvils travel a long distance away from the center. What happens when something moves above the surface of the earth for a long distance? The Coriolis effect bends its motion, right? That happens to the tops of the anvils. As the tops of the anvils spread out more and get bent more, they start to rotate. Eventually, the whole system of clouds starts rotating as well. At that point, meteorologists say that we have a **tropical cyclone**, as illustrated on the left.

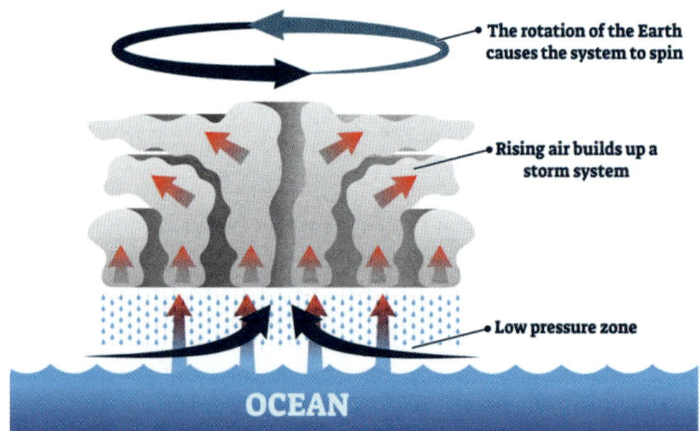

The Coriolis effect can turn a tropical disturbance into a tropical cyclone.

<u>Tropical Cyclone</u> – An organized system of thunderclouds that rotate over warm ocean waters

The word "tropical" refers to the fact that it must form over ocean waters near the equator, and the word "cyclone" refers to the rotating motion of the cloud system.

Please understand that the term "tropical cyclone" can refer to *any* organized system of thunderclouds that rotate, so it is a very general term. There are, however, specific levels to the severity of a tropical cyclone. The weakest tropical cyclone is called a **tropical depression**. The "depression" refers to the fact that the atmospheric pressure is low (depressed) in the area. If the winds in the cyclone get to the point where they are 63 kilometers per hour (39 miles per hour), the cyclone is called a **tropical storm**. In some ways, that's a misnomer, because even a tropical disturbance can produce storms. Nevertheless, at this point, the tropical storm is officially given a short, simple name by the World Meteorological Organization. The first tropical storm of the year is given a name that begins with "A" (like "Anne"), the next one is given a name that starts with "B," etc. The letters "Q," "U," "X," "Y," and "Z" aren't used, so there are only 21 possible names for tropical storms in a given year. If more than 21 tropical storms form in a year, the rest are named after the letters in the Greek alphabet (alpha, beta, gamma, etc.).

If the winds continue to pick up speed and end up reaching 119 kilometers per hour (74 miles per hour), the tropical cyclone is given a new designation. Oddly enough, the designation it is given depends on where it formed. If it forms in the Atlantic or the Northeast Pacific Ocean, it is called a **hurricane**. If it forms in the South Pacific or Indian Ocean, it is called a **cyclone**. If it forms in the Northwest Pacific Ocean, it is called a **typhoon**. They are all the same thing: a tropical cyclone whose winds have reached 119 kilometers per hour. They are just referred to differently because of historical biases. There was a lot of Spanish influence in North America in the 15th and 16th centuries, and the Spanish word for a cyclone is *huracán*. Since tropical cyclones formed in the Atlantic Ocean were important to North American sailors at that time, they were called hurricanes. On the other hand, the Arabic, Persian, and Hindi refer to such storms with the word *tufan*. When the same kinds of cyclones formed in Northwest Pacific Ocean, they were of interest to people who spoke those languages. From now on, I will simply refer to all such cyclones as "hurricanes," since I live in North America.

When a tropical cyclone becomes a hurricane, it has a specific structure, as shown in the drawing on the right. The spinning motion has produced a cloudless region in the center of the hurricane. It is called the **eye** of the hurricane, and there are only very light winds there. Because it is cloudless and not windy, it is very calm and pleasant. Of course, all the other parts of the hurricane are rainy, windy, and chaotic. Most hurricanes are about 200 kilometers (125 miles) across, while their eyes are, at most, 50 kilometers (30 miles) across.

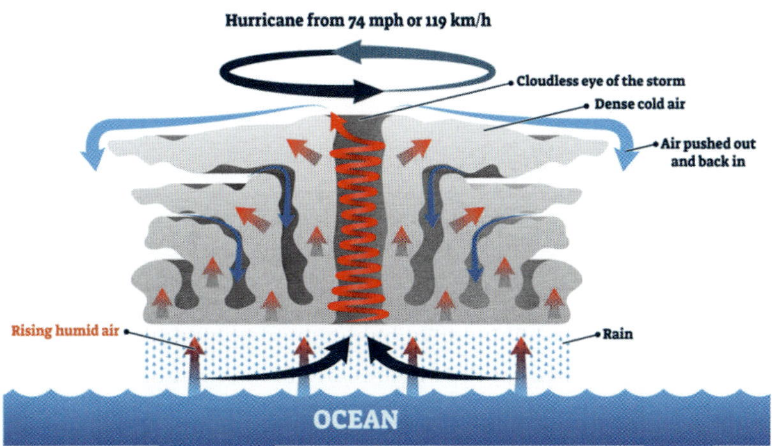

A hurricane (or typhoon or cyclone) has a distinct structure, with thunderclouds rotating around a cloudless eye.

Of course, a hurricane doesn't stand still while it develops; it is pushed along by the winds. If you remember from the previous chapter, around the equator, the winds curve so that they are blowing to the west. Thus, as a tropical cyclone develops, it tends to move west. However, they also move towards the pole in their hemisphere because of a complicated process called the "beta drift," which I don't want to go into. Just accept the fact that the general path of a hurricane is west and away from the equator. Thus, hurricanes in the North Atlantic travel northwest, while cyclones in the South Atlantic travel southwest. As long as there is warm water, a hurricane can become more powerful as it moves, because the warm, moist air can continually rise, "feeding" the clouds. However, once it actually hits land, it can't be "fed" with warm, moist air, so it quickly weakens. Nevertheless, it can cause a great deal of destruction before it dies out.

CATEGORY	mph	km/h	Knots (kn)	
1	79-95 mph	119-153 km/h	64-82 kn	Minimal Damage
2	96-110 mph	154-177 km/h	83-95 kn	Moderate Damage
3	111-129 mph	178-208 km/h	96-112 kn	Extensive Damage
4	130-156 mph	209-251 km/h	113-136 kn	Extreme Damage
5	≥157 mph	≥252 km/h	≥137 kn	Catastrophic Damage

Hurricanes are grouped into categories based on their wind speeds. ("Knots" are a nautical measurement of speed. 1 knot = 1.15 miles per hour)

Just like tornadoes, hurricanes can have different levels of severity, based on their wind speeds. Those levels are called "categories," and as you can see in the drawing on the left, the higher the category number, the larger the wind speeds and the more destructive the hurricane. Often, the category is referred to with "cat" and then the number. Hurricane Andrew in 1992, for example, was a cat 5 hurricane that hit Florida and caused an estimated 26.5 billion dollars of damage. It is considered the most expensive natural disaster in United States history.

Because we have satellites that orbit the earth, we can take pictures of hurricanes from above to see their structures. In the left picture below, for example, you see a hurricane that formed in the South Atlantic in 2004. Because it was formed in the South Atlantic, it is technically called a cyclone, but nevertheless, you can easily see the eye at the center of the hurricane and the clouds surrounding the eye. The shape even shows you that it is rotating clockwise. Next to it, you can see a hurricane that formed in the North Atlantic also in 2004. Once again, you can see its eye, and you can see that it is rotating. However, this one is rotating *counterclockwise*.

A hurricane in the Southern Hemisphere (left) and Northern Hemisphere (right). Notice that they rotate in opposite directions.

Why do these two hurricanes rotate differently? Because of the Coriolis effect. Remember, the Coriolis effect causes the rotation of the cyclone. Since the effect is opposite in the two hemispheres, hurricanes rotate opposite in the two hemispheres. In case you are wondering what would happen if a hurricane traveled from the Northern Hemisphere into the Southern

Hemisphere, remember that hurricanes move towards the poles and away from the equator, so that would never happen.

Is Severe Weather Getting Worse?

One common thing you will hear from many news sources is that severe weather is getting worse because the globe is warming. At first glance, this makes a lot of sense. After all, what produces a hurricane to begin with? Warm ocean water. If the earth is getting warmer, that should mean warmer ocean water and more powerful hurricanes. Also, we seem to hear a lot more about hurricanes nowadays than we did 20 years ago. Thus, it makes sense that the number of hurricanes has been growing over the years.

It might "make sense," but it's not true. Consider, for example, the graph below. It shows the number of hurricanes that have been formed on earth over time, starting in 1971 and going to the end of 2017. The squares on the top of the graph show all hurricanes, regardless of their category. The bottom squares show only the cat 3 – cat 5 hurricanes (what the graph calls "Major Hurricanes"). Notice that the number of hurricanes varies widely from year-to-year, and there is no overall pattern. If the earth's warming were causing more hurricanes, you should see a definite upward trend to the graphs. There is a very small upward trend to the major hurricanes, but a very small downward trend to all hurricanes. Thus, there is no real pattern.

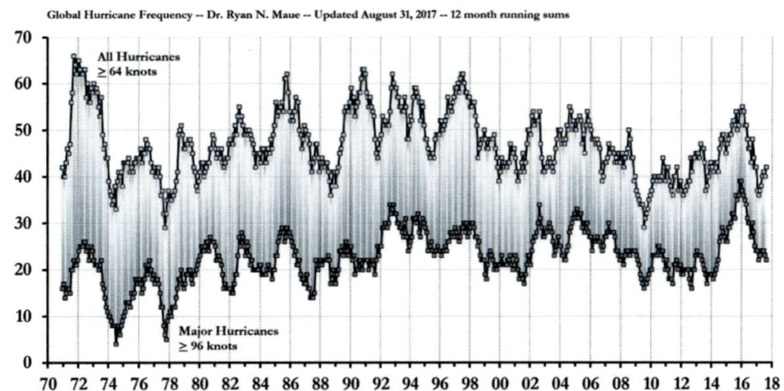

This graph shows the total number of hurricanes (top squares) and major hurricanes (bottom squares) that formed over the past 46 years.

Well, perhaps the number of hurricanes hasn't changed, but maybe their destructiveness has. After all, the warmer the waters, the stronger the hurricanes, right? Not necessarily. Meteorologists measure how strong a cyclone is using the "accumulated cyclone energy," which is abbreviated ACE. It simply measures how much energy accumulates in each cyclone over the course of its existence. That is shown in the graph to the right. The top squares are for all tropical cyclones that formed over the entire earth, and the bottom squares are for tropical cyclones that formed in the Northern Hemisphere. Once again, we see no definite trend. The amount of energy in cyclones has not been increasing over the 46 years covered in the graph.

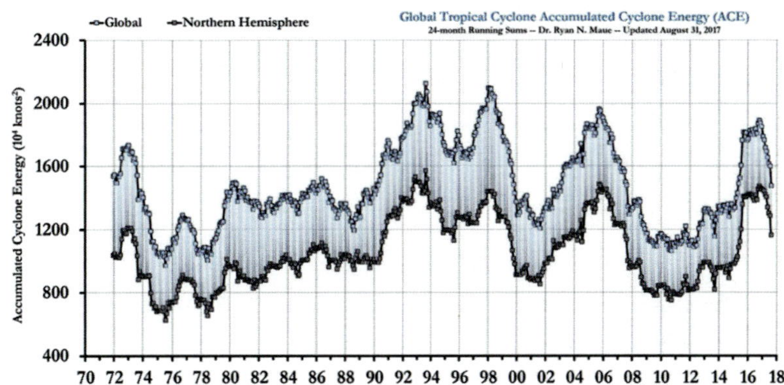

This graph shows the energy in all cyclones on earth (top squares) and all cyclones in the Northern Hemisphere (bottom squares) over the past 46 years.

But what about tornadoes? Those are also severe weather events, and they are formed when warm, humid air rises in response to a cold front moving in. Are tornadoes increasing in frequency? Well, unlike hurricanes, it is difficult to track tornadoes worldwide. After all, they are a lot smaller, and if they touch down in uninhabited regions of the earth, there won't be many consequences. As a result, we might not even know they occurred. However, the United States has been tracking its tornadoes since 1954. The results of this tracking are shown on the left. In the top graph, we have all tornadoes (EF1 or higher). Notice that once again, there is no pattern. We don't see the number of tornadoes going up or down as time goes on. On the bottom, there is a similar graph for just the stronger tornadoes (EF3 or higher). There does seem to be a trend here, but it is down, not up. If anything, the number of stronger tornadoes has been decreasing as the earth has warmed.

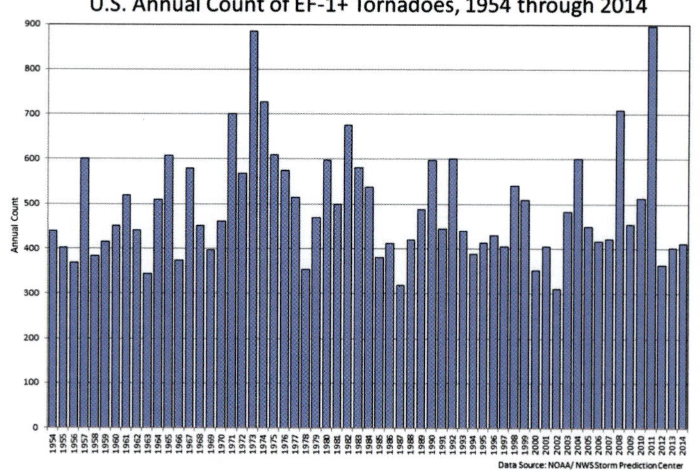

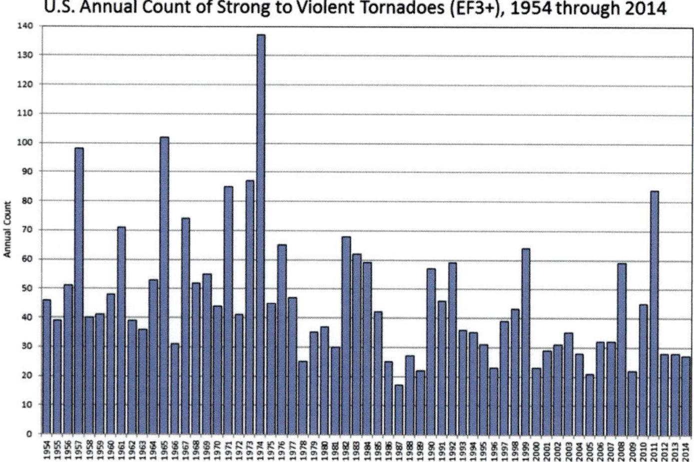

What does this mean? Scientists can't say for sure. Perhaps the earth has to get a lot hotter before severe weather increases in intensity. Perhaps severe weather events are so complex that there is no simple way to relate global temperature and severe weather. Perhaps the slight increase we see in temperature around the earth is caused by the urban heat island effect and thus wouldn't have an effect on severe weather over a large scale. What we can say for certain is that there has not been an increase in hurricanes or tornadoes over the last several decades. Thus, if someone says that there has been, you know that he or she is terribly misinformed!

Comprehension Check

14.14 Which has faster winds: a tropical depression, a tropical storm, or a tropical disturbance?

14.15 A hurricane is currently a category 4. If it hits land, will its category stay the same or change? If it changes, will the category number go up or down?

14.16 Hurricane Carl is about to hit Florida. How many hurricanes formed that same year before Carl?

14.17 You see a satellite image of a tropical cyclone but there are no landmarks to figure out where it is. How can you determine the hemisphere it is in?

Answers to the Comprehension Check Questions

14.1 It is increasing. The cloud pattern is indicative of a warm front, so warmer air should be moving in.

14.2 It is decreasing. Cold fronts replace warm air (which has low pressure) with cold air (which has high pressure). Since that's how the pressure changed, this means a cold front passed through, so the temperature should get colder.

14.3 In central United States. Notice that the western part of the warm front has a cold front behind it, and the triangles indicate that the eastern part of the cold front is headed towards the warm front. Cold fronts travel more quickly than warm fronts, so it might catch up with the warm front, making an occluded front.

14.4 Alabama. Blues and greens represent less intense rain than reds and yellows.

14.5 Collision coalescence makes water drops heavier near the bottom of the cloud, and the Bergeron process makes them lighter near the top. Collision coalescence requires that the water be liquid, which happens in the lower part of the cloud, where it is warm. The Bergeron process causes water drops to lose weight, because the water evaporates and then deposits onto ice crystals. That must happen high in the cloud, where the temperature is low.

14.6 It is more likely to become empty on the second day. The lower the relative humidity, the faster water evaporates. For the same temperature, then, 30% relative humidity results in faster evaporation than 75% humidity.

14.7 It could hit the ground as rain, freezing rain, or sleet. It cannot hit the ground as snow, because snow crystals must be formed in a cloud.

14.8 They are colder than the parts that get more snow. Remember, colder air is less humid, which produces less precipitation. The coldest parts of Antarctica receive only about 2 inches of precipitation a year, compared to as much as 12 inches in the warmer parts.

14.9 Look for snow crystals. If you see snow crystals, it must be snow. If not, it is rain that froze, so it is sleet.

14.10 You will see the lighting first. They must both travel to you from where they were made. Since light travels faster, it will get to you first. The farther you are away, the longer the time between them. For every second that elapses between seeing the lightning and hearing the thunder, you are roughly 340 meters (1,100 feet) from where the lightning formed.

14.11 The air pressure at the center would be lower. Remember, it's making an updraft, so air pressure has to be lower where the air is rising. That's also why a tornado can lift things up.

14.12 No, it will not. A funnel cloud must be produced in order for the mesocyclone to touch the ground. No funnel cloud means no tornado.

14.13 Ask him if it was clear or stormy. Dust devils form on clear, sunny days, while tornadoes form in thunderstorms.

14.14 A tropical storm. Remember that as it gets more severe, its wind speeds increase. A tropical disturbance is the first to form, followed by a tropical depression, followed by a tropical storm. Since that one is the latest of the three listed, it must have the fastest winds.

14.15 Its category number will decrease. Once a hurricane hits land, it gets weaker because it no longer has warm ocean water to power it.

14.16 Two others formed before it. Its name starts with "C," which means it is the third hurricane of the season.

14.17 Check to see which way it is rotating. The clouds show you the rotation, and that tells you the hemisphere.

Chapter Review

1. Define the following terms:

a. Weather front	d. Relative humidity	g. Wind shear
b. Cohesion	e. Absolute humidity	h. Mesocyclone
c. Saturated	f. Storm cell	i. Tropical cyclone

2. You see cumulonimbus clouds forming. What kind of weather front is probably moving in?

3. You see cirrus clouds form, followed by stratus clouds. What kind of weather front is probably moving in?

4. What do we call a weather front in which neither air mass really moves?

5. You see cirrus clouds form, followed by stratus clouds. Then, cumulonimbus clouds start to form. What kind of weather front is probably moving in?

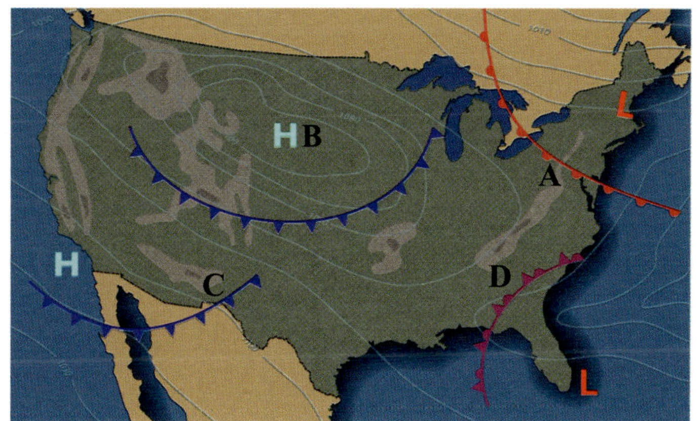

6. In the weather map on the right, at which point (A-D) is the atmospheric pressure highest?

7. In the weather map on the right, which point (A-D) is about to experience an occluded front?

8. In the weather map on the right, which point (A-D) just experienced a decrease in temperature?

9. In the weather map on the right, which point (A-D) is about to experience an increase in temperature?

10. In a cumulonimbus cloud, water drops are growing by the collision-coalescence process. Is this happening high in the cloud or low in the cloud?

11. An ice crystal in a cloud is growing. What is happening to the size of the water drops in the same part of the cloud?

12. One day, the temperature is 29 °C, and the relative humidity is 90%. The next day, the temperature is still 29 °C, but the relative humidity is 20%. Which day feels more uncomfortably hot?

13. Snow crystals are falling out of a cloud. Does that mean snow is falling on the ground? Why or why not?

14. Snow is falling on the ground. Have the snow crystals always been solid since the time they were first formed in the cloud? Why or why not?

15. Rain is falling, but the roads are icy. What do we call this kind of precipitation?

16. Sleet is falling. Has the sleet been ice since it left the cloud?

17. What are the three stages of a thunderstorm? At which stage are there both updrafts and downdrafts in the cloud?

18. In a cumulonimbus cloud, what kind of electrical charge is found near the top?

19. What are the two stages of a lightning strike, and which one is more powerful?

20. In a cumulonimbus cloud, balls of ice are seen traveling upwards. What will this cloud produce in addition to rain?

21. A supercell has formed. What must it make before it can make a tornado?

22. What do we call a tornado that forms on water?

23. If you see a rotating column of air rising from the ground into the sky on a sunny day, what are you witnessing?

24. Consider the terms "tropical cyclone," "tropical storm," "typhoon," and "tropical depression." Which of these terms can refer to the other three? For those other three, list them according to increasing wind speeds.

25. What two other terms not listed in the question above can be used to refer to a typhoon? What determines which term is used?

26. A hurricane is getting stronger. Has it hit land, or is it still on the ocean?

Chapter 15: Earth's Solar System

Introduction

I have already discussed the earth's orbit around the sun as well as a bit about the moon's orbit around the earth. However, there is so much more to learn about the **solar system** that we need to spend an entire chapter on it. To get started, look at the picture on the right. It was taken by a robotic spacecraft that was on its way to an asteroid. At the time, it was 400,000 kilometers (250,000 miles) from the earth, which is in the lower left portion of the picture. The moon is in the upper right. Please note that these are two separate images that have been put together in their proper positions. The spacecraft detected both of them, but the earth is *much* brighter than the moon. As a result, the moon is simply too dark to see when the camera is adjusted to get a good view of the earth. Thus, the picture was taken twice, once at a setting that gave a good image of the earth, and again at a setting that gave a good picture of the moon. They were then overlayed so that you see their proper positions from the spacecraft's perspective.

This picture shows the earth (lower left) and moon (upper right) as seen from a robotic spacecraft that was 250,000 miles away at the time. The brightness of the moon is adjusted to make it visible with the earth.

There are two things I want you to get from the picture. First, can you tell where the sun was when the picture was taken? Notice that the earth's right side and the moon's right side are both illuminated. The left sides are dark. That tells you the sun is shining on them from the right. Second, notice that both the earth and the moon are simply "hanging" out there in space. They are not supported by anything. Why is that important? Because the Bible described this situation long before scientists did! Referring to God, Job 26:7 states, "He stretches out the north over empty space And hangs the earth on nothing." Most ancient cultures thought that the earth had to be supported by something. Ancient Hindus, for example, believed the earth was held up by elephants. The Bible, however, teaches that the earth hangs on nothing, as you can see in the picture above.

How Does Gravity Keep the Earth in Orbit Around the Sun?

If I asked you what holds the earth in orbit around the sun, you would immediately say "gravity," right? But wait a minute. Gravity *attracts objects to one another*. If I drop a rock, what happens? It falls straight towards the ground, because gravity attracts the rock and the earth. As a result, the rock falls to the earth. Why doesn't the earth fall straight into the sun? The gravity that exists between the earth and the sun acts the same way. It pulls the earth to the sun, but the earth never reaches the sun. Why? Perform the following experiment, which will help you understand the answer.

Experiment 15.1: Twirling around

Supplies:
- A length of sewing thread that is about 61 cm (24 in) long
- A small plastic bag, like a sandwich bag (It can be zippered but doesn't have to be.)
- The mass scale from the laboratory kit made for this course
- The graduated cylinder from the laboratory kit made for this course (Make sure it is dry on the inside.)
- Rice, small beans, or something else made of small grains (You can use dirt, sand, flour, etc., but there is a possibility that the bag will break, so definitely do the experiment outside if you decide to use something like that.)
- Someone to help you
- A large room with nothing breakable in it or (better yet) an open area outside

Instructions:
1. Turn the mass scale on and make sure it is reading grams.
2. Put the graduated cylinder on the scale and hit "tare" to make it read 0.
3. Pour rice into the graduated cylinder until the mass reads somewhere between 45 grams and 55 grams. If you spill rice on the scale, blow it off the scale so that the reading comes just from the rice in the graduated cylinder.
4. Pour the rice from the graduated cylinder into the plastic bag.

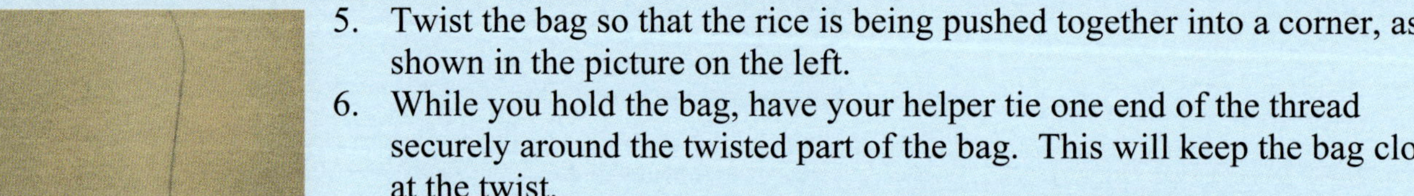

5. Twist the bag so that the rice is being pushed together into a corner, as shown in the picture on the left.
6. While you hold the bag, have your helper tie one end of the thread securely around the twisted part of the bag. This will keep the bag closed at the twist.
7. Grab the other end of the thread and wrap it around your index finger a few times. Then pinch your thumb against the thread on your finger so the thread stays securely twisted around your index finger.
8. Go to the large room or outside.
9. Hold the thread out at arm's length so that the bag of rice dangles in front of you.
10. Lift your hand up and watch the bag rise up as well. Not all that impressive, is it?
11. Slowly start twirling the bag of rice in a circle above your head. You don't want it to travel too fast at first. Just have it orbiting your head in a circle.
12. Now start to increase the speed of the bag. Try to concentrate on what the thread feels like as you increase the speed of the bag.
13. Continue to increase the speed of the bag, trying to get it moving as fast as possible. What happens?
14. If the thread didn't break, causing the bag to go flying away from you, either try to twirl it faster or add another 45-55 grams of rice and try again.
15. Clean up your mess.

'hink about what was happening when the bag was dangling out in front of you. Gravity was 'e bag down, but it didn't fall to the ground because the thread was holding it up. In other thread supplied a force that worked against the force of gravity. When you lifted the thread up, that increased the force with which the string was pulling up on the bag, so the bag lifted up as well. In the end, then, you were controlling where the bag was by using the string to exert a force on the bag.

What changed when you started twirling the bag around your head? The string was still pulling on the bag, but because of your twirling, the bag was moving as well. In what direction was the bag moving? Where did it move when the thread broke? It moved away from you, right? That's the direction it was moving. The thread kept that from happening, pulling the bag towards you. It wasn't pulling strongly enough to bring the bag any closer to you. It could only keep the bag from moving farther away from you. Because of this situation, the bag orbited your head.

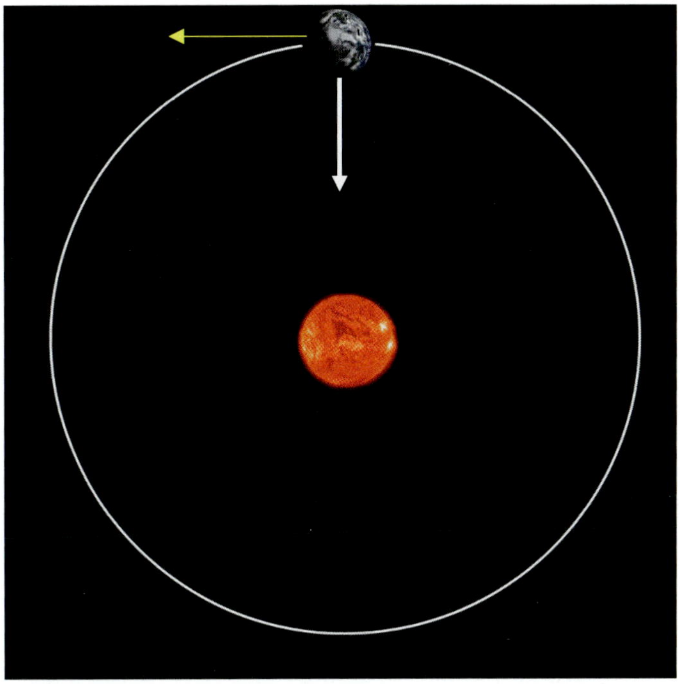

The same thing happens with the earth and the sun. The earth has a velocity that causes it to move away from the sun, as shown in the drawing on the right. However, the gravity between the earth and the sun pulls the earth towards the sun. It isn't strong enough to pull the earth any closer to the sun, but it is strong enough to keep the earth from moving any farther from the sun. Thus, the earth orbits around the sun.

The earth has a velocity that would cause it to move away from the sun. This doesn't happen because gravity attracts the earth to the sun. It isn't a strong enough attraction to pull the earth any closer, but it does keep the earth from moving any farther away.

One other thing to learn from the experiment is what caused the thread to actually break. The faster the bag moved, the more force the string had to use to keep it from moving farther away. Eventually, the force needed to keep the bag from moving away was more than the thread could supply, so it broke. What does that tell you about the planets orbiting the sun? If they are moving slowly, they don't need a lot of force to keep them in their orbit. If they are moving quickly, they need a lot of force to keep them in their orbit.

Do you remember the most important thing that affects gravity's strength? You learned about it when you learned about the tides. It's the distance between the bodies. When the bodies are close to one another, gravity is strong. When they are far from one another, gravity is weak. What does that tell you about the speed with which the planets orbit the sun? The closer the planet is to the sun, the stronger gravity is. Since that's the force keeping the planets from moving farther from the sun, the closer the planet is to the sun, the faster it can travel. We will discuss that more the next time you do science.

Comprehension Check

15.1 It takes one year for the earth to travel around the sun. Mercury is the closest planet to the sun. Does it take Mercury a longer time to travel around the sun, a shorter time, or the same time as the earth?

Our Solar System

As I am sure you are aware, the earth "shares" the sun with many other planets and, in fact, with many other bodies as well. All the things orbiting the sun make up what we call the **solar system.** To give you a general idea of what it looks like, the illustration below shows the eight planets that orbit the sun in their relative positions:

This simplified illustration shows the sun (left) and the planets that orbit it.

It's important to realize that there is no way to make an accurate representation of our solar system using an illustration in a book, so all illustrations you see in textbooks are simplified representations, as is indicated in the caption above. For example, the drawing properly shows that Mercury is closest to the sun, Neptune is farthest, and the other planets are in their correct order in between. However, the actual distances vary so dramatically that if they were shown accurately in the drawing above, the earth would have to be placed 0.2 inches from the sun. In the drawing, it is about 2 inches from the sun. Venus, of course, would have to be even closer, and Mercury closer still. Obviously, that would make those planets nearly impossible to see! Thus, the planets are spaced roughly evenly in the drawing, even though that's not at all the case in the solar system itself.

As I said, however, the order of the planets is correct, and that's worth remembering. So, you need to remember the order of the planets in terms of their proximity to the sun:

Order of the planets from the sun: **M**ercury, **V**enus, **E**arth, **M**ars, **J**upiter, **S**aturn, **U**ranus, **N**eptune

While this might seem hard to remember, you can come up with a mnemonic that will help. Notice, for example, that I have put the first letter of each planet in boldface type. I remember the order by using the phrase:

My **v**ery **e**ager **m**other **j**ust **s**ent **us** **n**achos.

The phrase is memorable to me, because it refers to the first anniversary gift my wife and I got from my mother (a gift certificate to a Mexican restaurant). You can make up any mnemonic you want, or you can just memorize the order, but you will have to know the order of the planets in the solar system.

The other aspect of our solar system that cannot be properly illustrated in a drawing like the one on the previous page is the relative sizes of the planets and the sun. In that drawing, the sun is way, way, way too small. In addition, even though it does properly show the smaller planets as smaller and the larger planets as larger, they are not drawn true to scale. If you were to illustrate the planets and the sun according to their actual sizes, you would have to make a drawing like the one you see on the right. The sun is so large that it cannot fully fit in a picture in which planets like Mercury and Mars are visible.

This illustration correctly shows the relative sizes of the sun and the planets.

In addition to the fact that we cannot easily illustrate the relative sizes of the planets and their distances from the sun in a single drawing of the solar system, there are typically a lot of things left out of such a drawing. For example, there are all sorts of other things orbiting the sun besides just the planets. For example, there are a lot of **asteroids** (as' tuh roydz) that orbit around the sun.

<u>Asteroid</u> – A small rocky body that orbits the sun

As the definition indicates, asteroids are small – too small to be considered planets. They can be found throughout the solar system, as illustrated in the drawing below, which shows the solar system from the center out to the orbit of Jupiter. Each purple and yellow dot represents an asteroid, so as you can see, there are lots of them throughout the solar system. Nevertheless, the majority of them are found between Mars and Jupiter, but closer to Mars. That region of the solar system is called the **asteroid belt**. Once again, it is difficult to illustrate things like this in a single drawing, so the asteroid belt isn't nearly as "crowded" as the illustration on the right suggests. First, while asteroids can be more than 500 kilometers (320 miles) across, many are "only" 10 meters (30 feet) or so across. Thus, most of the dots in the drawing are too big for the asteroids they represent. If they were smaller, however, you couldn't see them, so they have to be incorrectly large. Also, the drawing is flat, but the solar system isn't. Thus, there is empty space above and below each dot that cannot be illustrated in such a drawing.

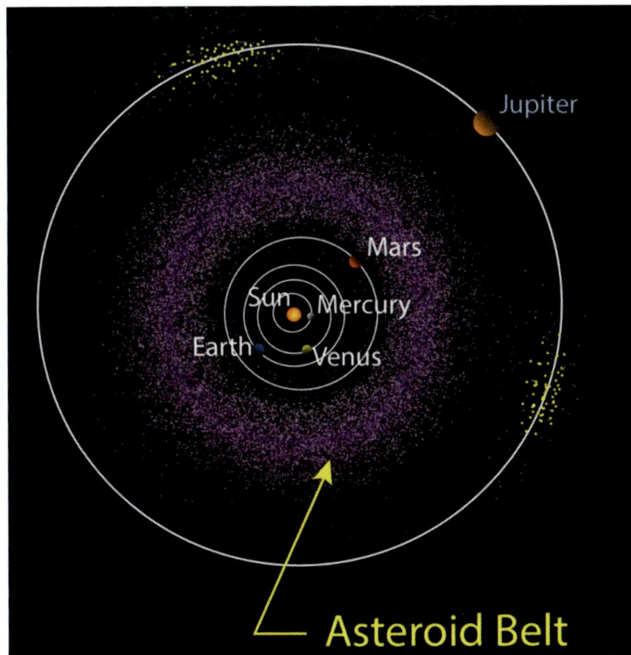

Each purple and yellow dot represents an asteroid.

Asteroids aren't the only small bodies that orbit the sun. There are also **comets**, which are quite different from asteroids. They are generally much smaller than asteroids, rarely reaching sizes of more than 750 meters (2,500 feet) across. More importantly, they have a rather different composition. While asteroids are made mostly of rocky material and some metals, comets are made of rock and ice. In fact, they are often called "dirty snowballs" to highlight that fact. Even though most of the ice in a comet is made of water, there are other frozen components to the ice. Often, there is frozen carbon dioxide. While you might find it odd to think of a gas like carbon dioxide being frozen, you have actually seen it if you have ever encountered "dry ice," which is frozen carbon dioxide.

Now remember that the planets orbit the sun in an ellipse. Asteroids do as well. However, those ellipses have low eccentricities, so they are mostly circular. The orbits of comets have high eccentricities. As a result, the difference between aphelion and perihelion is very large. In other words, they get very far from the sun during part of their orbit, but they get very close to the sun in another part of their orbit. In fact, they get so close to the sun when they are near perihelion that their ices turn into gases, which are expelled. This leads us to the definition of a comet:

Comet – A small, icy body that warms when near perihelion, losing some of its mass in the process

When the comet starts expelling gases, they tend to light up, producing a "tail" that can be quite spectacular, depending on the orbit of the comet and its composition.

The illustration on the left below shows you the path of a comet when it is near perihelion. The main body of the comet (the "dirty snowball") is called the **nucleus**. When the comet gets close enough to the sun, the ices start to turn into gases, and they get expelled from the nucleus. As they are expelled, dirt is liberated and gets caught up in the gases, being expelled from the nucleus as well. Instead of moving away from the nucleus in all directions, however, the gases move directly away from the sun. That's because the sun produces a lot of high-energy particles that produce a kind of wind (it's actually called the "solar wind") that blows the gases in the same direction. Thus, the gaseous tail of a comet is always pointed directly away from the sun. However, the dirt that is liberated is not as strongly affected by the gases, so it tends to produce a second tail that is usually easier to see. In a spectacular comet, such as comet Neowise shown in the picture to the right of the illustration, you can see both tails.

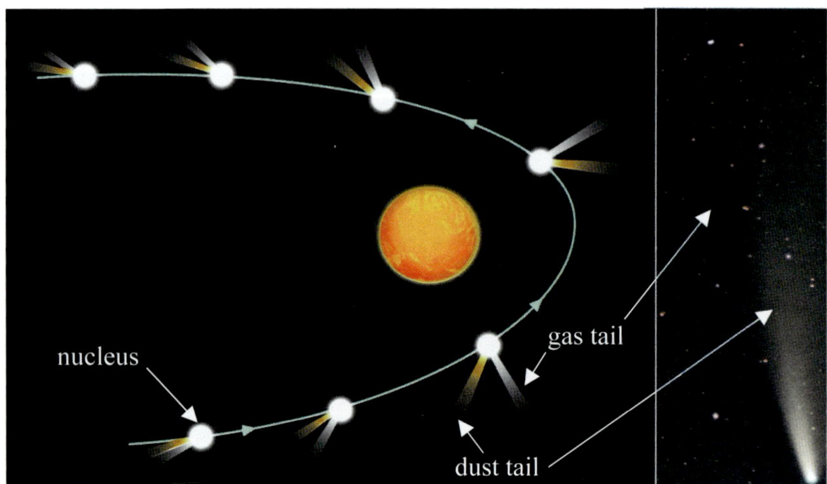

A comet develops tails when it is close to the sun. The dust tail is usually brighter, and the gas tail is the one that points directly away from the sun.

Because comets can make a spectacular sight in the night sky, they have been a curiosity to people throughout history. For example, ancient animal bones from China that were inscribed with writing more than 3,000 years ago describe comets. For most of history, comets were considered omens of terrible things to come, but as time went on, natural philosophers studied them to better understand their nature. In 1705, an astronomer named **Edmond Halley** (ha' lee) published a list of observed comets, and he noted that three of them (one observed in 1531, one in 1607, and one in 1682) behaved nearly identically in the night sky. He suggested that they were all the same comet, which took between 75 and 76 years to orbit the sun. Thus, he predicted that it should be seen again in 1758. He did not live to see it, but it appeared as he predicted, so it is now referred to as **Halley's Comet** (or Comet Halley). It has been regularly seen and recognized since then, with its latest appearance being in 1986. It is predicted to be seen again in 2061, when it once again gets near its perihelion.

The time it takes for any object to complete an orbit around the sun is called its **period**, so we can say that Halley's Comet has a period of 75-76 years (75.32 years to be more precise). It turns out that the periods of comets vary widely. Encke's Comet, for example, has a period of 3.3 years, while there is one comet whose speed and orbit indicate that its period would be 750,000 years. The difference in these periods is caused by the difference in the orbits. The farther a comet's aphelion is from the sun, the longer it takes the comet to orbit. Encke's Comet, then, is closer to the sun (even at aphelion) than is the comet whose period is estimated to be 750,000 years. Overall, however, we can separate comets into two broad groups. The first, **short-period comets**, have periods of 200 years or less. The second, **long-period comets**, have periods of longer than 200 years.

Now remember, each time a comet gets near the sun, gases and dirt are ejected from it. That means it loses mass. As a result, comets decay away with every orbit they make. If you are a uniformitarian, this presents a bit of a problem, because you need to believe that the earth and solar system are billions of years old. However, comets decay away. If a comet has a period of 200 years and can survive 100 trips around the sun, it should disappear after 20,000 years. Even a comet with a period of 750,000 years would decay away in 75,000,000 years if it could survive 100 trips around the sun. That's much less time than a uniformitarian requires for the age of the solar system.

As a result, uniformitarians must hypothesize that comets are constantly being generated from a source in the solar system. For a long time, uniformitarians taught that short-period comets were replenished from a belt of icy bodies called the **Kuiper** (ky' pur) **belt**, which exists just beyond the orbit of Neptune. The Kuiper belt is much like the asteroid belt, with its bodies orbiting the sun in a low-eccentricity orbit. For many years, uniformitarians taught that some of those bodies could be knocked out of their low-eccentricity orbits into high-eccentricity orbits, producing a short-period comet. We now know that this is not possible, because the bodies in the Kuiper belt are far too stable for that to happen to any significant extent. However, there is another set of icy bodies called the **scattered disc**, and those bodies have a more eccentric orbit. It is possible that they can be knocked into very eccentric orbits and become comets. However, we have never observed that happening, and we have observed only a few hundred scattered disc bodies, so it's not clear whether or not there are enough bodies there to allow for comets in a billions-of-years old solar system.

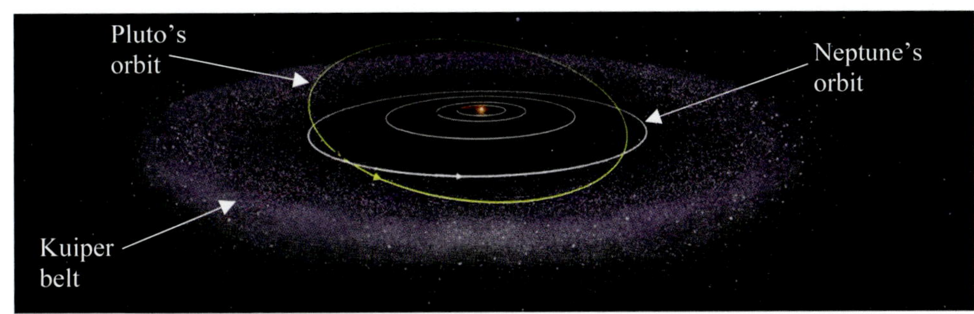

The Kuiper belt was once thought to be a source that could replenish short-period comets, but we now know that isn't correct.

Now remember, there are also long-period comets, and neither the Kuiper belt nor the scattered disc can be a source for those comets, because neither of those groups of icy bodies are far enough away from the sun to produce the long orbits required for long-period comets. Thus, uniformitarians must hypothesize the existence of yet another collection of icy bodies that can replenish the long-period comets we see. This hypothetical structure is called the **Oort cloud**, and it must be much farther from the sun than even the Kuiper belt. The problem, of course, is that we don't even know if it exists. However, it (or something like it) *must* exist if the solar system is billions of years old. If the earth and solar system are only thousands of years old, as YECs think, there doesn't have to be a large source to replenish comets. Thus, comets are much easier to understand if the earth and solar system are thousands of years old instead of billions of years old.

If you have a science book that is old, you might read about a ninth planet in the solar system. It is called **Pluto**, and its orbit is shown in the drawing on the previous page. Notice that its orbit is similar in size to Neptune's orbit, but it is tilted relative to the all the planets' orbits. While it was once thought that Pluto was a planet, scientists have discovered many other bodies that orbit the sun like Pluto and have many other characteristics in common with Pluto. As a result, scientists no longer call Pluto a planet. Instead it and several other bodies orbiting the sun are called **dwarf planets**. They are all smaller than the planets (that's why the term "dwarf" is used), and the way they orbit the sun is different from the way the planets orbit the sun. At least three of the dwarf planets are thought to have started out as members of the Kuiper belt but then were kicked out of the Kuiper belt by gravitational forces. They didn't become comets, however, because their orbits are not eccentric enough for them to get that close to the sun.

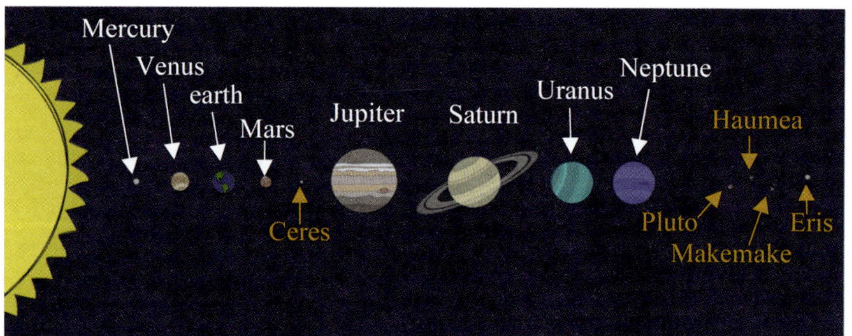

This illustration shows five dwarf planets (in gold), their positions in the solar system, and their sizes relative to the other planets.

Comprehension Check

15.2 Which planet travels faster as it orbits the sun: Mars or Saturn?

15.3 You are studying an asteroid and a comet. Which has the more eccentric orbit?

15.4 Suppose you observe a comet as it approaches perihelion, and you watch it for many nights in a row, until it is far from perihelion. How would the brightness of the comet change over that time period?

15.5 Looking at the drawing above, which of the dwarf planets almost certainly did not come from the Kuiper belt? Where did it most likely come from?

But It Doesn't Look Like That to Us!

Most likely, you have seen diagrams or animations of the solar system, so what I described to you in the previous section of this chapter probably didn't strike you as odd at all. However, it would have struck ancient people as very odd. After all, when we look up in the sky, we see the sun moving across the sky as well as up and down. From our point of view, then, the sun is moving, not the earth. In addition, when we observe the moon and the other planets in the night sky, we see that they also move. Indeed, the Greeks called them "planets" because the Greek word "*planetes*" means "wanderer," and the planets seemed to wander through the night sky. From our perspective, then, the sun and the planets move, but the earth does not.

Not surprisingly, the ancient Greeks thought that the earth was stationary, and the sun and planets moved around it. This is called a **geocentric** view, since "geo" refers to the earth, which is placed at the center of everything.

Geocentric – Having the earth at the center

A **geocentric solar system**, then, has the earth at the center, with the sun and the planets orbiting around the earth, as shown in the illustration below.

According to the ancient Greeks, the earth was surrounded by spheres made of an element found only outside of the earth. Like glass, this element was transparent, so you couldn't actually see the spheres. However, the sun, planets, and moon were each embedded in a sphere. All the stars were embedded in the outermost sphere. Those spheres rotated around the earth, causing whatever was embedded in them to move around the earth. The spheres were thought to make music as they turned. The music wasn't audible, but it could be "heard" by a person's soul, which in turn, affected the person's life. If you hear the phrase "music of the spheres," it refers to this view of the universe.

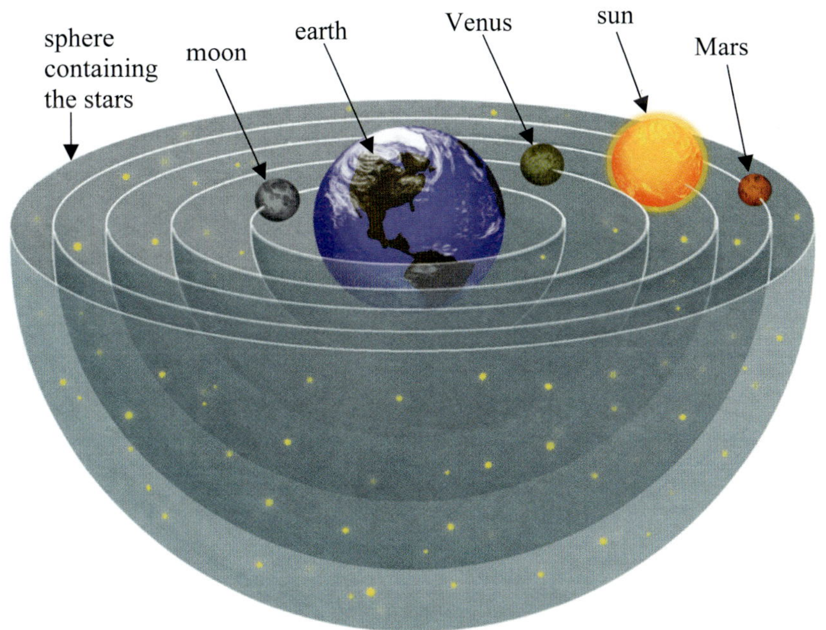

This illustration shows the ancient Greek view of the universe, with the planets, moon, and stars embedded in spheres that surrounded the earth and rotated around it, causing the motion seen in the sky. Please note that this view is incorrect.

Now, of course, such a view makes perfect sense given what we see. We see the sun, the moon, and the planets moving in the sky, so it makes sense that they move in the solar system. Also, having the earth at the center of the universe was consistent with the ancient Greek view that the earth was corrupt, but the heavens were perfect. If something started off perfect, then, it would be in the heavens. However, if it got corrupted, it would fall from the heavens. Well, if all the heavens surrounded the earth, then anything that got corrupted would fall to the earth. In other words, the earth was the heavens' garbage heap.

The ancient Greek philosopher Aristotle, and the ancient Greek natural philosopher Ptolemy were both proponents of the geocentric view and the idea that everything on earth was corrupt. They were so influential that natural philosophers strongly held to the geocentric view. Once the church became an influential voice in Europe, the geocentric view became even more well-established. The church didn't believe the earth was corrupt, like the Greeks, but it did believe that people were of special importance to God. Thus, it made sense that the earth would be at the center of the universe. Also, there are some passages in the Bible that can be interpreted as indicating that the earth doesn't move. Thus, most church leaders (both Protestant and Catholic) believed that the geocentric view was correct.

All that began to change when Nicolaus Copernicus wrote a book that was published in 1543, the same year he died. Copernicus argued that the earth could not be at the center of the universe. First, he said, the geocentric view of the universe was very messy. That's because many observations seemed inconsistent with the geocentric view, so to "fix" those inconsistencies, exceptions had to continually be made to the general rules established for the geocentric system. Since Copernicus thought the universe was made by God, who he called "the Best and Most Orderly Workman of all," the universe should not be so messy. It should work based on principles that had no exceptions. Thus, he constructed a view of the universe with the sun at the center, and everything orbiting around the sun. That is now called the **heliocentric** view, because "*helios*" is the Greek word for "sun."

Heliocentric – Having the sun at the center

Copernicus showed that the heliocentric view of the universe was more consistent with the observations that had been made, so there weren't as many exceptions to the rules. As a result, the universe was more orderly, as Copernicus would expect of something made by God.

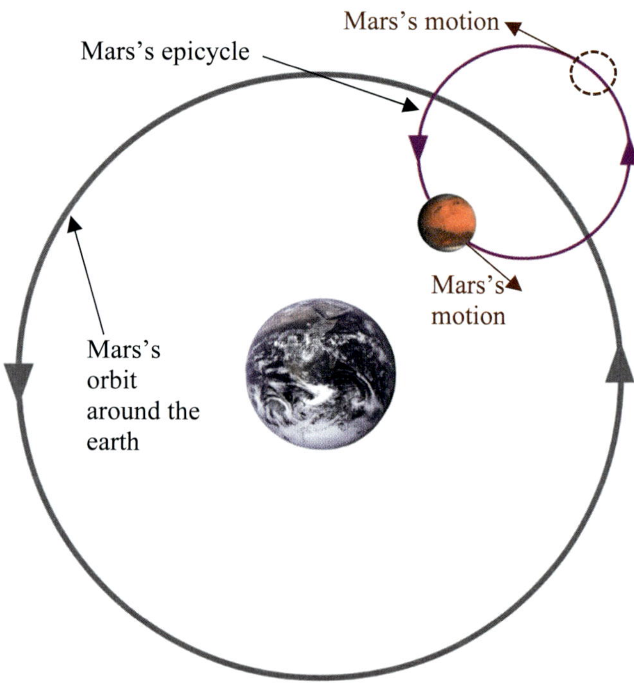

The geocentric view used epicycles to explain retrograde motion.

One of the most important observations that made the geocentric view so messy was that the planets moved in different directions depending on when you saw them. If you observed Mars over the course of several nights, you would see it moving consistently in one direction. However, if you continued watching, it would eventually turn around in the sky and start moving in the opposite direction! After a while, it would then turn back around and head back in the original direction again. This is called **retrograde motion**. If the planets orbited around the earth, this wouldn't be expected. They should always move in the direction of their orbit. However, those who supported the geocentric view "fixed" this problem by saying that the planets not only orbited the earth, but they also traveled in a circle around their orbits. That circle was called an **epicycle** (ep' ih sy' kul), which is illustrated on the left.

Notice how the epicycle can produce the retrograde motion seen in the night sky. Suppose you were on the earth looking at Mars. Where it is pictured right now, you would see it moving in the direction of the red arrow pointed down and to the right, because its epicycle makes it move that way. However, if you waited long enough, you would eventually see Mars when it is at the position of the dashed circle. At that point, it would be moving in the direction of the other red arrow, which is opposite the first red arrow! This was a great "fix" to the geocentric view, but the size of the epicycle or Mars's speed in the epicycle had to change from time to time to keep it consistent with observations that were being made. This was true of all the planets, which made the geocentric view so messy.

Now imagine what you would see in a heliocentric system. Remember that the planets move at different speeds, depending on their distance from the sun. The earth is closer to the sun than Mars, so it moves in its orbit faster than Mars does. So, there will be a time when the earth is behind Mars, but because it travels faster, it will catch up and then pass Mars. Now imagine what a person on the earth would see if he or she watched Mars for a while. A person on earth would have to look in the direction of the blue lines on the right in order to see it. Notice what happens from position 1 to position 2. Mars travels upwards in the sky. But what happens from position 3 to position 4? Mars moves down! From positions 4-7, it moves up again. Retrograde motion, then, is a natural consequence of the heliocentric view.

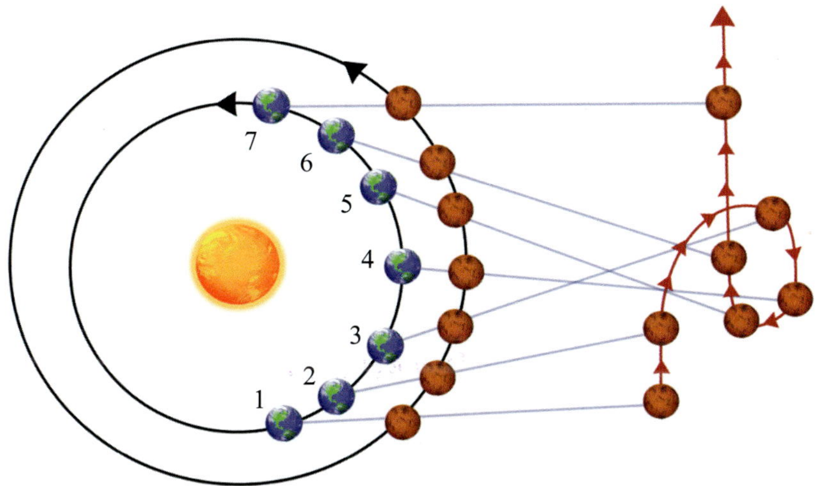

Rather than requiring a "fix" to understand it, retrograde motion (as seen on the right side of the illustration) is a natural consequence of the heliocentric view.

Now remember, all the planets move at different speeds, so in the heliocentric view, all the planets should show retrograde motion. In fact, they do. Because their speeds are different, the details of their retrograde motion vary, and what we observe is what we expect from the heliocentric view. But what about the moon? The moon orbits the earth, so its motion in the night sky shouldn't change direction. Thus, it should not have any retrograde motion, and, in fact, it does not.

Copernicus highlighted a couple of other observations that were expected in the heliocentric view but were inconsistent with the geocentric view. One had to do with the planet Venus. It is often called the "evening star," because it is so bright that it can be seen shortly after sunset. However, it can only be seen in the evening if you look in the western sky. In other words, you have to look in the direction of the sun to see Venus. Now because it is so bright, it can also be seen shortly before sunrise. Guess where you have to look to see it then? The Eastern sky, which is where the sun is rising. If you want to see Venus, then, you must look in the direction of the sun, because it seems to follow the sun.

Does this make sense in the geocentric system? No. Look at the drawing on the right. Venus and the sun travel in the sky at different speeds, so they should travel in their orbits at different speeds. So there should be times when Venus and the sun are on the same side of the earth, as shown in the drawing. During those times, Venus should be in the western sky at sunset, just as we see it. However, there should also be times when the sun is roughly in the same position, but Venus is on the opposite side of the earth, as shown with the dashed circle. During those times, you should have to look away from the sun to see Venus, so it should be in the Eastern sky at sunset. However, it never is.

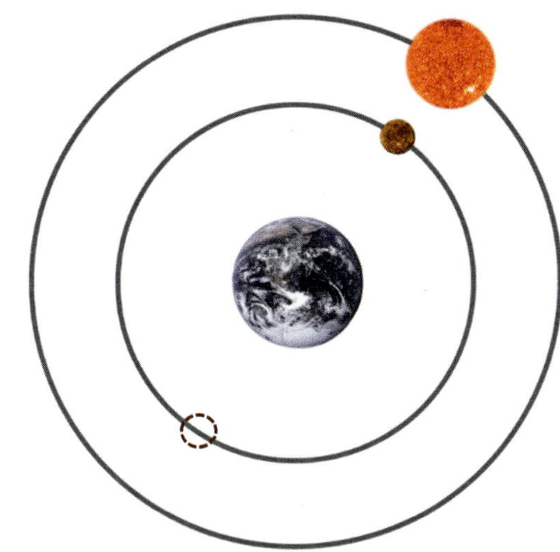
Venus shouldn't always be on the same side of the earth as the sun in the geocentric view.

One other observation Copernicus highlighted was the brightness of Mars. We see planets in the night sky because they reflect light that comes from the sun. In the geocentric system, a planet should be roughly the same distance from the earth all the time, so a planet's brightness should only depend on its distance from the sun. The closer it is to the sun, the brighter it should be, and the farther it is from the sun, the dimmer it should be. In the geocentric view, then, when should Mars be the brightest? It's when Mars and the sun are on the same side of the earth. Thus, Mars should be brightest when it is in the western sky at sunset. That's when it is near the sun in the geocentric system. It should be dimmest when it is on the opposite side of the earth compared to the sun, so it should be dimmest in the eastern sky at sunset.

Observations indicate exactly the opposite. Mars is brightest in the eastern sky at sunset, and it is dimmest in the western sky at sunset. Once again, this is completely understandable in the heliocentric system. Look at the drawing below. In the heliocentric system, the earth is closer to the sun than Mars. So imagine what happens when earth is between the sun and Mars, as shown on the right side of the drawing. At that point, Mars is close to the earth. When something is shining, it looks brighter the closer it is. Thus, you would expect Mars to be brightest when the earth is between Mars and the sun. To see Mars, you would have to look opposite the sun, so Mars should be brightest in the eastern sky at sunset. That's what we see. When the sun is between Mars and the earth, as shown in the left side of the drawing, Mars is farthest from the earth. Thus, it should be dimmest when it is on the same side of the earth as the sun. Thus, it should be dimmest in the western sky at sunset, and once again, that's exactly what we see.

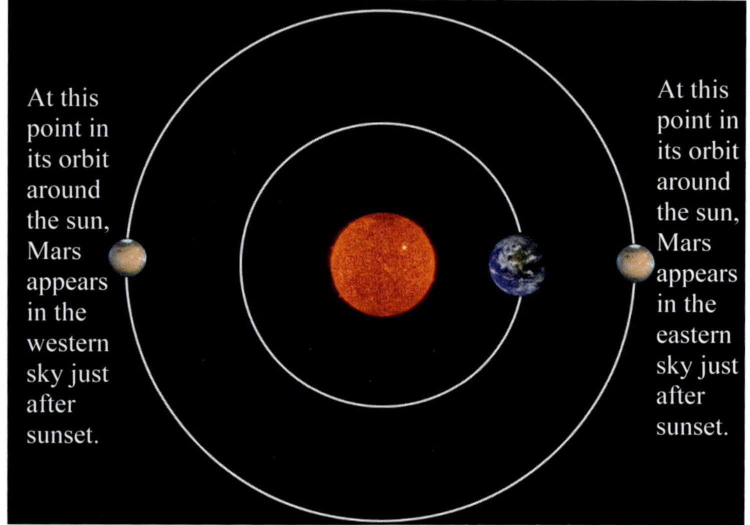

When it appears in the eastern sky after sunset, Mars is at its brightest, because it is closest to the earth. When it appears in the western sky after sunset, it is dimmer, because it is farther from the earth. This can only be understood in the heliocentric system.

When a planet is on the opposite side of the earth as the sun, we say that it is in **opposition** to the sun. Technically, then, we would say that Mars is at its brightest when it is in opposition to the sun. When a planet is in opposition to the sun, then, it appears in the eastern sky at sunset, because you have to look away from the sun in order to see it. If it is not in opposition to the sun, it is in the western sky at sunset, because you have to look towards where the sun would be in order to see it.

Since Copernicus, of course, lots of other observations have been made regarding the planets, and many of them are simply inconsistent with a geocentric solar system. However, they are all well-understood in a heliocentric solar system. Thus, from a scientific point of view, it is clear that the sun sits at the center of the solar system and the planets orbit around it. But that seems to be inconsistent with what the Bible says. For example, 1 Chronicles 16:30 says, "Tremble before Him, all the earth; Indeed, the world is firmly established, it will not be moved." Isn't that inconsistent with the heliocentric view? Not really. If you look at 1 Chronicles 16:30, this verse comes from a psalm of thanksgiving. A psalm is a song, which uses poetic language. Thus, the verse is simply a poetic way of saying that the earth will continue to do what God has created it to do. There are other verses that say the earth will not be moved, but they come from the book of Psalms (Psalm 93:1-2 and Psalm 96:10), which once again, are poetic in nature.

Now there is one place in the Bible that isn't poetic and has been interpreted to indicate that the sun is not at rest. That's Joshua 10:12-13:

> Then Joshua spoke to the Lord on the day when the Lord turned the Amorites over to the sons of Israel, and he said in the sight of Israel, "Sun, stand still at Gibeon, And moon, at the Valley of Aijalon!" So the sun stood still, and the moon stopped, Until the nation avenged themselves of their enemies. Is it not written in the Book of Jashar? And the sun stopped in the middle of the sky and did not hurry to go down for about a whole day.

Joshua is, indeed, commanding the sun to stand still, but notice that he tells it to stand still *at Gibeon* and the moon to stand still *at the Valley of Aijalon*. In addition, the verse later says that the sun stopped *in the middle of the sky*. Thus, Joshua's command doesn't imply that the sun moves *in the solar system*. It says that the sun moves in relation to Gibeon and the moon moves in relation to the Valley of Aijalon. Those are both places on the earth, and relative to them, the sun does move. In the same way, the sun does move in the sky. So these verses are simply saying that the sun moves relative to the earth, which is completely true.

In the end, then, a heliocentric view of the solar system is consistent with our observations, and it is also consistent with what the Bible says about the earth and the sun. So even though the church (both Protestant and Catholic) originally agreed with the ancient Greeks that the earth is at the center of the solar system, we now know that this idea is wrong, which shows that even the church can be wrong when it comes to science and how it relates to the Bible. This isn't because the Bible is inconsistent with science, but because man's interpretation of the Bible can be flawed.

Comprehension Check

15.6 When a theory needs to be adjusted to make it consistent with observations, scientists often say that epicycles are being added to the theory. Why?

15.7 Suppose you could stand at the surface of the sun and view all the planets. Would any of them have retrograde motion? What about the moon?

15.8 What two planets cannot be in opposition to the sun? You might want to go back to the drawing on page 436 to answer this question.

"Sampling" the Sun

I want to have a more detailed discussion of our solar system, and what better place to start than the sun? The gravitational attraction between it and every planet is what holds the solar system together, and its energy is what warms the earth, allowing life to flourish. Let's start with what the sun is made of, because it's quite different from the earth. Instead of being a rocky body surrounded by an atmosphere, the sun is a mixture of gases, mostly hydrogen atoms and helium atoms. In fact, 73.4% of the sun's mass comes from hydrogen, while 25.0% comes from helium. The remaining 1.6% of the sun's mass comes from carbon, nitrogen, oxygen, neon, magnesium, silicon, sulfur, and iron.

Now it's important to note that as of early 2021, the closest any spacecraft has been to the surface of the sun is 13.5 million kilometers (8.4 million miles). That same spacecraft, the Parker Solar Probe, is planning to get as close as 6.1 million kilometers (3.8 million miles) by the year 2025.

Nevertheless, that's not close enough to get a sample of the sun so that we can determine its composition. Fortunately, however, that's not necessary. We can determine what elements are in the sun by simply looking at the light the comes from it. How? Perform the following experiment to find out.

Experiment 15.2: A CD Spectrometer

Supplies:

- A tall, thin box, like a cereal box
- A CD or DVD that you don't mind ruining
- A serrated knife
- Scissors
- Cellophane tape
- Aluminum foil
- At least two of the following light sources: an incandescent light, a fluorescent light, or an LED light (An incandescent light is a standard bulb that heats up a filament to make light. You can use a light bulb from an uncovered lamp or an incandescent flashlight. There are compact fluorescent light bulbs that fit in a normal lamp, and some camping lanterns use fluorescent bulbs. Most cell phones have a flashlight app that's an LED light, and you can also use an LED flashlight or an LED light bulb made for lamps.)
- A room that can be made fairly dark

Instructions:

1. Make sure the cereal box is empty.
2. Turn the cereal box upside down.
3. Use the serrated knife to cut a slit into one edge of the box, starting about 5 centimeters (2 inches) from the bottom of the box. The cut should go through both sides of the box, and it should be at a shallow angle (about 30 degrees). It needs to be deep enough so that when you stick the CD in it, half of the CD will be inside the box. See the photo labelled "A."
4. Use the knife to cut a square out of the bottom of the box, right above the CD. That way, when you look into the square, you can see the part of the CD that is inside the box. See the photo labelled "B."
5. Use the knife to cut another hole in the box. This one should be on the edge opposite the CD so that you can look through the box and see the CD on the other side. It should start about 5 centimeters (2 inches)

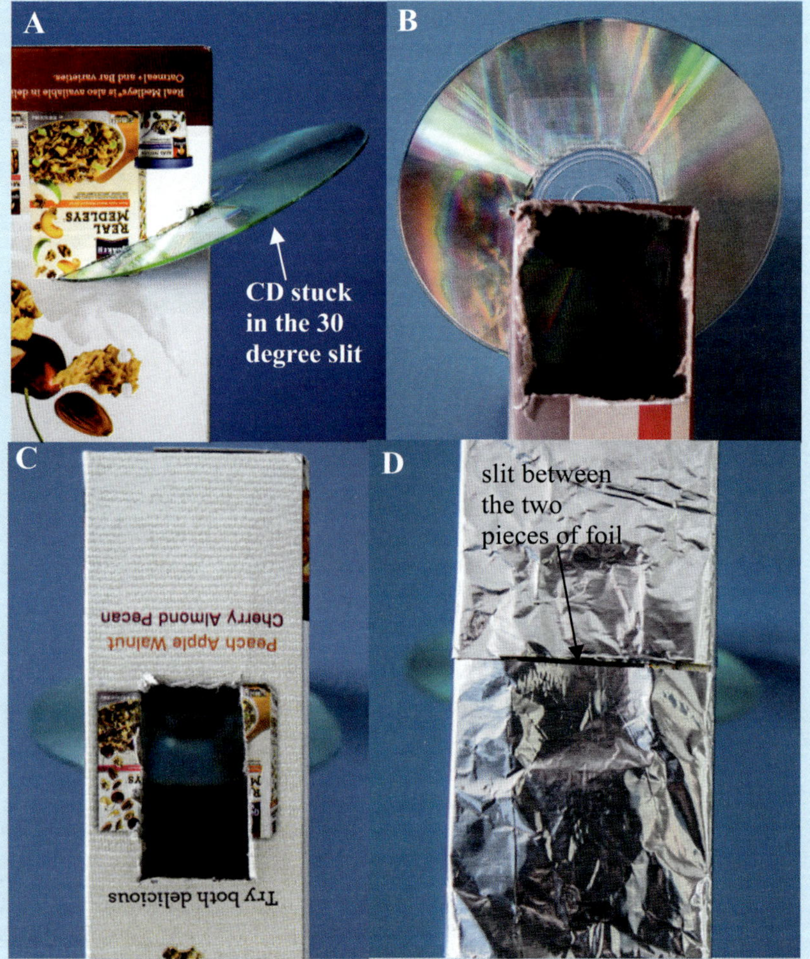

from the bottom of the box and go a bit farther down the edge of the box than the CD goes down inside the box. See the photo labelled "C."

6. Cover the hole you just cut with two pieces of foil so that there is just a tiny slit through which light can pass, as shown in the photo labelled "D." To do this, cut two pieces of foil that will each cover more than half of the hole you just cut in the box. Then tape them to the box so they cover the hole except for a tiny slit that is about the same height as the bottom of the CD that is sticking through the other end of the box. Be sure that the flattest edge of each piece of foil forms the slit, because it needs to be smooth.
7. You now have something you can use to analyze light in a specific way. You will shine a light through the slit you just made. That light will hit the CD, and you will look at the CD through the hole in the bottom of the box.
8. Take your light sources into the dark room.
9. Turn on one of your lights so that it is the only source of light in the room.
10. Hold the box so the light from the light source shines through the foil slit.
11. Look at the CD through the hole that you made in the bottom of the box.
12. Play with the tilt of the box and the position of your head until you see a nice rainbow on the CD. The blue color should be on the bottom edge of the CD. The red color should be on the top edge. If you can't see the entire rainbow at once, just move your head back and forth so that you "scan through" the rainbow on the CD. Note how the colors of the rainbow blend into one another.
13. Change the light source. If you used an incandescent light bulb, remember that it will be hot, so let it cool before you try to remove it from the lamp if you need to remove it.
14. Repeat steps 10-12 and look at the rainbow that this new light source gives you. What differences do you see now compared to what you saw in step 12?
15. Feel free to switch between the light sources a couple of times to see the difference between the rainbows.
16. Clean up your mess.

What did you see in the experiment? Most likely, the rainbows that you saw were different. If you used an incandescent light, you should have seen a smooth rainbow, with each color easily blending into the next. A fluorescent light bulb still should have produced a rainbow, but it shouldn't have been smooth. There should have been gaps in between some of the colors, and some of the colors were probably brighter than others. An LED light should have also made a rainbow with gaps.

Why were there differences in the rainbows? Well, remember that the color of visible light is determined by its wavelength. White light is a mixture of many different colors, and when the white light from each of your light sources hit the CD, those colors were separated by their wavelengths. An incandescent bulb uses a hot filament to make its light, and a hot filament produces lots and lots of wavelengths of all colors. When light from an incandescent source is split up according to its wavelengths, there is a smooth transition from one color to another.

A fluorescent light doesn't use a hot filament. The inside of the glass is painted with a chemical that glows when given energy, but it doesn't glow with as many colors as a hot filament. There is still a mixture of colors, so the light is still white, but since there aren't as many colors, the rainbow it produces has gaps in it. An LED bulb makes its light in a completely different way, and it often has even fewer colors. There is still a large enough variety for your eyes to perceive the light as white, but when it is split into its wavelengths, there are gaps between the colors. This is why a fluorescent bulb's light is not exactly the same kind of white as an LED bulb's light, and neither is the same as an incandescent bulb's light.

Now think about it this way. Once you got used to seeing those rainbows, you wouldn't have to see the bulb in order to know what kind it was. You could just look at the pattern of colors in the rainbow that each light produced, and you could determine the type of light bulb that way. Well, we can do the same thing with elements. When elements get excited, they emit light, but each element has its own pattern of wavelengths that it emits. We don't have to sample a mixture of excited elements to know what is in it. We can just analyze the wavelengths of light that are emitted, and we can determine what elements are in the mixture. This process is called **spectroscopy** (spek trah' skuh pee).

Spectroscopy – Identifying chemicals by studying the light they emit

Spectroscopy requires an instrument like the one you made in your experiment, which is called a **spectrometer**.

To determine what elements are in the sun, then, we simply do spectroscopy on the light that it is shining on the earth. Based on that spectroscopy, we can determine not only the elements that exist there, but also their relative amounts. We can do this for any star whose light we can see, so a lot of what we have learned about the stars in the night sky comes from spectroscopy.

Comprehension Check

15.9 Suppose you looked at a candle's flame using the spectrometer you made in your experiment. Would its rainbow look more like the one from an incandescent light, a fluorescent light, or an LED light?

Focusing on the Sun

The sun provides the earth with most of its energy, but how does the sun produce that energy to begin with? When you see pictures like the one below, the sun looks like a big ball of fire. However, that's not what it is. When something burns, it is chemically reacting with oxygen. There are a few oxygen atoms in the sun, but not many. In addition, they are not in the form that is necessary to participate in that kind of chemical reaction. So where does the energy from the sun come from? It doesn't come from chemical reactions; it comes from **nuclear reactions**. What's the difference between chemical reactions and nuclear reactions? In chemical reactions, atoms either exchange or rearrange their electrons. In nuclear reactions, atoms exchange or rearrange their protons and/or neutrons. Since the energy associated with protons and neutrons is much higher than the energy associated with electrons, nuclear reactions are much more powerful than chemical reactions.

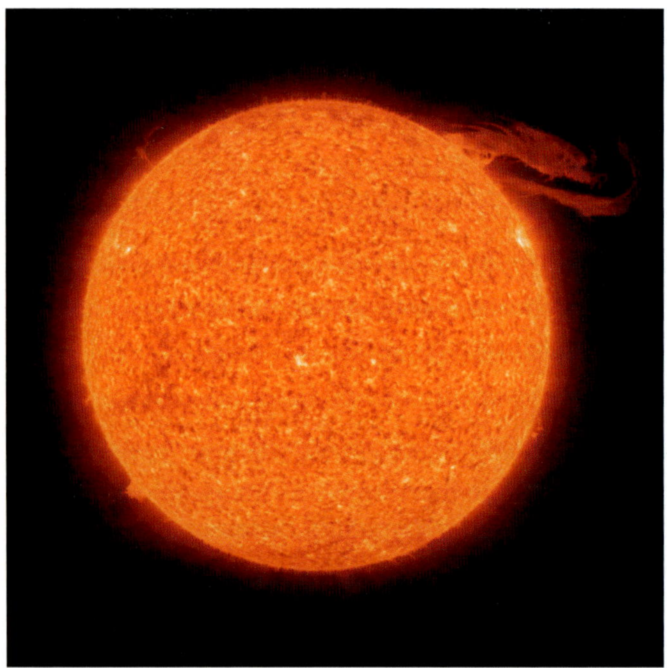

The sun may look like a big ball of fire, but it is not.

For example, I am sure that you know something about the atomic bomb. It is significantly more destructive than conventional bombs, because conventional bombs use chemical reactions to produce an explosion. The atomic bomb uses nuclear reactions to produce its explosions. The specific kind of nuclear reaction that powers a nuclear bomb is the same kind of reaction that runs a nuclear power plant: **nuclear fission**.

Nuclear fission – The process by which a nucleus with many protons and neutrons is transformed into two smaller nuclei

This isn't the kind of nuclear reaction the sun uses, however. Instead, it uses **nuclear fusion**, which is pretty much the opposite:

Nuclear fusion – The process by which two nuclei come together to form a single, larger nucleus

Depending on the nuclei, both kinds of reactions can produce enormous amounts of energy. While people have figured out how to use nuclear fission to make electricity, using nuclear fusion is more difficult. Currently, there is a lot of research being done trying to make viable electrical power plants based on nuclear fusion, but so far it has not been successful.

The nuclear fusion that goes on in the sun is illustrated in the drawing below. The "H" refers to hydrogen, and the "He" refers to helium. If a nucleus has one proton, it is a hydrogen nucleus. If it has two protons, it is a helium nucleus. The superscripts to the left of the letters count the number of protons and neutrons in the nucleus. A ^{1}H nucleus, then, is made of only one proton, while an ^{2}H nucleus has one proton and one neutron. In the same way, ^{3}He has two protons and one neutron, while ^{4}He has two protons and two neutrons.

In the sun, gravity pushes two protons together so strongly that they merge. The two-proton system isn't stable, however, so one of the protons turns into a neutron. That makes an ^{2}H nucleus. In the process, two other particles, a positron and neutrino, are made. Don't worry about those particles. Just follow what happens to the nuclei. The ^{2}H nucleus that was just made will then be pushed into another proton, resulting in an ^{3}He nucleus and a gamma ray (which is a high-energy electromagnetic wave). This happens over and over again in the sun, but the illustration shows it happening twice, producing two separate ^{3}He nuclei. Those two ^{3}He nuclei are pushed together, and two of the protons fly out of the system. That leaves two protons and two neutrons that join together to form a ^{4}He nucleus. There are a total of six protons involved in the reaction, but two protons are made in the end. Thus, the overall result is that four ^{1}H nuclei react with one another, forming a ^{4}He nucleus.

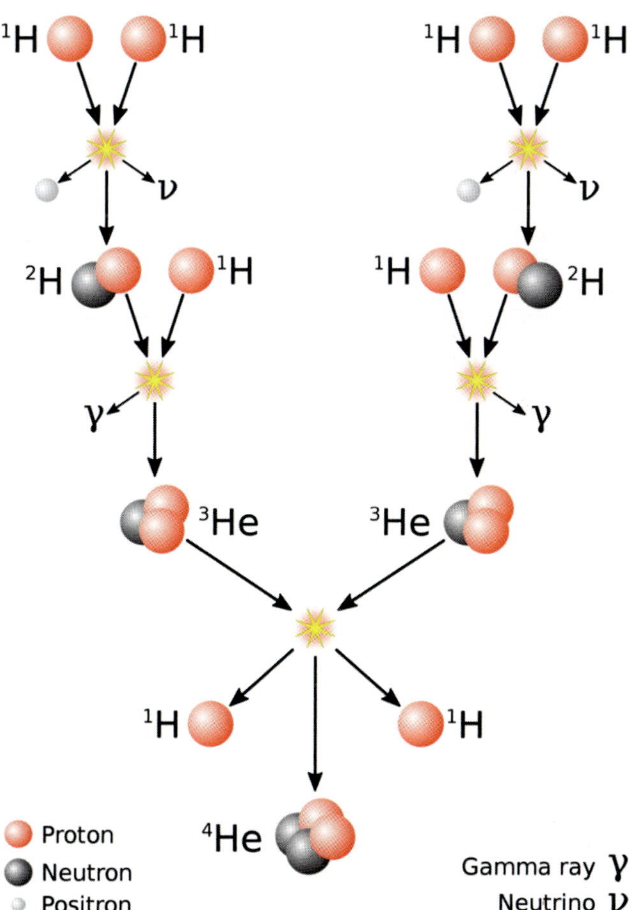

This illustration shows how four protons react to form a helium nucleus in the sun.

Now think about this process for a moment. Protons are positively charged, right? What happens when two positively-charged particles are brought close to one another? They repel each other, because that's what like charges do. This is why nuclear fusion is so hard to master. The protons must get very, very close to one another to join together, but they resist doing that, because they repel each other. Thus, *a lot* of energy has to be used to force the protons close enough to start this process. In the same way, when the ^{2}H and ^{1}H are pushed together, they repel one another, as do the two ^{3}He nuclei. Thus, each step of this process takes *a lot* of energy. In the end, the process releases more energy than it takes to make all the steps happen, so overall, energy is produced. However, without the energy needed to push all those positively-charged nuclei together, it simply can't happen.

This is what makes nuclear fusion so hard to master. Scientists haven't been able to find an efficient way to push the nuclei together. As a result, a lot of energy is wasted just making the process happen. If we could figure out how to get rid of the wasted energy, nuclear fusion would be a great way of producing energy in power plants. Unfortunately, there is so much wasted energy in human-made fusion systems that we end up putting in more energy than the process makes. Until this wasted energy problem is fixed, nuclear fusion will not be a viable way to generate power for people to use.

Now, of course, the sun has no problem with wasted energy. It is made up of an enormous amount of gas, which produces a lot of gravitational attraction. As a result, the gases on the outside of the sun are pulled in towards the center of the sun, which is called the **core**. The gases being pulled to the center of the sun produce so much pressure in the core that the steps of the fusion process I just discussed happen easily. That causes a lot of energy to be released by the core. As a result, the core has a temperature of about 15 *million* °C (28 million °F)!

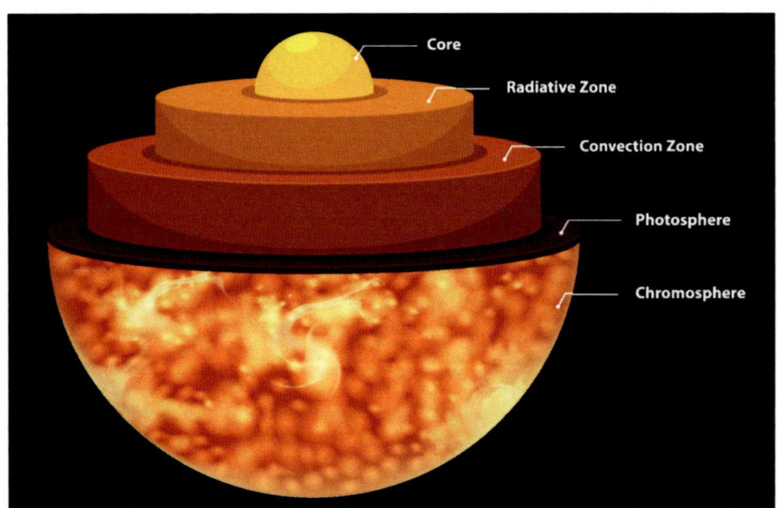

This illustration shows the major layers in the sun.

The energy produced by fusion in the core is mostly in the form of electromagnetic waves, and they start traveling outward towards the surface of the sun. Leaving the core, they then enter the **radiative zone** of the sun. The pressure isn't strong enough here to force nuclear fusion to occur, but there is still a lot of gas there, so the electromagnetic waves bounce around in the radiative zone, but they do eventually make their way out to the **convection zone**.

The gases that are near the bottom of the convection zone absorb the electromagnetic waves that reach them, heating them up. What does hot gas do? It rises. So the heated gases rise from the bottom of the convection zone to the top. When they reach the top of the convection zone, they cool off by emitting light. Once they cool, they begin to sink. Thus, the convection zone has currents of hot gases rising and cool gases sinking, much like the convection currents in the mantle. That's why it is called the "convection zone." Now remember, the gases cool by emitting light and that's the light we see coming from the sun. As a result, the region above the convection zone is called the **photosphere**. It is also referred to as the surface of the sun, and it has a temperature of about

5,800 °C (10,000 °F). Notice that the surface of the sun is much cooler than the core. That should make sense. The farther you travel way from a heat source, the cooler it gets.

Here's the part that doesn't make sense: There is a collection of gases above the sun's surface called the **chromosphere** (kroh' muh sfear). You can think of it as the sun's atmosphere. It is actually hotter than the surface of the sun, with a temperature of about 20,000 °C (36,000 °F). Scientists aren't sure why this is the case, and while there are some possible explanations, none of them have been fully supported by observations, at least not as of the time this book was published. The last major layer of the sun (not shown in the illustration) is the **corona**. It is even hotter than the chromosphere, which is once again not fully understood. It is composed mostly of charged particles and extends millions of kilometers beyond the surface of the sun.

The photosphere of the sun is always changing due to variations in how energy from the core reaches it. As a result, dark spots appear, move, and vanish on the photosphere. Not surprisingly, they are called **sunspots**. They are darker than the rest of the photosphere because they are cooler and are therefore emitting less light than their surroundings. The picture on the right gives you an interesting view of the sun's photosphere when the planet Venus was in between the sun and the earth. In that position, it blocked the light that would normally come from the sun, so it showed up as a black disk. The other dark patches are sunspots. Notice that they vary in size, but some of them are nearly as big as Venus!

This picture shows Venus blocking a portion of the sun's photosphere along with several sunspots.

Even though sunspots represent cooler areas on the photosphere, they are made when the sun is very active, so the more sunspots there are, the more energy the sun is producing and sending to the earth. As a result, the more sunspots there are, the warmer the sun is, and the fewer sunspots, the cooler it is. Do you remember the Little Ice Age discussed in chapter 12? Astronomers during that time period saw virtually no sunspots on the photosphere, which would mean the sun was cooler during that time period. Climate scientists disagree about how much of the Little Ice Age was caused by the fact that the sun was cooler, but it makes sense that a cooler sun was at least partially responsible for the earth being abnormally cool during that same time period.

Ancient Chinese astronomers had documented the existence of sunspots, but a German medical student named Johannes Fabricius first published a detailed analysis of them in 1611. That analysis was very important, because most natural philosophers taught that the sun was a perfect, unchanging body. Remember, the ancient Greek philosophers put the earth at the center of the universe because the heavenly bodies (like the sun) were perfect, while the earth was corrupt. That thinking was still guiding scientific study in the early 1600s, even though some natural philosophers at the time, like Galileo, were strongly arguing against geocentrism. When Fabricius showed that sunspots "blemished" the surface of the sun and that they changed over time, it caused more natural philosophers to question the assumptions made by the ancient Greeks, including the assumption of geocentrism.

Of course, we now know that the sun is neither perfect nor unchanging. Not only are there sunspots on the photosphere, but some of those sunspots actually lead to violent eruptions! Remember, a sunspot is a region where the photosphere is cooler. That means less energy is reaching the photosphere, so there is more energy "pent-up" beneath it. Sometimes, that "pent-up" energy rises to the surface quickly, causing a **solar flare**.

Solar flare – An eruption of intense, high-energy radiation from the sun's photosphere

Solar flares are the most powerful explosions that happen in our solar system. Most of the energy leaves the sun as electromagnetic radiation. If it is released in the direction of earth, it reaches the planet in eight minutes, which is the time it takes light to travel from the sun to the earth. Powerful solar flares can throw so much electromagnetic radiation at the earth that it disrupts radio communications. In September of 2017, for example, three hurricanes developed in the Atlantic Ocean. During one of the mornings where emergency teams were trying to coordinate their efforts to aid people being affected, a solar flare knocked out their ability to communicate for as long as eight hours, depending on the equipment that was being used. Four days later, another solar flare disrupted communications for up to three hours. Those disruptions hampered the relief efforts. The picture on the left shows a solar flare erupting from the photosphere.

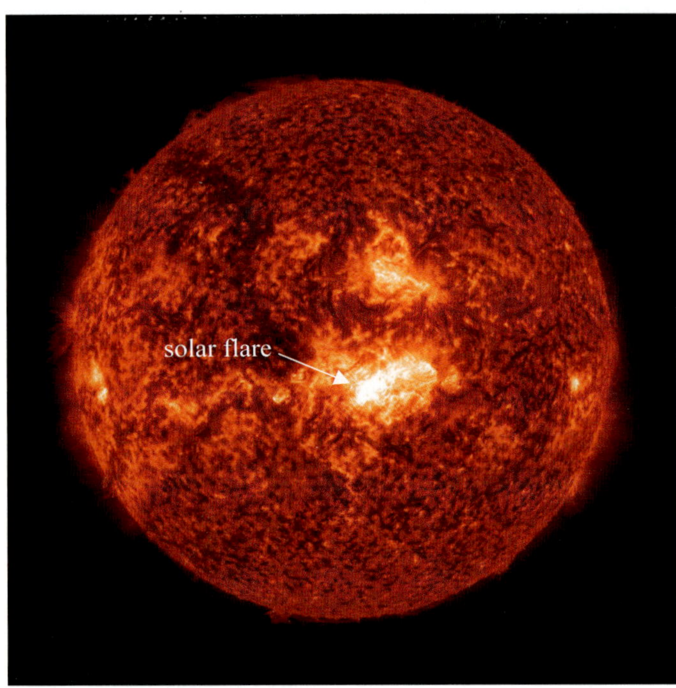

This picture shows a solar flare on the sun's photosphere.

Strong solar flares can cause another kind of eruption. Remember that the sun's corona contains very hot charged particles. A solar flare can cause those charged particles to be propelled away from the sun. This "blast" of charged particles is called a **coronal mass ejection (CME)**. Since CMEs are composed of particles, they travel slower than the electromagnetic waves that come from a solar flare. As a result, it can take up to three days for them to hit the earth. When they do, they can cause a lot more damage than the energy from a solar flare. They can disrupt power distribution, satellites, and satellite-based communication. In 1989, for example, a CME hit the earth and caused six million people in Canada to be without power for nine hours. If you look back at the image of the sun on page 448, you will see what looks like long clouds extending from the upper right part of the sun. That's a CME.

Comprehension Check

15.10 In a nuclear reaction, more nuclei are produced than what was used to get the reaction going. Is this nuclear fission or nuclear fusion?

15.11 Over time, what do you expect to happen to the amounts of hydrogen and helium in the sun?

15.12 Since 1990, the average number of sunspots on the photosphere has been decreasing. What do you think happened to the average number of solar flares over the same time period?

More Details About the Sun

Now if you think about it, the sun has two opposing forces working on it. Gravity is pulling all the gases that make up the sun towards the center. That produces the pressure that is strong enough to push protons together in the core and cause nuclear fusion to happen. So why don't all those gases get pulled to the center, making the sun much smaller than it is? Well, remember that nuclear fusion is happening in the core, and that produces a lot of energy, making the core really hot. What happens to gases when they get hot? They expand. So, gravity is pulling in on all the gases, which makes the sun smaller, but the heat generated by nuclear fusion is causing the gases to expand, which makes the sun bigger. What happens as a result? Perform the following experiment to get an idea.

Experiment 15.3: Hydrostatic Equilibrium

Supplies:
- A plastic bottle with a lid, like the kind water comes in (The thinner the plastic, the better.)
- A plastic straw
- A glass that is taller than a juice glass (See the picture labeled "B" below.)
- A few paper clips
- Scissors
- Water

Instructions:
1. Fill the glass mostly full of water.
2. Fill the bottle completely full of water.
3. Fold the straw at the middle so that it forms a "V."
4. Cut the straw about 2.5 centimeters (1 inch) from the bottom of the "V."
5. Pull the center loop of the paper clip out so that it forms a "V" as well.
6. Push the two ends of the paper clip into the two ends of the V-shaped piece of straw so that the paper clip keeps the two ends of the straw close to one another, as shown in the picture labeled "A."
7. Connect two unbent paper clips together and hang them from the paper clip that is holding the ends of the straw together, as you see in the pictures labeled "B" and "C."
8. Put the straw/paper clip contraption in the glass of water. It should just barely float, with just a little bit of straw's "V" poking above the surface of the water, as shown in the picture labeled "B." If it sinks, remove one of the hanging paper clips. If a lot of the contraption is above the water, add a third paper clip. If a paper clip touches the bottom of the glass, get a taller glass.
9. Once you have a contraption that floats as shown in the picture labeled "B," put it in the bottle that is full of water and screw on the lid so the seal is airtight. Your setup should look now like the picture labeled "C."

10. Hold the bottle in one hand and position it so you can easily see the contraption through the bottle.
11. Squeeze the bottle with the hand that is holding it. What happens to the contraption? It should sink when you squeeze. If water comes out the top, that means the lid isn't on tight enough. Refill the bottle and try again. If you don't see the contraption sink, squeeze as hard as you can. You won't break the bottle. If you can't get it to sink, add a paper clip to the bottom and try again.
12. Relax your grip on the bottle so that you are still holding it but not squeezing it. The contraption should float again.
13. Play with the system a bit, learning how hard you need to squeeze to just start it sinking and how much you need to relax your grip to make it float.
14. Try adjusting your grip so that the contraption stays suspended in the middle of the bottle, neither sinking to the bottom nor rising to the top. You should eventually be able to get it to stay anywhere in the bottle that you want it to stay.
15. Clean up your mess.

In your experiment, you made what is called a **Cartesian diver**, because it was a toy that is supposed to have been made by a famous philosopher named René (ren' ay) Descartes (day kart'). Based on what you already know, you might have figured out that when you squeezed the bottle, you changed the density of the contraption. Specifically, the squeeze you gave to the bottle was transmitted to the V-shaped straw, and the straw got smaller. When volume goes down, density goes up, which means that when you squeezed hard enough, the contraption's density got larger than the density of water, so it sank. When you relaxed your grip, the straw expanded, increasing the volume and lowering the density until it became lower than that of water, so it floated again.

Rather than thinking about density, however, I want you to think about this situation in terms of the forces involved. Without water in the bottle, the contraption would immediately fall to the bottom, because gravity is pulling it downwards. With water in the bottle, it doesn't fall. What does that tell you? The water must be pushing up on the contraption, overcoming the force of gravity. The force with which water is pushing on the contraption is called the **buoyant** (boi' yunt) **force**, and anything that is in water experiences this upward force. The larger the volume of the object, the larger the buoyant force it experiences.

So even though it is 100% correct to say that something floats in water when it is less dense than water, the *reason* that happens is because the force with which gravity is pulling the object down is less than the force with which the water can push it up. Thus, the buoyant force "wins," and the object doesn't sink. On the other hand, when an object is denser than water, the force with which gravity pulls it down is greater than the buoyant force. Thus, gravity "wins," and the object falls down through the water. When you squeezed on the bottle, you changed the volume of the contraption, which reduced the buoyant force that water could use to push it up. As a result, gravity "won," and the object sunk. When you relaxed your grip, you increased the contraption's volume, which increased the buoyant force, so the buoyant force "won," and it floated.

But what happened when you were able to get the contraption sitting in the middle of the bottle, neither falling toward the bottom nor rising toward the top? At that point, the force with which gravity was pulling on it *equaled* the buoyant force. As a result, neither force could "win," and the object simply stayed wherever it was when that balance of forces was achieved. Physicists call this balancing of forces an **equilibrium** between the forces.

Believe it or not, something very similar happens in the sun. Gravity is pulling all the gases in the sun together, which reduces the size of the sun. However, the heat being generated by fusion is pushing the gases away from each other, which increases the size of the sun. At some point, the force with which gravity is pulling will equal the force with which the heat is pushing, and the sun's size will not change. This is called the **hydrostatic** (hi' droh stat' ik) **equilibrium** point of the sun.

<u>Hydrostatic equilibrium</u> – The condition in which opposing forces in a fluid are balanced

But wait a minute. Isn't the sun made of gases? What's the word "fluid" doing in there? It turns out that gases can behave like very thin liquids, so physicists use the word "fluid" to refer not only to liquids, but also to gases that are behaving like thin liquids.

Now here's the really cool part. The sun's hydrostatic equilibrium is self-correcting. As hydrogen is fused to make helium, the gases in the sun become more compact. This actually increases the gravitational force pulling them together. However, as that force increases, the pressure pushing the remaining hydrogen atoms in the core together gets greater. That means more fusion happens. Of course, more fusion means more heat, which reestablishes hydrostatic equilibrium! As a result, the size of the sun stays pretty constant.

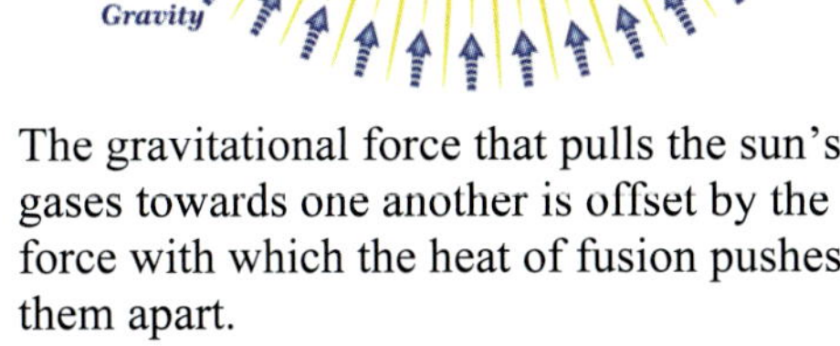

The gravitational force that pulls the sun's gases towards one another is offset by the force with which the heat of fusion pushes them apart.

Now if you think about it, while the sun's size stays relatively constant, the amount of light it produces actually increases over time. In other words, as time goes on, the sun slowly gets hotter. This is opposite from a fire, which gets cooler once it starts using up its fuel. As the sun uses up its hydrogen fuel, it actually gets hotter. The effect is small, so it is hard to see over a "short" timespan of a few hundred years. However, over a long time, it would become very important.

This presents a problem for uniformitarians, who think that life formed roughly 3.5 billion years ago. If we look at the fusion that would have been going on in the sun back then, it would have produced only about 75% of the light it produces now. That would result in an earth that is simply too cold to support life. Furthermore, since the water on earth would be ice, its albedo would be very high, which would make the earth even colder! According to uniformitarians, this high albedo would keep the earth far too cold for life, even after the sun got warmer. This is called the **young, faint sun problem**, and uniformitarians have been working to solve it for almost 50 years. So far, none of their proposed solutions are consistent with the data. Young-earth creationists don't have this problem, because they think the sun is only thousands of years old, and the change in the sun's output is too slow for it to be important over that time frame.

However, young-earth creationists have a problem of their own with the sun. According to what we know, the light generated in the core takes a long time to make its way to the photosphere, because of the way it interacts with the gases it travels through. Calculations indicate that it would take about 100,000 years for light to go from the core to the photosphere. If the sun is only a few thousand years old, how are we seeing any light from it at all? Most young-earth creationists simply say that when God created the sun, He made it fully functional, which means there was already light at the photosphere. That's a reasonable solution, but it indicates that the light we are seeing now was not actually made in the sun's core. Instead, it was made miraculously by God.

Comprehension Check

15.13 Suppose something happened that caused lots of gases to be added to the sun. What would happen to the amount of light coming from the sun?

The Moon

I have already discussed the moon to some extent, because it produces the largest effect on the ocean's tides. However, there are other things you need to know about the moon. First, you need to know that "moon" can also be used as a general term referring to any natural **satellite**.

Satellite – A small body that orbits a larger body

The planets are, in fact, satellites of the sun, since they are smaller than the sun and orbit it. The general term "moon," then, refers to a natural body that orbits a larger body. When you and I use the term "satellite," we are usually talking about a piece of technology that is put in orbit around the earth. Technically, that's an *artificial* satellite.

What we call "the moon" is earth's only natural satellite. It orbits the earth just like the earth orbits the sun, and for exactly the same reason. The moon is moving so that it should travel away from the earth. However, because gravity attracts the moon to the earth, the moon keeps getting pulled towards the earth. The pull isn't strong enough to bring it to the earth, but it is strong enough to keep it from getting farther away from the earth. Why doesn't the moon orbit the sun like the planets do? It is much closer to the earth, so it is more strongly attracted to the earth than it is to the sun.

Mercury and Venus don't have any moons, as far as we know. Why? Because they are closer to the sun. Remember, not only does gravity get stronger the closer two objects are, but it also gets stronger the more mass the objects have. Distance is more important than mass, but both play a role. Mercury and Venus are close enough to the sun that the sun's large mass overcomes the effect of distance, so anything that could orbit Mercury or Venus will be more strongly attracted to the sun. As a result, it will be pulled to the sun, not to Mercury or Venus. The earth is far enough from the sun that the sun's mass isn't large enough to overcome the effect of distance.

In this picture, taken with a telescope, the faint dots are the four largest moons that orbit Jupiter.

The other planets are even farther from the sun, and they all have moons. In fact, they have more than one. Galileo was the first to discover moons orbiting another planet. He built a telescope based on a design he had heard about, and he studied the planets and moon extensively with it. When looking at Jupiter, he saw four smaller bodies, and as he studied them over time, he saw that they orbited around Jupiter. This was astonishing, since no one thought any of the other planets had moons, and most people at that time thought that everything orbited around the earth. However, those four bodies clearly orbited around Jupiter, which meant that Jupiter had moons. Galileo also used it as evidence against geocentrism, since it showed that smaller bodies orbited bigger bodies, and the earth is much smaller than the sun. While Galileo is best known for discovering the moons, another natural philosopher, Simon Marius discovered them at roughly the same time, and he named them after lovers of the Roman God Zeus, who is the equivalent of Jupiter in Greek Mythology. Those lovers were Io, Europa, Ganymede, and Callisto, and those are the names that stuck.

Those four moons were just the tip of the iceberg. Based on current observations with modern telescopes, Jupiter has at least 79 moons! Saturn has even more: 82. Uranus has 27 and Neptune has 14. For planets other than the earth, Mars has the fewest moons, with only two. Despite the fact that moons are quite common throughout the solar system, there is something very special about the earth's moon – it is very heavy. Three of Jupiter's moons (Ganymede, Callisto, and Io) are heavier than the earth's moon, as is one of Saturn's moons (Titan). However, Jupiter and Saturn are much, much more massive than earth. If we compare the mass of the moon to the mass of earth, the moon has about 1% of the earth's mass. Ganymede, the heaviest of Jupiter's moons and the heaviest moon in the solar system, has only 0.008% of Jupiter's mass. Relative to the planet it orbits, then, the earth's moon is by far the heaviest!

It turns out that this is incredibly important. Remember, gravity's strength depends on the masses of the objects involved as well as the distance between the objects. Well, as the earth travels around the sun, the other planets are traveling around the sun as well. They are all traveling at different speeds, so the distance between the planets is constantly changing. For example, sometimes Venus is on the same side of the sun as the earth. This causes the planets to be closer together, which increases their attraction to each other. As a result, Venus's pull on the earth is harder. Other times, the two planets are on opposite sides of the sun, so they are farther apart, which means Venus's pull on the earth is weaker. The same thing happens with each planet in the solar system, so all the planets are constantly pulling on earth, but those pulls are always changing.

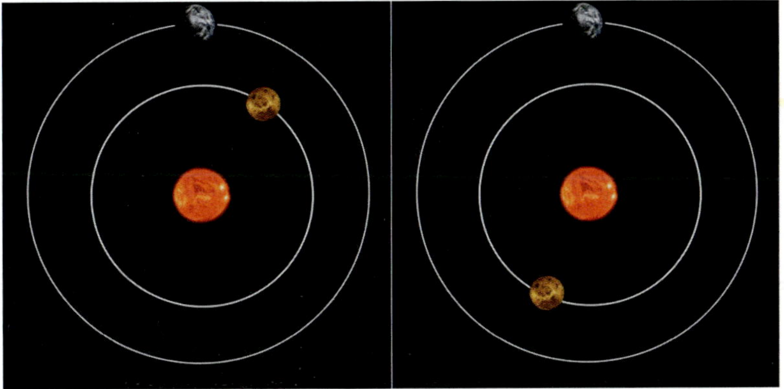
When Venus is closer to the earth (left), it pulls on the earth harder than when it is farther from the earth (right).

If that were the end of the story, the earth's axial tilt would change dramatically over time. Remember, the earth has an axial tilt of about 23.5 degrees, which is responsible for the seasons, how the winds form, etc. If the axial tilt were to vary significantly, it would dramatically change the climate, and the more dramatically the tilt changed, the more dramatic the climate changes would be. According to calculations, the varying pulls from the other planets could cause the earth's axial tilt to change by about 85 degrees. That means it could go from being hardly tilted at all to being flipped on its side! That would cause devastating climate changes.

Why doesn't this happen? Because the moon *stabilizes* the earth. Its gravitational force is strong enough to counteract most of the variation caused by all the different planetary pulls. As a result, the earth's axial tilt can only change by a few degrees, which isn't enough to cause much climate change. Now remember, the gravitational force depends on distance and mass. If the moon weren't so heavy compared to the earth's mass, or if it didn't have the right orbit, it wouldn't be able to stabilize the earth much. Thus, the reason the earth's climate stays relatively constant is because the moon has the right mass and the right orbit! This is another of the many indicators that the earth was designed by God specifically to be a haven for life.

Of course, the most obvious thing about the moon is that it changes in appearance over the course of about a month. Sometimes, the entire side of the moon facing the earth is illuminated. Sometimes, only a thin crescent is visible. These are referred to as **phases** of the moon, and they happen because the moon reflects the sun's light. As a result, the amount of the moon that we see lit up depends on the relative positions of the moon, earth, and sun.

Look at the diagram below. It shows the moon orbiting the earth. In the diagram, the sun is shining on both the earth and the moon from the right. Thus, the right side of the earth is lit up, because that's the part being hit by the sun's light. Now look at the moons embedded in the circle that represents the moon's orbit around the earth. Once again, in each case, the right side of the moon is lit up, because that's the part the sun's light is hitting.

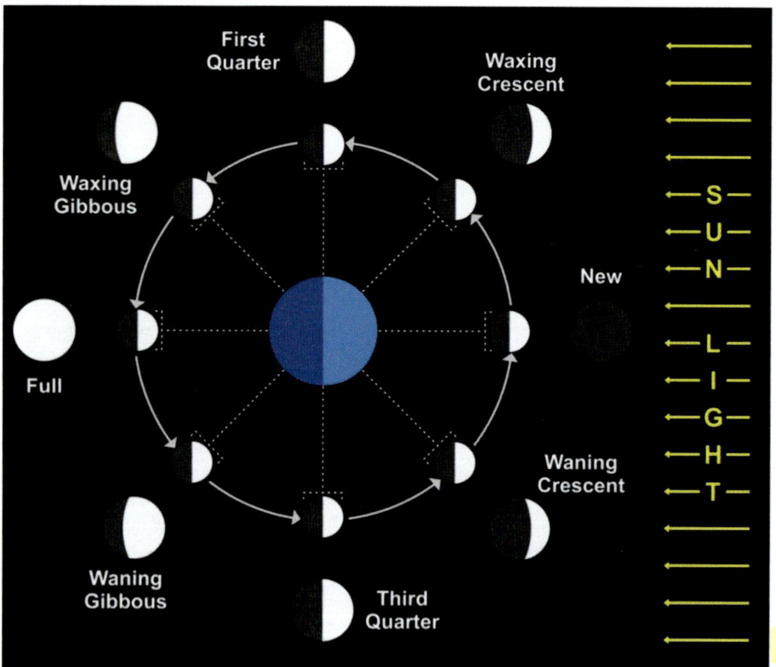

In this drawing, the blue ball at the center is the earth, and the sun is shining from the right. The white circle with arrows shows the moon's orbit around the earth. The gray and white balls that are in the orbit represent the moon at different points in its orbit. The gray and white ball beyond the orbit represent how the moon would appear to someone looking at it from the earth.

Now look at the drawing of the moon directly to the left of the earth. The right side of the moon is lit up, since that's where the sun's light is hitting it. A person looking up at the moon would be looking along the dashed line that goes from the earth to that drawing of the moon. Since the moon's right side is what the person sees, the person sees the entire moon lit up. In other words, the person sees the white circle of light labeled "full." That's called a **full moon**, and it happens when the side of the moon that is being illuminated by the sun is facing the earth.

Now look at the drawing of the moon directly above the earth. Once again, its right side is lit up, but a person looking at the moon won't see all that light. His or her gaze must be directed along the dashed line pointing to that drawing of the moon, so he or she sees only half of the moon lit up; the other half looks dark. That's called a **quarter moon**, and it happens when the relative positions of the earth, moon, and sun allow you to see only half of the sun's light reflecting from the moon. You might wonder why that's not called a "half moon." Well, at any given time, only half the moon is illuminated, so if you are seeing only half of the light, you are seeing one quarter of the moon illuminated. Since there are two times an observer from earth sees one quarter of the moon illuminated (the second time is represented by the drawing directly below the earth), there are two quarter moons.

Now look at the drawing of the moon directly to the right of the earth. Once again, half of it is illuminated, but a person looking at the moon would be looking along the dashed line going to the drawing. He or she would see *no light* at all. Thus, the moon would be very, very hard to actually see.

That's called a **new moon**. The other drawings represent other phases, but you don't need to get bogged down by their names. Just understand why the moon goes through these phases, and just remember the relative positions of the earth, moon, and sun for the full moon, new moon, and the two quarter moons. It takes just over 27 days for the moon to make one orbit, but the earth moves during that time, so it takes just over 29 days for someone to see the moon go through all its phases.

It's important to note that because all planets in the solar system reflect light from the sun, they all go through phases. However, the patterns of those phases are more complicated than the moon's phases, since the earth is moving relative to the other planets. Even so, when Galileo studied Venus with a telescope, he saw its phases, and he saw that they could be understood well in terms of heliocentrism but were incompatible with geocentrism. As a result, he used them as evidence for heliocentrism.

I told you previously that the moon's rotation on its axis is synchronized with its orbit around the earth. As a result, the same side of the moon is always facing the earth. However, that doesn't mean the moon looks the same everywhere on the earth. Look at the pictures of the moon below. The one on the left is not quite full, but the one on the right is. Disregarding the slight difference in phase, can you see the other big difference? Look at the structure pointed out in both pictures. It is a big crater that was named after **Tycho** (tee' koh) **Brahe** (brah' he), an important early astronomer. Notice that in the image on the left, the crater is on the upper right portion of the moon, but on the right, it is in the lower left. Once you see this, you should also see that the moon on the left seems upside down compared to the moon on the right. Why? The moon doesn't rotate in a way that would make that kind of change, and its rotation is synched with its orbit, so rotation has very little effect on how we see the moon.

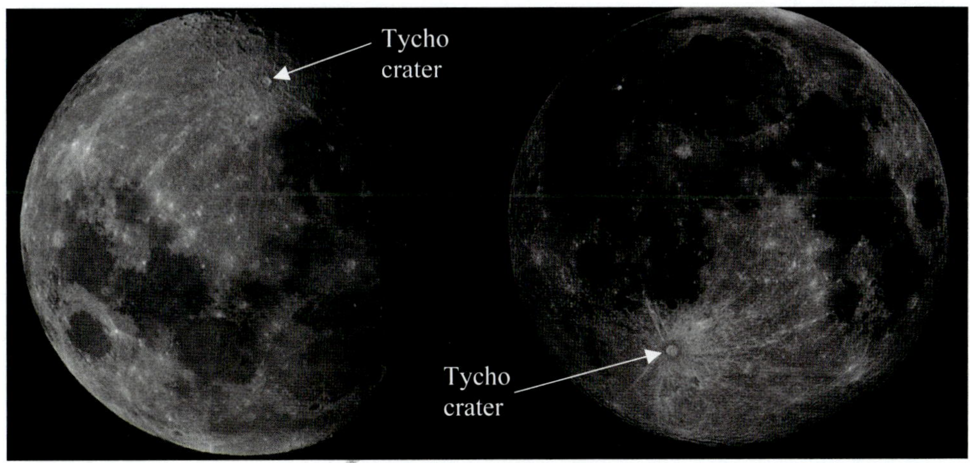

The moon on the left appears to be upside down compared to the moon on the right.

Why do the two pictures look so different? Because they were taken from different hemispheres of the earth. The one on the left was taken in Montevideo, Uruguay, while the one on the right was taken from Madison, Alabama in the U.S. Because the earth is roughly a sphere, the person taking the picture in Uruguay was standing upside down compared to the person taking the picture in Alabama. Thus, one sees the moon upside down as compared to the other! The view you get of the moon will probably be different from both of the ones in the picture, because you are on a different part of the earth, which will lead to a different perspective.

While you are looking at the pictures, notice that there are several dark spots on the moon. Those are called **maria** (mah' ree uh), which comes from the Latin word meaning "seas." Since the maria appear to be quite different from the rest of the moon's surface, ancient astronomers thought they might be oceans on the moon. Of course, we now know that there is no liquid water on the moon, but water in its solid phase can be found there. We now understand that the maria are large deposits of basalt that most likely are the result of volcanic eruptions that took place on the moon.

Astronauts brought back samples of rocks from a particular mare (singular of maria). When those rocks were analyzed, it was found that they had been magnetized. This is a problem for the uniformitarian view, since it says the moon shouldn't have ever had a magnetic field that could make basalt magnetic. Indeed, to this day, uniformitarian scientists do not have a successful theory that explains how the moon had a magnetic field when the maria formed but then lost it as time went on. However, do you remember the young-earth magnetic field theory I discussed in Chapter 9? It specifically explains how the moon's magnetic field originated and how it was lost later on.

The astronauts didn't just bring samples back from the moon. They also left things on the moon. In some cases, those things are just trash. However, other things were left on the moon so that they could be used to learn more. Look at the picture below. It shows the place where Apollo 11 landed the first astronauts on the moon. The picture is the result of merging images taken by the Lunar Reconnaissance Orbiter, an artificial satellite of the moon. As pointed out in the picture, the biggest thing you can see is the descent stage, which allowed the astronauts to land on the moon and then served as a launch pad when they left. It remains behind as trash. However, the two other white patches pointed out were left there for a purpose. The larger one, the passive seismic experiment, looked for seismic waves on the moon. While it lasted for only a short time, other moon landings left their own passive seismic experiments behind so that more data could be collected. The smaller one, the laser range retroflector, allows scientists to measure the distance to the moon very accurately by shining a laser beam at the reflector and measuring the time it takes for the light to return to earth.

This is a picture of the Apollo 11 landing site as taken from orbit around the moon.

Comprehension Check

15.14 Suppose we find another solar system with two planets. One is far from that solar system's star, while the other is near to it. Which planet is more likely to have at least one moon?

15.15 You observed a full moon one night, but heavy clouds obscured it for the next four nights. The clouds go away on the fifth night. When you see the moon on that night, will it be a quarter moon? If not, will you see more or less than half of it illuminated?

15.16 If you took a sample of rock from a part of the moon called the "Sea of Tranquility," would you expect it to be an extrusive igneous rock or an intrusive igneous rock? You might want to look back at page 98 to remind yourself what those terms mean.

Answers to the Comprehension Check Questions

15.1 It takes Mercury a shorter time to travel around the sun. Since it is closer to the sun, gravity is stronger, so it can move faster. Also, the distance it needs to travel to get around the sun is shorter, because the circle it orbits in is smaller. You don't need to know the second reason, but you should know that the closer a planet is to the sun, the faster it completes one orbit.

15.2 Mars. Remember from the first section that the closer the planet is to the sun, the faster it moves because the force holding it in orbit is stronger.

15.3 The comet's orbit is more eccentric. A comet must pass close enough to the sun at perihelion to expel some of its ice, so the orbit must be very eccentric.

15.4 At first, it would get brighter, but then it would reach a maximum brightness and begin to fade until you can no longer see it. As it approaches perihelion, it gets closer to the sun, meaning more gases and dirt will be ejected from the nucleus. That will make it brighter. However, once it passes perihelion, it will move farther away, reducing the amounts of gases and dirt ejected, making it dimmer.

15.5 Ceres probably didn't come from the Kuiper belt. It probably came from the asteroid belt. The Kuiper belt is beyond Neptune, so any dwarf planet that comes from there must be far from the sun. Ceres is between Mars and Jupiter, which is where the asteroid belt is.

15.6 Epicycles were used to adjust the geocentric view to fit the observations. While the word "epicycle" referred specifically to a circle in which a planet traveled, it is now a general term that means "an adjustment to make the theory fit the data."

15.7 The planets would not exhibit retrograde motion, but the moon would. Since the planets orbit the sun, they all move in the same direction in the sun's "sky." However, the moon is orbiting the earth, so it's orbit would act like an epicycle, causing its motion to change relative to the sun.

15.8 Venus and Mercury cannot be in opposition to the sun. Since their orbits are closer to the sun than the earth's orbit, they are never on the opposite side of the earth as the sun. They appear only in the western sky at sunset and the eastern sky at sunrise, because they are always in the general direction of the sun from the earth's point of view.

15.9 It would look most like the light from an incandescent light. Remember, an incandescent light heats up a filament to make light, and a candle also uses heat to make light. Thus, the lights are very similar. Most candles produce more yellow light than an incandescent light, so the yellow in the rainbow would probably be stronger, but that would be the main difference.

15.10 This is nuclear fission. Fission involves a single nucleus becoming two nuclei, which increases the total number of nuclei. Fusion involves two nuclei becoming a single nucleus, which reduces the total number of nuclei.

15.11 The amount of helium should increase, while the amount of hydrogen should decrease. The fusion that is going on in the sun causes hydrogen to be converted into helium. Overall, four H's become one He, so over time, the He should build up, while the H should dwindle away. Now, while the sun uses up a lot of hydrogen, it also has a lot of it. As a result, there is no worry about it running out of hydrogen any time soon. Based on what we have seen, the sun has five billion years' worth of hydrogen to use.

15.12 <u>The average number of solar flares has also been decreasing</u>. Remember, a solar flare is caused by certain types of sunspots. If the number of sunspots decreases, the number of solar flares should decrease as well.

15.13 <u>The amount of light would increase</u>. If a lot more gases were added to the sun, the sun's gravity would increase. That would cause the gases to be pulled more closely together. However, when the gases got pulled more closely together, the amount of fusion in the core would increase, due to the increased pressure. That would make more light.

15.14 <u>The one farther from its star will be more likely to have at least one moon</u>. Remember, Mercury and Venus have no moons because they are so close to the sun that the sun would pull their moons to it.

15.15 <u>It will not be a quarter moon. More than half of it will be illuminated</u>. It takes the moon over 29 days to go through its phases. That means to go from full moon to new moon will take half that time. To go from full to quarter would take one-fourth that time. That's more than seven days. Since you are looking at it 5 nights after the full moon, it will be in between full and quarter, which means more than half of it will be illuminated.

15.16 <u>It will be an extrusive igneous rock</u>. In fact, it will be basalt. Remember that "maria" means "seas." Thus, if it has the name "sea of," it is one of the maria. They are the result of volcanic eruptions, which means lava freezes, forming an extrusive igneous rock. Intrusive igneous rocks are made from magma freezing underground. There is intrusive igneous rock on the moon, but it is not in the maria.

Chapter Review

1. Define the following terms:

a. Asteroid	d. Heliocentric	g. Nuclear fusion	j. Satellite
b. Comet	e. Spectroscopy	h. Solar flare	
c. Geocentric	f. Nuclear fission	i. Hydrostatic equilibrium	

2. What keeps gravity from pulling the earth into the sun?

3. List the planets in the solar system in order based on their distance from the sun. Start with the one that is closest. How does the speed at which they orbit the sun vary with their distance?

4. What is the largest body in the solar system? What is the largest planet? What is the smallest planet?

5. What do we call the structure that holds most of the solar system's asteroids? Where is it?

6. What makes up the two tails of a comet? Which way do they point relative to the sun?

7. What are the two broad classes into which all comets can be placed?

8. What do we call bodies like Pluto that orbit the sun and are bigger than asteroids but smaller than planets?

9. What is an epicycle, and in which view of the solar system (geocentric or heliocentric) was it used?

10. What does the phrase "retrograde motion" refer to? Which view of the solar system (geocentric or heliocentric) explains it better?

11. When is Mars brightest in the night sky – when it is on the same side of the earth as the sun or when it is on the opposite side?

12. Does the Bible indicate that the sun moves in the solar system?

13. What are the two main gases that make up the sun? Which one is present in larger quantities?

14. What process does the sun use to produce energy? What is used up, and what is made?

15. What happens to the amount of energy the sun puts out as time goes on?

16. What is the hottest: the sun's core, radiative zone, convection zone, or photosphere? Why is it so hot there?

17. What do we call the surface of the sun?

18. What are sunspots, and what do they tell us about how active the sun is?

19. What would happen to the size of the sun if the amount of fusion happening in the core increased? What would that change in size do to the amount of fusion happening in the core?

20. Which planets don't have at least one moon? Why?

21. What does the moon do for the earth's axial tilt? How does that help make the earth a place where life can flourish?

22. In the diagram on the right, indicate the letters that are positioned where the new moon, the full moon, and the two quarter moons would be.

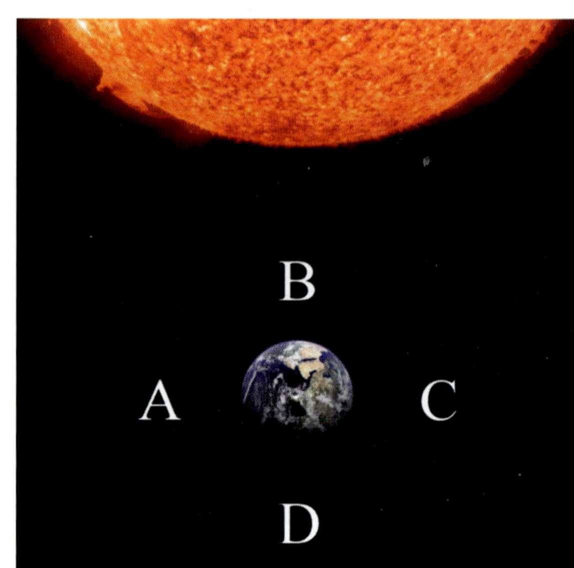

23. You take a picture of the full moon when you are at a latitude of 45 degrees north. You then take a picture of the full moon six months later when you are at a latitude of 45 degrees south. How will the pictures compare to one another?

24. What are the dark gray patches on the moon's surface called? What are they made of?

25. Scientists have used seismic waves to determine the moon's internal structure. How did they detect those seismic waves?

26. How do we get the most accurate information about the distance between the earth and the moon?

Chapter 16: Earth's Solar System and the Universe

Introduction

In the previous chapter, I showed you a diagram of the solar system. I made the point that the diagram cannot be drawn to scale, because the distances between the planets and the sun are incredibly different. Neptune, for example, is *100 times* farther from the sun than Mercury. In other words, if you drew a diagram of the planets with Neptune 8 inches (20.3 cm) away from the sun, Mercury would have to be drawn 0.08 inches (0.2 cm) away from the sun! But how do we know that? How have we measured the distances between the planets and the sun?

Nowadays, we use radar to determine the distance between the earth and all the planets. We send electromagnetic waves from earth and wait for them to bounce off the planet and come back to the earth. The time it takes for that to happen tells us how far the earth is from the other planets. Of course, the distance between the earth and the other planets keeps changing, since each planet (including the earth) moves in its own orbit. If we take enough radar measurements over a period of time, however, we can map out how the planets are moving and use that map to determine each planet's distance from the sun.

Interestingly enough, however, scientists had a really good idea of the distances between the planets long before radar was invented. They used a method that doesn't have the precision of radar, but it still produced excellent results, which allowed scientists to determine the layout of the solar system. It depends on a well-known phenomenon called **parallax** (pair' uh lacks).

Parallax – The effect by which the position of an object appears to change when viewed from different locations

While we no longer use this method to study distances in the solar system, we still use it to measure distances outside the solar system. As a result, it is an important technique to learn.

Measuring Distance with Parallax

You have probably experienced parallax before, even though you might not be familiar with the term. To understand the effect and see how it is sensitive to distance, perform the following experiment.

Experiment 16.1: Parallax and Distance

Supplies:

- A sheet of paper that is roughly 8.5 in x 11 in
- Two colors of marker or crayon (You want to leave thick marks on the paper. I use red and blue in my discussion, but you can use any colors you want.)
- A ruler
- A meterstick or some other long, thin stick
- Tape
- A long hallway or large room
- Someone to help you
- Something you can position at the end of the hallway or room and tape the paper to, like a chair with a flat back.

Instructions:

1. Turn the sheet of paper on its side so that the widest part is horizonal.
2. Use the ruler and one color of marker/crayon to draw a vertical line down the center of the paper.

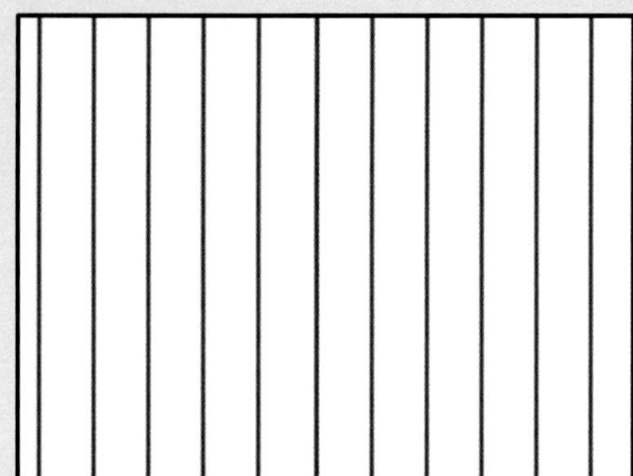

3. Use the ruler and the other color of marker/crayon to make vertical lines that start one inch from the line you just made and are each one inch apart. In the end, your paper should look something like the sketch on the left. Your colors might be different, but I will refer to this sketch in the instructions, so when I say, "red line" I just mean the line at the center. When I say, "blue line" I just mean the other lines on the paper.
4. Put the chair or other object at one end of the room/hallway and tape the paper to it so the lines are vertical. The paper shouldn't be higher above the floor than the meterstick/long, thin stick is tall.
5. Have your helper stand about six feet away from the piece of paper and hold the meterstick/long, thin stick so that the thinnest part of the stick is facing the paper.
6. Stand a couple of feet from the stick your helper is holding so that the stick is between you and the paper.
7. Close one eye. It doesn't matter which one.
8. Position yourself so that the stick is directly in front of the red line on the paper when one eye is closed. As you look at the stick, then, it should be blocking the red line from your view.
9. Once you have it lined up, close the eye you have opened and open the other eye. What line is the stick in front of now? It isn't in front of the red line anymore, is it? It may not be directly in front of any line, but find the line it is closest to and count the number of blue lines between the red line and the blue line it is closest to.
10. Have your helper keep holding the stick where it is. You need to back up so that you are about twice as far from the stick as you were before.
11. Repeat steps 7-9.
12. Have your helper keep holding the stick where it is. Once again, you need to back up so that you are about twice as far from the stick as you were before. If you don't have that much room, just get as far from the stick as you can.
13. Repeat steps 7-9.
14. If you have more room, repeat steps 12-13.
15. Clean up your mess.

What did you see in the experiment? You had lined up the stick so that it was right in front of the red line, at least as far as your open eye could tell. However, when you closed that eye and opened the other one, what did you see? You saw that the stick's position relative to the red line had moved. It was no longer in front of the red line but was near one of the blue lines. That's what parallax is. Your eyes are probably 4-6 centimeters (1.5-2.4 inches) apart from one another. As a result, they each viewed the stick from a slightly different position. Because of that, the stick's position relative to the lines on the paper was different for each eye.

What did you notice about how this difference depended on your distance from the stick? Each time you moved farther away, the number of blue lines you had to count from the red line should have decreased. In other words, the farther you were from the stick, the less its position changed when you switched from one eye to another. Thus, the amount of parallax you experienced with your eyes was smaller the farther you got from the stick.

What does this tell you? Parallax is sensitive to the distance between the observer and what is being observed. The closer you are to an object, the more parallax you get. The farther away you are, the less parallax you get. While your experiment didn't show this, it also depends on the distance between the two spots where the object is being observed. Your eyes stayed the same distance from each other the entire time (thankfully), but if you had been able to increase the distance between your eyes, you would have also seen the amount of parallax increase.

While your experiment only demonstrated that parallax is sensitive to the distance between you and what you are viewing, detailed geometry can be used to actually calculate that distance. All you need to know to use the equation is the distance between the two places from which the observations took place as well as the change in the observed position. All this geometry has been known for quite some time, so back in the late 1600s, two astronomers, Giovanni (jee oh vah' nee) Cassini (kuh see' nee) and Jean (zahn) Richer (ree' shay), arranged to view Mars on the same night in two different locations: Cassini stayed in Paris, and Richer went to French Guiana in South America. Remember, the farther apart the two observations, the more sensitive parallax is to distance. That's why Richer traveled so far from France. They each reported how they saw Mars relative to a specific star. Just as was the case in your experiment, the position of Mars relative to that star was different for each observer. They had chosen to view Mars when it was closest to earth, and when they used the geometry related to their observations, they calculated the distance to be 140 million kilometers (87 million miles). That is within 7% of the distance measured today with radar!

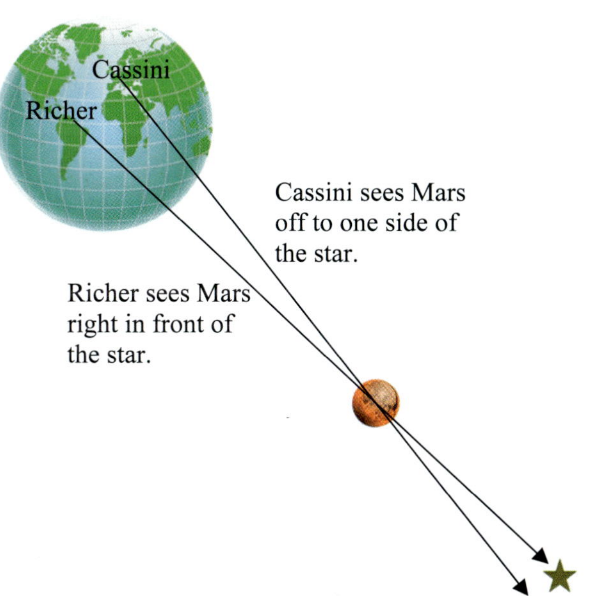

Cassini and Richer used parallax to accurately measure the closest distance between Mars and the earth.

As I said, we no longer use this method to determine the distance to the planets, but we still use it to determine the distance to many stars. In this case, however, we need a lot more distance between the two observations, so we take advantage of the earth's rotation around the sun. Essentially, we observe the star at one time of the year and then observe it again six months later. That way, the earth has traveled halfway around the sun, so the distance between the two points of observation is equal to the diameter of the earth's orbit. That allows the parallax to be very sensitive to distance.

Of course, God created a really big universe, so there are many stars that are too far away to be measured using parallax. As a result, other distance measuring techniques must be used. I will talk about that in an upcoming section of this chapter.

Comprehension Check

16.1 Looking at the drawing above, how could Richer and Cassini have improved their observation to make it even more sensitive?

Let's Get Some Perspective

For most of the history of science, we had no idea how big God's creation is. When Copernicus and his supporters were arguing for heliocentrism, for example, astronomers already knew about parallax. They said that if the earth traveled around the sun, the stars should show parallax over the course of a year, since we would be looking at them from different positions throughout the year. However, astronomers in those days saw no parallax in the stars. But remember from your experiment that the amount of parallax you see depends on the distance between you and what you are observing. Astronomers back then knew that, but they could not imagine that the stars were so far away that their equipment wasn't sensitive enough to actually detect parallax. It took almost 300 years for astronomers to develop good enough equipment to detect it, and by that time, Copernicus had already been vindicated.

My point is that most astronomers throughout the course of history were simply unable to think realistically about how big God's creation is, and it took a long time for them to develop the proper perspective. It's time for you to get the proper perspective. Let's start with the fact that the solar system I have been describing is only a tiny, tiny speck compared to all of God's creation, which is properly called the **universe**.

Universe – The sum total of all the natural components of God's creation

God created supernatural things as well, but it is difficult to study them scientifically. Science can find evidence for their existence, but it cannot study them. Thus, when talking about the things that scientists study, I will limit my discussion to the natural world.

To begin my discussion of the universe, let me show you a diagram that shows our solar system to scale when it comes to the distances between the planets and the sun. The relative sizes of the planets and sun are not correct, but at least the relative distances are. Notice how hard it is to see Mercury, Venus, the earth, and Mars. That's because compared to Neptune and Pluto, they are incredibly close to the sun. Because they are so close, they are often called the **inner planets**. By contrast, Jupiter, Saturn, Uranus, and Neptune are called the **outer planets**, and the asteroid belt is what separates the two groups.

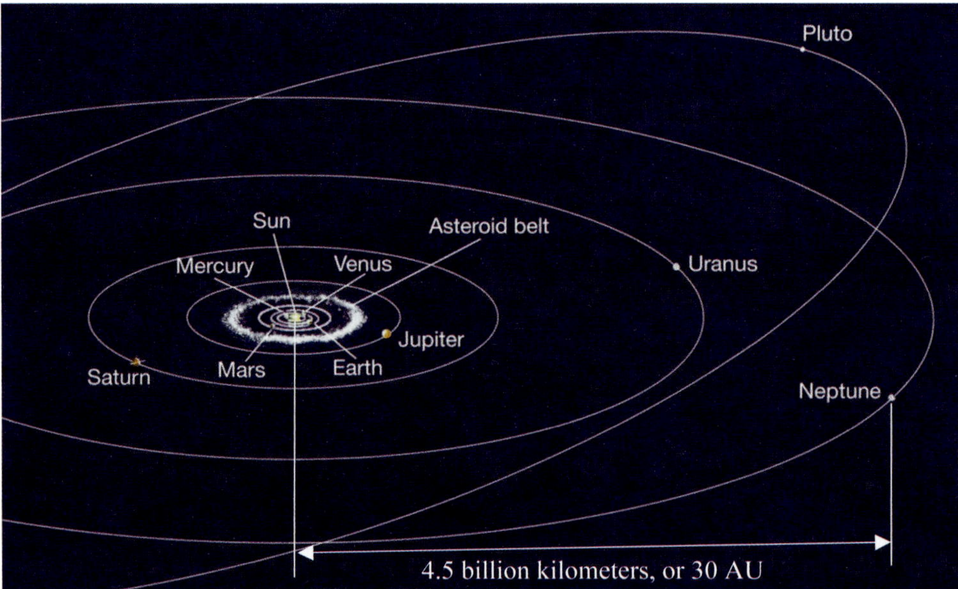

This diagram of the solar system shows the orbits of the planets (and Pluto) to proper scale

At the bottom of the diagram, I show the average distance between the sun and Neptune. It's 4.5 *billion* kilometers, which is the same as 2.8 *billion* miles! That's a mind-bogglingly-long distance. The distance around the earth's equator is "only" 40,000 kilometers. Thus, the distance between the sun and Neptune is the same as about 112,000 trips around the earth's equator! When numbers get big like that, they get annoying to write, so astronomers have defined a distance unit called the **astronomical unit (AU)**, which is equal to the distance between the earth and the sun.

Astronomical unit – The average distance between the earth and the sun (150 million kilometers)

Now please understand that the astronomical unit is just another way to measure distance. In other words, it is just another distance unit. Thus, you can convert any other distance unit (like kilometers) into astronomical units. For example, the distance between the sun and Neptune can be converted from kilometers to AU the way you learned to convert any units. Since 1 AU = 150,000,000 km:

$$\frac{4{,}500{,}000{,}000 \text{ km}}{1} \cdot \frac{1 \text{ AU}}{150{,}000{,}000 \text{ km}} = 30 \text{ AU}$$

The distance between the sun and Neptune, then, is 4.5 billion kilometers, which is the same as 30 astronomical units.

Since 30 AU is the distance between the sun and the farthest outer planet, is that the size of our solar system? No! The solar system extends beyond Neptune, because the solar system is really defined by the sun, and the sun emits radiation that travels much farther than the orbit of Neptune. Specifically, the high-energy charged particles that the sun emits travel quite far. They form a "bubble" around the solar system, protecting it from radiation that is trying to enter from the rest of the universe. This "bubble" is called the **heliosphere**, and it is what actually defines our solar system.

Heliosphere – The region of space protected by the sun's charged particles

The heliosphere extends to about 180 AU from the sun. So once you get to Neptune, you are only one-sixth of the way to the edge of the solar system!

Obviously, then, the solar system is *huge*. However, it is very small compared to the rest of the universe. In fact, while the sun is stationary at the center of the solar system, it is hurtling through the universe at a speed of 830,000 kilometers per hour (520,00 miles per hour)! This motion actually deforms the heliosphere, so the drawing on the right gives you a view of what the solar system looks like as it travels through space. It also shows you why we know so much about the limits of the solar system. In 1977, two robotic spacecraft (Voyagers 1 and 2) were launched to study the solar system. They studied the outer planets and then continued to travel farther and farther from the sun. Their instruments continue to send back data, and scientists have learned a lot from them. Based on the data, scientists concluded that both spacecraft went beyond the heliosphere in 2018. They are the only human-made devices that exist outside our solar system!

This shows what our solar system looks like as it travels through space. **Note**: The Voyager crafts are not drawn to scale, and the deformation of the heliosphere is exaggerated.

Outside Our Solar System

Even though the Voyager spacecraft are the only human-made devices that exist outside our solar system, there are a wealth of natural things found there. The most obvious things are the stars. When you look up at the night sky, you see lots of points of light. Most of them are stars. Depending on the time of year and where you are looking, however, some of those points of light might be planets in our solar system. We see stars because they make their own light, but we see planets because they reflect light from the sun.

How can you tell the difference between planets and stars in the night sky? Ancient astronomers noticed the difference by seeing how their positions changed. If you watch the night sky for a few hours, you will see that the points of light move in the sky, just like the sun moves in the sky. Of course, that's because of the earth's rotation. The picture on the left shows this rather dramatically. It was taken by pointing a camera at the night sky and leaving the shutter open for a long time. This causes each point of light in the sky to trace out its motion in the picture. Each curve, then, is produced by a single star as it travels in the night sky. The small, bright curve at the center is made by the north star, which is called **Polaris**. Its position changes the least over the course of the night, since it is almost right above the earth's axis of rotation.

This long-exposure picture taken of the night sky in Oregon shows the motion of the stars that is caused by the earth's rotation. The center of all those circles is right above the North Pole.

The positions of the lights in the night sky also change from day to day, because the earth orbits around the sun, which causes the position from which we view the rest of the universe to change every day. However, when the earth completes its orbit around the sun, we return to the original position from which we were viewing. Because of this, the stars return to their original position in the night sky. As a result, at a given time and day of the year, the positions of the stars in the sky are always the same the next year. However, that's not true for the planets, because they are also moving in their orbits. As a result, the planets seem to "wander" around the night sky, which as you learned in the previous chapter, is why ancient Greeks called them planets (*planetes*).

There is a much easier way to distinguish between planets and stars, however. If you look at a star long enough, you will eventually see it twinkle. That's because all the stars are very far away compared to the planets, so the points of light they make in the night sky are much, much smaller than the points of light that the planets make. You might not notice it, but it's true. As a result, even a small amount of dust floating in the air can block a star's light from hitting your eyes. That makes the star darken or maybe black out altogether. However, the dust is constantly moving, so it doesn't block the star's light for long. As a result, a star will appear to flicker off quickly and then back on again. That's what we call "twinkling." A planet, on the other hand, is much closer, so it makes a larger light in the night sky. That larger light isn't blocked effectively by dust, so a planet doesn't typically twinkle.

If you study the stars in the night sky carefully, you will find they come in a wide variety of colors. Look at the picture on the previous page. Those colors aren't photoshopped in. Those are the natural colors of the stars. That may surprise you, since you probably see most stars in the night sky as white. There are many reasons for this, and I will discuss two of them in a moment. For right now, however, look at the picture on the right. It shows a telescope's view of a portion of the night sky. Each of those points of light is a star (or multiple stars, as you will learn later), but notice the colors! There is a big, bright orange star near the bottom left, there are several yellow stars, some white ones, and a couple of blue ones. It turns out that the color of a star can tell you something about its makeup. For right now, however, you just need to understand that stars come in many colors.

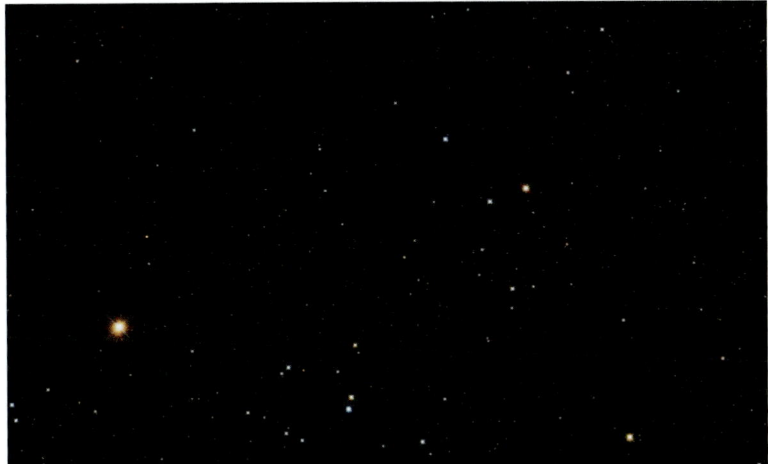

This picture shows a portion of the night sky as seen through a telescope. The bright orange star near the bottom left is called Aldebaran.

Why don't we usually see these colors when we look up at the night sky? One big reason relates to the way our eyes are designed. In order to make sure we can see under many different sets of conditions, God created our eyes with two different kinds of cells to detect light: rod cells and cone cells. The cone cells allow us to see color, but they need a certain amount of light in order to do that. The rod cells, on the other hand, can detect light even when it is very dim. However, rod cells do not detect different colors. You can think about it this way: rod cells allow you to see in black and white, while cone cells allow you to see in color.

When there is plenty of light, your cone cells are very active, and you see lots of color. However, if the light starts to dim, your cone cells can't gather as much information from the light, so they don't detect colors well. As a result, the rod cells start to take over, and the colors become muted. If the light dims even more, there will come a time where your cone cells cannot detect any colors. At that point, your eyes rely solely on your rod cells, and you see only black, white, and shades of gray. Depending on where and when you are viewing the night sky, you might be receiving so little light from the stars in the sky that only your rods cells respond, so you see a light, but you do not see a color associated with it.

That's why the long-exposure camera shot on the previous page shows the colors so well. Since the shutter was open for a long time, it was able to collect a lot of light. Now, of course, the camera doesn't have rod and cone cells, but the process that turns what the camera sees into the image requires more light to produce color images than it does to produce black-and-white images, so the effect is really the same. Since the camera collects more light, it can produce images with better colors. This is why photographers spend so much time making sure the lighting is right.

Surprisingly, the explanation is the same for pictures taken through telescopes. While most people think a telescope just magnifies what it looks at, it does much more than that. It takes a large amount of light from the sky and focuses it all into the lens you are looking through. Of course, since it is taking a large amount of light and focusing it down to a small lens, it ends up magnifying whatever it is pointed at. More importantly, however, that large amount of light allows you to see what you are looking at much better, including being able to see the colors more clearly. While there are other issues that govern how a telescope works, one of the more important ones is simply the size of the telescope's opening. The larger the opening, the more light it can collect, so the better it can produce accurate images.

The other issue that needs to be considered is the fact that in order to hit your eyes, the light must travel through the atmosphere. What does light do as it travels through the atmosphere? It bounces off the molecules in the air, right? That's what makes the sky blue and sunsets reddish orange. Well, the same thing happens to starlight. The light can bounce off molecules in the air, distorting what you see, including the color. It can also keep light from hitting you altogether, so you don't see the really dim stars.

How can this problem be fixed? Initially, astronomers would go to tall mountains in order to view the night sky. The higher you go, the less atmosphere there is, so the less distortion you experience. This is why many telescopes are built on tall mountains like Mauna Kea in Hawaii. Of course, the best thing to do is get rid of the air altogether, which is why many telescopes are launched into orbit around the earth, where there is very little atmosphere. The result is spectacular. Consider, for example, the photo on the left. It comes from the Hubble space telescope, which was put in orbit back in 1990. Notice all the vibrant colors. They are not the result of editing. They are the real colors with which the stars actually shine!

This picture was taken by the Hubble telescope, which is in orbit around the earth. Since the stars' light doesn't have to travel through the earth's atmosphere to reach the telescope, the picture is sharp and vibrant.

What causes these amazing colors? You will learn about that later. For right now, it's time to review what you have already learned. Before you finish, however, please note that the next experiment requires something to sit overnight, so you should do steps 1-5 of Experiment 16.2 before you finish your school day.

Comprehension Check

16.2 As I said, the heliosphere extends to about 180 AU from the sun. How many kilometers is that?

16.3 If you observe a planet in the night sky over many months, will its size be constant?

16.4 You have a bright red balloon in your hands. If you take it into a windowless room and darken the room until you can just barely see the balloon, what color will it appear to be?

16.5 Suppose you are using a telescope on a cold night where you need a fire to stay warm. For the best views, should the fire be between the telescope and what you are viewing or behind the telescope?

Star Light, Star Bright

As you look at stars in the night sky, you should notice something else rather obvious: they shine with different brightnesses. Look, for example, at the picture below. It shows stars in a portion the night sky as seen from the Northern Hemisphere. Some stars are so dim you can barely see them, and others are incredibly bright. The brightest one (which is low in the sky) is named "Sirius," but it is often called the "dog star." It is, in fact, the brightest star in the night sky. The orange star near the top of the picture is called "Betelgeuse" (bee' tuhl joos) and the bright stars near it form the constellation that is called "Orion." Why is Sirius so much brighter than the rest? Why are the stars in the constellation Orion brighter than most of the other stars in the picture? There are actually two main reasons. To understand the first reason, perform the following experiment.

This picture of the night sky shows the constellation Orion in the upper right. The orange star at the top of the constellation is Betelgeuse, and the brightest star in the night sky (Sirius) is low in the center of the picture.

Experiment 16.2: Distance and Brightness

Supplies:
(For the first part of the experiment, you need only the first three items.)

- A white card, like an index card (It can have lines on it, but it needs to be mostly white.)
- A paper towel
- Vegetable oil (Any plant-based oil will work.)
- A room that is a bit dim when the lights are turned off (It needn't be dark – just dim.)
- Two lamps from which the shades can be removed
- Three bulbs for the lamps, two of which have the same brightness (listed in watts) and one with a different brightness (They should be either all normal (incandescent) bulbs, all compact fluorescent bulbs, or all LED bulbs. If you use LED bulbs, make sure they all produce the same shade of white light, because some are bluer than others. If you can't find a bulb with a different brightness, use three lamps instead.)

Instructions:
1. Lay the paper towel on a counter or table.
2. Lay the index card on the paper towel.
3. Take the lid off the oil bottle and pour a small amount of the oil into the lid. You don't want much.
4. Pour some oil from the lid onto the center of the index card so that you make a circle of oil at the center.
5. Let the index card sit overnight. You can move it, along with the paper towel, to a place where it won't be disturbed if you like. The oil should soak into the card.
6. The next day, you should have an index card with an oil spot in the center.
7. Take the index card, the lamps, and the bulbs into the room that is a bit dim.

8. Set the two lamps so that they are a meter (3 feet) apart from one another and are plugged in.
9. If the lamps are of different heights, set the shorter one on a stack of books or have someone hold it so that the two bulbs are at the same height.
10. Make sure the bulbs in both lamps are the same brightness, and turn on both lamps.
11. Stand roughly at the midpoint of both lamps, but stand to one side so that you don't block the light traveling between the two bulbs.
12. Hold the index card between the light bulbs at the level of the light bulbs, so that it blocks the light traveling between the light bulbs.
13. Move the index card so it is only a few inches away from one of the bulbs.
14. Look at the card on the side that is facing you, not the side nearby the bulb.
15. Concentrate on the oil spot. You should see a lot of light from the nearby bulb shining through it.
16. Keeping your eyes on the oil spot and continuing to look at the same side of the index card, slowly move the card so that it stays at the same height but gets farther from the nearby bulb and closer to the bulb that is farther away. Notice how the oil spot changes in brightness.
17. When the index card passes the halfway point between the bulbs, you should notice a definite difference. Don't look at it from the other side. Continue to look at it from the same side.
18. Continuing to look at the oil spot from the same side, move the index card until it is just a few inches from the light bulb it was initially far from.
19. Now that you know how the oil spot changes, go back to the center and move the index card back and forth near the center. You should eventually find a point where the oil spot becomes nearly invisible. If the light bulbs are the same brightness, that should be halfway in between.
20. If you are using three lamps, put the third one right next to one of the other two and turn it on; then skip to #24.
21. If you are using bulbs of different brightness, turn off one of the lamps. If you are using incandescent lights, let the bulb cool down before you continue.
22. Remove the bulb from the light that is turned off and put in the bulb of a different brightness.
23. Turn on the lamp.
24. Repeat the experiment, and once again, try to find the place where the oil spot becomes almost invisible. It shouldn't be anywhere near the center. Is it closer to the brighter light source or the dimmer light source?
25. Clean up your mess.

What happened in the experiment? When light hit the oil spot on the card, some of the light passed through the card, and some reflected back off the card. Both bulbs were shining light on the oil spot, so both bulbs had light passing through the oil and reflecting back off the oil. When you started, the card was near one bulb, and since there was a lot of light from that bulb hitting the oil spot, you saw a lot of light traveling through the oil spot. When you had the card close to the other bulb, there wasn't a lot of light traveling through the oil spot anymore, because it was far from the bulb whose light passed through it. However, there was a lot of light reflecting off the card from the nearby bulb, so you saw that reflected light.

When the card was directly in between the two bulbs, there was light from one bulb traveling through the oil spot to hit your eyes, and there was light from the other bulb reflecting off the oil spot to hit your eyes. When the amount of light traveling through the spot was the same as the amount of light reflecting off the spot, the spot suddenly looked like the rest of the card, so it mostly disappeared. In other words, the oil spot was a means by which you could dctect how much light was hitting the card. If you could see the spot, one side of the card was getting more light than the other. When the spot nearly disappeared, each side of the card was getting the same amount of light.

So what happened in the second part of the experiment, where one light source was brighter than the other? In order for the oil spot to mostly disappear, you had to have the card farther from the brighter light and closer to the dimmer light. Only then was the card getting equal amounts of light on both sides. This tells you that when you are looking at a distant object, more light will hit your eyes the closer you are to it. That's true of anything that is emitting or reflecting light, including stars. If all other things are equal, then, the farther the star is away from the earth, the dimmer it will be. The closer it is, the brighter it will be.

In fact, that's why the sun is so bright. It is a star, and there are some stars in the night sky that are similar to the sun. However, they are much farther away from the earth. As a result, they are very dim compared to the sun. In fact, this is why the other stars are only visible at night, or in the late evening or very early morning. The sun is so close to the earth that it shines incredibly brightly in the sky. The other stars are always shining in the sky, but once the sun's light starts hitting the earth, it overwhelms the light coming from all the other stars, so you do not see them. The only time you see the other stars is when the sun's light is not hitting your side of the earth or just barely hitting your side of the earth. At that point, the sun's light is dim enough or gone completely, so it doesn't overwhelm the light coming from the other stars.

This effect is well-illustrated by the picture below, which was taken by astronauts on the moon. Look first at the sky. Notice that it is jet black, but there isn't a single star there. Why? Well, look at the big rock. One side of it (the side against which the astronaut is leaning) is lit up. The rock is casting a shadow on the other side. That means the sun is in the sky, out of the picture on the left. That light is reflecting off the rock, astronaut, the moon's surface, and the vehicle. This allows the camera to detect them, which allows you to see them in the picture. There are also stars in the moon's sky, but the camera can't detect them, because the sun's light overwhelms the star's light. It doesn't matter that the sky is black. As you have already learned, it's black because the moon has no atmosphere. Thus, even when the sun is shining, the sky is black. Because the sun is shining, however, you can't see the stars in the black sky. Their light is being overwhelmed by the sun's light.

You don't see any stars in the moon's black sky, because the sun's light is overwhelming their light, just like it overwhelms the stars' light in the earth's sky during the day.

Comprehension Check

16.6 Suppose you are looking at the stars in the night sky from your backyard, which is in the center of town. Now suppose you decide to leave the city and take a drive into the country that same night. Once you get far from all the buildings, street lights, etc., will stars in the night sky look any different? If so, how?

Distance Isn't the Only Factor

The brightness of a star depends not only on the distance between the earth and the star, it also depends on how quickly the star is producing energy. Different stars have different physical characteristics, which lead to different amounts of energy being produced. Thus, we have to make a distinction between the brightness we see and the actual brightness of the star itself. That means we have to define some terms. First, astronomers usually refer to a star's brightness as its **magnitude**.

Magnitude – (astronomy) A measure of a star's brightness

Because of the details of how it is defined, magnitude is a bit like a golf score: the *lower* the magnitude, the *brighter* the star.

Since magnitude depends on two things (distance and energy produced), we have to define two types of magnitude.

Apparent magnitude – The magnitude of a star as it is measured from earth

In other words, when we look at the night sky, the brighter the point of light, the lower the magnitude. Consider, for example, the picture shown below. There are a lot of stars in it, but just concentrate on the four I have labeled, which form the constellation known as the Southern Cross, which can only be seen in the Southern Hemisphere. Which is the brightest? You should be able to see that it's Acrux, which has an apparent magnitude of 0.77. The next brightest is Mimosa, with an apparent magnitude of 1.25. Gacrux is a bit dimmer, with an apparent magnitude of 1.63, while Iami is the dimmest of the four, with an apparent magnitude of 2.78. Notice that the brighter stars have lower magnitudes. In fact, a star can be so bright in the night sky that its apparent magnitude is negative. Sirius (the brightest star in the night sky) has an apparent magnitude of -1.46. The sun's apparent magnitude is -26.74, which is why its light overwhelms the light from all the other stars.

The four stars labeled in this picture form the Southern Cross. The brighter the star, the lower its apparent magnitude.

Now remember, distance plays a big role in a star's apparent magnitude. If we could get closer to a star, its apparent magnitude would go up. When studying a star, then, it is important to figure out how far it is from the earth. How do we do that? For stars that are "nearby," parallax works. If we view the star once and then wait 6 months to view it again, we are looking at it from opposite ends of the earth's orbit around the sun. Using geometry, we can then determine how far it is from the earth. What do I mean by "nearby"? I will get to that in a moment.

We need to define more distance units in order to make it easy to discuss distances to stars. The first one is based on how light travels. We know that it goes from one place to another very quickly. The sun is 1 AU away from the earth, but light makes that trip in just over 8 minutes. While the AU is already a big unit for distance, it is tiny compared to the universe as a whole. Thus, one distance measurement astronomers frequently use is the **light-year**.

Light-year – The distance light would travel in a year, based on how it behaves on the earth

Since light takes just over 8 minutes to travel one AU, a light-year is obviously a big unit! In fact, 1 light-year is the same as 63,000 AU! Now please understand that light-year *does not* measure time. It has the word "year" in it, but it is a distance unit, not a time unit. This will become important later on. There is one more distance unit that astronomers use. It's called the **parsec**, and it is just over 3.26 light-years. While I won't require you to know its definition, I will require you to know that it is a distance unit, since it is used to define a star's **absolute magnitude**.

Absolute magnitude – The apparent magnitude of a star if it is viewed from a distance of 10 parsecs

Absolute magnitude, then, gets rid of the effect that distance has on a star's brightness. It compares all stars assuming that they are 10 parsecs away from the person observing them.

Let's think about the difference between absolute magnitude and apparent magnitude. The sun has an apparent magnitude of -26.74. Would its absolute magnitude be bigger or smaller? Remember, the sun is really close to the earth ("only" 1 AU). To get its absolute magnitude, we would have to travel 10 parsecs from the sun and see how bright it is. That's really far away! The sun would be much dimmer that far away. Since dimmer stars have higher magnitudes, the sun's absolute magnitude is higher than its apparent magnitude. In fact, the sun's absolute magnitude is 4.83.

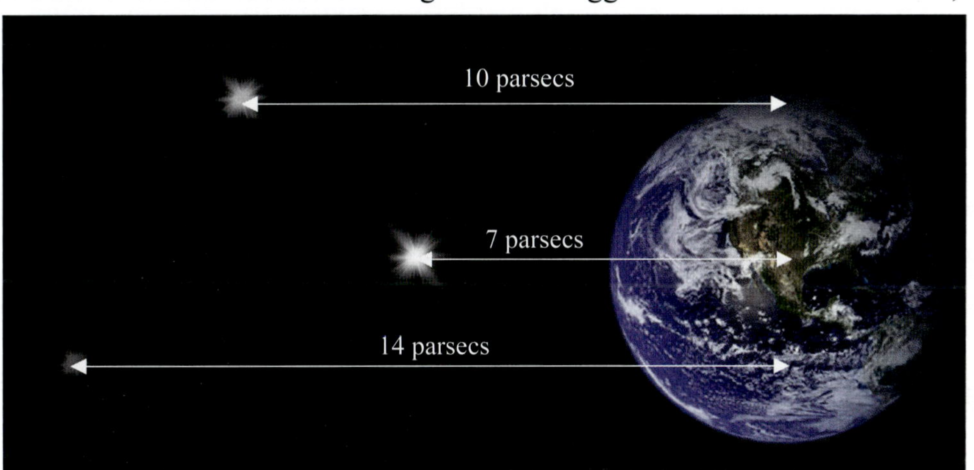

If a star were 10 parsecs (32.6 light-years) from earth, its apparent magnitude would be its absolute magnitude. If that same star were closer, the apparent magnitude would be a lower number, because the star would appear brighter. If it were farther than 10 parsecs from us, its apparent magnitude would be a higher number than its absolute magnitude, because it would appear dimmer.

As another example, consider Acrux, one of the stars pictured on the previous page. Its apparent magnitude is 0.77. Its absolute magnitude is -3.77. What does that mean? If we were 10 parsecs from Acrux, it would be *brighter*, because its absolute magnitude is lower than its apparent magnitude. So, is Acrux closer than 10 parsecs from earth, or is it farther than 10 parsecs from earth? If we viewed it from 10 parsecs, it would be brighter, which means 10 parsecs is closer. Thus, Acrux is more than 10 parsecs away from the earth. So, one way you can get an idea of how far a star is from earth is to compare its apparent and absolute magnitudes. If the absolute magnitude is *less than* the apparent magnitude it is farther than 10 parsecs from the earth, like Acrux. If the absolute magnitude is *greater than* the apparent magnitude, it is closer to the earth than 10 parsecs, like the sun.

But wait a minute. How can we determine a star's absolute magnitude? The farthest any human technology has traveled is just over 180 AU, which isn't even close to a parsec. How could we know the brightness any star would have if we were 10 parsecs away from it? Well, we can determine the distance to many stars by parallax. Once we know how far away the star is, there is a simple geometric relationship that tells us what percentage of the star's light we are actually seeing. We can use that to calculate the amount of light the star actually produces. Once we have that, we can use the same geometric relationship to calculate how much light we would see from the star if we were 10 parsecs away from it. In other words, if we know the distance to a star, we can use that distance and its apparent magnitude to calculate its absolute magnitude.

As I already said, absolute magnitude allows us to compare stars directly to one another, without worrying about the effect distance has on their brightness. When we do that, we find that some stars are naturally very bright, while other stars are naturally very dim. As I already mentioned, for example, our sun has an absolute magnitude of 4.83. Some stars are so dim that their absolute magnitudes are about 17. Some are so bright that their absolute magnitudes are about -12. So while the sun looks really bright to us, when it comes to its actual brightness, it is nowhere near the brightest nor the dimmest star in the universe.

What Happens When Parallax Doesn't Work?

So far, I have talked about measuring distance by parallax. While it is a very useful way to measure the distance to stars (and planets in the solar system), it has its limits. Remember from your experiment that the farther away you are, the less parallax you see. Some stars are so far from the earth that even if we view them from opposite sides of the earth's orbit around the sun, we cannot measure any change in their position. At that point, then, we cannot use parallax to measure how far away they are.

At what distances does this happen? It depends on the precision of the telescope being used. The more precise the telescope, the farther away a star can be when it comes to measuring distance by parallax. Telescopes that are on earth can use parallax to measure the distance to a star that is up to roughly 350 light-years away. However, the telescopes that are in orbit around the earth have a sharper view of the stars, so they can use parallax to measure distances to stars that are up to roughly 36,000 light-years away.

While 36,000 light-years seems like a long distance (more than 2 billion times the distance between the sun and the earth), it's still pretty short when it comes to the universe as a whole! To measure distances farther than that, astronomers make use of a particular kind of star. It's called a **Cepheid** (sef' eyed) **variable** star, and its apparent magnitude changes as time goes on. For example, the picture on the left shows two images of a star named "L Carinae" taken at different times. Notice how much dimmer it is in the first image. That's why it is called a "variable" star – its apparent magnitude varies. Surprisingly,

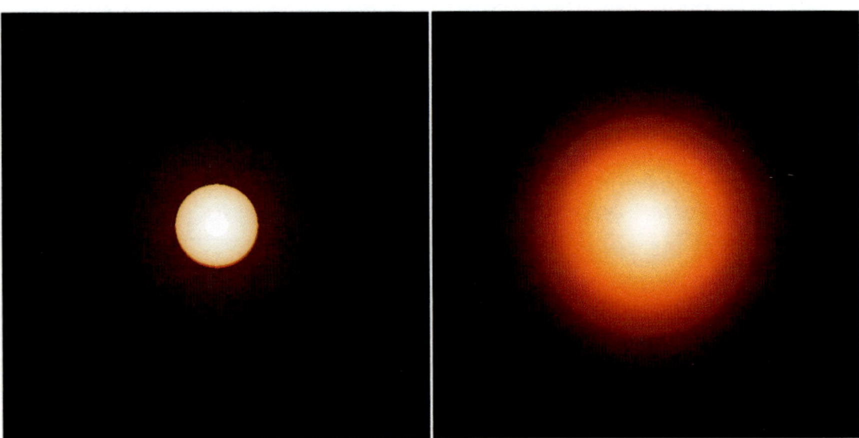

These images of the same star were produced at two different times.

there are a lot of variable stars in the universe, but Cepheid variable stars are very bright, and their apparent magnitudes change regularly, dimming and brightening over a period of 100 days or less.

How does this help us measure distance? Well, astronomers have studied the Cepheid variable stars that are close enough to us to measure with parallax. Since they know the distance to those stars, they can determine their absolute magnitudes. They found that the longer a Cepheid variable star takes to cycle from very dim to very bright, the smaller the absolute magnitude. In fact, they developed a mathematical formula that could relate the time of the cycle to the absolute magnitude of the star. For every Cepheid variable within the range of parallax, the equation works.

Think about what that means. When we see a Cepheid variable that cannot be measured with parallax, all we have to do is measure the time it takes to cycle from very dim to very bright. Using the formula that was developed, we can then determine the star's absolute magnitude. Once we have the absolute magnitude, we can use the star's apparent magnitude to determine how far away the star is! Using Cepheid variables, astronomers have been able to "measure" distances much longer than the distances measurable with parallax. I put the word "measure" in quotes because it's not quite a measurement. It is a measurement based on an assumption. It assumes that all Cepheid variables behave the same way – they all have a direct relationship between their cycle and their absolute brightness. We know that's true for Cepheid variable stars that are close enough to measure using parallax, but we can only *assume* it is true for all other Cepheid variable stars. If that assumption is wrong, then the distances measured using that relationship are also wrong.

Astronomers often refer to Cepheid variable stars as **standard candles**. While the name sounds a bit odd, it simply refers to something for which the absolute magnitude can be determined without knowing how far away it is.

Standard candle – A visible stellar object whose absolute magnitude can be determined without knowing the distance to it

There are other things visible with telescopes that can be used as standard candles, so astronomers can use more than just Cepheid variables to measuring long distances in the universe. However, *all* of those distances require some sort of assumption regarding how to determine the absolute magnitude of the standard candle.

Classifying Stars

Assuming all these standard candles work, we can determine the distance to many, many stars, and using that distance, we can calculate their absolute magnitudes. However, remember there is another difference among the stars: color. But remember what determines the color of light: its wavelength. So stars don't just produce different amounts of light. They produce different wavelengths of light. Some stars produce lots of long-wavelength light. If you remember the electromagnetic spectrum, red light has the longest wavelengths when it comes to the light we can see, so those stars tend to appear red to us. Other stars produce lots of short-wavelength light. They tend to appear blue to us.

Why do stars produce different colors of light? Because they have different temperatures. Think about your own experience with heat. When something is "red hot," it has a high temperature. But what about something that is "white hot?" It has a higher temperature, right? In a similar way, the color of a star is determined by its temperature. In fact, when astronomers measure all the wavelengths of electromagnetic radiation coming from a star, they can estimate its temperature. Once again, just as there are stars of many different absolute magnitudes, there are also stars of many different temperatures.

Interestingly enough, when you look at both the temperature and the absolute magnitude of the stars in the universe, you start to see a very interesting pattern, which is illustrated in the graph below. In the graph, the absolute magnitude is on the y-axis, while the temperature is on the x-axis. Just to make sure you see it, the temperature axis is "backwards," with hotter stars on the left and cooler stars on the right. Notice also that the temperature roughly correlates with color: the hotter stars tend to be blue, while the cooler stars are red.

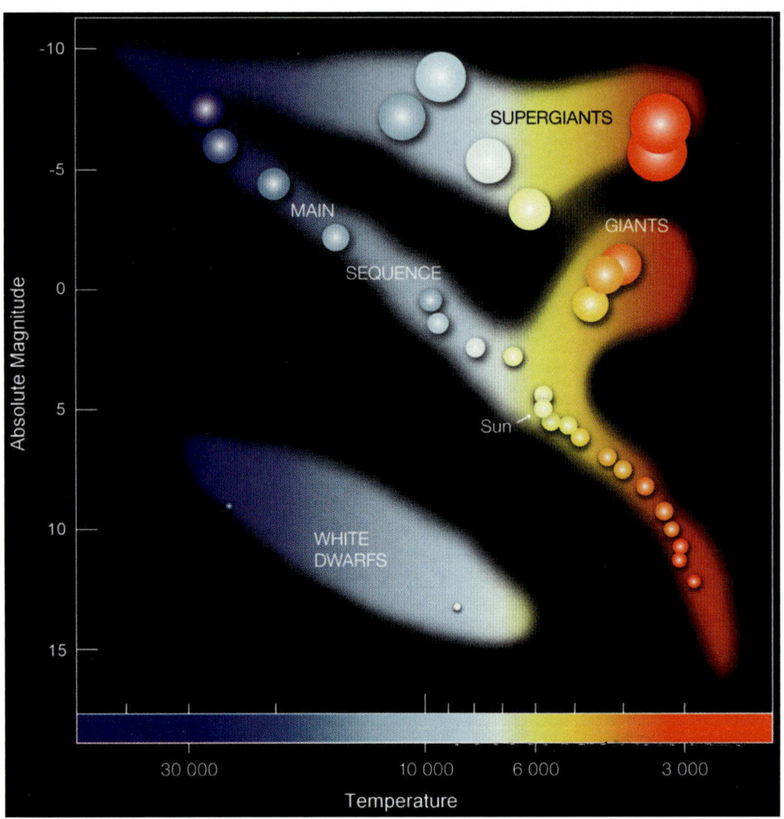

This Hertzsprung-Russell diagram helps to classify stars.

When you put lots of different stars on the graph, you see that they form groups. There is an oval near the bottom center. Those are hot stars that are dimmer than most (remember, the higher the magnitude, the dimmer the star). We call those stars **white dwarfs**, because they tend to be white and small. There is a diagonal stripe that goes from the top left to the bottom right, and the stars that fall in it, including the sun, are called **main sequence** stars. There is a "bulge" of stars that rises from the main sequence stars, which are called **giants**, because they tend to be very big. However, there is also a group of bigger stars, which are called **supergiants**.

The graph given above is called a **Hertzsprung** (hurt' sprung)-**Russell Diagram** (**HR diagram**), and it was created independently by two different scientists, Ejnar Hertzsprung and Henry Norris Russell, more than 100 years ago. Nevertheless, it is still considered the best general way to classify stars. The next time you do science, I will discuss the classifications that the HR diagram produces.

Comprehension Check

16.7 A star has an apparent magnitude of 3.4 and an absolute magnitude of 5.0. Is it more or less than 10 parsecs from the earth?

16.8 Which of the two pictures of the Cepheid variable on page 478 (the left one or the right one) shows the star at a higher apparent magnitude?

16.9 Looking at the HR diagram above, if a star has an absolute magnitude of -7 and a temperature of 6,000, is it a white dwarf, main sequence, giant, or supergiant?

Differences Among the Stars

As you already learned, the stars in the diagonal strip beginning at the top left of the HR diagram are called main sequence stars. About 90% of all stars that astronomers have analyzed belong to this group, including our sun. While there are a lot of differences among main sequence stars, there is one very important similarity they all share: They produce energy by fusing hydrogen atoms to form helium atoms in their cores. However, as you can see in the HR diagram, main sequence stars can have very different magnitudes and temperatures. The very bright, hot main sequence stars are at the top of the strip, while the dim, cool main sequence stars are at the bottom. Notice that the sun is somewhere near the middle – not the brightest, and not the hottest.

What accounts for the differences among the main sequence stars? The most important thing is the star's mass. The more mass a main sequence star has, the more fusion it can perform in its core, so the brighter and hotter it is. This is what the HR diagram on the previous page is trying to show with the relative sizes of the circles drawn in the main sequence strip. The brighter stars are represented with larger circles, because they have more mass, and the dimmer stars are represented by smaller circles, because they have less mass.

Near the center of the main sequence strip there is a "bulge" of stars that rises up and to the right on the HR diagram. That means they are brighter than main sequence stars, because they have lower magnitudes. They are not main sequence stars because they have no hydrogen in their cores. Because there is no hydrogen for fusion in the core, hydrostatic equilibrium cannot exist, so gravity crushes the core, making it very hot. This heat allows some fusion to take place outside the core. The heat produced by fusion pushes the other gases out, making the stars big. That's why they are called giant stars. However, they are cooler than main sequence stars of the same mass, so they are often called "red giants," because their cooler temperatures give them a red hue.

Supergiant stars are extreme versions of giant stars. They have a lot more mass and are able to do nuclear fusion on helium, making even heavier atoms. This produces a lot of energy, which is why they are brighter than giant stars. On the other end of the size scale, the white dwarfs are very compact stars that don't do nuclear fusion at all. Their cores are compacted by gravity, which is why they are hot enough to shine. However, without the ability to do nuclear fusion, they will grow dimmer as time goes on.

To give you an idea of just how much difference there is among the various stars, the drawing on the right shows you three supergiant stars (Betelgeuse, Antares, and Rigel), one giant star (Aldebaran), a main sequence star (the sun), and a white dwarf (Wolf 489). You might have trouble seeing the white dwarf in the drawing, since at this scale, it is a tiny white dot. Remember, the colors are a result of their temperatures, so even though the sun and Antares are very different in size, they are similar when it comes to temperature.

This drawing allows you to compare specific stars.

Stars Group Together

As you can see, there is a lot of variety among the stars. However, they tend to share an important characteristic: They group together to form galaxies.

Galaxy – A system of many stars held together by gravity

The sun, for example, is part of the **Milky Way** galaxy, which is shown in the image below. All of the light you see in the picture comes from stars that belong to the galaxy. Astronomers estimate that there are about 100,000,000,000,000 (one hundred thousand million) of them. Notice that they form a pattern of spiral arms. Not surprisingly, then, the Milky Way is called a **spiral galaxy**, and it is very big – about 100,000 light-years across.

This image of the Milky Way is based on the observations of the stars. The position of the sun is given by the arrow.

The size of the Milky Way should immediately tell you something about the image on the left. Remember, the Voyager spacecraft are just over 180 AU from the sun, which is "only" about 0.003 light-years! Thus, no human-made object has ever come close to leaving the Milky Way. As a result, we know that the image isn't a picture, because there isn't any camera outside the galaxy. Instead, this is a computer-produced "map" that is based on our observations of the stars around us. While we have no way of checking to see if it is accurate, it is consistent with all our observations, so it is at least a reasonable guess.

When you look at a galaxy, the brightness corresponds to where there are a lot of stars. Thus, you can see from the image above that the Milky Way has a lot of stars at its center and in a bar that extends outward from the center. In addition, there are more stars in the arms that make up the spiral than there are in between. What you can't see from the image is that all of these stars are orbiting around the center of the galaxy, just like the earth orbits around the sun. Indeed, if you could see a video of the Milky Way, you would see the arms rotating around the center.

Interestingly enough, however, individual stars do not necessarily orbit the center of the galaxy with the spiral arms. In fact, there is only one place in the galaxy where that happens, at a specific distance away from the center of the galaxy called the **corotation** distance. Guess where the sun is? It's at the corotation distance. This means that the sun (and its system of planets, including the earth), never has to interact with the arms. That's important, because the arms contain a lot of stars, and their gravitational pulls could interfere with the planets' orbits around the sun. In other words, we are at a very special place in the Milky Way!

You might wonder if there is anything at the center of the galaxy, since all the stars are orbiting around it. Well, the stars are so concentrated at the center (notice the bright bulge) that we can't say for sure. However, most astronomers think that there is a **black hole** at the galaxy's center. In fact, they think that most galaxies have a black hole at the center. What is a black hole? It is something so massive that its gravitational pull is strong enough to keep light from traveling away from it. Since no light can escape it, we can't see it, so it is impossible to detect it directly. However, if astronomers examine the stars in the bright bulge, the way they are moving is consistent with them orbiting a black hole. It also makes sense that there must be something very massive at the center of the galaxy. After all, it needs to be something that has strong enough gravity to hold the stars of the galaxy together over an incredibly huge distance!

God's creation is diverse, so you shouldn't be surprised to learn that there are galaxies in the universe different from the Milky Way. In fact, astronomers recognize three other types of galaxies, which are shown below.

Elliptical Galaxy Lenticular Galaxy Irregular Galaxy

An **elliptical galaxy** has a smooth appearance, because while there are more stars near the center of the galaxy, the stars are pretty evenly dispersed throughout. A **lenticular galaxy**, on the other hand, looks like a cross between a spiral galaxy and an elliptical galaxy. There are some circular areas where there are fewer stars, but there are no discernable arms like you see in a spiral galaxy. Of course, there are other galaxies that defy a general description, so we call them **irregular galaxies**.

As crazy as it sounds, galaxies also group together in **clusters**. Our galaxy, for example, is a part of a cluster known as the **Local Group**, which contains over 30 galaxies that are all attracted to one another by gravity. However, the Local Group is actually part of an even bigger structure called a **supercluster**, which contains over 700 galaxies that are all attracted to one another. When the universe as a whole is surveyed, we find that most galaxies are part of a cluster, and most clusters are part of a supercluster. In between these clusters and superclusters, there are areas with few stars at all. They are called **voids**.

Our Sun Is a Special Star

With all this talk about clusters and superclusters of galaxies, you have probably figured out that there are *a lot* of galaxies in the universe. Indeed, astronomers estimate that there are 100,000,000,000 (one hundred billion) of them. This is, of course, an estimate, because we don't really know how big the universe is, and we can't see all of it. Nevertheless, that number gives you a pretty good idea of the universe's scale. In addition, it is estimated that there are about a hundred billion stars in the average galaxy. As a result, a "rough figure" for the number of stars in the universe is 1,000,000,000,000,000,000,000 (a billion trillion). So, the star that we call our sun is just one among a billion trillion others, right? Not exactly!

When we study the sun and compare it to the other stars in the universe, we see that the sun is different from most stars. Consider, for example, the dog star, Sirius. If you look up at it, you see one very bright star. However, when a German astronomer studied the star with an improved telescope back in 1844, he saw that it's actually composed of *two stars*. The image on the left is from the Hubble telescope, which is in orbit around the earth. The big, bright light in the image is one of the two stars, and we call it Sirius A. It is a main sequence star. The small dot below it is the second star, a white dwarf called Sirius B. These two stars orbit one another and make up what is called a **binary star system**. Since these stars are part of the same solar system, we often say that they are **companion stars**.

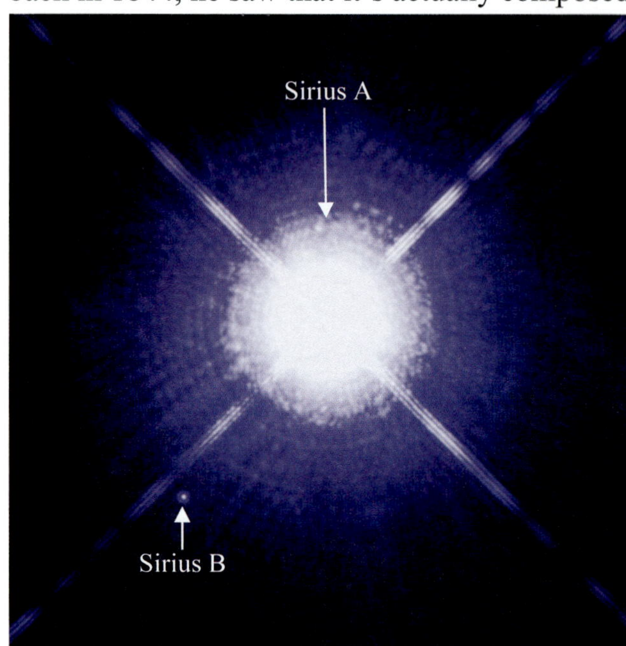

Sirius is actually a binary star system. Please note that Sirius B is very dim, so to see it, the exposure of the camera had to be long, which makes Sirius A look very bright. The "X" of light is an artifact of that exposure.

While you might not have imagined that two stars can occupy the same system, the fact is that there are many star systems with more than two stars in them. There are more than 150 known star systems with three stars (trinary star systems), and there are several with more. In fact, there is a star system "only" 470 light-years from earth that contains *seven* stars. While it is impossible to know for sure, most astronomers think that the majority of stars in the universe have at least one companion star. However, the sun does not. It is a solitary star.

Now remember, the sun is a main sequence star, which means it belongs to a group that contains most of the stars in the universe. However, there is a wide variety of masses and temperatures among main sequence stars, and most of them are very different from the sun. In addition, the sun has several planets that orbit it, and one of them (the earth) is at a distance that is ideal for getting just the right amount of energy to support life. If the earth were closer to the sun, it would get too hot to support life (as is the case with Venus and Mercury), and it if were too far from the sun, it would be too cold to support life (as is the case with Mars and the outer planets). Dr. Martin Beech did a study in which he tried to estimate how likely it would be to find another star like the sun. He concluded, "...if one picked a star at random within our galaxy, then there would be a 99.99% chance that it will <u>not</u> have the same intrinsic characteristics as our Sun and (basic) Solar System." (M. Beech, "Since When Was the Sun a Typical Star?," *Journal of the Royal Astronomical Society of Canada*, **105:**232, 2011, emphasis his).

But that's not the end of the story! Even when we find stars that have the same basic characteristics as the sun (we call them **sun-like stars**), our sun is still special among them. Why? Because it is very calm compared to other sun-like stars. Remember that the sun has sunspots – dark blemishes on its photosphere. As you learned previously, the more sunspots on the sun, the more active it is. When there are a lot of sunspots, the sun is a bit warmer than usual, and when there are only a few sunspots, the sun is cooler than usual. Well, all sun-like stars have spots on their photosphere, so all sun-like stars vary in their brightness. While we cannot actually see the spots on other stars (they are simply too far away), we can measure a star's variation in brightness. When we do that, we find that sun-like stars vary in brightness a lot more than the sun.

Think about what that means. Remember, when the sun varies in brightness, it's because of sunspots. Thus, it only makes sense that when a sun-like star varies in brightness, it's for the same reason – because of spots on its photosphere. Well, remember what can happen at sunspots – violent eruptions called solar flares. Those produce *a lot* of energy that can cause damage. Even worse, some solar flares produce coronal mass ejections, which can cause even more damage. If the sun were as variable in its brightness as other sun-like stars, there would be five times as many solar flares and coronal mass ejections! That would be devastating. However, because the sun is so "gentle" compared to other stars like it, we don't have to deal with such issues!

Think about what you have learned in this section of the chapter. The sun is at the corotation distance in the Milky Way galaxy. That means it never has to encounter the arms of the spiral, which are crowded with stars that could interfere with the planets' orbits. It also has characteristics that make it different from at least 99.99% of all other stars in the galaxy. Even when you consider the few sun-like stars that do exist, it is extraordinarily gentle. What does that tell you? It tells me that the sun was specifically designed to meet the needs of the life that God created on the earth!

Comprehension Check

16.10 Looking at the image on page 481, which is the hottest supergiant: Betelgeuse, Antares, or Rigel?

16.11 What kind of galaxy is pictured on the right (spiral, elliptical, lenticular, or irregular)?

16.12 You are studying a star with a telescope. Based on all that you have seen, you think it is a single star. However, an astronomer says you must be skeptical of any star that appears to have no companions. Why?

<u>It's a Dusty Universe</u>

Stars aren't the only things we see in the night sky. Look at the picture on the right. It was taken on a clear night, far from any cities or other sources of artificial light. You can see lots of stars, but what is that "stripe" running diagonally across the sky? It's what the spiral arms of the Milky Way look like from your perspective. Remember, you are inside the Milky Way. When you look up at the night sky, then, you see the spiral arms from the side, so they form stripes in the sky. There are clearly a lot of stars in the arms, but notice that it almost looks like there are glowing clouds in the arms

This picture shows the spiral arms of the galaxy from our perspective.

as well. It turns out that they are clouds, but they are not made of water. They are clouds of gas and dust that are glowing with energy. Each glowing dust cloud is called a **nebula** (neb' you luh), and the plural is "nebulae."

Now I do need to pause here and say something about the picture you just saw. There was a time where the night sky looked much like that on every clear night. However, you have probably never seen it look that way. Why? Because you live in an age where there is a lot of artificial light shining all night. Streetlights, lights shining from the windows of houses and businesses, headlights on automobiles, etc., all interfere with you seeing the dimmer objects (like nebulae) shining in the night sky. Remember, the stars are there even during the day, but the bright light of the sun overwhelms them, so you don't see them. Well, *any* light that is brighter than an object in the sky will overwhelm it as well. Light that interferes with our ability to see things in the night sky is often called **light pollution**. Thus, if you want to see the night sky as shown in the picture on the previous page, you need to do what the photographer did – go out on a clear night far, far from any light pollution. In many parts of the country, that is simply impossible to do, since there are so many cities, buildings, automobiles, etc. However, there are some places that are far from such things, and the night sky is breathtaking to behold there.

Because of light pollution, it is difficult to see nebulae with the naked eye. However, with the help of telescopes, astronomers can study them in exquisite detail. The picture below, for example, is of the **crab nebula**. When it was first seen through a telescope, it looked like a crab, but that was a result of the fact that the telescope was on the surface of the earth, so the image wasn't very clear. The picture on the left was put together from several images taken by the Hubble telescope, and it gives you a more realistic view of what the nebula actually looks like.

This is a Hubble Space Telescope image of the crab nebula.

This nebula is important because astronomers think they know what produced it. Chinese astronomers in 1054 recorded a new star appearing in the night sky where no star had been observed before. It got brighter and brighter, eventually it was so bright that it could be seen during the day. Then, it faded away. The crab nebula is there now. Astronomers think that the "new star" had actually been there all along, but it surged in brightness because it exploded! The nebula is made from the remains of that explosion. When a star explodes, it is called a **nova**, which just means "new star," because the surge in brightness might make it visible, even though it was too dim to be visible before. When a nova produces an incredibly violent explosion that makes it shine with the light of billions of stars, it is called a **supernova**. The explosion recorded in 1054 was clearly a supernova, since it was bright enough to be seen during the day.

Now remember, in order for us to see something, light must come from it in some way. In the case of the crab nebula, the supernova that produced it also produced a variable star known as a pulsar. That pulsar emits energy, which excites the dust and gases in the nebula. They eventually release that energy as electromagnetic radiation, and that's what makes the crab nebula visible. There are other types of nebulae in the universe, but I don't want to bog you down with all of them. Just realize that most of the diffuse "clouds" in the night sky are nebulae, and some of those nebulae are produced by novas and supernovas.

The fact that novas and supernovas were seen in ancient times serves as an example of how scientists can cling to ideas despite the fact that they are known to be incorrect. Chinese astronomers recorded the appearance of several "new stars." European astronomers also recorded them in the years 1006, 1572, and 1604. This means that astronomers knew that the heavens were not unchanging, as the Greeks believed them to be. After all, if new stars appeared in the night sky and then faded away, something *had* to be changing. Nevertheless, even as late as 1611, most astronomers still clung to the idea that the heavens are unchanging. Even after Fabricius detailed sunspots and their changes, there were still lots of astronomers who believed that the heavenly bodies were unchanging. Even though direct observations contradicted the idea, astronomers could not let it go, because it had been entrenched in the study of the heavens for too long. Unfortunately, this still happens today. Scientists often cling to ideas that are wrong, even when the facts say otherwise. The next section gives you another example of this unfortunate situation.

It's Getting Bigger

Throughout most of the history of science, it was considered scientific fact that the universe has always been the same size. It didn't matter that no one could measure the size of the universe. It was simply considered "obvious" that the universe is all there is, and thus its size could never change. This was so obvious that it guided all the scientific research on astronomy.

For example, Albert Einstein, one of the greatest scientists to have ever lived, spent a lot of his career trying to understand gravity. In 1917, he ended up developing a set of equations that explained how gravity shapes the universe, and those equations showed that the universe is expanding. However, he was so indoctrinated into the concept that the universe's size never changed that he "fixed" his equations by throwing in a made-up term he called the "cosmological constant." That term allowed his equations to be consistent with a universe that never changed size.

This is Albert Einstein 4 years after he made his greatest scientific blunder.

In 1929, that all changed. Sir Edwin Hubble (after whom the space telescope is named) demonstrated rather convincingly that the universe had to be expanding. Einstein was so surprised by Hubble's data that he visited Hubble in 1931 and ended up concluding that Hubble was right – the universe is actually expanding. This made him say that the cosmological constant he made up was his biggest scientific blunder. Even though his equations clearly said the universe is expanding, he clung to the misguided idea that the universe never changes in size. Fortunately, he was at least able to admit that error when he saw Hubble's evidence.

What was Hubble's evidence? Remember that atoms emit their own specific wavelengths of electromagnetic radiation when they are excited. As a result, we can look at the wavelengths emitted by stars and figure out what atoms the stars contain. Well, Hubble documented the fact that when he looked at stars that were far from the earth, the wavelengths emitted by the elements were longer than what came from nearby stars. In other words, the light coming from distant stars was "shifted" to longer wavelengths. In visible light, the longest wavelengths correspond to red, so this phenomenon is known as **redshift**. Hubble found that most stars show redshift, and the farther the stars are from the earth, the more redshift they show. It's almost as if the light waves are "stretched" when they travel from the star, and the farther away the star is, the more the light waves are stretched.

How can this be explained? Well, imagine a light wave leaving a star. Once it leaves the star, it takes time to travel to the earth. The farther the star is away, the longer its light should take to reach earth. If the universe is stretching the entire time the light wave is traveling, the light wave will be stretched along with the rest of the universe, as shown in the illustration on the left. The longer the light wave travels, the more it will be stretched. Hubble showed that the relationship between the distance to a star and its redshift is exactly what you would expect under that scenario. That was (and still is) considered very strong evidence that the universe is expanding.

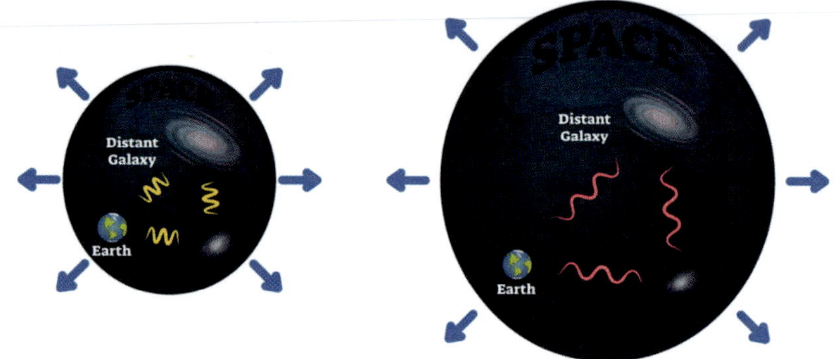

As the universe stretches, any light that has already left a star or galaxy will be stretched as well, giving it a longer wavelength. For visible light, that will make it redder.

Interestingly enough, while science taught that the universe was constant in size until Hubble's evidence was compiled, the Bible has always taught that the universe is expanding. Consider, for example, Job 9:8. It says that it is God "Who alone stretches out the heavens, And tramples down the waves of the sea." Psalm 104:2 also says that God is "Stretching out heaven like a tent curtain." This is just another example of how the Bible has scientific knowledge in it that was simply unavailable to the people who actually penned the words.

A Word About Distance and Time

Even though it is impossible to measure the size of the universe, we know it is a *really big* place. After all, parallax can be used to identify stars that are 36,000 light-years away. Assuming the standard candles we use are correct, we can observe galaxies that are 13.4 *billion* light-years away. Of course, there is probably a lot beyond what we can observe, since the farther away objects are, the harder they are to see with a telescope. Nevertheless, we can say with some confidence that the universe is many billions of light-years across.

Most astronomers will take distances like those and make what seems to be an obvious conclusion. If a star is 36,000 light-years away, the light that is reaching us now must have left the star 36,000 years ago. If it is 100 million light-years away, the light we are seeing must be 100 million years old. While that sounds perfectly reasonable, we know that it's not really true. Why? Because Einstein's equations (with or without the cosmological constant) tell us that time passes differently depending on where you are in the universe. While we like to think of time as something that marches on constantly no matter where we are, science tells us that's not true.

How does science tell us that? There are many ways, but the easiest way to see it is by looking at the global positioning system (GPS). It makes use of artificial satellites that orbit the earth about 20,000 kilometers above the surface of the earth. Remember that the force of gravity decreases the farther apart the objects in question are. Thus, the satellites that make up the GPS experience less of earth's gravity than we do on the earth's surface. According to Einstein's equations, the lower the force of gravity an object experiences, the faster time passes for that object. Thus, according to Einstein, time passes more quickly on the GPS satellites than it does on the surface of the earth.

Well, the GPS determines your position by seeing how long it takes a signal from your GPS device to reach the GPS satellites. If the GPS didn't use Einstein's equations to adjust for the difference in the way time passes on the satellites as compared to how it passes on the surface of the earth, it would not give you the correct position. However, when it does use Einstein's equations, the GPS always gives you the correct position. In other words, the GPS tests Einstein's equations every time it is used, and Einstein's equations always pass the test. Thus, we know that time passes differently depending on the gravitational fields in the area.

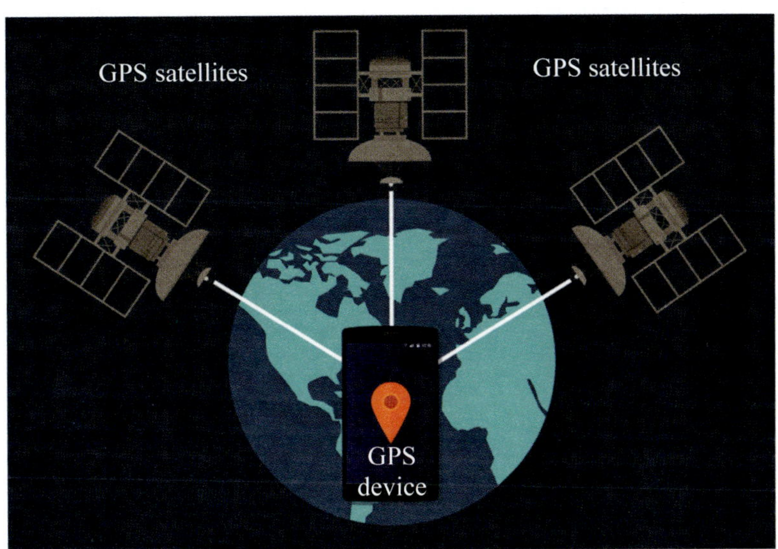

The global positioning system (GPS) determines your position by measuring how long it takes for a signal to travel from your device to different satellites in the system.

In order to assume that light from a star 100 million light-years away left the star 100 million years ago, you have to assume that the gravity it experiences is the same throughout its entire journey. However, we know that is wrong. The universe is made of clusters, superclusters, and voids. As a result, the gravity a light wave experiences along its journey varies considerably. Also, remember that the universe is expanding, and that expansion also affects the gravity in any given part of the universe. Thus, even if we are confident in our standard candles and can say for certain that the most distant galaxy observed is 13.4 billion light-years away, the light that we are observing from it is probably not 13.4 billion years old.

This is important, because it relates to the whole uniformitarianism versus catastrophism argument. If the light we are seeing from distant galaxies really is billions of years old, then the most obvious conclusion is that the universe must be billions of years old as well. In the uniformitarian view, the earth isn't as old as the entire universe, but a billions-of-years-old universe would imply that the earth is very ancient. Thus, most uniformitarians are eager to turn a light-year measurement into a time measurement as well.

Young-earth creationists, of course, don't agree with that. Some say that since God created the lights in the sky for our use, He simply created a beam of light from the star all the way to the earth. That way, all the stars could be seen as soon as they were created. The problem with that, of course, is that several novas and supernovas have been seen over the course of time. One that was observed in 2018 was measured to be 10.5 billion light-years away. That would mean that God didn't actually create the supernova. He created a nebula, but He sent light from it to indicate that a supernova happened. However, He didn't create the light from that fake supernova to initially reach the earth.

Instead, He created it so that it would arrive several thousand years after the earth was created. That doesn't make much sense.

Most young-earth creationists say that the Bible discusses things in the context of earth, so the entire universe is only a few thousand *earth* years old. However, a few thousand earth years can be billions of years in another part of the universe, depending on the relative strength of gravity in each place. So, if the earth started out in part of the universe with strong gravity, earth time would move slowly. If other parts of the universe, especially the more distant parts, experienced weak gravity, time would move quickly. Thus, a few thousand earth years could easily equal a few billion years in another part of the universe, depending on the difference in gravity in both regions. While the details are far beyond the scope of this course, there are young-earth creationist theories that have reasonable, mathematical explanations for how this can happen.

Obviously, there is a lot more to learn about the universe, but I think I have given you a good introduction to some of the most important issues related to what exists beyond our planet. At some point, one has to take a breath and review all that has been learned, and I think that time is now.

Summing It All Up

For better or for worse, you have reached the end of this earth science course. I hope you have learned a lot about this planet you call home and all that surrounds it. More importantly, however, I hope you have come to realize the truth expressed in Psalm 24:1-2, "The earth is the Lord's, and all it contains, The world, and those who live in it. For He has founded it upon the seas And established it upon the rivers." While we call the earth home for our short lifetimes, it does not belong to us. It belongs to God.

It is also a truly remarkable place, with many well-designed systems that keep it safe and hospitable to us and the other life it holds. When we look at these systems, it should be obvious that, as Nicolaus Copernicus said nearly 500 years ago, that the world "has been built for us by the Best and Most Orderly Workman of all." (Nicholas Copernicus, *De revolutionibus orbium coelestium*, 1543, in the preface and dedication to Pope Paul III)

Comprehension Check

16.13 A friend tells you about seeing an amazing nebula, and he indicates where you should look in the sky to see it for yourself. The next night, you look as hard as you can with the same equipment your friend has, but you don't see anything. What is the most likely explanation for the discrepancy?

16.14 A star is 511 light-years from earth. If the distance to that star had been measured 2,000 years ago, which would be a possible result: 500 light-years, 511 light-years, or 520 light-years?

16.15 Suppose twins are born on earth. At an early age, one of them moves to a space station orbiting the earth and stays there most of her life. When she comes back to the earth, how would her age compare to the age of her twin?

Answers to the Comprehension Check Questions

16.1 They could have made the observations from positions that were even farther apart. Richer could have been on the west coast of South America, for example, and Cassini could have been in Asia. The problem, of course, is that travel was difficult back then, and lots of places were hostile to those from other countries.

16.2 27,000,000,000 km. That's 27 *billion* kilometers! The definition of an AU says:

$$1 \text{ AU} = 150{,}000{,}000 \text{ km}$$

Put the number you want to convert over 1, and then multiply by a fraction made from the conversion unit. 1 AU needs to be on the bottom of the fraction, however, so that AU will cancel.

$$\frac{180\ \cancel{\text{AU}}}{1} \cdot \frac{150{,}000{,}000 \text{ km}}{1\ \cancel{\text{AU}}} = 27{,}000{,}000{,}000 \text{ km}$$

This is why we use AU when discussing the solar system. It's a lot easier to write the distances.

16.3 It will not be constant. The farther something is away, the smaller it appears. Well, the planets are moving in their orbits, while the earth is moving in its orbit. That means the distance between a planet and the earth will change. That will change its size. The closer the planet is, the bigger it will appear to be.

16.4 It will be dark gray or black. You can try this yourself. Once the light becomes too dim for your cones to operate, you will only see in black and white, so any color will appear to be some shade of gray.

16.5 It needs to be behind the telescope. Remember, the light bounces around in the air. What happens when you have a heat source? It makes air rise. Thus, if the fire is between the telescope and what you are viewing, there will be a constant stream of air rising in front of it, which will cause even more bouncing.

16.6 They will look very different. They will appear to be brighter, and there will be more of them. Remember, the light from stars can be overwhelmed by the light from the sun, because it is so close to you. Well, the lights in town are even closer. Thus, they can be bright enough to overwhelm some stars, and they make all the stars look just a bit dimmer. The less light you are exposed to, the brighter all the stars will seem, and you will actually be able to see stars you hadn't been able to see in town, simply because there is less light around you to overwhelm the stars' light.

16.7 It is closer than 10 parsecs. Remember, the larger the magnitude, the dimmer the star. Thus, the star is dimmer at 10 parsecs. That means 10 parsecs is farther from the earth. Thus, the star is closer.

16.8 The one on the left. Remember that high magnitude means dim, while low magnitude means bright. So when the star is dimmer, it has a higher magnitude.

16.9 It is a supergiant. A magnitude of -8 puts it near the top of the y-axis. A temperature of 6,000 puts it to the right on the x-axis. The stars near the top and to the right belong to the group labeled "supergiants."

16.10 Rigel is the hottest one. Remember that a star's temperature is given by its color. Red stars are cool, yellow stars are warmer, and blue stars are even warmer (see the HR diagram and notice the colors on the x-axis). Since Rigel is blue, it is the hottest.

16.11 It is a lenticular galaxy. You might be tempted to call it a spiral, but there are no real arms. There are just dark regions that separate the white regions. That makes it a lenticular galaxy.

16.12 Most stars have companions, so if a star doesn't seem to have a companion, it is unusual. Also, as Sirius shows, it is easy for one star's brightness to overwhelm its companion's brightness. Thus, you might have just missed the companion.

16.13 Most likely, you live in an area with more artificial lights than your friend. Remember, light pollution obscures what you can see in the night sky.

16.14 500 light-years. The universe is expanding. That means the stars and galaxies are moving farther away from one another. In the past, then, they would have been closer. 500 light-years is the only distance that is closer than the currently-measured distance.

16.15 She would be older than her twin. Time passes more quickly the weaker the gravity. Thus, she would have experienced more time passage than her twin, making her older. Now, of course, while the effect is measurable, it is still small. If this space station were at the same altitude as the GPS satellites, and she spent 20 years on it, she would only be about 7 hours older than her twin. Of course, the higher the orbit of the space station, the larger the effect would be.

Chapter Review

1. Define the following terms:

a. Parallax
b. Universe
c. Astronomical unit
d. Heliosphere
e. Magnitude
f. Apparent magnitude
g. Light-year
h. Absolute magnitude
i. Standard candle
j. Galaxy

2. There are two main factors that affect the ability to measure distance by parallax: the distance between the two observations being made and the distance to the star itself. How does an increase in each of them affect the sensitivity of parallax?

3. When we measure the distance to a star by parallax, what is the timespan between the first and second observation? Why?

4. Name the inner planets and then the outer planets. What separates these two groups?

5. At perihelion, Saturn is 9.0 AU from the sun. How many kilometers is that? (1 AU = 150,000,000 kilometers)

6. When it is farthest away, Mars is 401,000,000 kilometers from the earth. How many AU is that? (1 AU = 150,000,000 kilometers)

7. What human-made technology exists outside the heliosphere?

8. At a given time and day of the year, do the stars occupy the same position in the sky? What about the planets?

9. Why do telescopes make it easier to see stars?

10. Why do we build telescopes on tall mountains or put them in orbit around the earth?

11. Why don't you see the stars during the day?

12. Why don't you see stars in the black sky of pictures taken on the moon?

13. If one star has an apparent magnitude of 1.1 and another has an apparent magnitude of 5.6, which is brighter?

14. A star has an apparent magnitude of -0.7 and an absolute magnitude of 4.1. Is it more or less than 10 parsecs from the earth?

15. What is a Cepheid variable and how can it be used as a standard candle?

16. What determines the color of a star?

17. If a star is in the main sequence, how does it produce its energy?

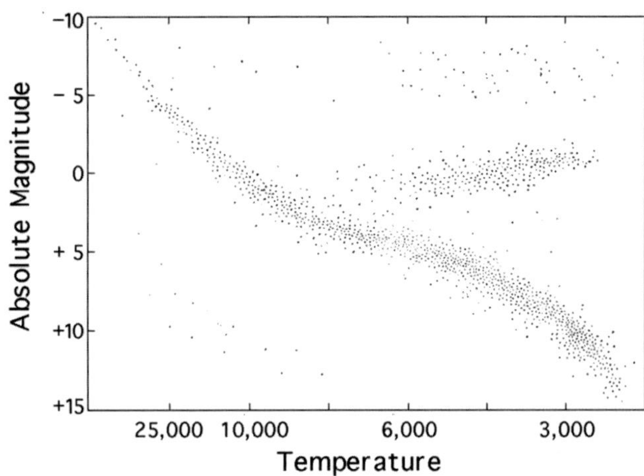

18. Using the HR diagram on the left, suppose a star has an absolute magnitude of -8 and a temperature of 5,000. What kind of star is it?

19. Using the HR diagram on the left, suppose a star has an absolute magnitude of 0 and a temperature of 10,000. What kind of star is it?

20. Using the HR diagram on the left, suppose a star is a white dwarf and its absolute magnitude is 10. Which is a possible temperature: 5,000 or 15,000?

21. What type of galaxy (spiral, lenticular, elliptical, or irregular) is the Milky Way?

22. What is the name of the cluster to which the Milky Way belongs?

23. What are companion stars and how common are they?

24. What is a nebula? What did you learn about in this chapter that can produce one?

25. What is redshift, and how does it give evidence for the idea that the universe is expanding?

26. How is the passage of time affected by gravity?

Glossary

Absolute humidity – The mass of water vapor found in a given volume of air

Absolute magnitude – The apparent magnitude of a star if it is viewed from a distance of 10 parsecs

Acid – A type of chemical that acts opposite of a base

Air mass – A large body of air with uniform temperature, pressure, and humidity at a specific altitude

Air pressure – The strength with which the air pushes on everything that it touches

Air resistance – The molecules in air must be moved out of the way in order to move through it, but that requires work, which means that air is hard to move through

Albedo – A measure of how much light is reflected from a surface instead of being absorbed

Alpha decay – Radioactive decay in which a nucleus ejects two protons and two neutrons as a single particle

Angiosperms – Plants that make enclosed seeds using flowers

Angular unconformity – An unconformity between two sets of strata that are tilted relative to each other

Aphelion – The point in a body's orbit when it is farthest from the sun

Apogee – The point at which the moon is farthest from the earth

Apparent magnitude – The magnitude of a star as it is measured from earth

Asteroid – A small rocky body that orbits the sun

Astronomer – A scientist who studies the sun, moon, and stars

Astronomical unit – The average distance between the earth and the sun (150 million kilometers)

Astronomy – The scientific study of the objects in the sky, such as the stars, moon, planets, and sun, as well as the universe as a whole

Atheist – A person who believes there is no God

Atom – The smallest chemical unit of matter

Atomic Mass – The average mass of the atoms that make up an element

Atomic number – The number of protons in the nucleus of an element

Atmosphere – The collection of gases that surround a planet

Bacteria – Microscopic creatures that are not plants or animals, considered the simplest form of life on earth

Base – A type of chemical that acts opposite of an acid

Basic Law of Magnetism – In magnets, opposite poles attract each other and like poles repel each other.

Bedding plane – A distinct separation between two rock strata

Beta decay – Radioactive decay in which a neutron emits an electron to become a proton

Biogeography – The study of the geographical distribution of animals on this planet

Buoyant – Able to float

Calibration – The process by which a measurement is compared to results with values that are known to be correct and adjusted so that the measurement is also correct

Cambrian Explosion – The abrupt appearance of many different kinds of animals in a short period of time during the Cambrian era

Catalyst – A chemical that speeds up a reaction without being used up in the reaction

Catastrophism – The view that most of the crust's features are the result of large-scale catastrophes such as floods, volcanic eruptions, etc.

Cell – The smallest functional unit of life

Centripetal force – A force that acts on an object moving in a circle, directed toward the center of the circle

Chemistry – The study of matter and how it changes

Chemical Energy – The energy stored in chemical bonds, such as food or oil

Chemical Weathering – The process by which chemical reactions break down rocks

Circumference – The distance around a circle or sphere

Classification – The act of grouping things together based on their similarities

Cleavage – (geology) The tendency of a mineral to break along parallel flat surfaces

Climate – The general weather conditions prevailing in an area over a long period of time

Climate proxy – A preserved physical characteristic used to determine past climate conditions

Climate sensitivity – The change in temperature that results from a doubling of the concentration of carbon dioxide in the atmosphere

Cohesion – A measure of how strongly molecules of a substance are attracted to one another

Combustion – A chemical reaction that occurs when oxygen combines with other substances to produce heat and usually light

Comet – A small, icy body that warms when near perihelion, losing some of its mass in the process

Compass – A device with a magnetic pointer that always points north

Compound – A substance that can decompose into simpler substances

Compression Fossil – A fossil preserved in sedimentary rock that has undergone compression

Concentration – The amount of a substance in a defined volume

Condensation – The process by which a gas turns into a liquid

Conduction – The process by which electrical charge moves from one place to another

Conductor – A substance through which electrical charges can move

Conservation of Mass – When substances change form, the total mass is neither created nor destroyed. It stays constant.

Convection – The process of transferring energy through the movement of heated or cooled fluid

Convergent boundary – A boundary between two plates that are moving toward each other

Convergent evolution – The process by which animals that are not closely related develop similar traits

Coprolite – Fossilized animal feces

Coriolis effect – The deflection of the path of any object that changes latitude over a long distance and is not firmly attached to the earth

Covalent bond – A chemical bond that is made by sharing electrons

Crest – The part of a wave that is highest

Decompose – To break a substance down into simpler substances

Density – The mass-to-volume ratio of a substance

Derived unit – A unit of measurement produced by the mathematical combination of simpler units

Detect – To discover or become aware of

Diamagnetic mineral – A mineral that is repelled by a magnet, no matter how the magnet is oriented

Diameter – The length of a line running from one edge of a circle or sphere to the other while passing through the circle or sphere's center

Diffusion – The movement of molecules from a region of high concentration to one of low concentration

Divergent boundary – A boundary between two plates that are moving away from each other

Deist – A person who believes in God, but believes that God does not reveal Himself to people

Disconformity – An unconformity between parallel strata of rock

Distillation – The process by which a solution is boiled and the vapor is then collected and condensed

Dome mountain – A mountain formed by magma pushing up on the lithosphere without breaking through it.

Effervescence – (geology) The bubbling of a mineral when it is exposed to acid

Electron – A negatively-charged component of the atom that orbits the nucleus

Electron capture – Radioactive decay in which a proton absorbs an electron to become a neutron

Electroplating – Using electricity to coat one metal in a thin layer of another metal

Electrophysiology – The study of how living things use electricity

Element – The simplest component of a substance that cannot be broken down

Emulsify – To disperse a fluid that normally doesn't dissolve so that it spreads throughout a mixture

Energy – The ability to do work

Epicenter – The point on the earth's surface directly above the focus

Equilibrium – A state of balance between opposing actions

Equinox – The two times each year when the length of the day is equal to the length of the night

Erosion – The process by which rocks and soil components are transported to a different location

Evaporation – The process by which a liquid changes into a gas

Extant – Something that exists presently

Extinction – The state in which a previously-existing species no longer exists on earth

Factor – A number used to divide or multiply another number

Fahrenheit – A temperature scale in which water freezes at 32 degrees and boils at 212 degrees

Fault – A fracture between two blocks of rock that allows the rocks to move relative to each other

Fault-block mountain – A mountain formed as a result of the motion of rocks along a dip-slip fault

Ferromagnetic mineral – A mineral that is strongly attracted to a magnet and can become a magnet if exposed to a magnet for a sufficient amount of time

Fluid – A substance that has no definite shape and can be easily influenced by pressure

Focus of an earthquake – The place where the earthquake began

Folded mountain – A mountain formed by the folding of the earth's crust caused by the collision of tectonic plates

Fossil – Evidence of past life preserved within the geological record

Fossil cast – A fossil formed when sediments enter a fossil mold and lithify

Fossil mold – A fossil formed when sediments lithify around an organism and it decays afterwards, leaving behind a cavity in the rock

Frequency – The number of times a crest or trough of a wave hits a specific point per second

Galaxy – A system of many stars held together by gravity

Gas – One of the three phases of matter – it occupies the most volume of the phases, and it fills and takes the shape of any container in which it is stored

Geocentric – Having the earth at the center

Geosphere – The solid components of the earth

Geyser – A vent in the earth's surface that ejects a column of hot water and steam

Glacier – A slowly-moving mass of ice caused by the accumulation of snow

Gram – A unit that measures mass

Gymnosperms – Plants like evergreens that make seeds which are open to the air

Gyre – A giant ocean current that forms an oval

Half-life – The time it takes for half a sample of radioactive atoms to decay

Heat – Energy that is exchanged because of a difference in temperature or a change in phase

Heat capacity – The amount of heat required to increase the temperature of an object by 1 degree

Heliocentric – Having the sun at the center

Heliosphere – The region of space protected by the sun's charged particles

Humidity – A measure of how much water vapor is in the air

Hydrogen bond – A weak bond between certain kinds of molecules caused by the attraction of a hydrogen atom in one molecule to an unshared pair of electrons on the other molecule

Hydrologic cycle – The constant motion of water throughout the earth's hydrosphere

Hydrosphere – All the water on the earth, regardless of its phase

Hydrothermal vent – An opening in the seafloor that emits hot, mineral-rich water

Hypothesis – An educated guess that attempts to explain an observation or answer a question

Humidity – A quantity that describes the amount of water vapor in the air

Hydrostatic equilibrium – The condition in which opposing forces in a fluid are balanced

Ice sheet – A permanent layer of ice covering a very large area of land

Ice shelf – A floating sheet of ice that is permanently attached to land

Igneous rock – Rock formed from the freezing of magma or lava

Imperial units – Measurement units defined by Great Britain in 1825

Index fossil – A fossil that can be used to temporally correlate strata of sedimentary rock

Indicator – A chemical that changes color depending on the amount of acid or base present

Indivisible – Cannot be broken down into smaller parts

Inflammable – Easily ignited and capable of burning rapidly

Infrared light – Light that is just beyond the red portion of visible light. It is not visible to the human eye

Insolation – The amount of solar radiation received by the earth

Invertebrate – An animal that does not have a vertebral column

Ion – An atom that becomes electrically charged because it has an imbalance of electrons and protons

Ionic compounds – Substances that are composed of positive and negative ions

Isomers – Compounds with the same chemical formula but different arrangements of atoms, which leads to different chemical properties

Isotopes – Atoms that have the same number of protons but different numbers of neutrons

Lava – Magma that emerges onto the earth's surface as a liquid

Law of Charge Conservation – Electrical charge cannot be created or destroyed.

Law of Mass Conservation – In any non-nuclear process, the total mass of everything involved must remain the same.

Latent heat – The heat associated with a change in phase

Light-year – The distance light would travel in a year, based on how it behaves on the earth

Linear scale – A scale that uses equal divisions to represent equal values

Liquid – One of the three phases of matter – it takes the shape of the container in which it is stored, and for most substances, it occupies more volume than the solid phase but less than the gas phase

Lithification – The process by which sediment becomes sedimentary rock

Lithosphere – The crust and the portion of the mantle directly below the Mohorovičić discontinuity

Loam – An easily-crumbled mixture of sand, silt, and clay

Logarithmic scale – A scale that uses equal divisions to represent multiplying by a specific number

Longitudinal wave – A wave that vibrates parallel to the direction it travels

Luster – The way a mineral shines when light hits it

Macroevolution – The idea that natural processes can, over a long period of time, transform an organism into a completely different kind of organism

Macroscopic organism – A living thing that is large enough to be seen with the unaided eye

Magma – Extremely hot liquid or semi-liquid rock located beneath the surface of the earth

Magnetic reversal – An event in which the earth's magnetic field switched its orientation to the opposite of what it is today

Magnitude – (astronomy) A measure of a star's brightness

Mantle hotspot – A relatively stationary plume of magma rising up from the mantle into the crust

Mass – A measure of how much matter is in an object

Mass extinction – The extinction of a large percentage of animals over a short geological time

Mass spectrograph – An instrument that measures the mass of individual atoms or molecules

Matter – A general term for any physical substance

Medium – A substance through which something travels or in which something is contained

Melting – The process by which a solid turns into a liquid

Mesocyclone – A rotating air mass at the center of a supercell

Mesosphere – The layer of the atmosphere above the stratopause and stratosphere

Metamorphic rock – Rock formed from existing rock that was subjected to heat and/or pressure

Metamorphosis – A change from one form to a very different form

Microevolution – The theory that natural processes can, over time, transform an organism into a more specialized version of that organism

Mineral – A naturally-occurring, inorganic substance usually found in rock that has its own chemical and physical properties, such as crystal structure, color, and hardness

Model – A representation of a process or an object in nature that allows us to more easily understand it.

Molecule – Two or more atoms linked together to make a substance with unique properties

Morphology – The form and structure of an organism or any of its parts

Mutation – An abrupt and marked change in the DNA of an organism compared to that of its parents

Natural selection – The process by which traits that make an organism more likely to survive and reproduce are preserved in subsequent generations, while traits that do the opposite are not preserved

Neap tide – A tide in which the sun's gravity fights the moon's gravity, reducing the extremes

Neutron – A component of the atom that has no charge and is found in the nucleus

Nonconformity – The separation between igneous or metamorphic rock and sedimentary rock above it

Nonpolar molecule – A covalent molecule that has an even distribution of electrons so there are no net charges in the molecule itself

Nuclear fission – The process by which a nucleus with many protons and neutrons is transformed into two smaller nuclei

Nuclear fusion – The process by which two nuclei come together to form a single, larger nucleus

Nuclear Process – A process in which an atom's nucleus undergoes a change in its number of protons and/or neutrons

Nucleus – (i) The center of the atom, where there is a concentrated, positive charge
(ii) An organelle in the cell which holds most of its DNA

Ocean acidification – The reduction of the ocean's pH due to increasing carbon dioxide in the air

Ore – A naturally-occurring solid material from which a useful mineral can be extracted

Organic compound – A covalent compound containing carbon atoms

Organism – A general term that refers to anything that is alive, be it plant, animal, or something else

Osmosis – The movement of solvent molecules across a semipermeable barrier from an area of low solute concentration to one of high solute concentration

Oxygen – A gas that people and animals breathe in and is necessary for most living things – plants produce it through photosynthesis

Ozone layer – The layer of the atmosphere that contains the highest concentration of ozone

Paleomagnetism – The study of the magnetic properties of rocks that were determined by the earth's magnetic field at the time they were formed

Paraconformity – A bedding plane that is thought to be an unconformity but doesn't look like one

Parallax – The effect by which the position of an object appears to change when viewed from different locations

Paramagnetic mineral – A mineral that is weakly attracted to a magnet and cannot become a magnet when exposed to one

Parts per million (ppm) – The amount of the chemical of interest found per million units of a mixture

Percolation – The movement of water through the pores in soil or rock

Perihelion – The point in a body's orbit when it is closest to the sun

Petrifaction – The process by which an organism's remains are replaced by minerals, preserving them

Phase – (of matter) One of the three ways matter appears: solid, liquid, or gas; (for the moon) A distinctive shape of the moon as seen in the night sky

Photosynthesis – The process by which a plant makes food for itself from water, carbon dioxide, and sunlight

Physical correlation – The matching of sedimentary strata based on their physical characteristics

Physical weathering – The process by which physical forces break down rocks

Polar molecule – A molecule with an uneven distribution of electrons, which results in slight electrical charges within the molecule

Pore – A small opening on a surface

Precipitation – (i) The process by which a solid forms from a solution
(ii) Water in its liquid or solid phase falling from clouds

Pressure – The force applied to an object divided by the area over which it is applied

Principle of Faunal Succession – The strata in which certain fossils are found can have a predictable position relative to the strata that contain other fossils

Principle of Superposition – In a series of rock strata, the oldest stratum is at the bottom, and the rocks get progressively younger the higher they are in the series.

Proton – A positively-charged component of the atom that is found in the nucleus

Pyroclastic material – Hot pieces of solid rock emitted during many volcanic eruptions

Radiant Energy – Energy transmitted in the form of light

Radiation – Energy that is emitted from a source

Radioactive isotope – An isotope that is unstable and eventually decays to become a different atom

Radiometric dating – The process by which radioactive decay is used to determine the age of a specimen

Ratio – The relationship determined by dividing the first quantity by the second quantity

Reflect – To bounce off an object and change direction

Refraction – Deflection of a wave due to a change in the material through which it is traveling

Relative humidity – The amount of water vapor in the air as a percentage of its saturation level

Residence time – The average amount of time a substance stays in a specific reservoir

Respiration – The act of breathing

Respiratory system – The body system that contains the mouth, nose, bronchial tubes, and lungs – it brings oxygen into the body and gets rid of waste gases

Rodent – A general term used to refer to animals like rats, mice, squirrels, etc.

Rock cycle – The process by which the three basic types of rock can be converted into one another

Salinity – A measure of how much salt is dissolved in water

Satellite – A small body that orbits a larger body

Saturated – Completely full of a substance so that no more can be accepted

Science – A systematic study of the natural world that allows us to understand it better

Scientific Law – A basic statement about how the world works

Scientific Method – A process of doing science in which you make observations, form a hypothesis, and test the hypothesis

Sedimentary rock – Rock formed from small particles accumulating and then sticking together

Semipermeable – Allowing passage of some substances but not others

Spring tide – A tide in which the sun's gravity adds to the moon's gravity, enhancing the extremes

Solar flare – An eruption of intense, high-energy radiation from the sun's photosphere

Solar system – The system in which the earth, the planets, their moons, and asteroids orbit the sun

Solid – One of the three phases of matter – for most substances, it occupies the least volume of all the phases

Solstice – The day at which the sun reaches its maximum or minimum height in the sky

Solute – A substance being dissolved in a solvent

Solution – A mixture of at least one solute in a solvent

Solvent – The substance a solute is being dissolved into

Species – The lowest level of biological classification that contains the most similar individuals

Spectroscopy – Identifying chemicals by studying the light they emit

Sphere – A round solid object in which all points on its surface are the same distance from the center

Standard candle – A visible stellar object whose absolute magnitude can be determined without knowing the distance to it

Storm cell – The smallest unit of a storm-producing system

Stratosphere – The layer of the atmosphere above the troposphere and tropopause

Streak – (geology) The color of a mineral when it is a powder

Subduction – The downward motion of a plate into the mantle as it moves under another plate

Sublimation– The process by which a solid turns directly into a gas

Symbiosis – A close, long-term relationship between different types of organisms

Tectonic plate – A massive, irregularly-shaped slab of lithosphere

Temperature – A measure of the energy associated with the random motion of a substance's molecules

Temporal correlation – The matching of sedimentary strata based on the time they were formed

Theologian – A person whose career is devoted to studying and understanding Scripture

Theory – A hypothesis that has been tested with a significant amount of data

Thermal energy – The energy associated with heat

Thermohaline current – An ocean current produced by changes in the salinity and temperature of seawater

Thermometer – A device that measures temperature

Thermosphere – The layer of the atmosphere above the mesopause and mesosphere

Tides – The rising and falling of an ocean, usually twice each day, caused by the sun and the moon

Trace fossil – A fossil that is formed by a trace of the organism rather than the organism itself

Transform boundary – A boundary between two plates that are moving alongside each other

Transverse wave – A wave that vibrates perpendicular to the direction it travels

Tropical Cyclone – An organized system of thunderclouds that rotate over warm ocean waters

Troposphere – The lowest region of the atmosphere, which contains almost all of the weather we experience

Unconformity – A separation between two strata caused by a period of erosion or no sediment deposition

Uniformitarianism – The view that most of the crust's features are the result of slow, gradual processes that have been at work for millions or even billions of years

Universe – The sum total of all the natural components of God's creation

Varve – A pair of thin sedimentary strata of contrasting color that were deposited in one year

Vertebrate – An animal that has a vertebral column

Visible light – Light that can be seen with the human eye

Void – An empty space

Volcanic mountain – A mountain formed by a volcano

Volume – A measure of how much room something takes up

Water table – The level below which the soil is saturated with water

Wavelength – The distance between crests in a wave

Weather front – A boundary that separates air masses with different characteristics

Wind shear – A variation in wind velocity that tends to produce a rotating motion in the air

Photo and Illustration Credits

Photos and illustrations by Dr. Jay L. Wile:
2, 19, 22, 28, 46, 47, 77, 107 (using images by nito, T.Thinnapat, www.sandatlas.org, Fokin Oleg, Tyler Boyes at www.shutterstock.com), 130, 133, 139, 147, 154 (bottom), 155 (top), 161, 170 (top – man is from stockunlimited.com), 275, 296, 311 (both), 319 (both), 320 (top), 325, 339, 340, 353 (data from the EPA), 360, 361, 434, 435 (using images in the public domain), 442-444 (using images in the public domain), 453, 457 (using images in the public domain), 464 (using images in the public domain), 467 (using images in the public domain), 477 (using images in the public domain)

Photos by Kathleen J. Wile:
257, 298, 312 (both), 358, 375, 382 (all), 446 (all)

Photo by Dr. Steven Austin:
118

Photo by Dr. Don R. Patton:
202

Illustrations by Julia Marie Mangeri:
24, 438 (left), 441

Photo by Danilo Pivato, Scientific Photographer of the Vatican Museums:
374 (top)

Illustrations by Scott Iblings:
156, 167, 170 (bottom), 203 (bottom), 259, 310, 323, 324, 374 (bottom), 385, 401, 402 (both), 403 (top), 421 (top)

Photo from alamy.com:
157 (Chris Sleby – top)

Image from thinkstock.com:
314 (earth image; moon from public domain; other elements added by Dr. Jay L. Wile)

Photos and illustrations from www.shutterstock.com (Copyright holder in parentheses):
xiii (Denis Belitsky – bottom), 4 (Mircea Maties), 5 (StudioMolekuul), 8 (G-Stock Studio), 9 (MSSA), 12 (Mockingjaz), 13 (Marzolino), 20 (Belinda Pretorius), 21 (OSweetNature), 26 (Suto Norbert Zsolt), 27 (Dmitry Lobanov), 35 (Soleil Nordic), 37 (Vadim Sadovski), 40 (Ellen Bronstayn), 45 (mahey), 48 (VectorMine), 50 (shoricelu), 52 (Gertjan Hooijer), 53 (seaonweb), 54 (Johnny Adolphson), 55 (vyasphoto), 56 (THONGCHAI.S), 57 (Juancat), 58 (Anton Foltin), 59 (Tricia Daniel – top), 60 (Fer Gregory), 65 (Victor Moussa), 66 (Nenad Randjelovic – left), 67 (Arzakae – top left, vvoe – top right, Epitavi – bottom left, Coldmoon Photoproject – bottom right), 70 (Marco Fine – top left, Albert Russ – top right, Madlen – bottom left, Lapis2380 – bottom right), 71 (Breck P. Kent), 73 (Moha El-Jaw – left, Albert Russ – right), 74 (Andriy Kananovych), 78 (Bjoern Wylezich), 79 (Aleksandr Pobedimskiy), 80 (Nastya22), 81 (IvanDonHuan – top left, Marco Fine – top right, MarcelClemens – bottom left, bonchan – bottom right), 82 (Ana-Maria Tegzes – top, MaraZe – bottom), 83 (Albert Russ), 84 (Puslatronik), 85 (Matteo Gabrieli – top, Bjoern Wylezich – bottom), 87 (Elzbieta Sekowska), 88 (Gorodenkoff), 89 (photo33mm – top, Imfoto – bottom), 90 (elen_studio – left, TinaImages – right), 95 (michal812 – left, Aleksandr Pobedimskiy – right), 96 (Bragin Alexey), 97 (Vladimirkarp – left, Breck P. Kent – right), 98 (Sarit Richerson – left, Amit kg – right), 100 (Breck P. Kent – top left and right, Bjoern Wylezich – bottom left, TR_Studio – bottom right), 101 (Leene), 102 (LiuSol), 103 (Yes058 – top left, sonsart – top right, vvoe – bottom left, anat chant – bottom right), 104 (Gerry Bishop), 105 (Yes058 – left, www.sandatlas.org – middle, Lu Mikhaylova – right), 108 (Tina Hankins), 109 (Michael Conrad), 110 (Jude Black), 111 (corlaffra), 112 (Horus2017), 113 (Mark Godden), 114 (Matauw), 116 (Bjoern Wylezich), 119 (LegART), 120 (SHTRAUS DMYTRO – top left, Heather Lucia Snow – top right, realfarbled – bottom left, losmandarinas – bottom right), 125 (OSweetNature), 129 (robin2), 131 (Designua), 135 (robin2, arrows added), 140 (Andrea Danti), 141 (Peter Hermes Furian), 142 (NoPainNoGain – top two, lichtmaster – bottom), 143 (Christoph Burgstedt), 144 (stihii – top, Designua – bottom), 148 (VectorMine), 153 (Chris Curtis), 154 (Barks – all top), 155 (Barks – bottom), 157 (BlueRingMedia – bottom), 158 (VectorMine), 160 (VectorMine), 162 (VectorMine), 163 (VectorMine), 164 (Fouad A. Saad), 165 (Oleksandr Khoma), 171 (corbac40), 172 (jmubalde), 173 (Daniel Prudek), 174 (Dominic Gentilcore PhD – top, Jim Parkin – bottom left), 175 (picturin), 176 (Kitnha), 177 (Deni_Sugandi – left, Tanguy de Saint-Cyr – right), 178 (Richard Whitcombe – top, eagledwarf – second from top, Chris Overby – third from top,

Wollertz – bottom), 179 (deLoon), 185 (MyImages – Micha), 187 (Bjoern Wylezich), 189 (Popiel – top, Karla Mtz – bottom), 190 (Breck P. Kent – both), 191 (Andreas Wolochow – top, Matthew R McClure – bottom), 193 (Igor Karasi – top, Breck P. Kent – bottom left, Carolina Jaramillo – bottom middle, AlicanA – bottom right), 195 (Jeffrey M. Frank), 197 (Ibe van Oort), 198 (Zoran Milosavljevic – bottom), 199 (COULANGES), 200 (Breck P. Kent – bottom), 209 (edited version of VectorMine), 216 (Designua), 217 (oorka), 218 (ed. from oorka), 219 (ed. from oorka), 220 (ed. from oorka), 225 (Hortimages), 226 (Craig Hanson), 227 (edited version of VectorMine), 228 (Dr Ajay Kumar Singh), 229 (Rubi Rodriguez Martinez – top, Dotted Yeti – bottom), 230 (NoPainNoGain), 233 (edited version of VectorMine), 235 (michal812), 237 (montage of sensee sriyota, zahradles, moi, IgorZh, Damsea, Steffen Foerester Photography), 239 (BoJe10), 249 (Peter Hermes Furian), 251 (Triff), 254 (JenJ_Payless), 255 (Daniel Eskridge), 256 (Rahnea Heard), 258 (Elza_R), 260 (photokai), 262 (Wor_K_Simkul), 266 (Anom Harya – left, Agnieszka Bacal – right), 269 (rozbeh – top, Samoli – bottom), 270 (Dotted Yeti – bottom), 271 (montage of Nikita Chisnikov, nwdph, Eric Isselee, Peter Fodor, Dennis van de Water, Axel Bueckert), 272 (Susan Flashman – top, Wonderly Imaging – middle, Agami Photo Agency – bottom), 281 (StudioMolekuul – bottom), 282 (Liaskovskaia Ekaterina), 285 (Liaskovskaia Ekaterina), 291 (Rachel Doreen), 292 (Rapax – top, NARUDON ATSAWALARPSAKUN - bottom), 293 (Andy Korteling), 294 (gritsalak karalak), 295 (francesco de marco – top, Islandjems - Jemma Craig – bottom), 299 (vvvita), 300 (Good_Stock), 301 (Artenex), 302 (karamysh – top, Tatsuo Nakamura – bottom), 303 (sedan504), 309 (vladimir3d – left, wewi-photography – right), 313 (grayjay), 315 (Anna L. e Marina Durante), 316 (Richard P Long), 329 (Tupungato – top, Goldilock Project – bottom), 334 (iSpyVenus), 341 (Vadim Sadovski), 343 (montage of Kwak Inhwan, Top Vector Studio, Sabelskaya), 344 (Travel Faery), 345 (VectorMine), 347 (Kirill Skorobogatko), 348 (Sylvie Corriveau), 350 (Juli Hansen), 352 (iamlukyeee), 355 (Ziablik), 357 (trgrowth), 370 (Biro Emoke), 371 (Ivan Smuk), 372 (sebikus), 377 (J. Marini), 378 (Peter Hermes Furian), 380 (Designua), 383 (mangostock), 386 (Designua), 387, (Designua), 390 (Mila Drumeva), 391 (irin-k), 392 (AleksD – top, CE Photography – middle, GiulianiBruno – bottom), 393 (Michael Dorogovich – top, Vectormine - bottom), 395 (MP kullimratchai – top, TAKENINE - bottom), 404 (NickJulia), 407 (Pixel Embargo and Pavel Talashov, montage), 408 (Alexey Kljatov, edited), 410 (Jorg Hackermann), 414 (Lisitskiyfoto – top, adamikarl – bottom), 415 (Albert Russ – top), 418 (VectorMine), 419 (both), 422 (desdemona72), 424 (VectorMine – both), 425 (VectorMine) , 426 (VectorMine– top), 431 (NickJulia), 436 (Christos Georghiou), 437 (Naeblys – top), 438 (EvhK – right), 450 (BlueRingMedia), 455 (Fouad A. Saad), 456 (David Hajnal), 458 (udaix), 459 (Fernando Da Rosa – left, Gregory H. Revera – right), 471 (Tragoolchitr Jittasaiyapanz), 473 (Erkki Makkonen), 476 (Alexei Koulikov), 481 (Aliona Ursu), 485 (Denis Belitsky – bottom), 488 (VectorMine), 489 (akhtiar Zein)

Illustration courtesy NASA/JPL-Caltech, Google:
128 , 381

Photos and illustrations from the public domain:
xii (top), xiii (top) 1, 7, 36, 39 (modified by Dr. Wile), 132, 136-138, 145, 174 (bottom right), 203 (top), 232, 234, 240, 241, 248, 261, 276, 281 (top), 288, 317, 318, 320 (bottom), 321, 322, 328 (both), 330, 346, 363, 388, 391, 405, 415 (bottom), 416, 420, 423, 426 (bottom two), 428 (both), 433, 437 (NASA/Caltech – bottom), 439, 440, 448, 449, 452, 460, 468, 469, 470 (Courtesy: Joshua Bury), 472, 475, 482 (NASA/JPL), 483 (all), 484, 485 (top), 486, 487

Images published under the Creative Commons Attribution 2.0 Generic license: (http://creativecommons.org/licenses/by-sa/2.0/):
115 (James St. John), 192 (Ryan Somma), 200 (James St. John – top), 263 (putneymark), 331 (Boris Radosavljevic), 451 (Brocken Inaglory)

Images published under the Creative Commons Attribution 3.0 Generic license: (http://creativecommons.org/licenses/by-sa/3.0/):
xii (Justin1569 – bottom), 66, (Robert M. Lavinsky – right) 201 (Hannes Grobe/AWI), 204 (© N. Tamura), 253 (Eurico Zimbres), 264-265 (Nobu Tamura), 268 (Dwergenpaartje – top), 274 (Sgbeer), 326 (Mariceti), 279 (Jfishburn), 403 (-xfi- – bottom), 412 (Pbroks13)

Images published under the Creative Commons Attribution 4.0 Generic license: (http://creativecommons.org/licenses/by-sa/4.0/):
59 (Larry D. Moore – bottom), 76 (MysteriousLion), 117 (Pierre Stromberg), 166 (Benjamin J. Burger), 188 (Cyclonaut), 196 (Wiemann, J., Fabbri, M., Yang, T. et al., "Fossilization transforms vertebrate hard tissue proteins into N-heterocyclic polymers," *Nat Commun* **9**:4741 (2018), 198 (Dwergenpaartje – top), 205 (Ghedoghedo), 206, 207, 236 (Paavo01), 238 (brewbooks), 250 (Graham Pearson), 268 (Sanjay Acharya – bottom), 270 (Maulucioni – top), 413 (Alexey Kljatov – left, Fedegrasso – right), 421 (Jason Weingart – bottom), 427 (Dr. Ryan M. Maue – both), 478 (ESO) , 479 (ESO)

Appendix A

Tables, Figures and Information for Reference

Metric Prefixes and Their Meanings

Prefix	Abbreviation	Meaning	Prefix	Abbreviation	Meaning
mega	M	1,000,000	**centi**	**c**	**0.01**
kilo	**k**	**1,000**	**milli**	**m**	**0.001**
hecto	H	100	micro	μ	0.000001
deca	Da	10	nano	n	0.000000001

Relationships Between Some Metric and U.S. Units

Physical Quantity	Metric Unit	U.S. Unit	Relationship
Distance	centimeter (cm)	inch (in)	1 in = 2.54 cm
Mass	gram (g)	slug (sl)	1 sl = 14,594 g
Volume	liter (L)	gallon (gal)	1 gal = 3.785 L

Soil Components:

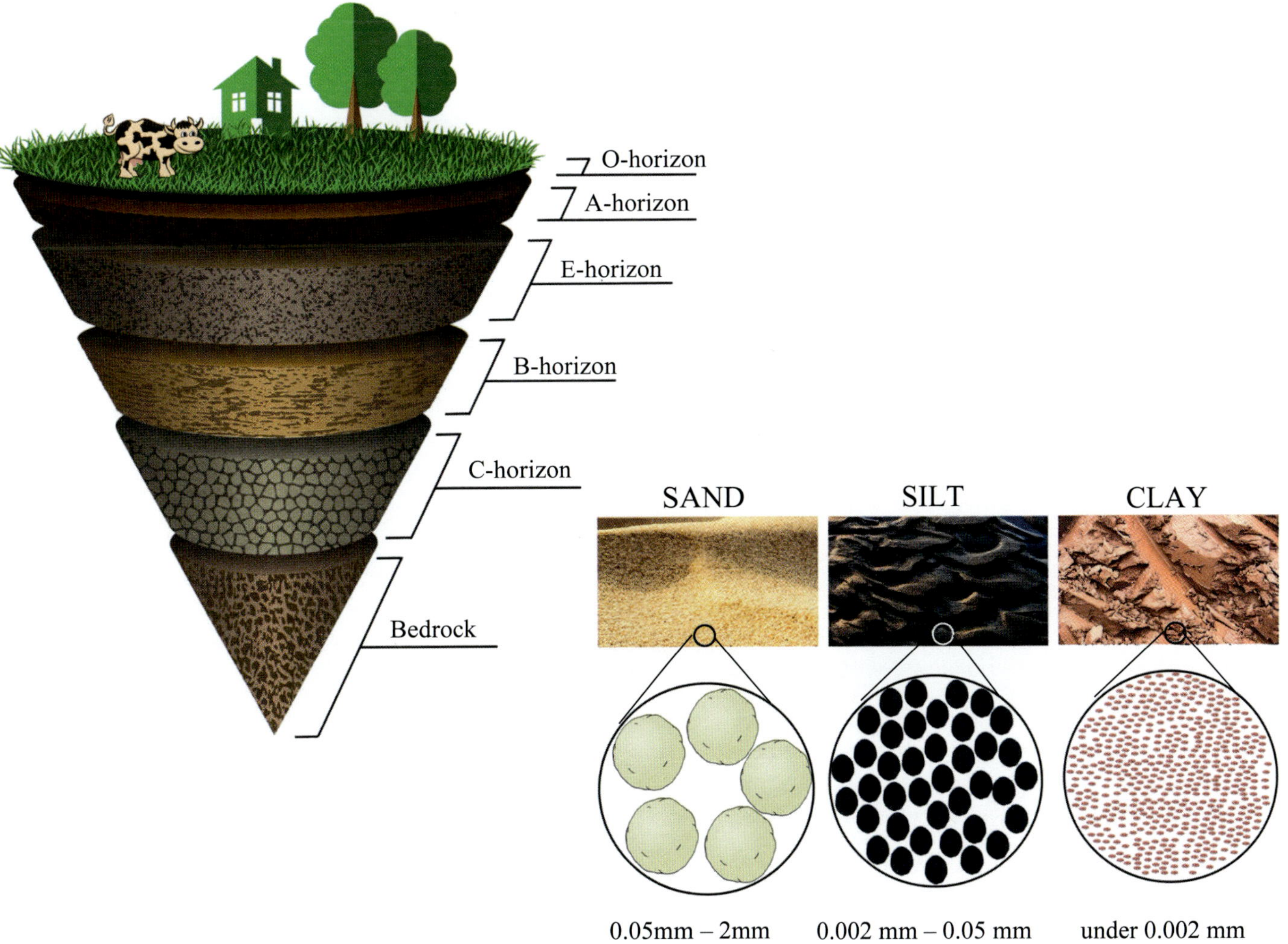

The Mohs Hardness Scale

DEFINITION

Mineral	Hardness	Mineral	Hardness
Talc	1	Potassium feldspar	6
Gypsum	2	Quartz	7
Calcite	3	Topaz	8
Fluorite	4	Corundum	9
Apatite	5	Diamond	10

SCRATCH TESTING

Scratch Tool	Hardness
Fingernail	2.5
Penny	3.0
Iron Nail	4.5
Glass plate	5.5
Streak plate	6.5

Divisions of the Earth:

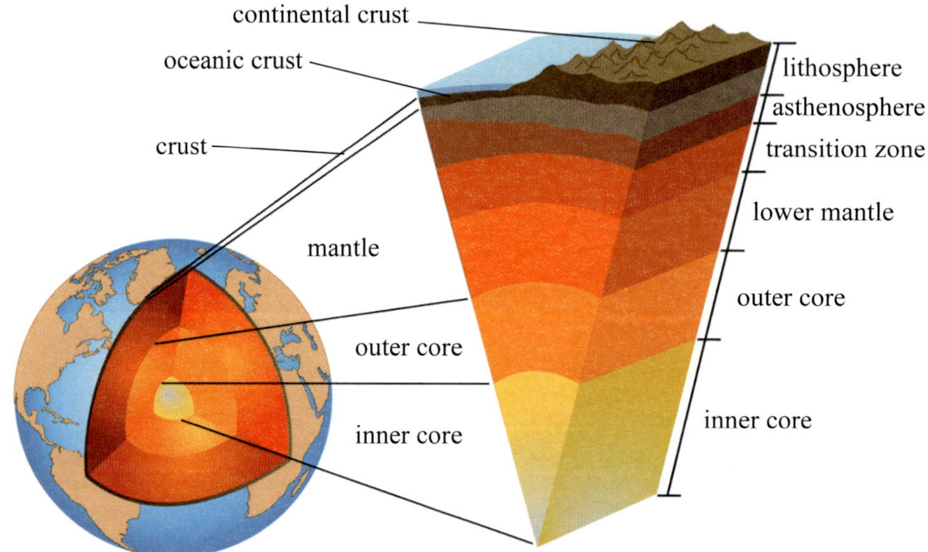

Major Plates of the Earth:

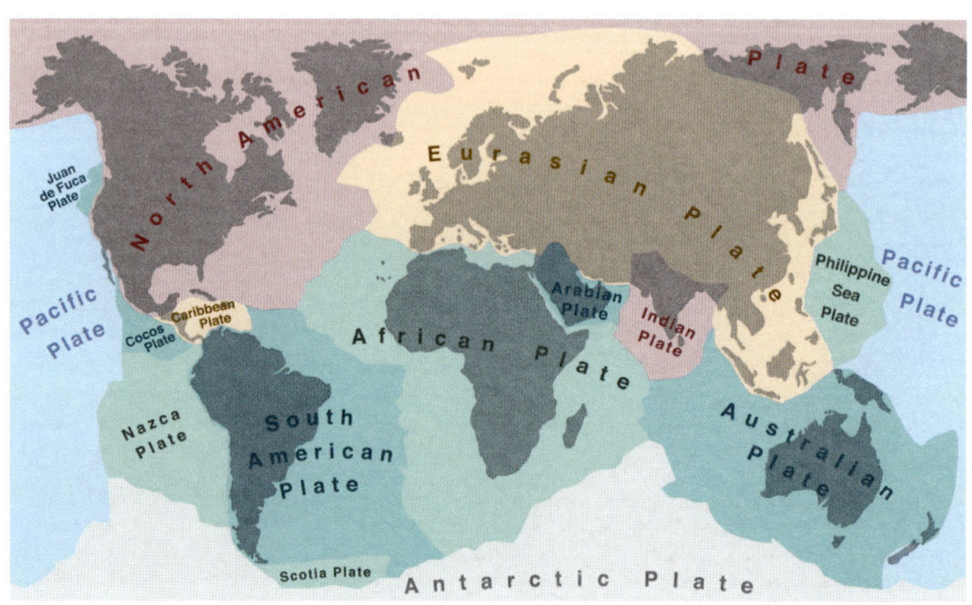

Richter Scale:

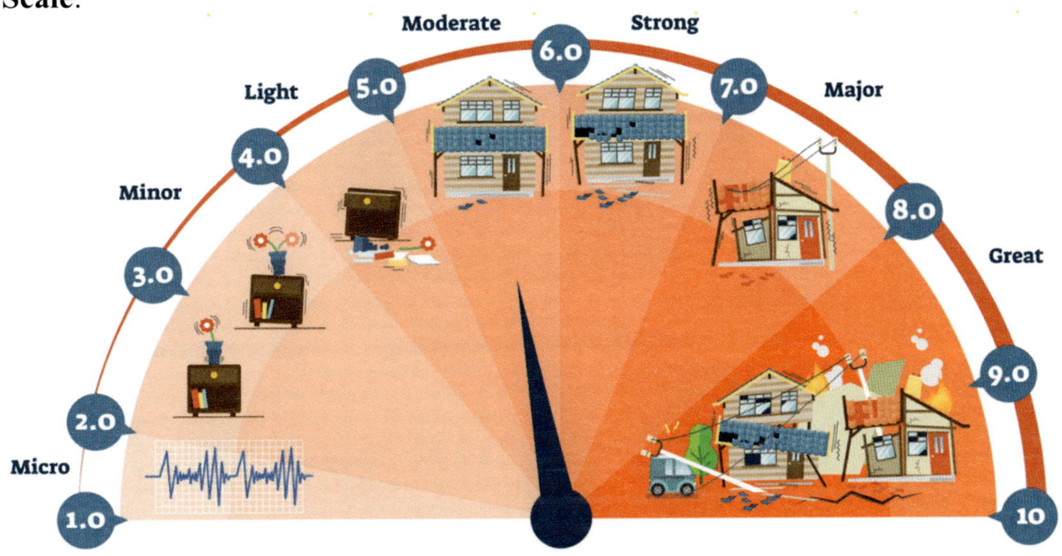

Geological Column:

Period	Fossils	Uniformitarian	YEC
QUATERNARY	mammoths, many things we see today	< 2.4 million years	Post Flood
TERTIARY	large mammals, grasses, apes	65-2.4 million years	Post Flood
CRETACEOUS	dinosaurs, deciduous trees, pteranodons	145-65 million years	Flood
JURASSIC	dinosaurs, cycad trees, birds	200-145 million years	Flood
TRIASSIC	dinosaurs, conifer trees, small mammals	252-300 million years	Flood
PERMIAN	synapsid and sauropsid reptiles	299-252 million years	Flood
PENNSYLVANIAN	ferns, reptiles, lycopod trees	318-299 million years	Flood
MISISSIPPIAN	amphibians, sharks	359-318 million years	Flood
DEVONIAN	armored fish, lobed-fin fish	416-359 million years	Flood
SILURIAN	nautilus, corals, bryozoans	423-416 million years	Flood
ORDOVICIAN	brachiopods, jawless fish	488-423 million years	Flood
CAMBRIAN	trilobites, sea worms, arthropods	540-488 million years	Flood
PRECAMBRIAN	algae, bacteria, protozoa, jellyfish	>540 million years	Pre-Flood

Water in the Hydrosphere:

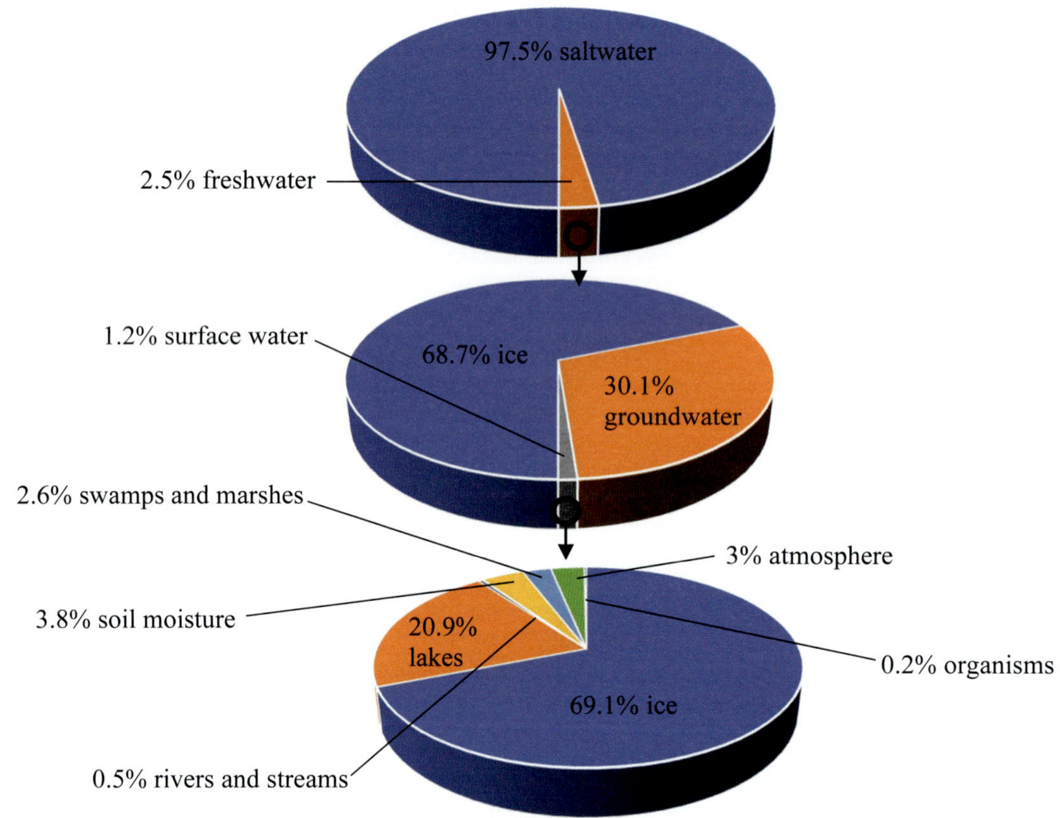

Reservoir	Residence Time	Reservoir	Residence Time
Antarctica	20,000 years	**Shallow groundwater**	100-200 years
Oceans	3,200 years	**Deep Groundwater**	10,000 years
Glaciers	20-100 years	**Lakes**	50-100 years
Seasonal Snow	2-6 months	**Rivers**	2-6 months
Soil Moisture	1-2 months	**Atmosphere**	9 days

Composition of Dry Air:

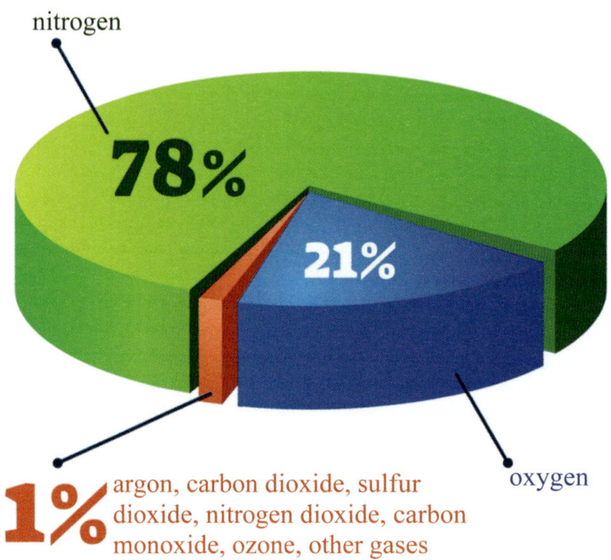

The pH Scale:

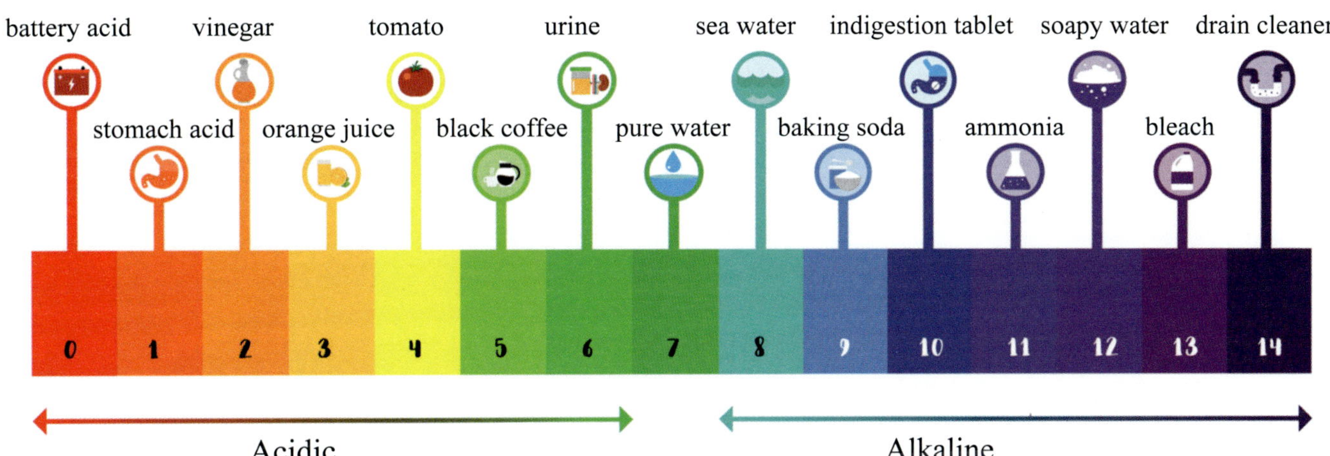

The Electromagnetic Spectrum:

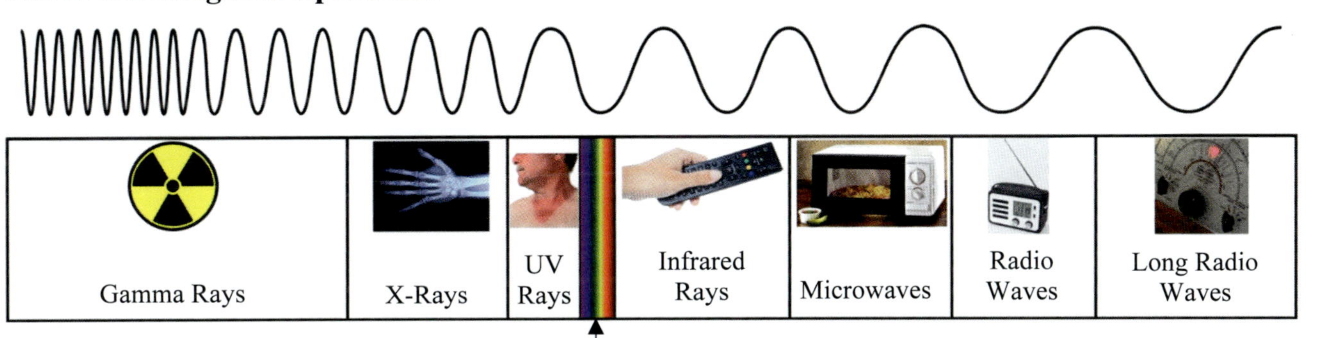

Tornado and Hurricane Ratings:

Rating	Wind Speed	Damage
EFO	65-85mph	minor roof, branches
EF1	86-110	broken windows
EF2	111-135	roofs off, large trees
EF3	136-165	homes damaged
EF4	166-200	homes leveled
EF5	200+	incredible damage

TORNADO RATING
Enhanced Fujita Scale

CATEGORY	mph	km/h	Knots (kn)	
1	79-95 mph	119-153 km/h	64-82 kn	Minimal Damage
2	96-110 mph	154-177 km/h	83-95 kn	Moderate Damage
3	111-129 mph	178-208 km/h	96-112 kn	Extensive Damage
4	130-156 mph	209-251 km/h	113-136 kn	Extreme Damage
5	≥157 mph	≥252 km/h	≥137 kn	Catastrophic Damage

Phases of the Moon:

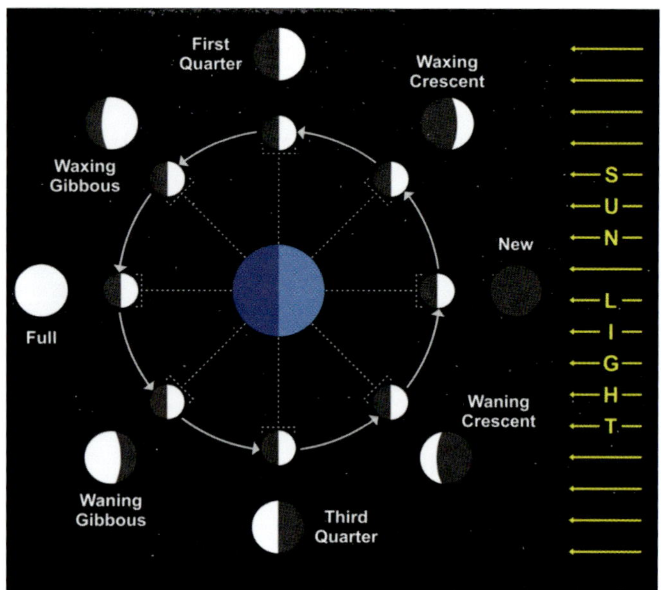

Hertzsprung-Russell diagram:

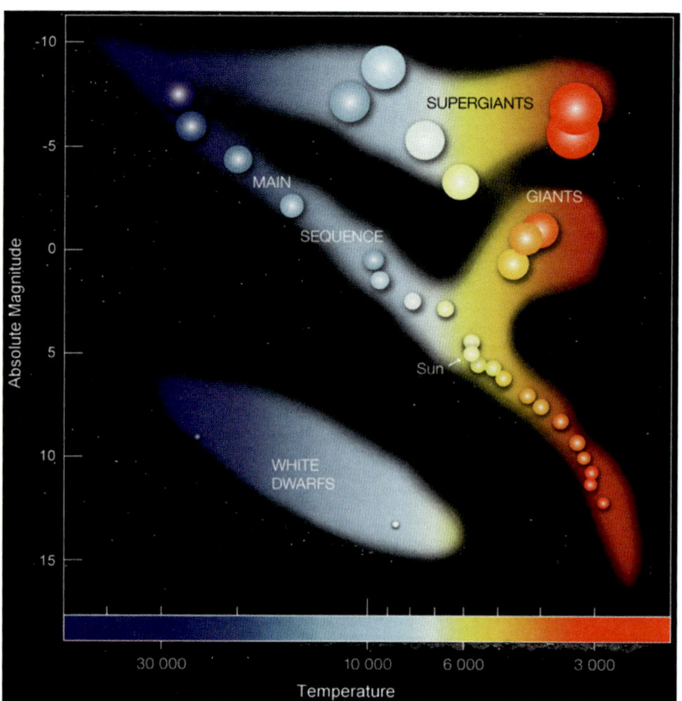

Equations and Facts That Must Be Remembered

$$\text{Area} = (\text{length})\cdot(\text{width})$$

$$\text{Volume} = (\text{length})\cdot(\text{width})\cdot(\text{height})$$

$$\text{density} = \frac{\text{mass}}{\text{volume}}$$

Oceanic crust is denser than continental crust.

1 ppm = 0.0001 percent

Order of the planets from the sun: **M**ercury, **V**enus, **E**arth, **M**ars, **J**upiter, **S**aturn, **U**ranus, **N**eptune

Appendix B

Experiment Supplies Needed for Each Chapter

In order to do the experiments in this course, you need to have the kit that is discussed on page iv of this book. In addition, you need the following common items:

Chapter 1:

- Five sheets of plain paper
- Tape
- Many books of different thicknesses
- A flat table or desk that is at least as long as four of the sheets of paper laid end-to-end
- Play-Doh or modeling clay
- Aluminum foil
- A toothpick
- Active dry yeast (Yeast for a bread machine will also work.)
- Water
- Ice
- Two smaller (like "snack sized") plastic bags that can be zipped shut (like Ziploc bags)
- Two larger (like "quart sized") plastic bags that can be zipped shut (like Ziploc bags)
- A measuring cup
- A measuring tablespoon
- Somewhere outside that you can get messy
- Someone to help you
- M&M candies (You need two each of four different colors. See the photo below.)
- A pan for boiling
- A bowl that holds roughly as much water as the pan (They just need to be close.)
- Two plates that are either metal or ceramic and can sit comfortably on top of the pan and bowl (The closer to white they are, the better.)
- A large serving spoon
- A stove or hotplate

Chapter 2:

- A place in a yard or park where you can dig. (If you don't have access to such a place, find a potted plant and use some of the soil in the pot, skipping steps 1-3. Soil from outside is better.)
- Two clear, plastic bottles, one of which has a lid (They should both be the size of a standard water bottle – 500 mL. Some sports drink bottles have wide mouths, which make the experiment easier.)
- A hand spade or a strong spoon that you can use to dig
- A raw potato (It doesn't have to be big. In fact, it could be just a piece of a potato.)
- A sharp or serrated knife
- A cutting board
- Paper towels
- A tall glass
- Water
- Three plastic water bottles (You can also use funnels or a combination of bottles and funnels.)
- Sheer hosiery that can be cut up (A couple of coffee filters will also work.)
- Tape

- Scissors
- A Styrofoam or paper cup
- A small balloon
- A 1-cup measuring cup
- A ½-cup measuring cup
- A ¼-cup measuring cup
- A spoon for stirring.
- A freezer-safe bowl

Chapter 3:
- White vinegar
- A metal paper clip (It must be bare metal, not covered in plastic.)
- A paper towel

Chapter 4:
- A blank white sheet of paper with no lines
- Paper towels
- Cleaning gloves
- A sink that is **not** stainless steel and has a faucet

Chapter 5:
- A pencil
- A small glass, like a juice glass
- Water
- A sewing needle that is strongly attracted to the magnet (most metal ones will be)
- Waxed paper or a plastic sandwich bag
- Scissors
- A casserole dish or a large bowl
- A pair of tongs that are wooden, plastic, or have plastic covers on its ends
- Aluminum foil
- Cellophane tape
- A standard battery (AA or C work best)
- A metal paper clip (It cannot be covered in plastic. It must be bare metal.)
- Someone to help you
- A tall, transparent glass
- Dark food coloring (Blue works best.)
- A rubber band
- A single ice cube
- A sink with a faucet that can deliver both hot and cold water

Chapter 6:
- 20 sheets of paper (They need to be 8.5x11 inches, and they can be blank or already used.)
- A transparent jar with a lid (If you don't have one, use a transparent glass and cover it with plastic wrap that is secured by a rubber band. You just want to limit any water that might spill out.)
- Water
- An empty metal can, like the kind vegetables come in

- A can opener
- Plastic wrap
- Tape
- Pepper
- A sink that can be plugged and filled with water
- Someone to help you
- A quarter or other large coin
- A bowl that is not transparent and doesn't have much of a rim at the top
- A container that will hold enough water to fill the bowl but can be held in one hand (like a pitcher)
- A chair and table

Chapter 7:
- A small glass, like a juice glass
- Vegetable oil (Olive oil or any other liquid cooking oil will work.)
- Active, dry yeast (available in the baking department of any supermarket)
- A ¼-teaspoon measuring spoon
- A spoon for stirring
- A sink
- Water
- A bowl
- An oven
- A baking dish, like a casserole dish
- Toilet paper
- A measuring tablespoon
- A ½-cup measuring cup
- A spoon for stirring
- Oven mitts or pads

Chapter 8:
- Baking soda
- Water
- Two small glasses, like juice glasses
- A measuring tablespoon
- A measuring teaspoon
- A 1-cup measuring cup
- A ¼-cup measuring cup
- Two spoons for stirring
- An oven
- Oven mitts
- A glass baking dish
- A napkin or paper towel
- 60 coins (You can actually use 60 of any small objects that have two easily-distinguished sides, like M&M candies. When the experiment and text refer to "heads" and "tails," you can just substitute the two sides of whatever object you are using.)
- A plastic container with a lid that is wider than it is tall and would easily hold hundreds of the items you are using.

- Any kind of carbonated beverage like soda pop, seltzer water, etc.
- A small pot or pan for boiling
- A stove

Chapter 9:

- Vinegar (White is best, because you can see through it, but any will work.)
- Two small glasses, like juice glasses
- A paper towel
- A small pot for boiling (One that has a pour spout is best.)
- A stove
- A freezer
- A teaspoon (not a measuring teaspoon)

Chapter 10:

- Four standard-size metal paper clips (You may want a couple more in case things don't go well at first.)
- Two bowls, like soup bowls
- A couple of squares of toilet paper
- A fork
- Water
- Vegetable oil (Any liquid cooking oil will work.)
- A freezer with enough space for the graduated cylinder from your kit to stand upright inside
- A stick of butter or margarine
- A means by which to melt the butter (a microwave and an appropriate container or a stove and a pan)
- Two small glasses, like juice glasses
- Paper towels
- A serrated knife, like a steak knife
- Two small cups or mugs for hot drinks (Styrofoam cups, for example, or coffee mugs)
- A pan for boiling water
- A stove
- Kitchen tongs
- Ice
- Salt
- Two small slices of raw potato
- Two small containers that can go in the freezer (Ideally, this would be two rectangles in an ice cube tray. Otherwise, two pill bottles or something of that size.)
- A small pot or pan for boiling water
- A pot lid with a handle on top (It doesn't have to fit the pot you are using.)
- A wide bowl that isn't as tall as the pot
- A Ziploc bag
- A measuring tablespoon

Chapter 11:

- Water
- A straw

- A rectangular baking pan that is at least 28 cm (11 in) long and 5 cm (2 in) deep (glass is best)
- Food coloring (Blue works best. The gel-based ones will work, but not as well.)
- A flat, level surface
- Someone to help you
- A pie pan
- The cardboard tube from the center of a roll of toilet paper
- A marble or small ball that easily fits in the tube
- Scissors
- Two paper or Styrofoam cups
- A glass baking dish or bowl that is wide enough for the two cups to sit side-by-side in the dish (The dish shouldn't be wide enough to fit three of the cups side-by-side. See the picture below.)
- Two different colors of food coloring
- A pen or other object that can make a hole in the cups
- Ice
- A small, clear plastic bottle, like the ½-liter bottles water comes in, and its lid (The flimsier the plastic bottle, the better.)
- Matches

Chapter 12:

- A stick that is at least 30 centimeters (1 foot) long and would be easy for you to break in two with your hands (It can be a dead stick from a tree or a thin piece of wood, but it must be much longer than it is wide.)
- Two full sheets from a newspaper, or two of any sheet of paper that is about 60 centimeters x 60 centimeters (23 inches x 23 inches). You can also tape pieces of paper together to make two sheets that size
- A table where the newspaper can be completely spread out
- A balloon
- A plastic bottle (or jug) that has a small enough opening for the balloon to fit over it (see picture on the next page)
- The nail from your kit
- A facial tissue or a couple of squares of toilet paper
- Something to cover your eyes, like safety glasses or goggles
- An empty thin aluminum can with a small opening, like the kind soda comes in (More than one is even better.)
- A stove
- A pan (Make sure it's an old pan, because this experiment can damage its finish.)
- Kitchen tongs
- A bowl
- Water
- Ice
- A plastic bottle, like the kind water comes in
- A bendy straw
- Play-Doh or other soft modeling clay
- Two small glasses, like juice glasses
- Toilet paper
- A small plate

- A spoon for stirring
- Vinegar
- Baking soda
- Ammonia (Sold in supermarkets with the household cleaners. You need only a drop or two.)

Chapter 13:
- Two standard sheets of white copy paper (without any lines)
- A 15-cm (6-inch) length of string that is stronger than thread
- A ruler
- A pencil
- Matches or a lighter
- A plate made out of something that will not catch fire
- Scissors
- A tall glass (It can be plastic, but it must be clear so that you can see through it.)
- An LED flashlight (If you don't have an LED flashlight, use a flashlight app on a smartphone. Ideally, the face of the flashlight should be smaller than the bottom of the glass.)
- Black construction paper, scissors, and tape if the flashlight face is bigger than the bottom of the glass.
- Milk
- Water
- A spoon for stirring

Chapter 14:
- A pan for boiling water. The flatter the bottom of the pan, the better, because the bottom of the pan needs to be evenly heated. Please note that this pan will be heated without anything in it for a while, which can be bad for some pans. Check with your parents to make sure you have a pan that is okay to use in that way. An oiled pan will not work for this experiment.
- A cup
- Water
- A stove
- A plastic bottle, like the kind water or soda comes in (the smaller the better)
- Two small glasses, like juice glasses
- Someone to help you
- Rubbing alcohol (If you have denatured alcohol, that's even better.)
- A balloon
- Clean, dry hair

Chapter 15:
- A length of sewing thread that is about 61 cm (24 in) long
- A small plastic bag, like a sandwich bag (It can be zippered but doesn't have to be.)
- Rice, small beans, or something else made of small grains. (You can use dirt, sand, flour, etc., but there is a possibility that the bag will break, so definitely do the experiment outside if you decide to use something like that.)
- Someone to help you
- A large room with nothing breakable in it or (better yet) an open area outside
- A tall, thin box, like a cereal box

- A CD or DVD that you don't mind ruining
- A serrated knife
- Scissors
- Cellophane tape
- Aluminum foil
- At least two of the following light sources: an incandescent light, a fluorescent light, or an LED light (An incandescent light is a standard bulb that heats up a filament to make light. You can use a light bulb from an uncovered lamp or an incandescent flashlight. There are compact fluorescent light bulbs that fit in a normal lamp, and some camping lanterns use fluorescent bulbs. Most cell phones have a flashlight app that's an LED light, and you can also use an LED flashlight or an LED light bulb made for lamps.)
- A plastic bottle with a lid, like the kind water comes in (The thinner the plastic, the better.)
- A straw
- A glass that is taller than a juice glass (See the picture labeled "B" below.)
- A few paper clips
- Water

Chapter 16:
- A sheet of paper that is roughly 8.5 in x 11 in
- Two colors of marker or crayon (You want to leave thick marks on the paper. I use red and blue in my discussion, but you can use any colors you want.)
- A ruler
- A meterstick or some other long, thin stick
- Tape
- A long hallway or large room
- Someone to help you
- Something you can position at the end of the hallway or room and tape the paper to, like a chair with a flat back.
- A white card, like an index card (It can have lines on it, but it needs to be mostly white.)
- A paper towel
- Vegetable oil (Any plant-based oil will work.)
- A room that is a bit dim when the lights are turned off (It needn't be dark – just dim.)
- Two lamps from which the shades can be removed
- Three bulbs for the lamps, two of which have the same brightness (listed in watts) and one with a different brightness (They should really be either all normal (incandescent) bulbs, all compact fluorescent bulbs, or all LED bulbs. If you use LED bulbs, make sure they all produce the same shade of white light, because some are bluer than others. If you can't find a bulb with a different brightness, use three lamps instead.)

Index

~B~

~C~

~D~

~E~

~F~

~G~

~H~

~I~

~N~

~O~

~P~

~Q~

~R~

~S~

~T~

~X, Y, Z~